Springer Complexity

Springer Complexity is an interdisciplinary program publishing the best research and academic-level teaching on both fundamental and applied aspects of complex systems — cutting across all traditional disciplines of the natural and life sciences, engineering, economics, medicine, neuroscience, social and computer science.

Complex Systems are systems that comprise many interacting parts with the ability to generate a new quality of macroscopic collective behavior the manifestations of which are the spontaneous formation of distinctive temporal, spatial or functional structures. Models of such systems can be successfully mapped onto quite diverse "real-life" situations like the climate, the coherent emission of light from lasers, chemical reaction-diffusion systems, biological cellular networks, the dynamics of stock markets and of the internet, earthquake statistics and prediction, freeway traffic, the human brain, or the formation of opinions in social systems, to name just some of the popular applications.

Although their scope and methodologies overlap somewhat, one can distinguish the following main concepts and tools: self-organization, nonlinear dynamics, synergetics, turbulence, dynamical systems, catastrophes, instabilities, stochastic processes, chaos, graphs and networks, cellular automata, adaptive systems, genetic algorithms and computational intelligence.

The three major book publication platforms of the Springer Complexity program are the monograph series "Understanding Complex Systems" focusing on the various applications of complexity, the "Springer Series in Synergetics", which is devoted to the quantitative theoretical and methodological foundations, and the "SpringerBriefs in Complexity" which are concise and topical working reports, case-studies, surveys, essays and lecture notes of relevance to the field.

In addition to the books in these two core series, the program also incorporates individual titles ranging from textbooks to major reference works.

Springer Series in Synergetics

Founding Editor: H. Haken

The Springer Series in Synergetics was founded by Herman Haken in 1977. Since then, the series has evolved into a substantial reference library for the quantitative, theoretical and methodological foundations of the science of complex systems.

Through many enduring classic texts, such as Haken's *Synergetics and Information and Self-Organization*, Gardiner's *Handbook of Stochastic Methods*, Risken's *The Fokker Planck-Equation* or *Haake's Quantum Signatures of Chaos*, the series has made, and continues to make, important contributions to shaping the foundations of the field.

The series publishes monographs and graduate-level textbooks of broad and general interest, with a pronounced emphasis on the physico-mathematical approach.

For further volumes:
http://www.springer.com/series/712

Andrei Ludu

Nonlinear Waves and Solitons on Contours and Closed Surfaces

Second Edition

Springer

Dr. Andrei Ludu
Embry-Riddle Aeronautical University
Department of Mathematics
S. Clyde Morris Blvd. 600
32114 Daytona Beach Florida
USA
ludua@erau.edu

ISSN 0172-7389
ISBN 978-3-642-44051-9 ISBN 978-3-642-22895-7 (eBook)
DOI 10.1007/978-3-642-22895-7
Springer Heidelberg Dordrecht London New York

Softcover reprint of the hardcover 2nd edition 2012

Printed on acid-free paper

Springer is part of Springer Science+Business Media (www.springer.com)

To Delia, Missy, and Nana, for everything.

Preface to the Second Edition

Everything the Power of the World
does is done in a circle. The sky is
round and I have heard that the earth
is round like a ball and so are all the stars.
The wind, in its greatest power, whirls.
Birds make their nests in circles,
for theirs is the same religion as ours.
The sun comes forth and goes down
again in a circle. The moon does the
same and both are round. Even the
seasons form a great circle in their
changing and always come back again
to where they were. The life of a man
is a circle from childhood to childhood.
And so it is everything where power moves.

Black Elk (1863–1950)

Nonlinear phenomena represent intriguing and captivating manifestations of nature. The nonlinear behavior is responsible for the existence of complex systems, catastrophes, vortex structures, cyclic reactions, bifurcations, spontaneous phenomena, phase transitions, localized patterns and signals, and many others. The importance of studying nonlinearities has increased over the decades, and has found more and more fields of application ranging from elementary particles, nuclear physics, biology, wave dynamics at any scale, fluids, plasmas to astrophysics. The soliton is the central character of this 167-year-old story. A soliton is a localized pulse traveling without spreading and having particle-like properties plus an infinite number of conservation laws associated to its dynamics. In general, solitons arise as exact solutions of approximative models. There are different explanation, at different levels, for the existence of solitons. From the experimentalist point of view, solitons can be created if the propagation configuration is long enough, narrow enough (like long and shallow channels, fiber optics, electric lines, etc.), and the surrounding medium has an appropriate nonlinear response providing a certain type of balance between nonlinearity and dispersion. From the numerical calculations point of view, solitons are localized structures with very high stability, even against collisions between themselves. From the theory of differential equations point of view, solitons are cross-sections in the jet bundle associated to a bi-Hamiltonian evolution equation (here Hamiltonian pairs are requested in connection to the existence of an infinite collection of conservation laws in involution). From the geometry point of view, soliton equations are compatibility conditions for the existence of a Lie group. From the physicist point of view, solitons are solutions of an exactly solvable model having isospectral properties carrying out an infinite number of nonobvious and counterintuitive constants of motion.

The progress in the theory of solitons and integrable systems has allowed the study of many nonlinear problems in mathematics and physics: nonlocal interactions, collective excitations in heavy nuclei, Bose–Einstein condensates in atomic physics, propagation of nervous pulses, swimming of motile cells, nonlinear oscillations of liquid drops, bubbles, and shells, vortices in plasma and in atmosphere, tides in neutron stars, only to enumerate few of possible applications. A number of other applications of soliton theory also lead to the study of the dynamics of boundaries. In that, the last three decades have seen the completion of the foundation for what today we call nonlinear *contour dynamics*. The subsequent stage of development along this topic was connected with the consideration of an almost *incompressible* systems, where the boundary (contour or surface) plays the major role.

Many of the integrable nonlinear systems have equivalent representations in terms of differential geometry of curves and surfaces in space. Such geometric realizations provide new insight into the structure of integrable equations, as well as new physical interpretations. That is why the theory of motions of curves and surfaces, including here filaments and vortices, represents an important emerging field for mathematics, engineering and physics.

The first problem about such compact systems is that shape solitons, which usually exist in infinite long and shallow propagation media, cannot survive on a circle or sphere. That is because such compact manifolds cannot offer the requested type of environment (long and narrow), even by the introduction of shallow layers and rigid cores. However, there is another basic idea which supports, in a natural way, the existence of nonlinear solutions on compact spaces. Because of its high localization, a soliton is not a unique solution for the partial differential system. Its position in space is undetermined because, far away from its center, the excitation is practically zero. On the other hand, all linear equations provide uniqueness properties for their solutions. It results that strongly localized solutions, and almost compact supported solutions can be generated only within nonlinear equations. There is an exception here: the finite difference equations with their compact supported wavelet solutions, but in some sense a finite difference equation is similar to a nonlinear differential one.

Despite the many applications and publications on nonlinear equations on compact domains, there are still no books introducing this theory, except for several sets of lecture notes. One reason for this may be that the field is still undergoing a major development and has not yet reached the perfection of a systematic theory. Another reason is that a fairly deep knowledge of integrable systems on compact manifolds has been required for the understanding of solitons on closed curves and compact surfaces.

The goal of the second edition of this book is to analyze the existence and describe the behavior of solitons traveling on closed, compact surfaces or curves. The approach of the physical problems ranging from nuclear to astrophysical scales is made in the language of differential geometry. The text is rather intended to be an introduction to the physics of solitons on compact systems like filaments, loops, drops, etc., for students, mathematicians, physicists, and engineers. The author assumes that the reader has some previous knowledge about solitons and nonlinearity in general. The book provide the reader examples of systems and models where the interaction between nonlinearities and the compact boundaries is essential for the existence and the dynamics of solitons.

We focused on interesting and recent aspects of relations between integrable systems and their solutions and differential geometry, mainly on compact manifolds. The book consists of 17 chapters, a mathematical annex, and a bibliography. First part contains the fundamental differential geometry and analysis approach. To render this book accessible to students in science and engineering, Chap. 2 recalls some basic elements of topology with emphasis on the concept of being compact. In Chap. 3 we review the representation formulas for different dimensions. The formulas express how a lot of information about the evolution of differentiable forms and fields inside a compact domain can be recovered only from its boundary. Chapter 4 introduces the reader to the calculus on differentiable manifolds, vector fields, forms, and various types of derivatives. We take the reader from map all the way to the Poincaré lemma. Next we introduce different types of fiber bundles, including the Cartan theory of frames, and the theory of connection and mixed covariant derivative (for immersions). Without always presenting the proofs, we tried though to keep a high level of rigorousness (relying on classical mathematical textbooks) all across the text while we still introduce intuitive comments for each definition or affirmation. Chapter 5 lays the basis for the differential geometry of curves in $\mathbf{R}_3$. We devote here special sections to closed curves and curves lying on surfaces. Complementary, in Chap. 6 we introduce elements of differential geometry of the surfaces with applications to the action of differential operators on surfaces. In Chap. 7 we derive the theory of motion of curves, both in two-dimensions, and in the general case. We devoted a section on the axiomatic deduction of the theory of motions based on differentiable forms and Cartan connection theory. We relate these motions with soliton solutions and find the nonlinear integrable systems that can be represented by such motions of curves. In Chap. 8 we discuss the theory of motion of surfaces, and we also relate it to integrable systems.

The second part of the monograph contains an exposition of the basic branches of nonlinear hydrodynamics. The working frame of hydrodynamics is the main content of the first part of the monograph, namely Chap. 9. In Chap. 10 we discuss problems on surface tension effects and representation theorems for fluid dynamics models. Chapter 11 concentrates with one-dimensional integrable systems on compact intervals, and their periodic solutions. Chapters 12 and 13 deal with nonlinear shape excitations of two-dimensional and three-dimensional liquid drops and bubbles. Chapter 14 is devoted to various applications of three-dimensional nonlinear drops, and also to compact supported solitons.

In the third part of the book, as a final goal for the first two parts, we present additional physical applications of nonlinear systems and their soliton solutions on various systems of different scales. In Chap. 15 we study the vortex filaments and other one-dimensional flows. In Chap. 16 we describe microscopic applications like elementary particles as solitons, instantons, exotic shapes in heavy nuclei, exotic radioactivity and quantum Hall drops. Chapter 17 deals with macroscopic applications like magnetohydrodynamic plasma systems, elastic spheres, nonlinear surface diffusion and neutron stars.

The book is closed by a mathematical annex including a section on nonlinear dispersion relations and their use for nonlinear systems of partial differential equations.

A legitimate question of the potential reader would be: "Why one more book on solitons?" First of all we have to acknowledge the importance of the interactions between compact boundary manifolds and the dynamics of particles and fields in mathematical in physical models. Historically the solitons are observed in sort of "infinite" systems like infinite long lines or curves, planes or open surfaces, or unbounded space. However, there is more and more evidence of the existence solitons or of localized patterns (like vortices) in compact lower dimensional spaces, like closed curves and/or surfaces. As examples, we can mention the unprecedent information technology advances in optical communication (light bullets and ultra-short optical pulses), solid-state spectroscopy, ultra-cold atom studies, soliton molecules, spinning solitons, quantum computers, spintronics and mass memory systems, femtosecond laser pulses, mesoscopic superconductivity, etc. Consequently, the reasons for writing this book are generated by a constantly increasing number of new challenges, vivid topics and hundreds of published articles.

If a substantial percentage of users of this book feel that it helped them to enlarge their outlook in the intersection between the fascinating worlds of nonlinear waves and compact surfaces and closed curves, its purpose has been fulfilled.

While writing the second edition of this book I have greatly benefited from discussions with my colleagues. I am particularly grateful to Ivailo Mladenov, Thiab Taha, Annalisa Calini who provided an inspirational and valuable help in the elaboration of this second edition. I am glad to mention the useful help from two of my students, Harry Wheeler and Tamika Thomas. For the best advices and uninterrupted encouragement I am indebted to my family.

Andrei Ludu
Daytona-Beach, Florida
December 2011

Contents

Symbols

In general the spaces ($\mathbb{R}^3$ for example), the vectors ($\boldsymbol{v}$), and the matrices are denoted with bold letters, and the dimension is represented as a subscript.

$A \lhd B$	$A \subset B$ and A has the same structure as B (it is a sub-structure)
A, dA	Area
$\boldsymbol{a}, \boldsymbol{A}, \boldsymbol{v}$	Vector field
$\alpha, \beta, \delta, \ldots$ (not γ)	Unspecified labels in general (or labels from 1 to 2)
$\boldsymbol{b}$	Binormal in the Serret–Frenet frame
$C_k(M)$	Class of differentiable functions of order k defined on M
$C_\infty(M)$	Class of infinite-differentiable functions defined on M, also called *smooth* in this book
$D_{\boldsymbol{x}}$	Directional derivative
$Diff(A, B)$, $Hom(A, B)$, etc.	Diffeomorphisms, homeomorphisms, etc. from A to B
▼	Covariant differential
∂M	Boundary of the domain M
E, F, G, e, f, g	Second fundamental form coefficients
f, df, f^*	Mapping, its differential, and the pull-back
g	Metrics
γ, Γ	Parametrized curve, or
Γ, ω	Connection, connection form
H	Mean curvature
$i, j, k, l, \ldots$	Labels in general (or labels in the $1, 2, \ldots, n \geq 3$)
$i = \sqrt{-1}$	If specified in the context
I	First fundamental form
K	Gaussian curvature
κ	Curvature of a curve

$\kappa_{1,2}$	Principal curvatures
κ_n	Normal curvature
κ_g	Geodesic curvature
M, X, Y	Manifold
ν	Viscosity
$\nabla_{\boldsymbol{v}}, \nabla_{\alpha'}$	Covariant derivative
$\nabla_\Sigma, \nabla_\Sigma \cdot, \nabla_\Sigma \times, \Delta_\Sigma$	Surface gradient, surface divergence, surface curl, surface Laplacean
$\boldsymbol{n}, \boldsymbol{N}$	Normal to a curve in the Serret–Frenet frame, normal to a surface
ODE, PDE	Ordinary or partial (system of) differential equation(s)
Ω, ω	Differential form
Π	Second fundamental form of a surface
Σ	Surface
$s,\ ds$	Arc-length
$TM, TX, TY, \ldots$	Tangent space
t	Time
$\boldsymbol{t}$	Unit tangent to a curve
$\boldsymbol{t}, \boldsymbol{N}, \boldsymbol{t}^\perp = \boldsymbol{N} \times \boldsymbol{t}$	Darboux frame associated to a given curve lying on a surface
τ	Torsion
$\hat{A}$	Tensor in general
τ_g	Geodesic torsion
u, v	Surface parameters
u, t	Curve parameters
V	Volume, only if results from context
$dV, d^n x, d^3 x$	Element of volume
$\boldsymbol{v}, \boldsymbol{u}, \boldsymbol{\omega}, \boldsymbol{V}, \boldsymbol{U}, \boldsymbol{W}$	Velocities or vorticities
$\boldsymbol{v}(\boldsymbol{w})$	Lie derivative of $\boldsymbol{w}$ with respect to (or along) $\boldsymbol{v}$
w.s.	Without summation
$x^i = \{x, y, z\}$	Specific three-dim coordinate notation
$x^\sigma = \{u, v\}$	Specific two-dim coordinate notation

Part I
Mathematical Prerequisites

In the first part of this book we study the topological, geometrical and algebraical prerequisites needed in the investigation of solitons traveling on bounded or compact manifolds. After introducing some basic elements of topology, with emphasis on compact spaces, we present the influence of boundary of a manifold over its interior points. We enumerate the representation theorems, namely those formulas, and their applicability ranges, providing the values of functions everywhere inside their domains of definition if their values on the boundaries are known. Further on, we introduce some elements of differential geometry on manifolds (vector fields, forms, derivatives) culminating with the Poincaré Lemma. A certain amount of space is devoted to the theorems of existence and uniqueness, both from the point of view of differential equations and from the point of view of geometry.

Many of the integrable equations of nonlinear science have essentially equivalent realizations in terms of differential geometry of curves and surfaces in space. Such geometric realizations provide new insight into the structure of integrable equations, as well as new physical interpretations. Therefore, we dedicate the last chapter of this part to the theory of motion of curves and surfaces. This theory also contains important tools in the study of forthcoming applications like kinematics and dynamics of filaments, interfaces, vortices, liquid boundaries of drops, bubbles, shells, nuclear surface, etc.

Chapter 1
Introduction

1.1 Introduction to Soliton Theory

Nonlinear evolution equations describe a variety of physical systems, at different scales from elementary particle models, to atomic and molecular physics, including fields like super-heavy nuclei, cluster radioactivity, atomic clusters, quantum hall drops, nonlinear optics, plasma and mesoscopic superconductor vortices, complex molecular systems, solid state, localized excited states, and Bose–Einstein condensates. At lab scale we have examples from fluid dynamics, pulses in nerves, swimming of motile cells and electric lines. Larger scale applications are related to tides in neutron stars or impact of stellar objects. It is of particular interest to examine the dynamics of localized solutions on compact domain of definitions like closed segments, closed curves, or closed surfaces, in one word on the boundaries of some compact domains.

The most useful nonlinear systems are of course the integrable ones, i.e. those solvable by inverse scattering. These particular systems have soliton solutions and infinite number of conservation laws. The traditional nonlinear systems, Korteweg-de Vries, modified Korteweg-de Vries, sine-Gordon, Schrödinger nonlinear equation and Kadomtsev-Petviashvili, were investigated in numerous works and books (see for example the following books and the references listed herein [2, 67, 71, 78, 79, 135, 169, 311]). In addition to these equations, there are other numerous other examples of integrable evolutionary systems in one ore more space dimensions. As a general property, all these systems have at least one dimension much larger than the other ones. For example, all models based on the two-layer configuration need the approximation of long channels, or long lines. In the present book we do not elaborate on such "long-scale" systems, and we do not review them in detail. We rather focus on compact physical systems modeled by nonlinear evolution equations. Some solutions derived in long systems may exist in the compact ones. The cnoidal waves which are periodical, or compact supported solutions. Some other solutions may be specific only to the compact

A. Ludu, *Nonlinear Waves and Solitons on Contours and Closed Surfaces*,
Springer Series in Synergetics, DOI 10.1007/978-3-642-22895-7_1,

systems, like we noticed in the theory of nonlinear oscillations of two-dimensional drops.

A *nonlinear evolution system* is a system of partial differential equations in variables time plus several space dimensions having the form

$$\frac{\partial u}{\partial t} = F(x, t, u, u^{(0,\tilde{p})}),$$

where $u(t, x^1, \ldots, x^n)$ is a complex vector function defined on a domain in $\mathbb{R} \times D \subset \mathbb{R} \times \mathbb{R}^n$, and where in the RHS the arbitrary functional F depends on the coordinates and derivatives of the function at spatial coordinates only ($\tilde{p}$ is a multiple index with n components). A *solitary wave solution* of the nonlinear evolution equation is a solution with the asymptotic form at $t \to \pm\infty$, $u \to u_\infty(t, x^1, \ldots, x^n) = f(x^1 - V^1 t, \ldots, x^n - V^n t)$, with arbitrary constant velocities V^i in all space directions. The definition does not exclude standing traveling waves with the same above form at all moments of time. A *soliton* is a solitary wave solution of a nonlinear evolutionary system which asymptotically preserves its shape and velocity against interactions with any other (linear or nonlinear) solutions of the same system, or against any other type of localized disturbance $\delta(t, x^1, \ldots, x^n)$ [2,67,71,79,135,169,242,311].

We define a *conservation law* of the nonlinear evolution system, a triple $(T(t, x^i, u, u^{(k,\tilde{p})}), \boldsymbol{X}(t, x^i, u, u^{(k,\tilde{p})}), \Xi)$, where the function T is the *conserved density*, the vector (X^i) is the *flux*, and Ξ is a linear first order partial differential *continuity equation* of the form

$$\Xi \Rightarrow \frac{dT}{dt} + \nabla \cdot \boldsymbol{X} = 0,$$

where the first term is the total time derivative, and T is such that

$$\frac{d}{dt} \int_D T(t, x^i, u_s, u_s^{(k,\tilde{p})}) d^n x = 0,$$

for any solution u_s of the nonlinear evolution equation.

1.2 Algebraic and Geometric Approaches

There are two exact mathematical approaches to a science problem: algebraic and geometric. Sometimes they provide similar results, but sometimes they reveal different features or relationships of the same object. Matrices represent a nice example of situation providing different results if we apply the geometrical or the algebraical approaches. For example, let us look at two 5×5 Heisenberg matrices (square matrices with entries 0, 1)

$$\begin{pmatrix} 0&0&1&0&1 \\ 0&1&0&1&1 \\ 1&1&0&0&1 \\ 1&0&1&1&0 \\ 0&1&1&0&0 \end{pmatrix} \quad \begin{pmatrix} 1&0&0&0&0 \\ 0&1&0&0&0 \\ 0&0&-1&0&0 \\ 0&0&0&1&0 \\ 0&0&0&0&1 \end{pmatrix}$$

The left one has no interesting geometrical feature, while its determinant is 5 which is the maximum possible value for such a Heisenberg 5×5 matrix since there are very few such maximal determinant matrices of this type. On the other hand, the matrix to the right in the above figure has a nice symmetrical structure but its determinant is -1, which algebraically is very common. These are differences between the algebraic and geometric points of view. Topological invariants, for example the Euler characteristics, or the rank of homotopy groups are calculated algebraically. The characteristics of curves, especially of loops, can be analyzed in terms of group theory, too. On the other side, the best efficiency of using groups and algebras is met when these algebraic objects have additional differentiable structure, and become geometrical objects like Lie groups, and fields defined on surfaces. When we study a physical problem we like to reveal both its algebraic and its geometric interpretation. Compacts systems, especially nonlinear compact systems like dynamical drops, closed shells, closed loops, etc., take profit of such dualities, because their differential structures are altered by periodic boundary conditions, by non-zero curvatures, or by the coupling between different terms of different orders or scales.

Another problem related to nonlinear compact systems is the need for compact supported solutions. Solitons have long tails which are not convenient for compact domains, unless one works in some approximations where the tail can be neglected to a certain extent. However, such pseudo-periodicity conditions introduce strong instabilities. Cnoidal waves type of solutions are better for compact domains because, on one hand, they can overlap over the same pattern by periodicity, and on the other hand, they offer enough exoticism in their shapes to match traveling isolated excitation like bumps or kinks. Nevertheless, a nonlinear system can generate even more localized solutions, like the compact supported solitons (e.g. compactons or peakons where the internal nonlinear dispersion structure can provide compactification of solutions). A nonlinear system is the natural frame for compact solutions, and a geometrically compact nonlinear system can take profit of that. On the contrary, a linear system has all its solutions uniquely determined by its initial conditions, so there is no freedom for a compact object to be placed in different initial positions, with the same effect on the general solution, like in the nonlinear cases.

There is, however, one exception. The multi-scale finite-difference linear systems like fractals or wavelets. These types of functions bring another interesting situation related to compact nonlinear problems: the hidden connection between nonlinear differential equations and finite-difference equations, via the infinite system of ordinary differential equations that represent both of them in some special cases.

In order to illustrate this point of view we mention the family of functions $f_\alpha(x) = \tanh(\alpha x)$ with the property $f_\infty(x) - f_\infty(x-1) \to 2\Phi_H(x)$, where $\Phi_H(x)$ is the Heaviside scaling function defined 1 on [0, 1] and 0 in the rest of the real axis. On one hand f_α is a solution of a nonlinear equation $f_\alpha' - \alpha^2 f_\alpha^2 - \alpha^2 = 0$, and on another hand, the limit $f_\infty(x)$ fulfils the two-scale finite difference linear equation $f_\infty(x) = f_\infty(2x) + f_\infty(2x-1)$.

1.3 A List of Useful Derivatives in Finite Dimensional Spaces

Throughout the chapters of this book we use calculus on finite dimensional manifolds, differential forms, and integral invariants. Why do we need so many diverse geometric objects for our applications? In the spirit of justifying the necessity of these mathematical tools we illustrate with a simple example about derivatives. In the following calculations we use several types of derivatives, among which we enumerate:

1. The partial derivative (in local coordinates)
2. The differential of a map
3. The directional derivative
4. The exterior derivative of a form
5. The Lie derivative of a geometrical object
6. The covariant derivative
7. Pseudo-differential operators

In the following we try to remember about their different ways of action and differences between them, so that the reader can figure out if they are useful or not.

1. *The partial derivative*
   These derivatives transform a scalar function (a 0-form) into a vector field (the dual of a 1-form), namely the gradient ∇f. We can build all sorts of symmetric or skew-symmetric linear combinations of partial derivatives acting on vectors or scalar fields (curl, divergence, Laplace operator, etc.), operators that form the object of vectorial analysis.
2. *The differential*
   The generalization of the partial derivative to calculus on manifolds is provided by the *differential of a map*. It is a generalization of the gradient operator. In local coordinates the differential of a map is the Jacobian matrix of that map. If we map a manifold into itself $F : M \to M$ we have actually a transformation or a flow of the points of M. These motions of points in M are integral curves of some vector field tangent to M. Then, the differential of this map measures the change of the position of the points along this transformation vector field.
3. *The directional derivative*
   It measures how a certain local quantity Q changes along a given direction $\boldsymbol{v}$, i.e. $D_{\boldsymbol{v}}Q$ of Q along $\boldsymbol{v}$. In the case of real three-dimensional manifolds the

directional derivative reduces to the scalar product between the gradient and a given direction.

4. *The exterior derivative*
   In $\mathbb{R}^3$ we have a hierarchy, called de Rham complex
$$0 \leftrightarrow 1 \leftrightarrow 2 \simeq 1 \leftrightarrow 3 \simeq 0.$$
   A differentiable covariant tensor field, i.e. a k-form ω can be mapped into a higher order form by repeated differentiation. However, the partial derivative will never produce the cyclic type of de Rham hierarchy, like the fact that the "curl" of a "gradient" is zero and the "divergence" of the "curl" is zero, and so on. The most important result of the exterior derivative is contained in the Stokes theorem and Poincaré lemma.
5. *The Lie derivative*
   It is the operator which in effect tells us the infinitesimal change of the geometric object ω when moved along integral curves of a given field $\boldsymbol{v}$, from one point x to a new point x'. The idea is to take the value of $\omega(x')$ at the new point, to pull it back towards the initial point x by using the dual F^* (or co-differentiation), and then compare the two values $F^*(\omega(x')) \backsim \omega(x)$. An example illustrates the importance of the Lie derivative. Let us have a fluid described in cartesian coordinates, and its volume element $dx\ dy\ dz$. How does the volume element change along the flow? If the flow is described by the Lagrangian trajectories of the fluid, i.e. curves of tangent field $\boldsymbol{V}$, then the directional derivative of the volume element along $\boldsymbol{V}$ is zero. However, the Lie derivative of the associated volume form $\Omega_{vol} = dx \wedge dy \wedge dz$ is $\div\, \boldsymbol{v}$ which is not zero, and we have $\boldsymbol{v}(\Omega_{vol}) = \nabla \cdot \boldsymbol{v}$, where the *exterior product operation* $\wedge$ will be defined in Sect. 4.2. Actually, it is a known fact that the volume is preserved during the flow only if the field $\boldsymbol{v}$ is solenoidal.
6. *The covariant derivative*
   When differentiating along a surface, the "inhabitants of the surface" can only see that part of the derivative lying in the tangent plane. Given a vector field $\boldsymbol{v}$ the covariant derivative $\nabla_{\boldsymbol{v}}$ is the projection of the directional derivative on the tangent space. As opposed to the Lie derivative which needs a vector field to exists, the covariant derivative can be defined only locally if we know the direction at a point, because we take profit of the connection.
7. *The covariant exterior derivative*
   Instead of the regular exterior derivative applied to a k-form with scalar components, we have a Lie algebra (of contravariant vectors) valued k-form. In this case the partial derivative with respect to the coordinates of the components of the form is substituted with the covariant derivative of the contravariant vector new components.
8. *Pseudo-differential operators*
   One can define the inverse of a differential operator as a formal series of partial derivatives with differential functions as coefficients

$$\sum_{n=-\infty}^{\infty} C_n(f(x))D_x.$$

We can present these observations in the diagram below, where by F we denoted a differential map between manifolds, and by ω a k-form or a vector field.

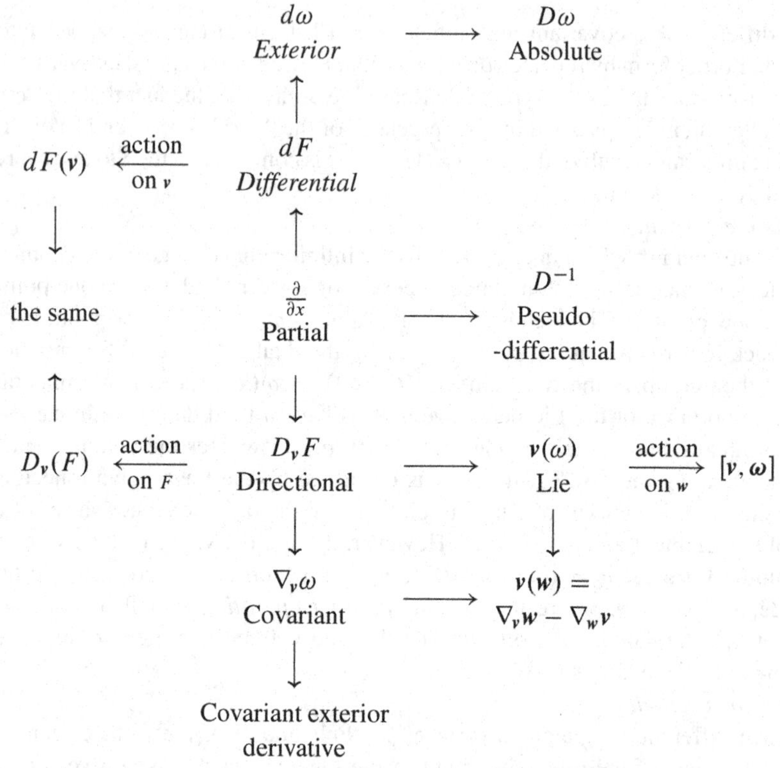

In addition to these types of derivatives working in finite dimensional spaces (and sometimes called "horizontal" derivatives [242]) we scientists occasionally need the so-called *functional derivative*, working as a generalization of the directional derivative except in infinite dimensional spaces. An example is given by the *variational derivative* whose action on a functional measures the infinitesimal variations of this functional when arbitrary small changes are applied to the dependent variables. If the working space has some topological and algebraic structures, the functional derivative can be defined more formally as either a Fréchet derivative (in Banach spaces) or a Gâteaux derivative (in locally convex spaces). Even if in this book we discuss deformations and motions of curves and surfaces we will not use in the following the functional derivative formalism. This happens mainly because we investigate these deformations from a more general aspect (the geometric one) than the restricted Lagrangian point of view.

Chapter 2
Mathematical Prerequisites

Before entering in the field of nonlinear waves on closed contours and surfaces we need to recall some useful mathematical concepts. The *cnoidal waves*, *solitary waves*, and *solitons* are solutions of nonlinear equations that could be partial differential (PDE), integro-differential, finite difference-differential, or even functional equations. They describe the evolution of the wave solutions in space and time. These nonlinear equations are usually coupled with linear or nonlinear boundary conditions (BC), initial conditions, or asymptotic conditions. The properties of solutions are dependent on the topological and geometrical structure of the space on which they are defined. In the following we assume for the reader to be familiar with the general concept of group, Abelian group, quotient group, rank of a group, and group homomorphism.

2.1 Elements of Topology

In this section we introduce some elements of topology related to the idea of boundary [68, 160, 274, 291, 344]. Some working theorems are very important and their generality raises sometimes the question: "how is this possible?" The following few sections try to reveal a little bit of the insights of such properties. When we investigate a space from the topological point of view, the basic questions are: how large, how dense, how tight, or how fuzzy is such a space? In Table 2.1, we present how topology addresses these questions. A topological space (X, τ) is a set X and a family $X, \emptyset \in \tau \subset \mathcal{P}_X$ of *open sets* stable against finite intersection and arbitrary reunions. The complement of any open set is *closed*. To any point $x \in X$ we can associate a family $\mathcal{V}$ of neighborhoods of x, $V(x) \in \mathcal{V}$ defined by the property $V(x) \in \mathcal{V}$ if $\exists A \in \tau, x \in A \subset V(x)$. A family of open sets in (X, τ) is called base if any open set of the topology is a reunion of sets in that family. A point $x \in X$ is called adherent if $\forall V(x), V(x) \cap A \neq \emptyset$.

A. Ludu, *Nonlinear Waves and Solitons on Contours and Closed Surfaces*,
Springer Series in Synergetics, DOI 10.1007/978-3-642-22895-7_2,

Table 2.1 Properties of topological spaces

Question	Topological property (or invariant)
How large?	Compactness
How fuzzy?	Separation
How many pieces?	Connectedness
How complicated?	Separability
How much measurable?	Metric space

A closed set contains all its adherent points. An adherent point is the rudiment of the concept of limit. We need the following definitions:

$$\text{int}\, A = \overset{\circ}{A} = \{x \in A |\ \text{exists}\ D \in \tau, x \in D \subset A\},\ \text{interior of}\ A,$$

$$\overline{A} = A \cup \{x \in X | x\ \text{adherent point to}\ A\},\ \text{closure of}\ A,$$

$$\partial A = \overline{A} - \text{int}\, A,\quad \text{boundary of}\ A.$$

The open property of a set is relative to the topology of the space. For example, the real interval (a, b) is open in the usual metric topology on $\mathbb{R}$, but it is neither open nor closed in the plane $\mathbb{R}^2$, while a loop is closed both in $\mathbb{R}^2$ and $\mathbb{R}^3$. A family $B_\alpha \in B \subset \tau$ with the property that $\forall D \in \tau, D = \cup B_\alpha$ is called a *base*. A set $A \subset X$ with the property $\overline{A} = X$ is called *dense* in X. A space with countable base is called *separable*. A topological space which is also a vector space such that the algebraic operations with vectors and scalars are continuous in the topology is a *linear topological space*. The space $C_0[0, 1]$ of continuous real functions defined on $[0, 1]$, for example, is separable because any such function can be the limit of a countable sequence of polynomials. Any harmonic complex function defined on the surface of the unit sphere in $\mathcal{R}_3$ can be expressed as a series of spherical harmonics Y_{lm}, so this space is separable, too.

A function defined on X with values in Y is *continuous* if the inverse image of any open set in Y is an open set in X. A bijective continuous function is called *homeomorphism*. Topological spaces are classified as modulo homeomorphisms and *topological invariants* (properties preserved by homeomorphisms). Topological properties of a space X can be investigated by choosing a test topological space S (known one) like $\mathcal{R}^n$ or $C_0(X)$, and building homeomorphisms $hom : S \to X$. When the image of a topological invariant in S is not anymore a topological invariant in X we know that X moved from a certain homeomorphism class into another [235]. The set of all homeomorphisms between two topological spaces X, Y is denoted by $\text{Hom}(X, Y)$. The property of homeomorphism, like any topological property, can be loosen up by using instead the property of *local homeomorphism*. A function is a local homeomorphisms if for any point of its domain of definition there is an open neighborhood of that point on which the restriction of this function is a homeomorphism onto its image. Obviously, homeomorphism implies local homeomorphism.

Definition 1. A *covering map* from a topological space C to another topological space X is a continuous surjective map $cov : C \to X$ such that $\forall c \in C$ and $\forall U(c)$ an open neighborhood of c we have $cov^{-1}(U) = \bigcup_\alpha V_\alpha$, $V_\alpha \bigcap V_\beta = \emptyset$ a disjoint union of open sets in C, and $cov \in \mathrm{Hom}(V_\alpha, U)$.

The "larger" space C is called the covering space, and the space X is called the base space. Traditional examples of covering maps are projecting a helix to its base circle, or by wrapping a plane around a cylinder.

2.1.1 Separation Axioms

The uniqueness property of solutions of a nonlinear partial differential system is not only important in itself, but it also provides the freedom to build solutions by any available methods. Uniqueness is mainly controlled by two mechanisms. One is related to the boundary, initial, asymptotic, regularity, or normalization conditions. The second is related to the internal constrains of the spaces for variables and parameters. Uniqueness is very strongly related to the topological property of *separation*. In topology there are several more refined definitions for the concept of *separation* [68, 160, 291, 344]. The various forms of separations, i.e., *separation axioms* introduce different types of topological spaces:

- $\mathcal{P}_1$. $x \neq y$. This is the weakest separation criterium.
- $\mathcal{P}_2$. $\mathcal{V}(x) \neq \mathcal{V}(y)$, the two points x and y do not have the same families of neighborhoods: they are *topologically distinguishable*.
- $\mathcal{P}_3$. $A \cap \overline{B} = \emptyset$, each set is disjoint from the other's closure; the sets are *separated.*
- $\mathcal{P}_4$. $\exists \mathcal{V}(x) \cap \mathcal{V}(y) = \emptyset$, points separated by disjoint neighborhoods. This form of separation is the most used in analysis, since it makes the transition from points to open sets.
- $\mathcal{P}_5$. $\exists \overline{\mathcal{V}(x)} \cap \overline{\mathcal{V}(y)} = \emptyset$, points separated by disjoint closed neighborhoods.
- $\mathcal{S}$. $A \cap B = \emptyset$, disjoint sets.
- $\mathcal{PS}$. $x \notin A$, the element does not belong to the set.
- $\mathcal{F}$. $\exists f \in C_0(X)$, $f(\xi_1) \neq f(\xi_2)$. There is a continuous function on X which takes distinct values in two disjoint quantities ξ that can be points and/or sets. This last form of separation is very useful when working with spaces of functions, e.g., in the Weierstrass approximation theorem.

According to the separation axioms there are four types of topological spaces:

1. Regular (R).
   A topological space is Kolmogorov (or T_0) if $\mathcal{P}_1 \to \mathcal{P}_2$, i.e., the space is such that any two distinct points have different families of neighborhoods (are topologically distinguishable). A topological space is *symmetric* (or R_0) if $\mathcal{P}_2 \to \mathcal{P}_3$, i.e., the space is such that any two topologically indistinguishable points have a disjoint neighborhood with respect to the other point (separated) (see Fig. 2.1). A stronger separation axiom defines X as a preregular space (or R_1)

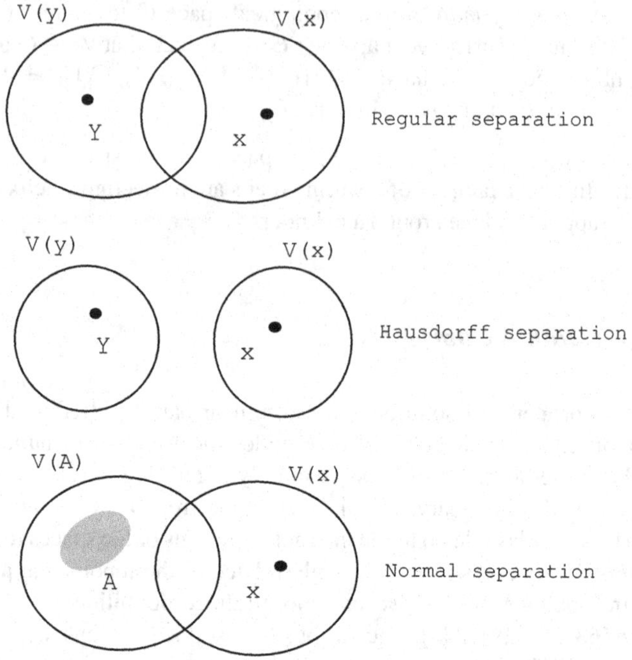

Fig. 2.1 Forms of separation axioms: regular, Hausdorff, and normal. Loops represent neighborhoods

if $\mathcal{P}_2 \to \mathcal{P}_4$, i.e., any two topologically indistinguishable points have disjoint neighborhoods. This axiom can be enhanced even more if we ask that any point x and disjoint closed set C, $x \notin C$ are separated by a continuous function, namely if $\mathcal{PS} \to \mathcal{P}_4$, and the space is called regular. As application, for example, any topological vector space is regular [160].

2. Hausdorff (H).
A topological space is Hausdorff separated (H or T_2) if $\mathcal{P}_1 \to \mathcal{P}_4$, i.e., its distinct points are separated by disjoint neighborhoods (see Fig. 2.1). The Hausdorff separation is the most used one in analysis and operator theory. For example, to build a Banach (commutative) algebra of functions defined on a base space X, we need this space to be Hausdorff (and compact). A very important application of H spaces is related to their property that the intersection of all closed neighborhoods of any point reduces to that point, $\forall x \in X, \cap\overline{\mathcal{V}(x)} = x$. This property is actually the basis of the uniqueness of the limit for the convergent sequences in H spaces. Moreover, this property plays the essential role in the proof of the Cauchy integral representation formula. There is interference between separation and compactness properties: the image of a compact through a continuous function $f : E \to F$ is compact, only if F is Hausdorff. The separation property is requested because we need to label the sets of a finite

covering of E (produced by reciprocal images of an open covering of F) by elements in E. So, if F is not separated, the images of two distinct such elements may belong to the same open set in F, which destroy the construction. As an example, the topology induced by a family of seminorms is in general Hausdorff.

Since the Hausdorff property is so essential to the uniqueness of solutions of equations, we give the following example of a non-Hausdorff space. Let us consider in $\mathbb{R}^2$ the sets $A_1 = \{(x, 0)|x \in \mathbb{R}\}$ and $A_2 = \{(x, 1)|x \in \mathbb{R}\}$, and let us introduce an equivalence relation $\sim$ between the points $(x, y) \in A = A_1 \cup A_2$ defined by $(x, y) \sim (x', y')$ if $x = x'$ and $y = y'$ or $x = x' < 0$ and $y \neq y'$. We organize the quotient set $X = A/ \sim$ as a topological space with the canonical interval topology on $\mathbb{R}$. The points $(0, 0)$ and $(0, 1)$ in X are distinct but have no disjoint neighborhoods.

3. Normal (N).
   In a normal topological space, any two disjoint closed sets are separated by neighborhoods, i.e., $\mathcal{S} \to \mathcal{P}_4$, or $\forall \overline{A} \cap \overline{B} = \emptyset, \exists \mathcal{V}(\overline{A}) \cap \mathcal{V}(\overline{B}) = \emptyset$ (see Fig. 2.1). For a Hausdorff space, this request becomes the Tietze–Uryson lemma. A topological space with the topology induced by a metric is normal, and a compact space is also normal [291]. Normal spaces are important in problems related to the partition of unity. Partitions are important in the theory of prolongation of continuous functions.
4. Completely separated (C).
   Here the separation criterium is the function separation. There are already several types of topological spaces completely separated as follows: completely Hausdorff spaces (CH or completely T_2) where $\mathcal{P}_1 \to \mathcal{F}$, completely regular spaces (CR) where $\mathcal{P}_5 \to \mathcal{F}$, and completely normal (CN) where $\mathcal{P}_3 \to \mathcal{P}_4$. We also have perfectly normal spaces (PN) if $\mathcal{S} \to \mathcal{F}$, etc.

In addition to these types of topological spaces, there are other spaces where separation is defined by combining different forms of separation. In Figs. 2.2 and 2.3, we represent some of the interconnections between all these spaces.

2.1.2 Compactness

The compactness property of a topological space (or set) tells if this space is "bounded" in some sense, without having a metric or a distance available. The compactness property is actually more powerful than boundedness, since the latter is not preserved by homeomorphisms. A topological space is a *compact space* if every open covering has a finite subcovering. In metric spaces (see Sect. 2.1.6) compact is equivalent with closed and bounded. Actually, it is easier to understand the concept of noncompact. The real axis is noncompact because if we cover it with the intervals $(n, n + 1)$ and $((2n + 1)/2, (2n + 3)/2)$, n integer, and we eliminate any of them the axis has at least one point uncovered. A compact Hausdorff space

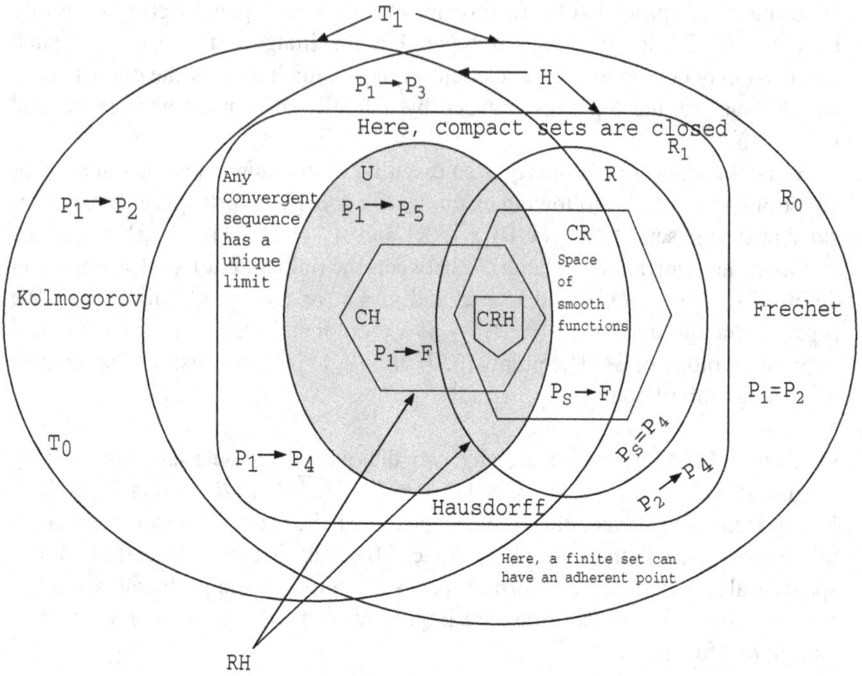

Fig. 2.2 Relationships between separation axioms presented in a Venn diagram. Part 1: the normal spaces are not included here. *Circles* represent classes of spaces fulfilling separation axioms, together with their inclusion and intersection properties. Each space is identified by an abbreviation (H = Hausdorff) and the text shows the corresponding axiom of separation. The *shaded area* represents the regular Hausdorff (T_3) space. The two *inside ovals* represent topological spaces where the separation axioms involve functional separation (definition F)

is usually called a *compact*, and a compact metric space is called *compactum*. An example of a compactum is any finite discrete metric space. A *continuum* is a connected compactum. The image of a compact set through a continuous function into a Hausdorff space is a compact set. As an immediate consequence, a continuous function defined on a compact space is bounded and has a maximum and a minimum.

Although compactness is a global property of a space, it can also be obtained starting from local level. We define a weaker request for compactness, i.e., a *local compact* space as a Hausdorff topological space with the property that any element has at least one compact neighborhood. A local compact space X can always be submerged into a larger topological compact space $\tilde{X}$ such that $X \sqsubset \tilde{X}$ and $\tilde{X} \setminus X = \omega$ (Alexandroff's compactification). The extra element ω is called the point at infinity. In the case of $\mathbb{R}^2 \simeq \mathbb{C}$, $C \cup \omega = \tilde{C}$ is called the extended complex plane. A local compact linear topological space has finite dimension. There are also refinements of the compactness property, like precompact, paracompact,

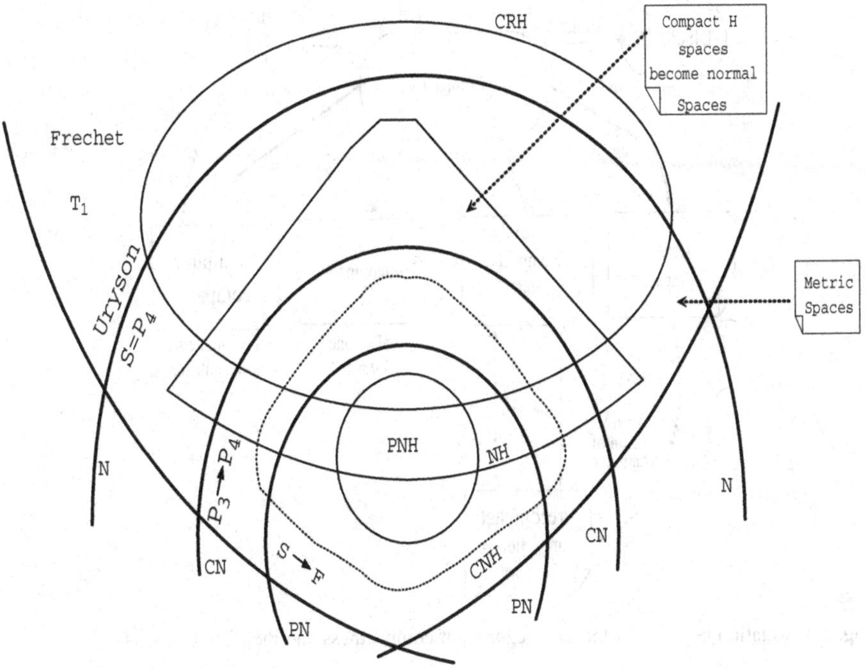

Fig. 2.3 Relationships between separation axioms in a Venn diagram, Part 2: the normal spaces are included. This figure is a zoom in of Fig. 2.2, and the space CRH has the same signification. The *thicker boundaries* represent topological spaces where the separation axioms involve functional separation (definition F)

relatively compact, countable compact, etc., but we do not need these concepts in our book. Basically, they occur whenever we relax one of the three properties defining compactness [68, 160, 291] (see Fig. 2.4).

An *open map* is a function between two topological spaces which maps open sets to open sets. Likewise, a *closed map* is a function which maps closed sets to closed sets. The open or closed maps are not necessarily continuous. A continuous function between topological spaces is called *proper* if inverse images of compact subsets are compact. An *embedding* between two topological spaces is a homeomorphism onto its image.

2.1.3 Weierstrass–Stone Theorem

How is it possible for the Taylor series to exist? That is, how is it possible to know all the values of a continuous function from just knowing a countable sequence of number, the coefficients of the Taylor series. The answer is related to the separation axioms and it is the Weierstrass–Stone theorem. This theorem is also the answer for

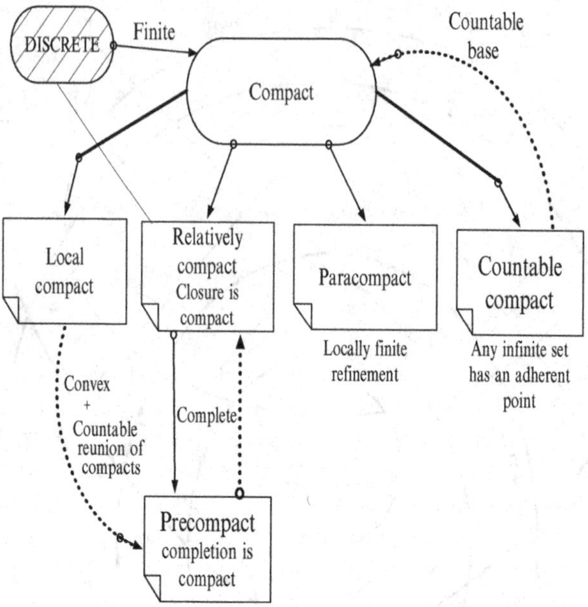

Fig. 2.4 Relation between different categories of compactness and their implications

the questions in Sect. 2.2, namely how is possible to find the values of a function in an n-dimensional domain, knowing only the values of the function in the $(n-1)$-dimensional boundary? Weierstrass proved that a real function defined on $[0,1]$ is the uniform limit of a series of polynomials. Later on Stone explained that the essential property of the polynomials that allow such a perfect approximation is that they form an algebra.

Theorem 1 (Weierstrass–Stone). *A subalgebra $\mathcal{A}$ of the Banach algebra of $C_0(X)$ continuous real functions defined on a Hausdorff compact space X, is dense in $C_0(X)$ if and only if:*

1. $1 \in \mathcal{A}$.
2. $\forall x \neq y \in X, \exists f \in \mathcal{A}$ *such that* $f(x) \neq f(y)$.

The first condition actually requires $\forall x \in X, \exists f \in \mathcal{A}$ such that $f(x) \neq 0$. We meet this condition if we try to generate a Hausdorff linear topological space. The algebraic structure of the functions $\mathcal{A}$ is required to have included in $\mathcal{A}$ the elements $Sup(f,g)$ and $Inf(f,g)$ for $\forall f, g \in \mathcal{A}$. The second condition requires that the algebra $\mathcal{A}$ "separates" points in X, in the sense of the $\mathcal{F}$ form of separation, like in the case for example when X is a completely regular Hausdorff (CRF) space. For details about the proof and Banach algebras one can consult, for example, [160] and references cited therein at page 516. Basically the idea is that any real continuous function defined on a Hausdorff compact X can be infinitely well approximated with other functions selected from a closed subalgebra of $C_0(X)$.

The Weierstrass–Stone theorem tells us that any vector-valued continuous function, no matter how complicated it is, can be infinitely well approximated with simpler functions g_α (where α is a label), as long as these simpler functions form a Banach algebra $\mathcal{A}$, i.e., $\mathcal{A} \ni g_\alpha \to f$. Moreover, if $\mathcal{A}$ is a separable space (to be defined later), then we have a countable basis of continuous functions, $\alpha \simeq n$, and consequently we can express f, for all $x \in X$, by a (maximum) countable set of coefficients associated with f approximating series. Since $\mathcal{A}$ is an abstract Banach algebra which $\mathcal{F}$ separates X, there is freedom to choose its elements, i.e., such a richness of examples: Taylor polynomials, orthogonal polynomials, trigonometric series, etc. The Weierstrass–Stone theorem can be equally applied to complex functions, with an additional request: $\forall g \in \mathcal{A}, \overline{g} \in \mathcal{A}$, where $\overline{g}$ is the complex conjugation.

We have two important corollaries. The space of polynomials defined on a compact $C \in \mathbb{R}^n$ with coefficients in a seminormed vector space $\mathbf{V}$ is dense in the space of continuous bounded functions defined on C with values in $\mathbf{V}$. The second corollary of the Weierstrass–Stone theorem allows us to approximate any complex vector-valued continuous function defined on the unit complex circle $S_1 \subset \mathbb{R}^2$ with trigonometric polynomials [291, Chap. XXII]. This corollary has important consequences for differential systems on closed curves and surfaces. Namely

Lemma 1. *Trigonometric polynomials with coefficients in* $\mathbf{V}$ *are a dense set in* $\{f : \mathbb{R} \to \mathbf{V} | f\, \textit{continuous, periodic}\}$.

2.1.4 Connectedness, Connectivity, and Homotopy

A topological space X is *connected* if it is not the disjoint reunion of two or more nonempty open sets. Connected spaces have a very interesting property: the only sets with empty boundary are the total space and the empty set. We can introduce a stronger type of connectedness through the concept of *arc* or *path*. Let $x, y \in X$ be two arbitrary points in a topological space. We have

Definition 2. A path from x to y is a continuous map $\Gamma : [0, 1] \to X$ such that $\Gamma(x) = 0, \Gamma(y) = 1$. An arc from x to y is a path which is also a homeomorphisms onto $\Gamma[0, 1]$.

So, an arc is a path which has also a continuous inverse.

Definition 3. The topological space X is *pathwise-connected* (*or arcwise-connected*) if any two of its points can be joined by a path (by an arc).

Some authors do not make a difference between path and arc in this context, and many references use the term path-connected instead of pathwise-connected, etc. Every path-connected space is connected, but not conversely. A traditional example is the graphics of the real function $\sin(1/x)$ which is in *one-piece* in $\mathbb{R}^2$ but there is no path between the points $(-1/\pi, 0)$ and $(1/\pi, 0)$ of its graphics. Any path-connected Hausdorff space is also arc connected, so again we want to emphasize the importance of axioms of separation. Connectedness is a topological invariant.

Finally, there is third type of criterion for connectedness. If any loop (closed smooth path) in the space is contractible to a point (can be smoothly deformed to a point) the space is called *simply connected* or 1-connected. Such a space is in one piece (connected) and has no "holes." The space is *n-multiply connected* if it is $(n-1)$ multiply connected and if every map from the n-sphere into it extends continuously over the $(n+1)$-disk. By *sphere* we mean here just the boundary of a sphere, for example in an n-dimensional normed space the $(n-1)$-sphere is the set $\{x/\ ||x|| = R\}$. The $(n-1)$-dimensional sphere is the boundary of an n-dimensional disk. The n-connectedness property is a generalization of pathwise connectedness, from paths to higher dimension surfaces.

Let X be a space and a function $f : X \to X$. An element $x_f \in X$ is a *fixed point* for the application f if $f(x) = x$. Also, a set $A \subset X$ is an *invariant set* if $f(A) \subset A$. Any continuous function defined on a real interval $[a, b]$ has at least one fixed point. The fixed point theorems [52] are successfully applied in field theory, biological problems and logistic equations, dynamics of population [327], and in mathematical economics. One of the most important applications is about iterated maps [93, 94]. A theorem due to Tikhonov [160, 312], enounces that compact and convex sets in a Hausdorff local convex space have the fixed point property.

If all the closed smooth curves (loops) in X can be continuously deformed one into another, we call this property *homotopy*. More rigorous

Definition 4. Let $\Phi : [0,1] \to M$, and $\Psi : [0,1] \to M$ be piecewise smooth closed paths on a manifold M. A *homotopy* from Φ to Ψ is a continuous function $\gamma : [0,1]^2 \to M$ such that $\forall t \in [0,1], \gamma(0,t) = \Phi(t), \gamma(1,t) = \Psi(t)$, and $\forall s \in [0,1]$, the path $\gamma(s,t)$ parameterized by t is closed and piecewise smooth.

All loops in X belong to the same equivalence class with respect to homotopy equivalence relation, so the group generated by the homotopy classes of X via the composition of curves is trivial identity. We call this group, *homotopy group* of X, and we denote it with $\pi_1(X)$. In algebraic topology one can prove that the groups of homotopy are topological invariants [235, 242].

An interesting result combining some of the concepts we introduced so far is this: any local homeomorphism from a compact space to a connected space is a covering, see Definition 1. The proof of this theorem is based on the fact that the local homeomorphism still preserves the property of being open, and the compactness of C insures that we can always choose a finite sub-cover from any open cover of it. Being finite, we can always choose its neighborhoods small enough to be pairwise disjoint, so all the conditions of being a covering map can be accomplished.

2.1.5 *Separability and Basis*

A metric space is *separable* if it has a countable dense subset Y, $Y \subset X, \bar{Y} = X$, where $\bar{Y}$ is the closure of Y, i.e., Y and all its adherent points (the boundaries).

Usually, the set Y is called *basis*, and if X is separable, members of Y can approximate any $x \in X$ as closely as we like. One of the Weierstrass theorems shows that the set of polynomials is a dense set in $C_0([0,1])$, so continuous real functions on a compact space can be approximated with polynomials to the best extent.

2.1.6 Metric and Normed Spaces

Metric spaces deal with *completeness* property. A metric topological space (M, τ, d) is a topological space (M, τ) endowed with a positive symmetric function $d : M \times M \rightarrow \mathbb{R}^+$ called *distance*, fulfilling the triangle inequality $\forall x, y, z \in M, d(x,z) \leq d(x,y) + d(y,z)$, and $d(x,y) = 0 \leftrightarrow x = y$. In a metric space M we can define an open ball (or disk) of center $x_0 \in M$ and radius $R \in \mathcal{R}^+$ as $B(x_0; R) = \{x | d(x, x_0) < R\}$. Any metric space is Hausdorff, by inheriting from the common real topology. In a metric space we can define *bounded* sets, if they can be enclosed in a certain ball. A compact metric space is separable. A linear space where we defined a nonnegative real function (a norm) $|| \cdot ||$ which is positively homogenous, subadditive and is zero only in the origin of the linear space is a *normed space*. A normed space is a metric space with the relation $d(x,y) = ||x - y||$, and consequently has all the properties of metric spaces. In a normed space the topology is normed induced and we have convergency in norm (the strong convergency). Any metric space M can be *completed* to $\overline{M}$ by adding to M the limits of all its Cauchy sequences. In a complete metric space all Cauchy sequences are convergent to a certain, unique limit. In a compact metric space any sequence contains a convergent subsequence. A complete normed linear space (where the metric is induced by a norm defined in the linear space) is called a *Banach* space.

A complex bilinear continuous symmetric form defined on a linear vector space $< \cdot, \cdot >: V \rightarrow \mathbb{C}$ is called a *scalar* or *inner* product. A space together with a scalar product, $(X, < \cdot, \cdot >)$ is *Euclidean*. For example on the linear topological space of integrable (in what ever sense integrability is needed) functions defined on a space X we define the scalar product

$$< f, g >= \int_X f(x) g^*(x) dx,$$

with g^* complex conjugated. The scalar product induces a norm, and obviously a distance $||f|| = \sqrt{< f, f >}, d(f,g) = \sqrt{< f - g, f - g >}$. A *Hilbert space* is a complete Euclidean space. The scalar product can measure the property of being *orthogonal* which generalizes the linear independence property in a geometric way. A maximal linear independent set of elements in X is a *basis* in X, and if X is Euclidean and the basis elements are mutually orthogonal and of unit norm, it is called *orthonormal basis*. Special functions, like orthogonal polynomials, spherical harmonics, etc. (Sect. 18.3), form orthonormal bases in spaces where the integral of the square magnitude of the functions are finite, $L_2(X)$.

The key theorem about representation of functions is the following:

Theorem 2. *Every separable Hilbert space H_s has a countable orthonormal basis $B_N \subset H_s$, i.e., $\bar{B}_N = H_s$.*

The following chapters, and all representation formulas theory, are entirely based on this result. It means that on a Hilbert space, any element can be approximated as good as we want with elements from this countable (discrete) basis. As strange as it may look, there are nonseparable Hilbert spaces in physics. For example in canonical quantum gravity, the space of functions defined on connections, A, modulo gauge transformations G, $L_2(A/G)$, is nonseparable [179].

2.2 Elements of Homology

The meaning of homology will become more transparent when we will use it in the Poincaré Lemma, and in compact boundary representation formulas (Sect. 3.1.4). For reference on the topics we suggest the bibliography [112, 235]. An oriented *p-simplex*, $p > 0$ integer, in $\mathbb{R}^n$ is generated by an ordered system of $p+1$ vectors, and it is the p-dimensional manifold

$$\sigma^p = [v_0, \ldots, v_p] = \left\{ v \in \mathbb{R}^n \mid \sum_{i=0}^{p} t_i v_i, \sum_{i=0}^{p} t_i = 1 \right\}.$$

Basically, the generalization of a segment (1-simplex), a triangle (2-simplex), and a tetrahedron (3-simplex) is to higher dimensions. A p-simplex is topologically homeomorphic with a p-ball. The subset $t^i = 0$ is an $(p-1)$-plane, or face, and the end points of the vectors are the vertices. A *simpliceal complex* is a set K of simplexes constructed such that all their faces also belong to K, and any two simplexes in K are either disjoint, or their intersection is a common face of each of them. A topological space homeomorphic to a simpliceal complex is called *triangulated.* In the following we work only on these triangulated spaces. Based on the triangulation K of a given manifold we can construct the Abelian groups $C_p(K)$, $p = 0, \ldots, n$ freely generated by the oriented p-simplexes of K, with integer coefficients, called the *group of chains* (not to be confounded to sets of continuity of order k !). We define the linear *boundary operators* as

$$\partial_p : C_p(K) \to C_{p-1}(K), \tag{2.1}$$

with the action $\partial_p \sigma^p = \sum_{j=0}^{p} (-1)^j [v_0, \ldots, v_{j-1}, v_{j+1} \ldots, v_p]$ creating thus a $(p-1)$-simplex. It is easy to verify that the boundary operator is a group homomorphism, $\partial_0 c_p = 0$, and

$$\partial_{p-1}\partial_p = 0, \tag{2.2}$$

which is the central property of homology, and somehow the main philosophy of the compact surfaces, contours, boundaries in general:

The boundary of a boundary is the empty set.
The immediate consequence in cohomology is that
the external derivative of order two is always zero.

Like we mention in Chap. 1, again a pure algebraic property like skew-symmetry of ∂_p provides a deep geometrical result. The kernel of the boundary operator, $Z_p(K) = \mathrm{Kerr}(\partial_p)$, is a subgroup of the group of chains, namely the group of boundary-less chains which are called p-cycles. Also the image of the boundary operator is called the group of the p-boundaries $B_p(K) = \partial_{p+1}(C_{p+1}(K))$. So, basically we have for each p the following succession of (normal) subgroups: $B_p \subset Z_p \subset C_p$. It is easy to notice that we can construct the quotient (factor) groups $C_p(K)/Z_p(K)$, $C_p(K)/B_p(K)$ and $Z_p(K)/B_p(K)$, and we have the group homomorphism $Z_p \sim C_p(K)/B_{p-1}(K)$. The quotient group

$$H_p(K) = Z_p(K)/B_p(K), \tag{2.3}$$

namely the *homology group* of order p of K. This factorization of p-cycles modulo p-boundaries over K introduces an equivalence relation in the group of cycles. In other words, two p-cycles of K are homologous if their difference is a p-boundary. Being Abelian freely generated, all the homology groups are isomorphic with some $\mathbb{Z}^n$ group. The rank of H_p group counts the number of p-dimensional holes of K. The rank of a group is smallest cardinality of its generating set. For example, $H_0(S_n) \sim H_n(S_n) \sim \mathbb{Z}$ and $H_p(S_n) \sim \{0\}$ for $p \neq 0, n$. A $T_2 \subset \mathbb{R}^3$ torus has the homology described by $H_0(T_1) \sim \mathbb{Z}$, $H_2(T_1) \sim \mathbb{Z}^2$, $H_3(T_1) \sim \mathbb{Z}$, and $H_p(T_1) \sim \{0\}$ for the rest of p.

We define the Euler characteristic χ of K the expression

$$\chi(K) = \sum_{p=0}^{n} (-1)^p \operatorname{rank} H_P(K), \tag{2.4}$$

which is one of the essential topological invariants for the Gauss–Bonnet formula (see Theorem 20) applied to closed Riemannian manifolds and for the Euler–Poincaré formula. For example $\chi(S_1) = 0, \chi(S_2) = 2, \chi(T_1) = 0, \chi(T_2) = -2$, etc. The Euler characteristic defines the genus g of a closed orientable surface by $g = (2 - \chi)/2$, which can be loosely understood as the number of "handles" of the surface.

2.3 Group Action

Let X be a topological space and G a topological group (that is a group which is also topological space and the two structures are reciprocal compatible). We say that G acts on X (from the left) if there is a continuous map $m : G \times X \to X$ such that

1. $m(g, m(h, x)) = m(gh, x)$ for $g, h \in G, x \in X$
2. $m(e, x) = x$, for $x \in X$

The entity (X, G, m) is called a *G-space*. For an efficient introduction in the theory of group actions from the differential geometry point of view we recommend the text [81], while for more technical details and applications we recommend [242]. We have the following definitions. The set $G_x = \{g \in G | m(g, x) = x\}$ is called *isotropy group* of x (or stabilizer subgroup of x). The set $O_x = \{m(g, x) | g \in G\}$ is called the *orbit* of x. The set of all orbits is denoted X/G and it is called orbit space and it is a topological space through the quotient induced topology with respect to the canonic projection $x \to O_x$.

- The action of G on X is *free* if the isotropy group is trivial for all x.
- The action of G on X is *proper* if the map $\theta : G \times X \to X \times X$ given by $(g, x) \to (x, m(g, x))$ is a proper function.
- The action of G on X is *transitive* if it possesses only a single group orbit, i.e. if all elements are equivalent. The G-space (X, G, m) is a *homogeneous space* if G acts in a transitive way.

The *principal homogeneous space* (or torsor) of G is a homogeneous space X such that the isotropy group of any point is trivial. Equivalently, a principal homogeneous space for a group G is a topological space X on which G acts freely and transitively, so that for any $x, y \in X$ there exists a unique $g \in G$ such that $m(g, x) = y$. If X is a G-space with proper action the quotient space X/G is Hausdorff. All these properties and definitions can be extended if the space X is a differentiable manifold, and G is a Lie group acting on X, case in which the structure (X, G, m) is called a G-manifold. Moreover, if the action of G is proper and free X/G has a differentiable manifold structure and the canonical projection $X \to X/G$ is a submersion.

Chapter 3
The Importance of the Boundary

How is it possible to describe any analytic or harmonic function on a compact set in terms of much simpler "construction blocks" like polynomials? Or, how is it possible to know the values of a function inside a compact domain, by knowing only its values on the boundary? Well, these simplifications are possible because the "bricks" are actually organized in complicated and versatile structures. For example the B^* algebras. And in addition, the compact domains are certainly among the simples ones, being always reducible to finite reunions. Actually, a complicated structure like a B^* algebra, defined by 24 axioms (out of which 13 axioms on commutative algebras, five axioms on norm, one for completeness, and five more specific axioms) can be realized by continuous functions defined on a compact set. It is not the only example. The space l_1 of complex sequences with norm given by the sum of the modules of the terms is isomorphic with the algebra of functions whose Fourier series is absolutely convergent. Also, a compact Hausdorff space, with topology induced by distance, is homeomorphic with a compact subset of $[0,1]^{\mathrm{N}}$. Any two separable Hilbert spaces are isomorphic, and so on. These similarities bring a unifying point of view: objects of apparently distinct nature, like Weierstrass–Stone theorem on function approximation, Wiener theorem on absolutely convergent Fourier series, spectral expansion of self-adjoint operators, the theorems of Tikhonov, Stone–Čech, or the fixed-point theorem of Brouwer, the Cauchy formula on complex functions, the Green representation theorem, the Poincaré Lemma, etc. actually provide the same fundamental truth: simplification by approximation is possible on compacts.

3.1 The Power of Compact Boundaries: Representation Formulas

The most fascinating analytical properties of compact boundaries embedded in differential manifolds are the representation formulas. We present in the following a review on the most important *representation formulas* for different dimensions of

A. Ludu, *Nonlinear Waves and Solitons on Contours and Closed Surfaces*, Springer Series in Synergetics, DOI 10.1007/978-3-642-22895-7_3,

the boundary. The general problem is the following: we have a domain $D \in \mathbb{R}^n$, and its boundary ∂D, in our case a compact surface. The representation formulas allow calculation of the values of a smooth and *harmonic* function (usually is enough to be of class $C_2(D)$, and the exact definition for harmonic will be specified for each dimension in particular) defined on $\bar{D}$ (closure of D) in all points of the interior of D, $\int D = D \; \partial D$, if we only know the values of the function, and of its partial derivatives, on the boundary ∂D. For $n = 1$ this representation is called Taylor series, for $n = 2$ is called Cauchy integral formula, for $n = 3$ it is called Green identity, etc., and in general all these are the expression of the Poincaré Lemma and the generalized Stokes theorem.

3.1.1 Representation Formula for n = 1: Taylor Series

If $n > 0$ is an integer and $f : [a,x] \subset \mathbb{R} \to \mathbb{R}$ is $C_n([a,b])$ and $C_{n+1}(a,b)$ function, then

$$\left| f(x) - \sum_{k=0}^{n} \frac{(x-a)^k}{k!} f^{(k)}(a) \right| \to 0, \text{ if } x \to a. \tag{3.1}$$

Let us retain the vital importance of the compact character of the $[a,x]$ interval. In other words, we know (with good enough precision) all the values of the function on a continuous interval, if we know just a discrete set of values, namely the derivatives of the function in one point. This theorem is a magic conversion of the continuous into countable and of the global into local. The truth beyond the power of representation of the Taylor theorem consists in the nature of the topology of both the real axis, and the Hilbert space of continuous functions. These Hilbert spaces are separable so they admit countable orthonormal bases by definition. From here we can represent any continuum through an at most countable set of numbers, which is nothing but the set of the coefficients of the Taylor series. So, in the real case, the representation formula is a consequence of the discrete/continuous play in the real topology. We also mention the important fact that a continuous real function on a compact real interval is bounded and attains its bounds.

3.1.2 Representation Formula for n = 2: Cauchy Formula

Let $D \subset \mathbb{C}$ and let $f : D \to \mathbb{C}$ be analytic. Let z_0, r such that $D_1(z_0, r) \equiv \{z \| z - z_0| < r\} \subset D$. For all $z \in D_1(z_0, r)$ we have

$$f(z) = \frac{1}{2\pi i} \int_{\partial D_1(z_0,r)} \frac{f(z')}{z' - z} dz'. \tag{3.2}$$

In other words, if the function is smooth enough on D (i.e., analytic) the values of the function inside any domain are known if we know the values of the function on its boundary [118, 312]. A complex function is *analytic* if it is differentiable and its derivative is continuous. Actually, this further guaranties the existence of all higher-order derivatives. A complex function $f(z) = g(z) + ih(z)$, with $g, h : D \to \mathbb{R}$ is differentiable if its components fulfill the Cauchy–Riemann conditions $g_x = h_y, g_y = -h_x$. The Cauchy–Riemann conditions actually can be written in vector form as $\nabla g \cdot \nabla h = 0$, in other words requesting the families of curves $g =$ const., $h =$ const., to be orthogonal on $\mathbb{C}$. In other words, if $V = (g, h, 0)$ is a flow, the Cauchy–Riemann conditions are equivalent to $\nabla \times V = 0$, i.e., an irrotational flow. The Cauchy–Riemann conditions are also equivalent to the existence of the complex derivative df/dz, or to the cancelation of the derivative with respect to the complex conjugation of the argument, i.e., $\partial f/\partial z^* = 0$. This last condition is equivalent to the request for harmonicity of the components, $\Delta g = \Delta h = 0$.

The power of the Cauchy representation formula is based on the special properties of analytic functions. If a function $f(z)$ is analytic in a domain D, then the contour integrals of f on any two homotopic loops are equal. We recall that two curves are homotopic (see Definition 4) if they can be deformed smoothly one into another. But what is beyond the Cauchy theorem? Actually the reason for the existence of the powerful Cauchy integral formula is double: on one hand the special topology of the plane, and on the other the continuity of the function. The traditional proof begins with a very simple structure, a triangle in the complex plane. One can prove that an analytic function on a triangle has zero integral along its boundary. This is because one can split any triangle into four smaller triangles, and so on, like in a fractal image. The topological limit of this construction exists, because all these triangles are closed sets in the plane topology. So, by a repeated process of division, we can reduce the perimeter of all these triangles to zero, and then the function, being continuous, will be forced to cancel over this boundaries.

3.1.3 Representation Formula for n = 3: Green Formula

Let us have a domain $D \subset \mathbb{R}^3$ with a boundary ∂D with smooth normal, and two functions $\Phi, \Psi \in C_2(D)$. The following integral relation exists (Green's second identity)

$$\iiint_D (\Phi \Delta \Psi - \Psi \Delta \Phi) d^3x = \iint_{\partial D} \left(\Phi \frac{\partial \Psi}{\partial N} - \Psi \frac{\partial \Phi}{\partial N} \right) dA, \tag{3.3}$$

where Δ is the three-dimensional Laplacian operator, and $\partial/\partial n$ is the directional derivative along the normal to ∂D, i.e., $N \cdot \nabla$. Then, if the function Φ is harmonic on the interior of D, $\Delta\Phi = 0$ then we have the Green representation formula

$$\Phi(\boldsymbol{r}) = \frac{1}{4\pi} \iint_{\partial D} \left(\frac{1}{|\boldsymbol{r} - \boldsymbol{r}'|} \frac{\partial \Phi}{\partial N'} - \Phi \frac{\partial}{\partial N'} \frac{1}{|\boldsymbol{r} - \boldsymbol{r}'|} \right) dA' \tag{3.4}$$

for $\forall \boldsymbol{r} \in D$. More generally, if

$$G(\boldsymbol{r}, \boldsymbol{r}') = \frac{1}{|\boldsymbol{r} - \boldsymbol{r}'|} + h(\boldsymbol{r}, \boldsymbol{r}'), \tag{3.5}$$

is the Green function associated with D and h is a harmonic function $\Delta' h = 0$ when $\boldsymbol{r}, \boldsymbol{r}' \in D$, then

$$\Phi(\boldsymbol{r}) = \frac{1}{4\pi} \iint_{\partial D} \left(G(\boldsymbol{r}, \boldsymbol{r}') \frac{\partial \Phi}{\partial N'} - \Phi \frac{\partial G}{\partial N'} \right) dA'. \tag{3.6}$$

If the Green function is chosen such that $G_D(\boldsymbol{r}, \boldsymbol{r}')|_{\boldsymbol{r}' \in \partial D} = 0$ we have a Dirichlet boundary problem, and if the Green function is chosen such that $(\partial G_N / \partial N')(\boldsymbol{r}, \boldsymbol{r}')|_{\boldsymbol{r}' \in \partial D} = -4\pi / S$, we have a Neumann boundary problem (where S is the area of ∂D). If ∂D is compact, then both the Dirichlet and Neumann problems provide unique and stable solution for elliptic partial differential equations on D, through the representation formula (3.6). These two conditions applied independently are too much constrain for hyperbolic or parabolic partial differential equations [64, 317]. The Green representation formula applies everywhere we have harmonic, or almost harmonic functions. In potential theory, and hence in potential flow, in electrostatics and magnetostatics, theory of minimal surfaces and application in surface tension driven systems, etc.

3.1.4 Representation Formula in General: Stokes Theorem

A more accurate mathematical approach on the Poincaré Lemma, based on homology (Sect. 2.2) and differential forms (Sect. 4.6), is done in Sect. 4.8. The generalized Stokes theorem is the coronation of all the representation formulas in the geometry of compact boundaries.

Let M be an m-dimensional manifold, and $B \subset M$, a compact, oriented b-dimensional submanifold (see Sect. 6.4 for details on definitions), with boundary $\Sigma = \partial B$. Let ω^{p-1} be a continuous differentiable $(p-1)$-form on M (Sect. 4.6). That is a $(p-1)$-covariant smooth tensor field $\omega_{i_1,\ldots,i_{p-1}}(x), x \in M$. Then we have

$$\int_B d\omega^{p-1} = \int_{\partial B} \omega^{p-1}, \tag{3.7}$$

where d is the *exterior derivative* acting on forms (Definition 23). We do not provide here the algebraic details (it can be found in Sect. 4.6) mainly because we are interested here to underline rather the geometric interpretation of the Stokes theorem, as a representation. In that, let us remember that we can triangulate B and ∂B (Sect. 2.2), and obtain the sequence of chain groups $C_p(B)$, $p = 0, \ldots, m$, and we can have the boundary operator ∂_p mapping one chain into another, like in the upper sequence in (3.8).

$$\begin{array}{ccccccccc} \cdots & \xleftarrow{\partial_{p-1}} & C_{p-1} & \xleftarrow{\partial_p} & C_p & \xleftarrow{\partial_{p+1}} & C_{p+1} & \xleftarrow{\partial_{p+2}} & \cdots \\ & & \downarrow & & \downarrow & & \downarrow & & \\ \cdots & \xrightarrow{d_{p-1}} & \Omega^{p-1} & \xrightarrow{d_p} & \Omega^p & \xrightarrow{d_{p+1}} & \Omega^{p+1} & \xrightarrow{d_{p+2}} & \cdots \end{array} \tag{3.8}$$

Now, for any given p-chain, and for any given differentiable $(p-1)$-form $\omega^{p-1} \in \Omega^{p-1}$ from the cotangent bundle associated to M we can calculate the integral

$$\int_{\partial_p C_p} \omega^{p-1}, \tag{3.9}$$

by decomposing the $(p-1)$-chain resulting from $\partial_p C_p$ into its constituent p-simplexes, and integrate ω^{p-1}, Lebesgue or Riemann, along each $(p-1)$-simplex of $\partial_p C_p$. This integration is a scalar product, a bilinear functional, defined on the $(p-1)$-chain space times the space of $(p-1)$-forms. Consequently, this scalar product maps the sequence of boundary operators acting toward the left in the upper sequence in (3.8), into a reverse sequence of operators, acting toward the right, in the sequence of corresponding spaces of form (cotangent bundles) Ω^p. See the bottom sequence in (3.8). Consequently we are in the possession of a splendid geometrical–algebraic tool, called the De Rham complex [112, 235], in which spaces of simplexes dually correspond to spaces of differential forms, and boundary operators correspond dually to exterior derivative operators, and this duality is actually represented by the generalized Stokes theorem. Indeed, a dual pair $(\partial_p C_p, \omega^{p-1})$ generates the integral in (3.9). If we move one step to the right in the De Rham complex (3.8), the differential form ω^{p-1} is mapped into its derivative $d\omega^{p-1} \in \Omega^p$, and the boundary $\partial_p C_p$ is mapped into its interior C_p. Since the boundary operator and the exterior derivative are dual, the geometrical fact that the boundary of the boundary is the null set has its dual into the *closure property* of the exterior derivative (4.15)

$$\partial_2 = \{\emptyset\} \leftrightarrow d^2 = 0.$$

All representation formulas presented earlier, or in other sections of the book, like Sect. 10.6, are based on this generalized Stokes equation. More details and examples on other special types of representations, especially those used in fluid dynamics, are provided in Sect. 10.6.

3.2 Comments and Examples

Geometrically, the concept of compact means closed and bounded, while algebraically compact means finite. Another example of duality is provided by the boundary of a boundary which is the empty space. A geometrical expression of this theorem is the Gauss–Bonnet theorem: the total curvature is constant no matter of smooth deformation of the surface. This geometric theorem has algebraic consequences in integrability and differential forms, i.e., in the "Poincarè Lemma." Finally, from the physical point of view, this boundary property relates to the existence of vortices or fields without sources on compact manifolds.

An interesting property of compact surfaces is the relation between the area of the surface and the number of dimensions of the embedding space. The area and volume of a sphere of radius R, $S_n = \{x \in \mathbb{E}^n \mid X_1^2 + \cdots + x_n^2 \leq R^2\}$, in an n-dimensional Euclidean space, like $\mathbb{R}^n$, are given by

$$\mathcal{A}[S_n] = \frac{2\pi^{\frac{n}{2}}}{\Gamma\left(\frac{n}{2}\right)} R^{n-1}, \quad \mathcal{V}[S_n] = \frac{\pi^{\frac{n}{2}}}{\Gamma\left(\frac{n}{2}+1\right)} R^n. \tag{3.10}$$

In Fig. 3.1, we plot the area and the volume of the unit sphere ($R = 1$) function of the number of dimensions n of the space. It is interesting to remark that a unit sphere has a maximum area in a space with seven dimensions, and a maximum volume in a space with five dimensions. It is also interesting to mention that the ratio between the area and the volume of the unit sphere, $\mathcal{A}[S_n]/\mathcal{V}[S_n] = n$, is just the dimensions of the space. In other words, when we increase the dimension of the space, more and more points of the interior of the sphere (and in general

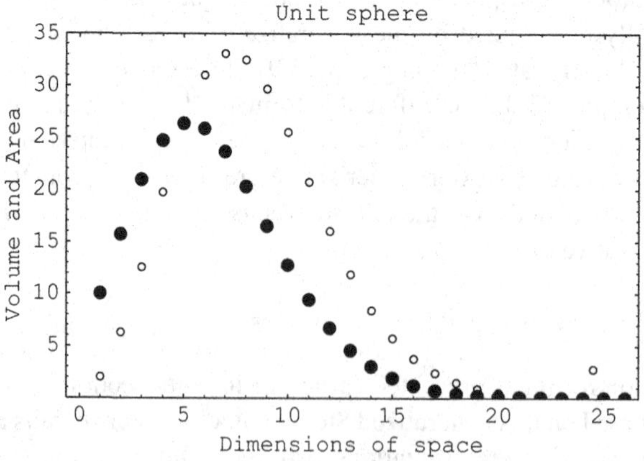

Fig. 3.1 Area (*white circles*) and volume (*black circles*) of the unit sphere, plotted in arbitrary units vs. the number of dimensions of the space

of any closed surface homeomorphic with a sphere) are concentrated toward the sphere surface. This is (see for example [207]) the most basic proof of existence of equilibrium temperature. In a statistic system of many free particles, where the phase space has a dimension of $n = 2 \times 3 \times 10^{26}$ or larger, almost all states of bounded energy are concentrated at the surface of a sphere of radius equal to the energy. So, almost all particles tend to have the same equilibrium temperature. On the contrary, in a space of any dimension, the ratio between the area and the volume decreases with increasing of radius. So, the larger the container, the less points are next to the surface. This fact may be an explanation of the fact that biochemical systems that require long time of slow transformations toward a final state, perform better in larger containers.

If we define the parameter *area over volume ratio* of a certain closed shape in an Euclidean space

$$AOV = \frac{Area}{Volume}(n, \sigma, \varsigma) \tag{3.11}$$

where n is the number of dimensions of the space, σ is a similarity parameter that measures "how large" is the object, and ς describes the shape. For $n = 3$ we have $AOV(3, \sigma, \varsigma) = C(\varsigma)/\sigma$, and for example $C(sphere) = 3$, $C(cube) = 6$, $C(cylinder) = 2(1 + \frac{R}{h})$, and so on. For the sphere we have

$$AOV(n, R, S_n) = \frac{2\Gamma\left(\frac{n}{2} + 1\right)}{R\Gamma\left(\frac{n}{2}\right)} = \frac{n+2}{R}, \tag{3.12}$$

so the AOV for the unit sphere is proportional to the number of dimensions of the space, and inversely proportional to the radius. That means that the larger the dimension of the space, the larger is the set of points in the area compared to those in the bulk.

A last interesting example is about unbounded smooth objects. Let us consider the function $f : [1, \infty) \to \mathbb{R}$, $f(x) = x^{-\alpha}$, $\alpha \in (0.5, 1)$. This function has an intriguing property. The surface of revolution produced by the rotation of the graphic of this function around Ox, between $x = 1$ and ∞, has infinite area, but its inside has finite volume. The infinite "funnel" obtained like that offer a paradox to the person who would like to paint it: one needs a fine amount of paint to fill it up, but it requests an infinite amount of paint to paint its surface.

Chapter 4
Vector Fields, Differential Forms, and Derivatives

The following results and some of the proofs, can be found in many excellent text books of differential geometry. For example [46] is a very readable and clear textbook with content based on theorems and proofs for geometrical objects in $\mathbb{R}^{2,3}$. Shifrin is also an excellent compact and short text rich in applications. For more abstract treatment (I was always puzzled by a book on geometry without any figures) especially on higher than three dimension differentiable manifolds we recommend the classic [158]. At the same level of abstraction, but more focused on specific topics we recommend [306] especially for applications on fiber bundles, [74] for applications concerning vector fields, and [119] for applications towards Lie groups and transformations. In between these levels of approach we also recommend for their wide range of *action* [19] for a very friendly general treatment of surfaces, or [162] as a very pictorial book on geometry with many applications.

The reason of using calculus on manifolds and differential geometry tools in physical applications is to solve physical problems as specific as possible in a mathematical frame as general as possible. By enhancing the mathematical generality of the approach one can increase the range for potential applications. In general the first attack on a physical problem is how to choose the appropriate working space. The next step is to choose an appropriate frame in that space. Choosing the space is basically a matter of topology, while choosing the correct frame is a matter of differential calculus on manifolds and differential geometry.

Topology, as theoretical physicist's primary tool, is mainly interested in objects whose properties are invariant under changes of the space. Topological objects like sets, neighborhoods, or curves are being classified according to their topological properties: closeness, compactness, connectivity, separability, etc., while topological spaces are classified by homeomorphisms. The topological properties which do not change under homeomorphisms are topological invariants. More specifically, if $\mathcal{X}, \mathcal{Y}$ are topological spaces, and $\psi : \mathcal{X} \to \mathcal{Y}$ is any homeomorphism (meaning f is bijective and bicontinuous function) those topologically invariant objects defined on $\mathcal{X}$ have equivalent counterparts on $\mathcal{Y}$ through f. Such theories based on classes of equivalence modulo homeomorphisms are useful not only to

A. Ludu, *Nonlinear Waves and Solitons on Contours and Closed Surfaces*, Springer Series in Synergetics, DOI 10.1007/978-3-642-22895-7_4,

investigate new objects in a given space (i.e., to check whether the new object belongs to such an invariant topological structure against homeomorphisms), but also to study new topological spaces. Let us exemplify. We start from a pair of homeomorphic topological spaces $(\mathcal{X}, \mathcal{Y})$ and some set of objects Π that form a topological invariant, i.e., $\Pi(\mathcal{X}) = \Pi(\mathcal{Y})$ or $h(\Pi(\mathcal{X})) = \Pi(\mathcal{Y}), h \in \hom(\mathcal{X}, \mathcal{Y})$. We choose a certain element $\alpha \in \Pi(\mathcal{X})$, and we begin to change the space $\mathcal{Y} \to \mathcal{Y}'$, while mapping $\alpha \to f(\alpha) \in \mathcal{Y}'$. If for a certain new space $\mathcal{Y}'$ the element $f(\alpha)$ is not anymore in the Π class, then $\mathcal{Y}'$ is not homeomorphic anymore with $\mathcal{X}$. For example, let us choose $\mathcal{X} = \mathbb{R}^3 - \{0\}$ (the punctured space) and let Π be the set of loops based on some point $x_0 \neq 0$ in $\mathbb{R}^3$ that can be smoothly deformed to a point (contractible loops). This set is a homotopy class in $\mathcal{X}$. For example the loop $\alpha = \{(\cos t, \sin t, 0) | t \in [0, 2\pi]\}$ belongs to this class, because we can always deform it to a point such that we can avoid the origin. Now, let us map this loop in the punctured plane $\mathcal{Y}' = \mathbb{R}^2 - \{0\}$. In this space this loop is not anymore contractible to a point, so it does not belong to the Π class anymore. Consequently, this map is not a homeomorphism, and hence $\mathbb{R}^3$ and $\mathbb{R}^2$ are not homeomorphic.

In differential geometry on manifolds we are interested in objects whose underlying geometrical properties are independent of any particular choice of a coordinate system. This request is very much related to the fundamental request of congruence in geometry: figures that differ only by rigid motions are congruent. The coordinate formulation of a certain object α can change from space to space, but the essential geometrical properties remain the same if the two spaces are connected by diffeomorphisms (i.e., infinitely differentiable functions with infinitely differentiable inverse). The concepts of smooth manifolds and differentiable maps on these manifolds create the most appropriate frame for such an approach. We note that in the following we will use the term *smooth* for a map (or function) which is of class C_∞ (called indefinite or infinite differentiable), and we will use the term *differentiable* for a map (or function) which is of class $C_k, k < \infty$.

4.1 Manifolds and Maps

The bottom model for a differentiable manifold M is a convenient topological space (for example one fulfilling certain decent separation axioms for the sake of the uniqueness of definitions based on limits and calculus) covered with partially overlapping local coordinate systems that can be changed from one another in a smooth manner. The only constraint is that both the degree of smoothness of the local coordinate transformations (e.g., continuous of a certain class k, or differentiable, or analytical, etc.), and the dimension of the local coordinate systems to be the same all over the manifold. Objects, for example, that begin in one end as a bounded two-dimensional surface (a stripe) and end up in the other end as a one-dimensional string, are not differentiable manifolds, though they may present a high interest for some physical studies.

Definition 5. We define an *n-dimensional real differentiable manifold* to be the pair $(M, \mathcal{A})$, where M is a Hausdorff topological space and $\mathcal{A} = \{(U_\alpha, \phi_\alpha)|\alpha = 1, 2, \dots\}$ is a countable *atlas* formed by *local coordinate maps*. Each such map consists in an open set $U_\alpha \subset M$ and a one-to-one function $\phi_\alpha : U_\alpha \to V_\alpha \subset \mathbb{R}^n$ onto an open connected subset V_α of $\mathbb{R}^n$, which satisfy the properties:

1. The atlas forms a countable open partition of M, $\bigcup_\alpha U_\alpha = M$.
2. $\forall \alpha, \beta, \quad \phi_\beta \circ \phi_\beta^{-1} : \phi_\alpha(U_\alpha \bigcap U_\beta) \to \phi_\beta(U_\alpha \bigcap U_\beta)$ is a smooth (infinitely differentiable C_∞) function.

A sketch of the definition is presented in Fig. 4.1. The coordinate charts induce in M a topological space structure inherited from $\mathbb{R}^n$. The degree of differentiability of the overlap functions $\phi_\beta \circ \phi_\alpha$ determines the degree of smoothness of the manifold: differentiable C_k-manifolds and smooth C_∞-manifolds also called analytic manifolds. Any Euclidean space is a smooth manifold with an atlas consisting of only one chart, the space itself $U_1 = \mathbb{R}^n$, and identity map $\phi_1 = \mathbf{1}$. Another useful example is provided by the unit n-dimensional sphere defined

$$S_n = \{(x^1, x^2, \dots, x^i, \dots, x^{n+1})|x^i \in \mathbb{R}, \sum_{i=1}^{n+1}(x^i)^2 = 1\},$$

realized as a hypersurface in $\mathbb{R}^{n+1}$. We can describe S_n as an n-dimensional real differentiable manifold with an atlas of two charts, namely:

$$U_1 = S_2 \backslash \{(0, \dots, 0, 1)\}, \quad U_1 = S_2 \backslash \{(0, \dots, 0, -1)\},$$

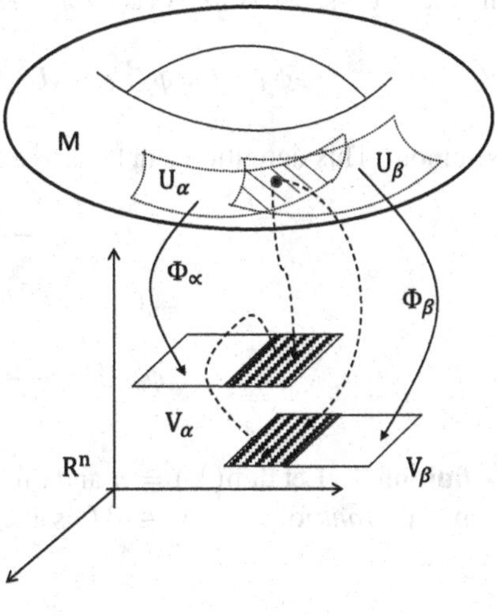

Fig. 4.1 A pictorial view of a smooth manifold

i.e., the unit sphere minus the north and south poles. The coordinate maps

$$\phi_\alpha : U_\alpha \to \mathbb{R}^n \simeq (y^i), \;\; \alpha = 1,2; \; i = 1,\ldots,n,$$

can be defined by the stereographic projections from the respective poles

$$\phi_\alpha(x^i) = \left(\frac{x^1}{1 \mp x^{n+1}}, \frac{x^2}{1 \mp x^{n+1}}, \ldots \right), \;\; i = 1,\ldots,n+1, \alpha = 1,2.$$

It is easy to check that $\phi_1 \circ \phi_2^{-1} : \mathbb{R}^n \backslash \{0\} \to \mathbb{R}^n \backslash \{0\}$ is a diffeomorphism (smooth bijective map), given by

$$\phi_1 \circ \phi_2^{-1}(y^1, y^2, \ldots, y^n) = \left(\frac{y^1}{\sum_{i=1}^n (y^i)^2}, \frac{y^2}{\sum_{i=1}^n (y^i)^2}, \ldots \right).$$

In addition to the defining atlas, one can always introduce more coordinate charts (U, ϕ) keeping the requirement that they are compatible with the given charts. This means that $\forall \alpha$, $\phi \circ \phi_\alpha$ is differentiable on the intersection $\phi_\alpha(U \cap U_\alpha)$. We can expand the atlas to include all compatible charts, and in this case we call the collection a *maximal collection* of charts. The maximal atlas is not any more countable, though.

Because the maps defining the local coordinates are one-to-one with the corresponding open sets in M, we can simplify the notation. While referring to certain local coordinates on a manifold we will ignore the explicit reference to the map ϕ_α defining the local coordinate chart.

Definition 6. A map $f : X \to Y$ between two smooth manifolds X, Y is *smooth* (or of class C_∞ called infinite differentiable) if its local coordinate expression is a smooth map in every coordinate chart, at any point of M.

In other words, $\forall x \in M$, $\forall (U_\alpha, \phi_\alpha)$ such that $x \in U_\alpha$, and $\forall (U_\beta, \phi_\beta)$, we have

$$\phi_\beta \circ f \circ \phi_\alpha^{-1} : \phi_\alpha(U_\alpha \cap f^{-1}(U_\beta)) \subset \mathbb{R}^m \to \mathbb{R}^m$$

is smooth. This definition can be also expressed as the diagram:

$$\begin{array}{ccc} U_\alpha & \xrightarrow[f]{} & U_\beta \\ {\scriptstyle \phi_\alpha}\downarrow & & \downarrow{\scriptstyle \phi_\beta} \\ \phi_\alpha(U_\alpha) & \xrightarrow[\phi_\beta \circ f \circ \phi_\alpha^{-1}]{} & \phi_\beta(U_\beta) \end{array} \tag{4.1}$$

Definition 7. Let $\dim(X) = n$ and $\dim(Y) = m$, and $f : X \to Y$ a differentiable map. The *rank* of f at $x \in M$ is the rank of the Jacobian matrix expressed in convenient local coordinates $(x^i), (y^j = f^j(x))$:

$$\operatorname{rank}(f) \equiv \operatorname{rank}(J) = \operatorname{rank}\begin{pmatrix} \frac{\partial f^1}{\partial x^1} & \frac{\partial f^1}{\partial x^2} & \cdots & \frac{\partial f^1}{\partial x^n} \\ \frac{\partial f^2}{\partial x^1} & \frac{\partial f^2}{\partial x^2} & \cdots & \frac{\partial f^2}{\partial x^n} \\ \cdots \\ \frac{\partial f^m}{\partial x^1} & \frac{\partial f^m}{\partial x^2} & \cdots & \frac{\partial f^m}{\partial x^n} \end{pmatrix}_{m\times n}$$

A *maximal rank* map on a set $A \subset X$ is a smooth function having its rank $= \min(n, m)$ for each $x \in S$.

There is another definition of the differential of the map in more geometrical terms [4]. This definition is valid for maps defined on real vector spaces, but it can be easily extended to manifolds by the local diffeomorphism provided by the atlas. Two functions $f, g : X \to Y$, where X, Y are vector spaces over $\mathbb{R}$ of dimensions n_X, n_Y, respectively, are *tangent* at $x_0 \in X$ if

$$\lim_{x\to x_0} \frac{||f(x) - g(x)||}{x - x_0} = 0. \tag{4.2}$$

Then we can define the differential $Df : X \to L(X, Y)$ as the map $Df(x)$ with the property that the function $g(x) = f(x_0) + L(x_0)(x - x_0)$ is tangent to f. Here L is the space of linear maps on $X \times Y$, i.e., $n_x \times n_Y$ matrices. In other words, the differential of f at x is given by the first-order terms in the Taylor expansion of f at x.

In Sect. 2.1.2 we defined embedding for topological spaces. For differential manifolds we define a *submersion* as a differentiable map $f : M \to N$ between differentiable manifolds whose differential is everywhere surjective. An *immersion* is a differentiable map between differentiable manifolds whose derivative is everywhere injective (an immersion does not need to be injective itself). The concepts of submersion and immersion are dual to each other. That is they are maximal rank maps such that if $\dim(M) < \dim(N)$ we have an immersion, while if $\dim(M) > \dim(N)$ we have a submersion. A stronger constraint is the smooth *embedding* which is an injective immersion and a topological embedding (i.e. homeomorphism onto its image) at the same time. An immersion (submersion) maps the coordinates in a faithful way, while an embedding is in addition topological or geometrical structure preserving.

4.2 Differential and Vector Fields

In Euclidean geometry we investigate spaces by using vectors (as subspaces of directions), we generalize vectors to tensors, and then to tensorial fields. In a similar matter, we can enrich the structure of a differentiable manifold with the help of curves defined on it. A curve defined on a differentiable manifold defines a direction, and a collection of such curves defines a linear space. Indeed, let us suppose that $\gamma : I \to X$ is a differentiable(C_k map) curve defined on the open $I \subset \mathbb{R}$ with values in a smooth n-dimensional manifold X. In the following we will call such a curve a *parameterized curve*. In local coordinates the curve is defined by n smooth differential functions $\gamma(t) = (x^1(t), \ldots, x^n(t))$ of the real variable $t \in I$. At each point $x = \gamma(t)$ the curve has an n-dimensional unit tangent vector defined by the derivative $\gamma'(t)$. In local coordinates we use to denote this tangent vector as

$$\mathbf{v} = \gamma'(t) = \left(\frac{dx^1}{dt}, \ldots, \frac{dx^n}{dt}\right) = \sum_{i=1}^{n} \frac{dx^i}{dt}\frac{\partial}{\partial x^i},$$

where we formally use the symbols $\partial/\partial x^i$ to represent a local basis for the components of this tangent vector in x.

Definition 8. The collection of all tangent vectors to all possible parameterized curves passing through a given point $x \in X$ is called the *tangent space* to X at x, and it is denoted $T_x M$.

The tangent space is isomorphic with an n-dimensional real vector space through the canonical application $\theta : T_x X \to \mathbb{R}^m$, $\theta_x(\xi^i(x)\frac{\partial}{\partial x^i}) = (\xi^i)$. The collection of all tangent spaces corresponding to all points of X is called *tangent bundle* and it is denoted as

$$TX = \cup_{x \in X} T_x X.$$

By the property of overlapping and differentiability of charts in the atlas, all the tangent spaces on a manifold can be smoothly connected.

Definition 9. A differentiable function $\mathbf{v} : X \to T_x X$ is called a *vector field* on the smooth manifold X.

In local coordinates $\mathbf{v}(x) = \sum_{i=1}^{n} \xi^i(x)\frac{\partial}{\partial x^i}$ where $\xi^i(x)$ are n differentiable real functions. A simple example is provided by the gradient field $\forall i, \xi^i = 1$ defined on the Euclidean space $X = \mathbb{R}^n$, i.e., $\nabla = \sum_{i=1}^{n} \frac{\partial}{\partial x^i}$. Any parameterized curve on a differentiable manifold has an associated vector field generated by its tangents existing in the tangent bundle. Conversely, for any vector in the tangent space at a point, we can define a unique parameterized curve (the integral curve or the flow) that passes through this point, and has its tangent equal to this vector. A vector field $\mathbf{v}$ is singular (non-singular) at a point x if $\mathbf{v}(x) = 0$ ($\mathbf{v}(x) \neq 0$).

Definition 10. Let X be a differentiable manifold, $I \subset \mathbb{R}$ open, and $\mathbf{v} \in TX$ a differentiable vector field on X. An integral curve of $\mathbf{v}$ at $x \in X$ is a parameterized curve $\gamma(u) : I \to X$ such that $\mathbf{v}(\gamma(u)) = \gamma'(u)$ for each $u \in I$.

We present in Sect. 4.3 a proof of the theorem of existence and uniqueness of integral curves for a particular case. For the general proof, especially related to dynamical systems applications we recommend [4, 68, 235, 242] books.

In Sect. 9.6.1 we present a local version of the theorem of existence and uniqueness of integral curves related to hydrodynamical systems.

An integral curve can be also interpreted as a one-parameter local group of transformations on X.

Definition 11. A set of vector fields $\mathcal{S}$ defined on a smooth manifold X is *rank-invariant* if the dimension of the linear space spanned by $\mathcal{S}$ along the flow of any of the vectors $\boldsymbol{v} \in \mathcal{S}$ is constant.

In the following we will use the mute convention for summation.

Definition 12. For a given vector field $\boldsymbol{v} = (\xi^i)$ defined on a differentiable manifold X, and a differentiable function $f : X \to \mathbb{R}^m$ we define *the action of* $\boldsymbol{v}$ *on the function* f in local coordinates by

$$\boldsymbol{v}(f)(x) = \xi^i \frac{\partial f}{\partial x^i}.$$

The quantity $\boldsymbol{v}(f)$ can be viewed as a linear operator acting on f, or as a function defined on the manifold X with values in $\mathbb{R}^m$, which generalizes the concept of derivative along a given direction. In some books this operator is also called the *directional derivative* (for example [299]) and is written as

$$D_{\boldsymbol{v}} f(x) = \nabla f(x) \cdot \boldsymbol{v}.$$

The above formula is obtained if we consider a parameterized curve γ on X and we identify $\gamma'(x(t)) = \boldsymbol{v} \in \mathbb{R}^m$. Then $(f \circ \gamma)'(x(t)) = \nabla f(x) \cdot \boldsymbol{v}$.

Definition 13. Any differential map $f : X \to Y$ induces a linear map

$$df : T_x X \to T_{f(x)} Y,$$

called the *tangent map* (or the *differential map*) and defined by the diagram

$$\begin{array}{ccc} T_x X & \xrightarrow[df]{} & T_{f(x)} Y \\ \theta_x \downarrow & & \downarrow \theta_{f(x)} \\ \mathbb{R}^n & \xrightarrow[(\phi \circ f \circ \phi^{-1})'(\phi(x))]{} & \mathbb{R}^m \end{array} \quad . \tag{4.3}$$

where $n = dim(X), m = dim(Y)$. That is $df = \theta_{f(x)}^{-1} \circ (\phi \circ f \circ \phi^{-1})'(\phi(x)) \circ \theta_x$.

Alternative notations for the tangent map are T_* [242] or f_*. There are three possible interpretations of the tangent map.

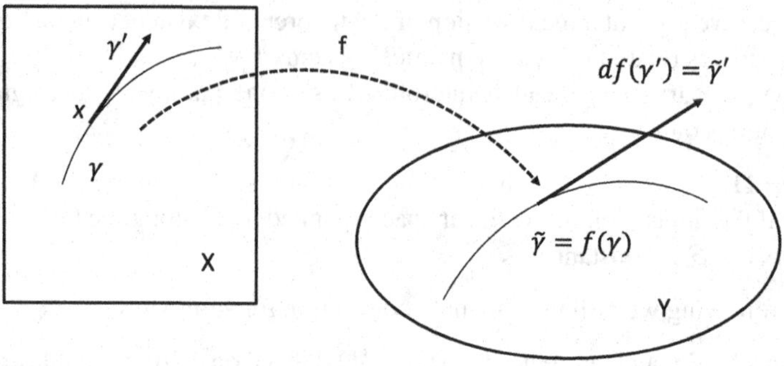

Fig. 4.2 The tangent map

The first one is related to curves: any parameterized curve $\gamma(t)$ on X is mapped by f into a parameterized curve $\tilde{\gamma}(t) = f(\gamma(t))$ on Y. Thus, f induces a map from the tangent vectors to γ at x to the corresponding tangent vectors to $\tilde{\gamma}$ at $f(x)$ (see also Fig. 4.2). The second interpretation of the tangent map is defined in terms of its action on tangent vectors $\boldsymbol{v} = (\xi^i) \in T_x X$ in the local coordinates:

$$df(\boldsymbol{v}) = \xi^i \frac{\partial f^j}{\partial x^i} \frac{\partial}{\partial y^j} \in T_{y=f(x)} Y, \tag{4.4}$$

where (x^i) and (y^j) are local coordinates in X and Y, respectively. In this context, the tangent map is the Jacobian matrix of the map f at x, acting as a linear transformation on the tangent vectors. If $\{(1, 0, \ldots, 0), \ldots, (0, \ldots, 0, 1)\}$ is a local basis in $T_x X$, then df transforms it into a basis in $T_{f(x)} Y$ of the form $\{\frac{\partial f}{\partial x^1}, \ldots, \frac{\partial f}{\partial x^n}\}$.

The third interpretation of the tangent map is in terms of the action of a vector field. In this context, if we have $f(x) = (f^1, \ldots, f^m)(x)$, then the action of the tangent map on a tangent vector, $df(\boldsymbol{v}(x)) = \boldsymbol{v}(f^j(x)) \frac{\partial}{\partial y^j}$, is nothing but the action of this vector, considered as a vector field in the tangent bundle, over the components of f in a local basis in Y. In other words, the tangent map is the directional derivative $df(\boldsymbol{v}) = D_{\boldsymbol{v}} f(x)$. Consequently, we write here one of the most useful equations in the differential geometry of surfaces, namely the relation between the tangent map of a map f, the action of a vector field $\boldsymbol{v}$, and its directional derivative

$$df(\boldsymbol{v}) = D_{\boldsymbol{v}} f(x) = \boldsymbol{v}(f). \tag{4.5}$$

Let us have a differentiable manifold X of dimension n. At every point $x \in X$ we can define the dual of the tangent space, the cotangent space $T_x^* X$. The space of skew-symmetric covariant tensors of rank 1 on X is a linear subspace $\Omega^1 T_x^* X \subset T_x^* X$ of the cotangent space. Its elements are called 1-*forms*, $\omega(x)$. In local coordinates (x^i) the 1-form is denoted $\omega = \omega_i dx^i$, where the dx^i form an abstract skew-symmetric local basis for the cotangent space. The 1-form is precisely defined by its action of differential vector fields

$$(\omega; v) = \left(\omega_j dx^j; \xi^i \frac{\partial}{\partial x^i}\right) = \sum_{i=1}^{n} \omega_i \xi^i \in \mathbb{C}.$$

This definition can be generalized to *differentiable k-forms*, namely skew-symmetric covariant tensor fields of rank $0 \leqslant k \leqslant n$ defined on X, $\omega = \omega_{i_1 i_2 \dots i_k} dx^1 \wedge dx^2 \wedge \dots \wedge dx^k$. Here the $\wedge$ "exterior" product represents the skew-symmetric property. The space of k-forms is denoted $\Omega^k T_x^* X = T_x^* \otimes T_x^* \otimes \dots \otimes T_x^*$, k times (or simply denoted $\Omega^k T_x^*$). It has dimension dim $(\Omega^k T_x^* X) = n!/(k!(n-k)!)$. The local basis at x, $dx^1 \wedge dx^2 \wedge \dots \wedge dx^k$, has the properties:

1. $dx^1 \wedge dx^2 \wedge \dots \wedge dx^k = 0$ if two indices are equal.
2. Permutation of two indices changes the sign.
3. The expression $dx^1 \wedge dx^2 \wedge \dots \wedge dx^k$ is linear.

In general, the local basis of differentials for a k-form is a generalized cross-product in more than three dimensions. A 0-form is a differentiable function on X, a 1-form is a covariant vector field, and a maximal dimension n-form is the volume element.

The properties of differentiable forms make them extremely useful and valuable for all sorts of geometry problems. In some applications we can generalize the above (traditional) representation of differential k-forms in terms of skew-symmetric $k \times k$ matrices of complex numbers. We can construct, for example, $k \times k$ skew-symmetric matrices with entries taken from a contra-variant tensor field. For example, we can write the object $\omega^i_{jk} dx^i \wedge dx^j$ where $i, j, k = 1, \dots, n$ which is a contra-variant vector with respect to the superscript i, and a 2-form with respect to j, k.

Let us have a differentiable map between two manifolds $f : X \to Y$.

Definition 14. The dual map of the tangent map at x, i.e., $df^* : T_{f(x)}Y \to T_x X$, is called the pull-back (or codifferential) of f. One generalizes the pull-back to k-forms by $\Phi^* : \Omega^k T^*_{f(x)} Y \to \Omega^k T_x^* X$ namely

$$\omega'_{i_1, i_2, \dots, i_k} = \omega_{j_1, j_2, \dots, j_k} \frac{\partial f^{j_1}}{\partial x^{i_1}} \frac{\partial f^{j_2}}{\partial x^{i_2}} \cdots \frac{\partial f^{j_k}}{\partial x^{i_k}} \tag{4.6}$$

The pull-back relates to the tangent map between X and Y by the following expression

$$(\omega; df(v) = (f^*(\omega); v), \tag{4.7}$$

meaning that k-forms in Y act on the derivative $df(v)$ of the vector field v on X in the same way as the pull-back $f^*(\omega)$ of the forms in X act on vector fields v on X, see Fig. 4.3.

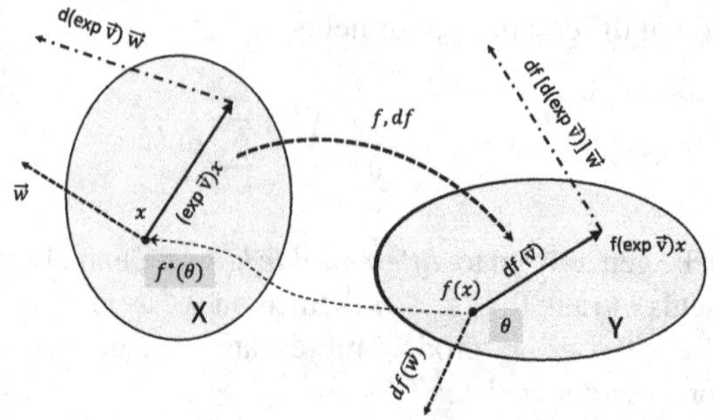

Fig. 4.3 We show two vector fields $\boldsymbol{v}, \boldsymbol{w}$ and a point $x \in X$ which are mapped into $f(x) \in Y$, and into $e^{\boldsymbol{v}}x \in X$ by the flow box of $\boldsymbol{v}$, respectively (we choose $\lambda = 1$ in this figure). The differential of the vector field acts in agreement with the pull-back. The shaded rectangles represent a k-form at x and at $f(x)$

4.3 Existence and Uniqueness Theorems: Differential Equation Approach

We showed that a vector field on a manifold M is a mapping $M \to TM$ that assigns to each point $x \in M$ a vector in T_xM. A vector field may be interpreted alternatively as the first-order system of partial differential equations (PDEs), i.e., a dynamical system [4].

In the following we introduce an important result from the theory of first-order PDEs, namely the fundamental theorem of existence and uniqueness of solutions under Cauchy conditions. Actually, we present here the general version for a system of coupled, nonlinear PDE of order 1 depending on two independent variables (to have a pictorial geometrical interpretation in terms of surface geometry). The extension of this theorem to many dimensions is a simple technical extension, and it does not introduce any new special insights. We begin with the fundamental theorem for one PDE in one unknown function $f(u, v)$ depending on two independent variables.

Definition 15. For a given function defined on the open sets $F : U \times V \times W \subset \mathbb{R}^2 \times \mathbb{R}^2 \times \mathbb{R} \to \mathbb{R}$, we define a partial differential equation (PDE) of order one, the equation

$$F(u, v, f_u, f_v, f) = 0. \tag{4.8}$$

The function $f(u, v) : U \to \mathbb{R}$ is called a solution of this PDE if the expression $F(u, v, \partial f(u, v)/\partial u, \partial f(u, v)/\partial v, f(u, v)) \equiv 0$ transforms the PDE, $F = 0$, into an identity on U.

Definition 16. We call the *Cauchy problem* (or *Cauchy condition*) associated to the function $f(u, v)$ and the PDE, the following set of three quantities:

1. A vector function $\boldsymbol{a}(\epsilon) : I \subset \mathbb{R} \to U \subset \mathbb{R}^2$.
2. A real function $\gamma(\epsilon) : I \subset \mathbb{R} \to W \subset \mathbb{R}$.

3. A constant vector $\boldsymbol{b}_0 \in V$ and a number $\epsilon_0 \in I$, and four constraints between the solution f and (4.8).
4. $F(\boldsymbol{a}(\epsilon_0), \boldsymbol{b}_0, \gamma(\epsilon_0)) = 0$.
5. $\frac{d\gamma}{d\epsilon}(\epsilon_0) = \boldsymbol{b}_0 \cdot \frac{d\boldsymbol{a}}{d\epsilon}(\epsilon_0)$.
6. $f(\boldsymbol{a}(\epsilon)) = \gamma(\epsilon)$.
7. $\left(\frac{\partial f}{\partial u}(\boldsymbol{a}(\epsilon_0)), \frac{\partial f}{\partial v}(\boldsymbol{a}(\epsilon_0))\right) = \boldsymbol{b}_0$.

Theorem 3. *The Cauchy problem $\boldsymbol{a}, \gamma \in C_2(I)$ for the PDE equation $F(u, v, f_u, f_v, f) = 0$, with the supplementary restriction*

$$\left|\begin{pmatrix} \frac{\partial F}{\partial u}\frac{du}{d\epsilon} & \frac{\partial F}{\partial u}\frac{dv}{d\epsilon} \\ \frac{\partial F}{\partial v}\frac{du}{d\epsilon} & \frac{\partial F}{\partial v}\frac{dv}{d\epsilon} \end{pmatrix}\right| = 0$$

has a unique solution $f(u, v) \in C_2(\mathcal{V}(\boldsymbol{a}(\epsilon)))$ on a neighborhood $\mathcal{V}(\boldsymbol{a}(\epsilon))$, fulfilling the Cauchy conditions (1–4) from the Definition 16.

We do not give the proof of Theorem 3 here (the reader can find a detailed proof of this theorem in [120]). We just introduced here Theorem 3 to comment on the geometrical interpretation in terms of surfaces.

We consider the independent variables $(u, v) \in U$ as parameters, and the solution $f : U \to \mathbb{R}$ of the PDE (4.8) as a parameterized surface S, $\boldsymbol{r}(u, v) \in \mathbb{R}^3$, defined by the graphics $f(u, v)$ (see Fig. 4.4).

The PDE (4.8) is integrable if there are solutions (i.e., surfaces) passing through every point of the working space $U \times W \subset \mathbb{R}^2 \times \mathbb{R} \subset \mathbb{R}^3$. The PDE defining equation, $F(u, v, \boldsymbol{b}_0, f) = 0$, provides a relationship between any given point $\boldsymbol{r} = (u, v, f) \in U \times W$, in the working space, and a vector $\boldsymbol{b}_0 \in V$ defined in some two-dimensional abstract vector space. Actually, the mathematical expression of (4.8) says that given a point $(u, v, f) \in U \times W$ and one component of a vector in V (f_u), we can get the other component $f_v(u, v, f, f_u)$. In local flat coordinates the geometric meaning is even simpler.

A solution $f(u, v)$ of (4.8) is a surface S parameterized by the local (flat) coordinates (u, v). We introduce a map from V to $T_{(u,v,f)}\mathbb{R}^3$ defined by

$$V \supset \boldsymbol{b} = (f_u, f_v) \to \\ \{(1, 0, f_u), (0, 1, f_v), \frac{(-f_u, -f_v, 1)}{\sqrt{1+f_u^2+f_v^2}}\} = \{\boldsymbol{r}_u, \boldsymbol{r}_v, \boldsymbol{N}\}$$

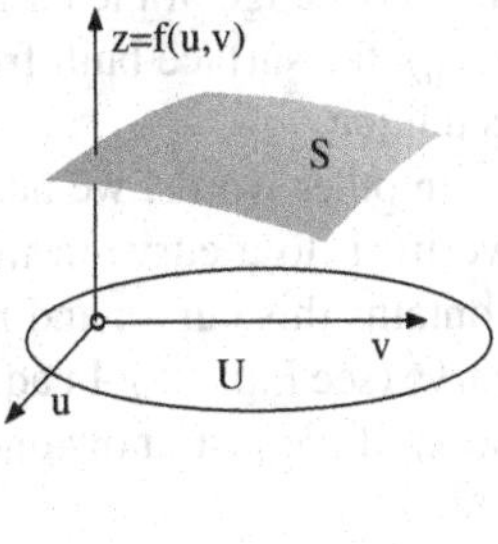

Fig. 4.4 The PDE solution $f(u, v)$ as a parameterized surface S

which provides the Darboux trihedron associated to the parametrization (u, v). In this geometric picture, the integrability of the PDE means that we have a relation which associates for any point and direction, a plane passing through that point and through that direction. This plane is actually the tangent plane to the graphics of the solution f, at the point (u, v). That is, we can write (4.8) in the form $f_v = f_v(u, v, f_u, f)$, and hence associate to any point (u, v, f), and to any direction $(1, 0, f_u)$, the other direction $(0, 1, f_v)$.

The Cauchy conditions (1–3) from Definition 16 assure uniqueness of the solution, and show how to actually construct it. The Cauchy conditions consist in a parameterized curve $\alpha : I \to U$ in the space of the parameters, defined by $\boldsymbol{a}(\epsilon) = (u(\epsilon), v(\epsilon))$, and a parameterized curve defined by $\gamma(\epsilon)$. In the following we assume that the parameter ϵ is the arc-length along α. The curve Γ defined by $(u(\epsilon), v(\epsilon), \gamma(\epsilon)) \in U \times W$ lies in the surface-solution S, by Cauchy condition (6), because if f is solution, then $f \circ \boldsymbol{a}(\epsilon) = \gamma(\epsilon)$. The Cauchy conditions (4,5,7) provide that the two components of the vector $\boldsymbol{b}_0$ are actually the components of the unit tangent of the curve Γ expressed in the basis associated with the (u, v) parametrization at the point $(u(\epsilon_0), v(\epsilon_0), f(u(\epsilon_0), v(\epsilon_0)))$. Indeed, on one hand $d\gamma/d\epsilon$ is the third component of the unit tangent $d\boldsymbol{r}/d\epsilon = D_{\boldsymbol{a}}\boldsymbol{r}$. On the other hand

$$D_{\boldsymbol{a}}\boldsymbol{r} = u_\epsilon(1, 0, f_u) + v_\epsilon(0, 1, f_v) = (u_\epsilon, v_\epsilon, u_\epsilon f_u + v_\epsilon f_v).$$

Consequently, if relation (5) from Definition 16 holds, then $d\gamma/d\epsilon = \boldsymbol{b}_0 \cdot (u_\epsilon, v_\epsilon)$, and $\boldsymbol{b}_0$ is actually equal to (f_u, f_v).

Now, we go for the geometrical interpretation of the existence and uniqueness theorem. The basic idea is simple: If we can build a plane passing through an arbitrary point of the space, we can smoothly extend this plane to an infinitesimal surface on a neighborhood of that point. The rest of the surface is just analytic continuation. A plane is generated by two directions. In the neighborhood of the given Cauchy curve Γ, we have one direction provided by the unit tangent of the curve, and the other direction provided by the PDE equation (starting with the coordinates of the point and the tangent direction). This will build the whole surface, hence the solution.

Indeed, the PDE equations tell us that for any point $(u, v, f) \in U \times W$, and for any direction through this point, $(1, 0, f_u)$, we are given a whole plane through this point and this direction. The solution $f(u, v)$ is the surface having this plane as a tangent plane at any $(u, v, f(u, v))$. In addition, the Cauchy condition provides that from any given parameterized curve $\Gamma : I \to U \times W$ (provided by $\boldsymbol{a}$ and γ), and the knowledge of the tangent plane in one of its points (generated by $\{\boldsymbol{b}_0, d\gamma/d\epsilon\}_{\epsilon_0}$ at ϵ_0), the surface built from the PDE as shown above, and containing the curve Γ, is unique.

In other words, we have a parameterized curve $\boldsymbol{a}$ in the parameter space U, and we lift it to a curve in the whole space $U \times W$. We want to find the surface that contains this curve, and has a prescribed tangent plane in one of the points of this curve (see Fig. 4.5). In addition, we can build the tangent plane to this surface at any point, if we just know one direction of this plane at that point. Now, it is obviously

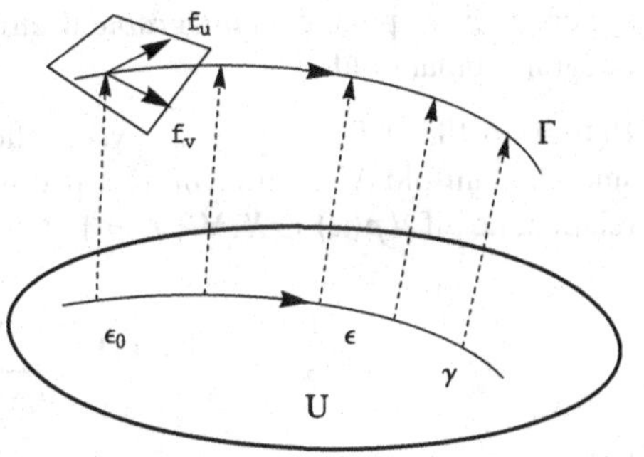

Fig. 4.5 The Cauchy condition for a two-dimensional first-order PDE

how the surface of the solution is built. The curve Γ is part of this surface, and the unit tangent of Γ is in the tangent plane of the surface. In any of the points of Γ, $(u(\epsilon), v(\epsilon), \gamma(\epsilon))$, we have one tangent direction $(u_\epsilon, v_\epsilon, \gamma_\epsilon)$, and the PDE provides the other direction, so we can built a tangent plane to the surface in any point of the curve Γ. Then, by analytic continuation, we can extend the surface from Γ toward the whole domain of F.

In the following we provide the general theorem for existence and uniqueness of solutions for (nonlinear) PDE of order m.

Theorem 4. *If the PDE equation of order m for $f(x_1, \dots, x_n) : D \subset \mathbb{R}^n \to \mathbb{R}$ can be written in the explicit form*

$$\frac{\partial^{|m|} f}{\partial x_1^{|m|}} = F\left(x_1, \dots, x_n, u, \frac{\partial f}{\partial x_1}, \dots, \frac{\partial^{|m|} f}{\partial x_1^{m_1} \dots \partial x_n^{m_n}}\right),$$

with $|m| = m_1 + m_2 + \cdot + m_n$, such that the derivative $\partial^{|m|} f / \partial x_1^{|m|}$ does not occur anymore in the RHS of the above expression, then the Cauchy problem

$$f(x)|_{x_1 = x_1^0} = g_0(x_2, \dots, x_n), \left.\frac{\partial f}{\partial x_1}\right|_{x_1 = x_1^0} = g_1(x_2, \dots, x_n), \dots$$

$$\dots, \left.\frac{\partial^{|m|-1} f}{\partial x_1^{|m|-1}}\right|_{x_1 = x_1^0} = g_{m-1}(x_2, \dots, x_n),$$

attached to this equation has one unique analytic solution $f(x_1, \dots, x_n)$: $V(x_1^0, \dots, x_n^0) \to V(u_0)$, if the function F is analytic on a neighborhood of the point $(x_2^0, \dots, x_n^0, u_0, \dots)$, and the functions $g_1, \dots, g_{m-1}$ are analytic on a neighborhood of $(x_2^0, \dots, x_n^0)$.

We can generalize the integrability concept for a general manifold.

Definition 17. Let $\mathcal{S} = \{v_1, v_1, \dots, v_n\}$ be a finite set of n vector fields defined on a smooth manifold X. We call *integral submanifold* of $\mathcal{S}$ a submanifold $Y \subset X$ whose tangent space T_pY is spanned by the system $\mathcal{S}$ at every point $p \in N$. The system at every point $\mathcal{S}$ is *integrable* if through every point $p \in X$ there passes an integral submanifold.

Definition 18. A finite system of vector fields $\mathcal{S} = \{v_1, v_2, \ldots, v_n, \}$, defined on a smooth manifold X, is in *involution* if it is algebraically closed under commuting relation, i.e., if $\forall p(x) \in X, \forall i, j = 1, \ldots, n$

$$[v_i, v_j] = \sum_{k=1}^{n} c_{ij}^k(x) v_k,$$

where $c_{ij}^k(x)$ are differentiable real functions on X, and $[,]$ is the Lie bracket defined by the action (see Definition 12) of two differential vector fields on functions $f : M \to \mathcal{R}$

$$[v, w]f = v(w(f)) - w(v(f)).$$

If the manifold X is only differentiable C_k, and the vector fields are differentiable of class C_{k-1}, then the Lie bracket is differentiable of class C_{k-2}. In coordinates the Lie bracket has the specific action

$$[v, w] = \left(v^j \frac{\partial w^i}{\partial x^j} - w^j \frac{\partial v^i}{\partial x^j} \right) \frac{\partial}{\partial x^i}$$

Of course, if the vector fields generate an n-dimensional Lie algebra $\mathcal{L}_{\mathcal{S}} \subset T_p Y$ at every $p \in X$, they are in involution. The concept of involution can be generalized to an infinite system of vector fields $\mathcal{S}_\infty$ by asking $\forall v, w \in \mathcal{S}_\infty$, we have $[v, w] \in \mathcal{S}_\infty$.

Theorem 5. *Frobenius theorem. A finite system of vector fields defined on a smooth manifold S is integrable if and only if it is in involution. If the system of vector fields is infinite, it has to fulfill in addition the rank-invariant condition, see Definition 11.*

For a proof of the theorem the reader can see [242, Chap. 1] and references herein. The dimension of the integral manifold is equal to the dimension of the linear space spanned by $\mathcal{S}$ at any point on X, which does not prevent this dimension to change from point to point. We can check the theorem by choosing a parameterized surface in the so-called local *flat* coordinates $r = (x, y, h(x, y))$. For any differentiable function $f : \mathbb{R}^3 \to \mathbb{R}$ we have the action

$$r_u(r_v(f(x, y, z))) = \left(\partial_x + \frac{\partial h}{\partial x} \partial_z \right) \left(\partial_y + \frac{\partial h}{\partial y} \partial_z \right).$$

It is easy to verify that the two tangent vector fields along the local coordinates fulfill $[r_u, r_v] = 0$.

Next important concept for integrability and symmetry of dynamical systems is the *Lie algebra*.

Definition 19. A Lie algebra is a vector space $\mathfrak{g}$ together with a bilinear operation $[\cdot,\cdot] : \mathfrak{g}\times \to \mathfrak{g}$, called Lie bracket or commutator, satisfying the axioms:

1. $[\boldsymbol{a},\boldsymbol{b}] = -[\boldsymbol{b},\boldsymbol{a}]$, skew-symmetry
2. $[\boldsymbol{a},[\boldsymbol{b},\boldsymbol{c}]] + [\boldsymbol{c},[\boldsymbol{a},\boldsymbol{b}]] + [\boldsymbol{b},[\boldsymbol{c},\boldsymbol{a}]] = 0$, Jacobi identity

$\forall \boldsymbol{a},\boldsymbol{b},\boldsymbol{c} \in \mathfrak{g}$.

The dimension n of the Lie algebra is the dimension of the vector space. Usually, the Lie algebras in use for physics are finite dimensional, but there are exceptions especially in field theory. An algebra homomorphism from $\mathfrak{g}$ into a Lie algebra of square matrices is a representation of the Lie algebra. A Lie algebra is uniquely determined by the basis $\{\boldsymbol{v}_i\}_{i=1,\dots,n}$ of its vector space, and by its *structure constants* $c_{ij}^k = -c_{ji}^k$ defined by

$$[\boldsymbol{v}_i,\boldsymbol{v}_j] = c_{ij}^k \boldsymbol{v}_k.$$

Usually, the above relation is given in tabular form, i.e., the commutator table of the Lie algebra. If a Lie algebra is generated by vector fields $\boldsymbol{v} \in TM$ defined on the tangent space of a differentiable manifold M, we can introduce the *exponential map* as a one-parameter smooth transformation $\exp(\epsilon\boldsymbol{v}) : M \to M$ and we call the subset $\{\exp(\epsilon\boldsymbol{v})x | x \in M, \epsilon \in (-\epsilon_{max},\epsilon_{max}) \subset \mathbb{R}\}$ the orbit of the one-parameter local Lie group generated by $\boldsymbol{v}$. Obviously $\exp(0\boldsymbol{v}) = \text{Id}$. Conversely,

$$TM \ni \boldsymbol{v}_{x\in M} = \left.\frac{d}{d\epsilon}\right|_{\epsilon=0} \exp(\epsilon\boldsymbol{v})x, \quad \forall \boldsymbol{v} \in \mathfrak{g},$$

is the Lie equation. Namely, given the (exponential) one-parameter Lie group of transformations based on some initial point $x \in M$, the tangent vector to the curve $\exp(\epsilon\boldsymbol{v})x \subset M$ at $\epsilon = 0$ is the infinitesimal generator of the transformation.

4.4 Existence and Uniqueness Theorems: Flow Box Approach

For a differential vector field $\boldsymbol{v}$ on the differential manifold X we define an integral curve $\gamma(\lambda) : I \to X$ a parameterized curve whose tangent vector at any point coincides with the value of $\boldsymbol{v}$ at the same point

$$\gamma'(\lambda) = \boldsymbol{v}(\gamma(\lambda)).$$

In local components these equations define a system of ordinary differential equations, where the integral curve is the solution. The existence and uniqueness of such an integral curve is guarantied locally by the general existence and uniqueness theorem (Theorem 4) exemplified in Sect. 4.3 [4, 68, 242]. However, this theorem is local and in general does not assure the existence of a global integral curve. To have an intuitive geometrical picture about integral curves, we discuss here the concept of "flow box" introduced in [4, Chap. II].

Definition 20. Let us have a differentiable manifold M. The flow box of a vector field $\mathbf{v}$ at $x \in X$ is a unique triple $(V(x), a, \{F_\lambda\}_{\lambda \in (-a,a)})$ where $V(x)$ is an open neighborhood of x, $a > 0$ and F_λ is a continuous family of differentiable functions $F_\lambda : V(x) \to X$ such that:

1. For any $y \in V(x)$, $F_\lambda(y)$ considered as a function of $\lambda \in (-a, a)$ is an integral curve of $\mathbf{v}$, i.e., $\partial F_\lambda(y)/\partial\lambda = \mathbf{v}(F_\lambda(y))$.
2. For any $\lambda \in (-a, a)$, the mapping $F_\lambda(x) \to F_\lambda(V(x))$ is a diffeomorphism.

In other words, at any point of X, and for a given "size" a of the flow box, we can find local integral curves filling up a neighborhood and mapping it diffeomorphic along X.

The existence of a flow box at any point is guaranteed by the general theorem of existence and uniqueness theorem (Theorem 4) applied on the homeomorphism provided by the charts overlapping $V(x)$, Fig. 4.6.

Theorem 6. *For any given vector field* $\mathbf{v}$ *on a manifold* X*, and for any* $x \in X$ *there is a flow box of* $\mathbf{v}$ *in* X*. This flow box* $(V(x), a, F_\lambda)$ *is unique in that any other flow box of the same point* $(V'(x), a', F'_{\lambda'})$ *has* $F_\lambda = F'_{\lambda'}$ *on* $V(x) \cap V'(x) \times [-a, a] \cap [-a', a']$.

In order to prove this theorem we notice that the uniqueness results from the fact that any two integral curves $\gamma_1(\lambda), \gamma_2(\lambda)$ of the same field, at x, are equal on the intersection of their domains of definition. Indeed, let be $\Lambda = \{\lambda \in (-a, a)| \gamma_1(\lambda) = \gamma_2(\lambda)\} \subset Domain(\gamma_1) \cap Domain(\gamma_2)$. By the definition Λ is closed (being obtained as the kernel of a continuous function). Also, for any $\lambda \in \Lambda$ there is a neighborhood $(\lambda - \epsilon, \lambda + \epsilon)$ contained in a chart (U, ϕ) of X such that the curves $\phi(\gamma_i(\lambda + t))$, $|t| < \epsilon$, $i = 1, 2$ agree for $t = 0$. Again, by the general theorem of existence and uniqueness theorem (Theorem 4), it results that the two curves, and consequently their inverse images agree on the whole neighborhood $(\lambda - \epsilon,$

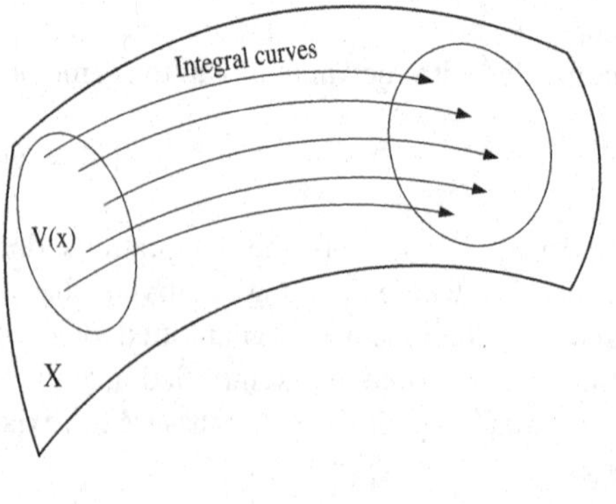

Fig. 4.6 Flow box generated by a vector field

$\lambda + \epsilon$). Consequently this neighborhood is contained in Λ, so Λ is also open. Since $(-a, a)$ is connected, it results that $\Lambda = (-a, a)$.

We have a comment about the size a of the flow box which gives a measurement of the degree of locality of the integral curves. From the very beginning the domain of definition of all integral curves is set to the same interval $(-a, a)$, contrary to the habit in differential geometry (where the parametrization of the curve is not essential and can be changed). Such fixed domain is needed to keep simple the proof for the uniqueness of the flow box vs. change of parametrization. This standard domain of definition does not introduce any limitation when we speak about maximal integral curves because we introduce this concept in a different way. That is, we defined the set $D_v = \{(x, \lambda) |$ there is an integral curve passing through this point$\} \subset X \times \mathbb{R}$. The vector field v is complete if $D_v = X \times \mathbb{R}$. For a complete vector field any integral curve can be extended so its domain becomes $(-\infty, \infty)$. For example, the velocity field of a potential flow past a rigid obstacle is not complete, because there are stream lines ending at stagnation points.

The set D_v can be partitioned by the unique mapping $F_v : D_v \to X$ (the integral of v), constructed such that the curves $\lambda \to F_v(x, \lambda)$ are integral curves at x, for all $x \in X$. Now we define a maximal integral curve to be $\lambda \to F_v(x, \lambda)$. If, in addition v is complete, the function F_v is called flow of the vector field v. The collection of all maximal integral curves for a given vector field is called a *foliation* of X, where the maximal integral curves themselves are called *leaves* of the foliation. Because of the properties of the flow box, and the transitive action of F_v on X, the family $F_v(x, \lambda)$ is called *one-parameter local group of diffeomorphisms* (for exact definition see for example [242]). If X is complete this family becomes a Lie group of diffeomorphisms acting on X. In terms of group theory the vector field is called group infinitesimal generator.

4.5 Compact Supported Vector Fields

The flow box plays an interesting role when the vector field has compact support [4]. Let us assume that X is compact and v is a vector field defined on X. For $x \in X$ let us consider a maximum integral curve γ through x, and let its domain be $(-b, b)$ with $b < +\infty$. We can always find a sequence $b_n \to b$ such that (by compactness and Hausdorff property of X) $\gamma(b_n)$ is convergent to some unique point $x_b \in X$. We can construct a flow box $(V(x_b), a, F)$, and for n larger than a certain limit $\gamma(b_n)$ points lie in $V(x_b)$. Consequently γ can be extended beyond b, and so on to infinity, and minus infinity. It gives the following result:

Lemma 2. *Any vector field defined on a compact manifold is complete. Moreover, vector fields with compact support are complete.*

The flow box is the main tool in the introduction of the Lie derivative, and it is useful for handling differential equations and global invariance.

4.6 Differential Forms and the Lie Derivative

If X is an n-dimensional differentiable manifold and $x^i, i = 1, 2, \ldots, n$ are the local coordinates, we can express formally the vector field $\boldsymbol{v}(x) \in T_x X$ in components

$$\boldsymbol{v} = v^i(x) \frac{\partial}{\partial x^i}, \tag{4.9}$$

and its action on functions $f : X \to \mathbb{R}$ becomes

$$\boldsymbol{v}(f)(x) = v^i(x) \frac{\partial f}{\partial x^i}. \tag{4.10}$$

We can express formally the flow box diffeomorphisms F_λ from Definition 20 as

$$F_\lambda(x) = e^{\lambda \boldsymbol{v}} x, \tag{4.11}$$

also called the exponentiation of the vector field, or *exponential map* [46, 242], because of the structure of its formal differential equation (Lie equation)

$$\left. \frac{dF_\lambda(x)}{d\lambda} \right|_{\lambda=0} = \boldsymbol{v}(x).$$

In reverse, for a given diffeomorphism $\phi : X \to X$, the vector field whose exponential provides this transformation (if it exists) is called the *infinitesimal generator* of ϕ. The set $\{F_\lambda(x)\}_{\lambda \in (-a,a)}$ represents a one-parameter local Lie group of transformations acting on $V(x)$. It is useful to mention the action of the vector field on differentiable functions defined on X. By using the formal Taylor series for f we have

$$f(e^{\lambda \boldsymbol{v}} x) = \sum_{k \geq 0} \frac{\lambda^k}{k!} \boldsymbol{v}^k(f)(x). \tag{4.12}$$

The value of the function in the transformed point is obtained by repeated action of the vector field on the function at x

$$\begin{array}{ccc} x & \xrightarrow{\ f\ } & f(x) \\ \downarrow{\scriptstyle \lambda} & & {\scriptstyle \boldsymbol{v}^k}\downarrow \\ e^{\lambda x} & \xrightarrow{\ f\ } & f(e^{\lambda x}). \end{array}$$

The generalization of a function, or of an infinitesimal surface or volume element, is the differential k-form (defined in Sect. 4.2 and simply called k-forms in the followings) defined as a differential skew-symmetric covariant tensor field on X, with entries in the k-times exterior product of the cotangent space of the n-dimensional manifold X [235, 242]

$$\omega = \sum_{\substack{i_1 < i_2 < \cdots < i_k \\ i_1 \ldots i_k = 1}}^{n} \omega_{i_1 i_2 \ldots i_k}(x)\, dx^{i_1} \wedge dx^{i_2} \wedge \cdots \wedge dx^{i_k} \in \Omega^k T^*_x X,$$

which can be written in a simpler form

$$\omega = \frac{1}{n!} \omega_{i_1 i_2 \ldots i_k}\, dx^{i_1} \wedge dx^{i_2} \wedge \cdots \wedge dx^{i_k},$$

because the quantities $\omega_{i_1 i_2 \ldots i_k}$ form a skew-symmetric tensor. For a set of k vector fields on X we have the action of the k-form on these fields given by

$$(\omega; v_1, \ldots, v_k) = \omega_{i_1 i_2 \ldots i_k} v_1^{i_1} \ldots v_k^{i_k}, \tag{4.13}$$

where we use the dummy index summation convention.

Definition 21. A k-form ω and an r-form θ can be combined into a new $(k + l)$-form by the exterior product, through the $\wedge$ operation

$$\omega \wedge \theta =$$

$$\sum_{\substack{i_1 < \cdots < i_k \\ i_1 \ldots i_k = 1}}^{n} \sum_{(j) \in P(i)} (-1)^{\nu} \omega_{\underset{<}{j_1 \ldots j_k}} \theta_{\underset{<}{j_{k+1} \ldots j_{k+r}}} dx^{i_1} \wedge \cdots \wedge dx^{i_{k+r}},$$

where the symbol $<$ means that the lower indexes are taken in increasing order, $P(i)$ means all the $k + l$ permutations of the i indexes, $(j) = (j_1, \ldots, j_{k+r})$ is a multi-index, and ν is the signature of each permutation in the sum. The exterior product is linear, distributive, and the pull-back map is linear under the exterior product, and $\omega \wedge \theta = (-1)^{kr} \theta \wedge \omega$. For example, if $\omega, \theta \in \Omega^1 = T^* X$ are 1-forms on the n-dimensional differentiable manifold X we have

$$\omega \wedge \theta = (\omega_i \theta_j - \omega_j \theta_i) dx^i \wedge dx^j,\ i, j = 1, 2, \ldots, n.$$

Here is another example for ω a 2-form, and θ a 1-form

$$\omega \wedge \theta = (\omega_{12}\theta_3 - \omega_{13}\theta_2 + \omega_{23}\theta_1) dx^1 \wedge dx^2 \wedge dx^3.$$

Definition 22. We define the interior product between a vector field and a k-form ω the $(k-1)$-form

$$(\boldsymbol{v} \perp \omega)_{i_1 \ldots i_{k-1}} = \sum_{\substack{l=1 \\ l \neq i_1 \ldots i_{k-1}}}^{n} v^l \omega_{\underset{<}{l;\, i_1 \ldots i_{k-1}}}.$$

For example $\partial_x \perp dx \wedge dy = dy$. The action of the interior product is given by

$$(\boldsymbol{v} \perp \omega; \boldsymbol{v}_1, \ldots, \boldsymbol{v}_{k-1}) = (\omega; \boldsymbol{v}, \boldsymbol{v}_1, \ldots, \boldsymbol{v}_{k-1}). \tag{4.14}$$

The last operator we need for our purposes is the *exterior derivative.*

Definition 23. For any k-form ω we define the linear operator $d : \Omega^k T_x^* X \to \Omega^{k+1} T_x^* X$ acting on ω and producing a $(k+1)$-form

$$d\omega = \sum_{i,I} \frac{\partial \omega_I}{\partial x^i} dx^i \wedge dx^I,$$

where I is the increasing ordered multilabel defining ω components.

The exterior derivative is linear, commutes with the pull-back map, and – most importantly – has the closure property

$$d(d\omega) = d^2\omega = 0. \tag{4.15}$$

We conclude this series of definitions with a set of useful relations between all these operators. In the following ω is a k-form, and θ is an r-form, $k, r \leq n$, and $\boldsymbol{v}$ is a vector field:

$$\boldsymbol{v} \perp (\omega \wedge \theta) = (\boldsymbol{v} \perp \omega) \wedge \theta + (-1)^k \omega \wedge (\boldsymbol{v} \perp \theta). \tag{4.16}$$

$$d(\omega \wedge \theta) = d\omega \wedge \theta + (-1)^k \omega \wedge \theta \tag{4.17}$$

$$\boldsymbol{v} \perp (\omega \wedge \theta) = (\boldsymbol{v} \perp \omega) \wedge \theta + (-1)^k \omega \wedge (\boldsymbol{v} \perp \theta). \tag{4.18}$$

For a given vector field $\boldsymbol{v}(x)$ on X and a certain geometrical object $\omega(x)$ defined on X (like another vector field or a k-form), it is natural to ask how does ω changes along the integral curves of $\boldsymbol{v}$. Since at different points $e^{\lambda \boldsymbol{v}} x$, the quantity $\omega(x)$ takes values in different spaces of a fiber bundle ΩX over X (example the tangent bundle TX, cotangent bundle T^*X, tensor bundle $T_k^j X$, etc.) we have to compare the values of $\omega(x) \in \Omega_x X$ with the pulled-back values of $\omega(e^{\lambda \boldsymbol{v}} x) \in \Omega_{e^{\lambda \boldsymbol{v}} x}$. This technique is known under the name of the *Lie derivative.*

Definition 24. We introduce the *Lie derivative* of ω at $x \in X$, with respect to $\boldsymbol{v}$, as the expression $\boldsymbol{v}(\omega)$ defined by

$$\boldsymbol{v}(\omega)|_x = \lim_{\lambda\to 0} \frac{\Phi_\lambda^*(\omega|_{\exp(\lambda \boldsymbol{v})x}) - \omega|_x}{\lambda} = \left(\frac{d\Phi_\lambda^*(\omega|_{\exp(\lambda \boldsymbol{v})x})}{d\lambda}\right)_{\lambda=0},$$

where $\Phi_\lambda x = e^{\lambda \boldsymbol{v}} x$ and Φ_λ^* is the pull-back of Φ between the corresponding tensor spaces (Definition 14).

For example the Lie derivative of a function $\boldsymbol{v}(f)$ is nothing but the expression (4.10). The Lie derivative of a vector field $\boldsymbol{w}$ is

$$\boldsymbol{v}(\boldsymbol{w}) = [\boldsymbol{v}, \boldsymbol{w}]. \tag{4.19}$$

In general the Lie derivative of a k-form $\omega = \omega_{i_1,i_2,\ldots,i_k} dx^{i_1} \wedge dx^{i_2} \wedge \cdots \wedge dx^{i_k} \in \Omega_k X$ is

$$\boldsymbol{v}(\omega) = \boldsymbol{v}(\omega_{i_1,i_2,\ldots,i_k}) dx^{i_1} \wedge dx^{i_2} \wedge \cdots \wedge dx^{i_k}$$
$$+ \sum_{j=1}^{k} \omega_{i_1,i_2,\ldots,i_j,\ldots,i_k} dx^{i_1} \wedge \cdots \wedge \boldsymbol{v}(dx^{i_j}) \wedge \ldots dx^{i_k}, \tag{4.20}$$

where we can use the formula $\boldsymbol{v}(dx^{i_j}) = dv^{i_j} = (\partial v^{i_j}/\partial x^k) dx^k$ from (4.9). For example if $k = 2$ we find the Lie derivative of a 1-form by knowing its action on vector fields

$$(\boldsymbol{v}(\omega); \boldsymbol{w}) = \boldsymbol{v}(\omega; \boldsymbol{w}) - (\omega; [\boldsymbol{v}, \boldsymbol{w}]) \tag{4.21}$$

In components, on $X = \mathbb{R}^2$ we can write the Lie derivative of a 2-form by using $\boldsymbol{v}(x, y) = \xi(x, y)(\partial/\partial x) + \eta(x, y)(\partial/\partial y)$, $\omega = \omega_{12} dx \wedge dy$ and we have

$$\boldsymbol{v}(\omega) = \left[\xi \frac{\partial \omega_{12}}{\partial x} + \eta \frac{\partial \omega_{12}}{\partial y} + \omega_{12}\left(\frac{\partial \xi}{\partial x} + \frac{\partial \eta}{\partial y}\right)\right] dx \wedge dy.$$

A very useful application of Definition 24 and (4.19) and (4.20) is provided in Sect. 9.2.6 where we introduce the concept of covariant (or convected, convective, material) time derivative related to the Eulerian–Lagrangian frame transformations in hydrodynamics.

Another direct application of the Lie derivative, introduced in [4], is based on the concept of flow box, see Sect. 4.4. Let $\boldsymbol{v}$ be a vector field on a manifold X, i.e., $\boldsymbol{v}(x) \in \cup_{y \in X} T_y X$ and a flow box (see Sect. 9.3) $(U(x), \alpha, F)$ at x. That is we have an open neighborhood $U(x)$ of x, and a continuous set of homeomorphic copies of $U(x)$, labeled by U_λ, $\lambda \in (-\alpha, \alpha) \subset \mathbb{R}$ mapped by the diffeomorphism $F_\lambda : U(x) \to U_\lambda$. Let also ω be a tensor field on X, for example $\omega(x) \in \cup_{y \in X} T^j_{y,i} X$, i.e., an i-order covariant and j-order contravariant tensor field. According to the definition of the flow box (Definition 20) we can choose an arbitrary $\lambda \in (-a, a)$ and an arbitrary point, $y \in U_\lambda$, and consider the value of the tensor field in that point, $\omega(y) \in T^{j}_{y,i}X$. Then pull-back this value into $T^{j}_{F^{-1}(y),i}X$ by using the dual of the diffeomorphisms F_{λ}^{-1*}. We obtain a tensor $F^{\lambda}_{-1*}(\omega(y))$ (see Fig. 4.7). Since this pull-back can happen for any value of λ we actually built a curve $\omega_{\lambda}(x)$ of tensors defined at x, lying in $T^{j}_{x,i}X$. Now we can apply the elementary concept of derivation along a curve for this curve of tensors, and the resulting object is the Lie derivative of an i-covariant, j-contravariant tensor at x. So we have

$$v(\omega) = \left. \frac{d\omega_{\lambda}(x)}{d\lambda} \right|_{\lambda=0}, \tag{4.22}$$

see also Fig. 4.8. So, the Lie derivative is not a different mechanism of differentiation of tensors on a manifold, but just involves a special way of choosing the curve along which we differentiate. That is a curve induced by the vector field v, but lying in the space of tensors over the point x (a curve in the fiber of an (i,j) tensor bundle of base X over x). In this sense, the Lie derivative is very similar with the covariant

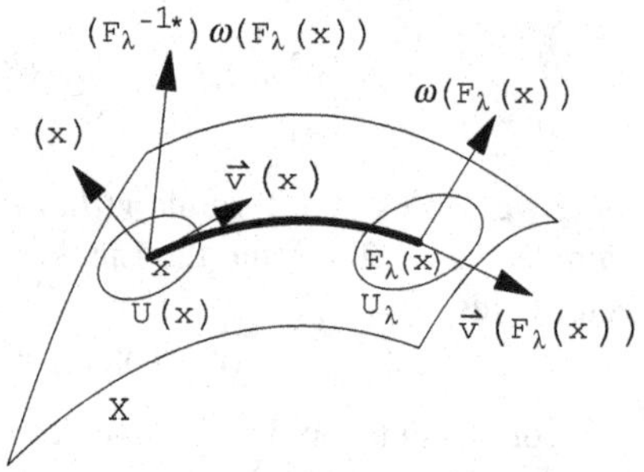

Fig. 4.7 The mechanics of the Lie derivative. The *thick line* is the integral curve of the vector field v

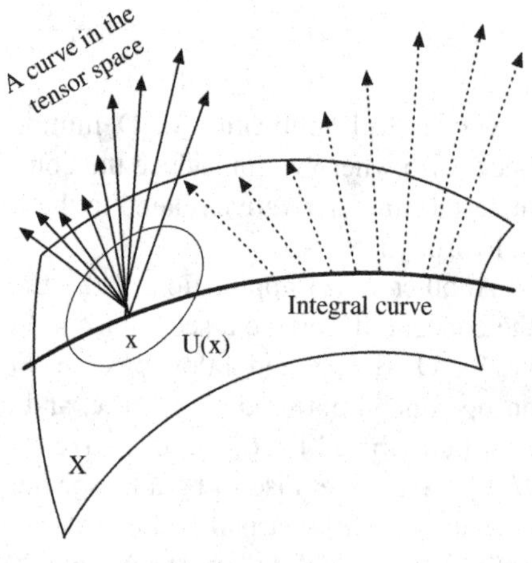

Fig. 4.8 The result of application of the procedure in Fig. 4.7. The tensors over different points along the integral curve are pulled back in the same tensor space, forming a parameterized curve of tensors which can be differentiated

derivative (see also the comment and the diagram about derivatives in Sect. 1.3). The *flow box* plays the same role for the Lie derivative, as the *connection coefficients* (Christoffel symbols) have for the covariant derivative (see Sect. 6.1).

The Lie derivative has the following properties with respect to the interior, and exterior algebra of forms

$$\begin{aligned} \mathbf{v}(\omega \wedge \theta) &= \mathbf{v}(\omega) \wedge \theta + \omega \wedge \mathbf{v}(\theta), \\ \mathbf{v}(d\omega) &= d\mathbf{v}(\omega), \\ \mathbf{v}(\mathbf{w} \perp \theta) &= [\mathbf{v}, \mathbf{w}] \perp \theta + \mathbf{w} \perp \mathbf{v}(\theta), \end{aligned} \tag{4.23}$$

and the important relationship between the Lie derivative and the exterior derivative

$$d(\mathbf{v} \perp \omega) = \mathbf{v}(\omega) - \mathbf{v} \perp (d\omega). \tag{4.24}$$

Equation (4.24) has an elegant generalization for k-forms

$$\begin{aligned} (d\omega; \mathbf{v}_1, \ldots, \mathbf{v}_{k+1}) = {} & \frac{1}{k+1} \sum_{i=1}^{k+1} (-1)^{i-1} \mathbf{v}_i (\omega; \mathbf{v}_1, \ldots, \hat{\mathbf{v}_i}, \ldots, \mathbf{v}_{k+1}) \\ & + \frac{1}{k+1} \sum_{i=1, i<j}^{j=k+1} (-1)^{i+j} \\ & \times (\omega; [\mathbf{v}_i, \mathbf{v}_j], \mathbf{v}_1, \ldots, \hat{\mathbf{v}_i}, \ldots, \hat{\mathbf{v}_j}, \ldots, \mathbf{v}_{k+1}), \end{aligned} \tag{4.25}$$

where the first sum in the RHS term represents the regular action of vector fields on scalar functions in the sense of Eq. (4.10), and the hat placed on a vector means that vector should be omitted from the counting. For $k = 1$ this equation has the form

$$(d\omega; \mathbf{v}, \mathbf{w}) = \mathbf{v}(\omega; \mathbf{w}) - \mathbf{w}(\omega; \mathbf{v}) - (\omega; [\mathbf{v}, \mathbf{w}]).$$

In particular this expression is very important in two cases. First one is when the 1-form is valued in a Lie algebra of a Lie group, and the two vector fields are invariant to this group. In this case the first two terms in the RHS are zero, and we have the famous Maurer-Cartan equation

$$d\omega + \frac{1}{2}[\omega, \omega] = 0, \tag{4.26}$$

where the bracket is the Lie bracket of the Lie algebra. In the second case the form $d\omega + \omega \wedge \omega$ represents the curvature 2-form of a linear connection, also called the first Cartan structure equation.

4.7 Differential Systems, Integrability and Invariants

In the following we will show how can we generalize the concept of differential equations on differentiable manifolds, and how we can use the elegant tool of k-forms the express the integrability conditions in the simplest way. We work on a differentiable manifold X of dimension n.

Definition 25. A differential system of dimension $m \leq n$ on X is a differentiable map D which associates to any point $x \in X$ a vector sub-space D_x of $T_x X$. A sub-manifold $X_D \subset X$ is an integral manifold of D if the pull-back $di^*(T_y X_D) \subset \mathrm{D}_{i(y)}$, for any $y \in X_D$, where $i : X_D \to X$ is the canonical injection of X_D on X.

A differential system on X is *completely integrable* if we have a maximal dimension integral manifold through each point of X. In nonlinear physics, where the differential system models some process, there is another term defining a completely integrable system, especially when X is infinite dimensional: *exactly solvable system* (or model).

Of course, there is a generalization of the Frobenius theorem (Theorem 5 of Sect. 4.2) in the following form. The differential system D on X is completely integrable if and only if the vector space generated by D at any point is a local Lie algebra, that is for any point of X, and any two vector fields of this subspace their Lie bracket also belongs to this subspace.

Among differential systems of physical interest we have the so-called *total differential systems* (or Pfaffian system) which are described by simultaneous vanishing of a number of linear independent 1-forms $\omega^j \in \Omega^1 T^* X$. We have the following integrability theorem

Theorem 7. *A total differential system of order m given by the equations* $\omega^j = 0$, $j = 1, \ldots, m \leq n$ *on X is integrable if*

$$d\omega^j \wedge \omega^1 \wedge \cdots \wedge \omega^m = 0, \; j = 1, \ldots, m.$$

For example if we have a total differential system of order 1 on $\mathbb{R}^3$ given by $\omega = Pdx + Qdy + Rdz = 0$ the above integrability condition $d\omega \wedge \omega = 0$ has the well known form

$$P\left(\frac{\partial R}{\partial y} - \frac{\partial Q}{\partial z}\right) + Q\left(\frac{\partial P}{\partial z} - \frac{\partial R}{\partial x}\right) + R\left(\frac{\partial Q}{\partial x} - \frac{\partial P}{\partial y}\right) = 0.$$

One of the most important geometrical concepts is the *invariance* under certain transformations, either of coordinates or through continuous groups. Consequently we have two basic types: invariants of a transformation $\Psi : X \to X$, and invariants of a vector field $\boldsymbol{v}$. If the transformation is the flow or an integral curve of the vector field, or the vector field is the infinitesimal generator of the transformation, the two types coincide.

Let X be a differentiable manifold, $\Psi : X \to X$ a differentiable map, and $Y \subset X$ a submanifold.

Definition 26. The submanifold is Ψ-invariant if $\Psi(Y) \subset Y$.

The definition can be extended to more than one mapping, which eventually forms a group of applications $\Psi \in G$, in which case we call Y G-invariant.

Definition 27. An application $f : X \to \mathbb{R}^n$ is Ψ-invariant if $\forall c \in \mathbb{R}^n$ the submanifold $\{x \in X \,|\, f(x) = c\}$ is Ψ-invariant.

Let X be a differentiable manifold, $\boldsymbol{v} \in TX$ a differentiable vector field on X, and $Y \subset X$ a submanifold.

Definition 28. The submanifold is an invariant manifold of $\boldsymbol{v}$ if $\forall y \in Y$, $\boldsymbol{v}(y) \in T_yY \subset T_yX$.

In other words, a submanifold is invariant of a vector field if the restriction of this field is tangent to the submanifold. From the uniqueness of the integral curves we have a sufficient criterium for the invariant set.

Lemma 3. *If the submanifold $Y \subset X$ is an invariant manifold of $\boldsymbol{v}$, $a > 0$, $y \in Y$, and $\gamma : [-a, a] \to X$, $\gamma(0) = y$ an integral curve of $\boldsymbol{v}$, then there is $0 < b < a$ such that $\gamma([-b, b]) \subset Y$.*

In addition, sufficient conditions are provided by Lemma 4.

Lemma 4. *The differentiable function $f : X \to \mathbb{R}$ is invariant of a vector field $\boldsymbol{v}$ on X if and only if $\boldsymbol{v}(f) = 0$. As a corollary, if a submanifold $Y \subset X$ is defined implicitly by a set of equations $\{f_j(x) = 0\}_{j=1,\dots,n}$, then Y is invariant of $\boldsymbol{v}$ if $\boldsymbol{v}(f_j) = 0, \forall j = 1, \dots, n$.*

Proof. By using (4.12) we note that if $\boldsymbol{v}(f_j) = 0$ the functions are invariants, and conversely. □

We make the following comment: the notation $\boldsymbol{v}(\dots)$ in general represents the Lie derivative of a certain geometrical object. If this object is a function, like above, its Lie derivative coincides with the action of the vector field on the function, so there is no danger of misunderstanding. However, with respect to k-forms invariance we deal with the Lie derivative. Let X be a differentiable manifold, $\boldsymbol{v} \in TX$ a differentiable vector field on X, and $\omega \in \Omega_k T^*X$ a differentiable k-form defined on X (Sect. 4.6).

Definition 29. ω is an *invariant* k-form of $\boldsymbol{v}$ if its Lie derivative has the property $\boldsymbol{v}(\omega) = 0$.

Sometimes an invariant form is also called constant of motion. The invariant form ω of $\boldsymbol{v}$ has the property that it is constant along the integral curves of $\boldsymbol{v}$. That is

$$\frac{\partial F_\lambda^*(\omega)}{\partial \lambda} = 0, \tag{4.27}$$

for any flow box (V, a, F_λ) of $\boldsymbol{v}$. An invariant form ω has a series of important properties:

1. On any oriented, compact k-dimensional submanifold $Y \subset X$ with boundary ∂Y we have the identity
$$\int_{\partial Y} F_\lambda^* \omega = \int_{\partial Y} \omega, \tag{4.28}$$
for any λ of any flow box in Y. In other words, an invariant form, considered a volume form, conserves the volume associated with a submanifold of the same dimension.
2. The exterior differential $d\omega$ is also invariant.
3. All interior and exterior algebraic operations between invariant forms of $\boldsymbol{v}$, or between invariant forms and $\boldsymbol{v}$ are also invariant forms of $\boldsymbol{v}$ (for a list of such operations see Sect. 4.6). More general, a set of invariant forms of $\boldsymbol{v}$ form a subalgebra of $\Omega_k T^* X$, closed under the interior and exterior operations (i.e., $d, \wedge, \top$).
4. The Lie derivative $\boldsymbol{v}(d\omega)$ is closed.

4.8 Poincaré Lemma

This section needs the elements of homology introduced in Sect. 2.2. It is also related to Sect. 3.1.4 where we emphasize the importance of this lemma for compact boundaries representation formulas, and especially for the generalized Stokes theorem (3.7).

Theorem 8. *If the manifold M is contractible to a point, then all closed forms on M are exact.*

A space is contractible is there is a deformation homeomorphism that contracts m to one of its points. A *closed form* ω is a differential form with the property that $d\omega = 0$. An exact p-form $\phi \in \Omega^p$ has the property that there exists a $(p-1)$-form $\psi \in \Omega^{p-1}$ such that $d\psi = \phi$. This lemma is a generalization of the fact that on simple connected domains a total exact differential is integrable, and its path integral does not depend on the path. For example in $\mathbb{R}^3$ the Poincaré Lemma is the cause for the existence of the following important vector analysis relations

$$\nabla \times \nabla \Phi = 0, \quad \nabla \cdot (\nabla \times \boldsymbol{V}) = 0,$$

$$\nabla \times \boldsymbol{V} = 0 \rightarrow, \exists \Phi, \ \boldsymbol{V} = \nabla \Phi, \quad \nabla \cdot \boldsymbol{V} = 0 \rightarrow \exists \boldsymbol{W}, \ \boldsymbol{V} = \nabla \times \boldsymbol{W}.$$

For example, in the classical real analysis on $\mathbb{R}^3$ we can write the De Rham sequences as follows

$$
\begin{array}{ccccccc}
\Phi & \xrightarrow{d=\nabla} & \nabla\Phi & \xrightarrow{d=\nabla\times} & \nabla\times(\nabla\Phi)=0 & & \\
0-\text{form} & & 1-\text{form} & & d^2=0 & & \\
 & & \boldsymbol{v} & & & & \\
 & & 1-\text{form} & \xrightarrow{d=\nabla\cdot} & \nabla\times\boldsymbol{w} & \xrightarrow{d=\nabla\times} & \nabla\cdot(\nabla\times\boldsymbol{w})=0 \\
 & & \text{dual vector} & & 2-\text{form} & & d^2=0 \\
 & & & & \boldsymbol{V} & \xrightarrow{d=\nabla\cdot} & \nabla\cdot\boldsymbol{V} \\
 & & & & 2-\text{form} & & 3-\text{form}
\end{array}
\quad (4.29)
$$

4.9 Fiber Bundles and Covariant Derivative

The general idea is to decompose a topological space or differentiable manifold E, called total space, in a Cartesian product. If E is inherently twisted this decomposition is only possible locally and we call this space a *fiber bundle*. Hence a Cartesian product is a trivial bundle. The general definition is the following [158, 306].

Definition 30. A a fiber bundle E is a collection $E(X, F, \pi, G)$ as follows:

1. E, X, F are a topological spaces called: the bundle (or total) space, the base space, and the standard fiber. The map π is a surjection $\pi : E \to X$ called canonic projection.
2. G is a topological group of transformation homomorphisms acting on F and called the structure group.
3. There is a family of coordinate neighborhoods $(U_\alpha, \Phi_\alpha) \in U$, with U_α open sets covering X, and $\Phi_\alpha : U_\alpha \times F \to \pi^{-1}(U_\alpha)$ homeomorphisms, such that $\pi\Phi_\alpha = Id\ X$.
4. The set of all $\Phi_\alpha \circ \Phi_\beta = g_{\alpha,\beta}(x)$ (called transition functions) for any fixed $x \in X$, is homeomorphic with G.

The way the coordinates are assigned to a fiber F at a point $x \in X$ is handled by the structure group of homeomorphisms of F. Nevertheless, on the top of any coordinate neighborhood $\pi^{-1}(U_\alpha)$, when we move along the fiber, we actually stay on the same base point x. The inverse image $\pi^{-1}(x) = E_x$ is called the fiber at x.

In other words a fiber bundle is a space E which can be projected onto another (simpler) space X, which is already partitioned in open coordinate neighborhoods. Locally, E can be expressed in terms of coordinates in X and coordinates in the fiber F. The total space looks locally like a simple direct product between a coordinate neighborhood and the fiber. However, globally this is not true in general. The maps $U_\alpha \times F$ are glued together (where the coordinate neighborhoods overlap) in different ways across X. The gluing maps taking $U_\alpha \times F$ to $U_\beta \times F$ are the transition functions $g_{\alpha,\beta}(x)$. Thus, the structure group controls the gluing operations between local parts of the total space of the fiber bundle. The definition of the fiber bundle equally works if the spaces E, X, F are differentiable manifolds, and the group G to be a Lie group.

A cross-section in a bundle is a differentiable injective map $\phi : X \to E$ so that $\pi\phi = Id\ X$. A cross-section is in a way a generalization of the graphic of a function defined on the base space with values in the fiber bundle. Different from a regular graphic, a cross-section takes different values at different points $x \in X$ in different spaces, namely homeomorphisms of the fiber.

Usually, in physics and geometry books the traditional example of fiber bundle is the Möbius band $E(S_1, [0, 1], \pi, O(2, \mathbb{R}))$, [235, 306], namely a compact two-dimensional manifold based on a circle which locally is $\mathbb{R}^2$ but globally is not a cylinder. We present here a nontraditional example of fiber bundle. We define the total space E_W as the set of all words $\in E_W$ spoken on Earth now, and we are interested in two features: the meaning and the language to which a word belongs. However, E_W is not the Cartesian product of the set of all meanings F_{mean}, and the set of all languages X_{lang} because languages and meanings have different structures. Indeed, the subsets of meanings for different languages are different for different cultures, so meaning and language combine in a twisted way. The base space is X_{lang} and the projection $\pi : E_W \to X_{\text{lang}}$ is defined by associating for any word a unique language. The fiber F_{mean} is the space of all possible meanings of all possible concepts. The local fiber π^{-1}(language) is the set of words of different meanings spoken in a given language. The structure group acting on F_{mean} is how the structure of meanings changes from language to language. Let us choose an open set $U_\alpha \subset X_{\text{lang}}$ of "related" languages. Then $\pi^{-1}(U_\alpha)$ is the set of words spoken in this family of related languages. The mapping $\Phi_\alpha : \pi^{-1}(U_\alpha) \to U_\alpha \times F_{\text{mean}}$ is defined by associating to each word a unique meaning, i.e., the map word $\to$ (language, meaning). The structure functions $g_{\alpha,\beta}$ are the dictionaries needed at the points where two different spoken languages overlap. A cross-section in E_W is a rule by which we associate to any word from each language a meaning, and a connection would be a way to associate to all words the same meaning. The covariant derivative at German$\in X_{\text{lang}}$ in the direction English would be to find out how words of the same meaning change between these two languages.

Definition 31. The fiber bundle $E(X, \mathbb{V}_n(\mathbb{R}), \pi, GL(n, \mathbb{R}))$ where $\mathbb{V}_n(\mathbb{R})$ is some $n-$dimensional real vector space, and $GL(n, \mathbb{R})$ is the general (real) linear group, is a vector bundle if any $x \in X$ has an open neighborhood $U(x)$ and a diffeomorphism $\phi : U(x) \times \mathbb{V}_n(\mathbb{R}) \to \pi^{-1}(U(x))$ which is an isomorphism between $\mathbb{V}_n(\mathbb{R})$ and E_x.

In other words any local fiber E_x is isomorphic to the standard fiber vector space $F = \mathbb{V}_n(\mathbb{R})$, and the corresponding isomorphisms depend smoothly on x in the base space. A typical example of vector bundle is the tangent bundle $T\Sigma$ of a parameterized differentiable surface Σ in $\mathbf{R}_3$. The base space is the surface itself, and the tangent bundle is the set of all tangent vectors at all points of the surface. The projection is the assignment for each vector of its initial point. The fiber at x is the tangent plane at x and is a topological vector space. Choosing a unique representative $F = \mathbb{R}^2$, linear correspondences $E_x \to F$ can be constructed, but not uniquely. In this case the structure group G is the full linear group operating on F. A cross-section here is just a differentiable vector field over the surface.

4.9.1 Principal Bundle and Frames

Many differential geometry objects originate directly for the theory of Lie groups and algebras. In the following $\mathfrak{g}$ will represent an n-dimensional Lie algebra associated to the Lie group G, and $\boldsymbol{A}, \boldsymbol{B}, \cdots \in \mathfrak{g}$. A function is called left invariant if it commutes with the left group translations, or with their adjoint representation. In a Lie algebra we can define two important objects which later on will become handy in the definitions of vector bundles and connections [158]. We introduce the property.

Definition 32. A left invariant 1-form ω defined on $\mathfrak{g}$ fulfils the equation of Maurer-Cartan

$$(d\omega; \boldsymbol{A}, \boldsymbol{B}) = -\frac{1}{2}(\omega; [\boldsymbol{A}, \boldsymbol{B}]) \tag{4.30}$$

for any $\boldsymbol{A}, \boldsymbol{B} \in \mathfrak{g}$,

see also (4.26). Then, we have

Definition 33. A canonical 1-form θ defined on G is a left invariant $\mathfrak{g}$-valued 1-form uniquely determined by the invariance relation $(\theta; \boldsymbol{A}) = \boldsymbol{A}$.

As a consequence, if $\{\boldsymbol{e}_1, \ldots, \boldsymbol{e}_n\}$ is a basis for $\mathfrak{g}$ we can write

$$\theta = \theta^i \boldsymbol{e}_i, \quad d\theta^i = -\frac{1}{2} C^i_{jk} \theta^j \wedge \theta^k, \tag{4.31}$$

where $[\boldsymbol{e}_i, \boldsymbol{e}_j] = C^k_{ij} \boldsymbol{e}_k, k = 1, \ldots, n$ define the structure relations (constants).

A Lie group G can act on a manifold and induce orbits, see Sect. 2.3. However, its Lie algebra $\mathfrak{g}$ is "local", and it cannot act at different points on the manifold like G does, except if the manifold is G itself. In order to globalize this locality we need to enrich the structure of the manifold and make it a fiber bundle. In a fiber bundle there is more "freedom", and we will introduce vertical and horizontal displacements by using the covariant derivative, and the connection form, respectively.

Definition 34. A principal bundle over the "base" space X with "structure" group G is a fiber bundle $P(X, G)$ on which G acts freely (on the right) and $X = P/G$.

We note that we simplified the notation of a fiber bundle and mention in the parenthesis only the base space and structure group. Every fiber $\pi^{-1}(x)$ of a principal bundle is diffeomorphic to G, and actually the base space is just the space of all orbits of the action of G on P. For any element $\boldsymbol{A} \in \mathfrak{g}$ we can construct a *fundamental vector field* $\boldsymbol{A}^* : X \to TX$ defined by $\forall x_0 \in X, \boldsymbol{A}^* =$

$d(e^{tA}x_0)/dt \in T_{x_0}X$, that is the vector field tangent to the one-parameter Lie subgroups generated by A. The fundamental vector field is tangent to each fiber at each point of P. The best example of principal bundle is the *bundle of linear frames* (or simply frames) over an n-dimensional manifold X. It is the principal bundle $\mathrm{F}X = P(X, GL(n, \mathbb{R}))$ which consists of ordered bases in T_xX defined at each x, namely linear frames.

Theorem 9. *If dimX=n a linear frame* $v \in FX$ *can be also understood as a linear mapping of some canonical basis of a vector space* $\mathbb{R}^3$ *in* TX, *i.e.* $u(e_i) = X_i, i = 1, \ldots, n$.

Moreover, by using the natural inner product of vectors in $\mathbb{R}^n$, we define the bundle $\mathrm{OF}X = P(X, O(n, \mathbb{R}))$ called *bundle of orthonormal frames* over X.

The bundle of frames explains how the frames at a given point of X change under the action of a group, but does not relate this to the possible change of the point x itself under the action of the group. In order to combine these two actions, if the manifold X is n-dimensional we need the concept of *associated vector bundle* to the principal bundle.P To construct it we begin with $P(X, G)$ and use a finite dimensional vector space called *standard fiber* F (F in isomorphisms with some $\mathbb{R}^n$). The new vector bundle is denoted $E(X, \mathbb{R}^n, \pi_E, G; P)$, its canonical projection is π_E, and its space is nothing but the quotient space $E = (P \times F)/G$. The tangent bundle $TX = \bigcup_{x \in X} T_xX$ is the associated vector bundle to the principal bundle FX of frames. We illustrate this construction in (4.32).

Space of frames over$X$: Space of directions in$X$:

$$\begin{array}{ccc} \mathrm{FX} = P(X, GL(n, \mathbb{R})) & \xrightarrow{\mathbb{R}^n} & TX = E(X, \mathbb{R}^n, GL(n, \mathbb{R}), \pi_E; \mathrm{FX}(X)) \\ \pi \downarrow & & \pi_E \downarrow \\ X & & X \end{array} \tag{4.32}$$

Now Theorem 9 can be better understood; the bundle $P = FX$ consists in frames, the bundle $E = TX$ consists in vectors placed in frames modulo action of G. The local character of each such element is given by the canonical projections. However, the manifold generated by a fixed frame (at a point) and al possible vectors (at the same point) is a fiber in TX and it is isomorphic to the generic fiber F. So, any frame $u \in FX$ generates an isomorphisms $\pi^{-1}(x) \ni u : F \to \pi_E^{-1}(x)$, that is, u gives to any abstract vector from FX a set of components and places it in a frame. The frame u maps this abstract vector into the tangent space TX and gives it geometrical meaning. This construction can be seen in parts of Fig. 4.10. If instead of tangent spaces we use affine spaces constructed upon the tangent spaces, the vector bundle of linear frames becomes the bundle of affine frames.

The quintessence of the vector/frame duality can be presented in a nut-shell by introducing the 1-form called the *canonical form* $\theta \in \Omega^1(\mathrm{FX})$ on the principal bundle of frames FX with values in the standard fibre F (see how the canonical form was introduced for a Lie algebra in Definition 33 and (4.31)). The action of the canonical form on a vector $X \in T\text{FX}$ is $(\theta; X) = u^{-1} \circ d\pi(X) \in F$.

If X is a n-dimensional affine space, then a point $x \in X$ is represented by a position vector $r = x^i e_i$, whose components are given in a certain frame $\{e_i\}_{i=1,\dots,n} = u \in \pi^{-1}(x) \in$FX. The question is: how does this position vector changes with dr by infinitesimally moving the frame. The answer is given by the canonical form, that is by

$$dr = (\theta; X) = (\theta^i, X)e_i, \tag{4.33}$$

where $X \in T_u$FX describes this infinitesimal motion of the frame in the tangent space to the bundle of frames.

4.9.2 Connection Form and Covariant Derivative

The bundle of frames does not provide a recipe of how frames transform when the base point moves through the base space. In order to provide such a law we need an extra construction which is the connection on X, see Fig. 4.9. A connection should provide the infinitesimal transformation of a point in the vector bundle when we perform an infinitesimal move in the base. Since the infinitesimal transformations are described by vectors in the tangent space, a connection should map a point (to

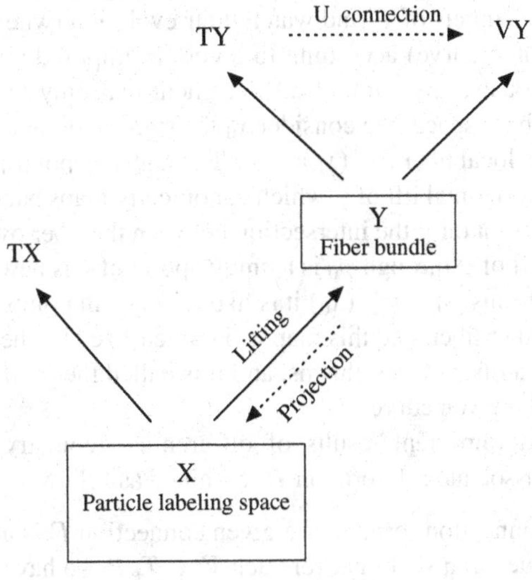

Fig. 4.9 From a base space X we can lift to the tangent bundle TX, and to any other fiber bundle Y. The resulting bundle can be also lifted to its tangent bundle, and there we can define the vertical space VY and the connection $\mathcal{U}$

be moved) in the vector bundle to a vector in the tangent bundle to the vector bundle (how this point transforms), map depending on a vector in the tangent space of the base (the direction of moving).

In the following we introduce the connection in a principal bundle $P(M, G)$.

Definition 35. A connection Γ in P is the assignment of an G-invariant subspace $H_p \lhd T_pP$, for any $p \in P$ and depending differentiable on p, called horizontal subspace.

Let us note that a connection does not see the base space, unless specified otherwise: it only works in the tangent space to each fiber of P. The orthogonal complement of H_p is called *vertical subspace*, is denoted by V_p, and we have $T_p = V_p \oplus H_p$. Any vector $V \in T_pP$ can be uniquely decomposed in two orthogonal components $V = vV + hV$ each in the corresponding sub space $vV \in V_p, hV \in H_p$. A horizontal lift of a vector field on X is the unique horizontal vector field on P such that the differential of the canonical projection on $d\pi : TP \to TX$ maps it to the initial vector field. Any parameterized curve in X, and any point $p \in P$ provide a *lift* of this curve to a unique horizontal (with horizontal tangent vectors) curve in P, to which it canonically projects. As an example, imagine P as the orthonormal frame bundle over $\mathbb{R}^3$, and a curve in this space. At any point in the base space we can choose a variety of frames, any frame from the local fiber. But there is only one such frame which is also a Serret-Frenet frame for that curve, and we denote it p_0. When we move along the curve in the base space, this Serret-Frenet frame transforms from fiber to fiber in a "parallel" way, following the lifted image of the curve.

The existence of a connection on a principal bundle P (the *Cartan connection*) allows us to "flag" elements of P and watch their evolution (when we move in the base space along some curve) according to a certain imposed law called *parallel displacement*. Obviously, any parallel displacement can only be defined along a certain curve in the base space. We consider x_0 the starting point of a parameterized curve $\gamma \subset X$, and its local fiber $\pi^{-1}(x_0) \subset P$. Through any point p_0 in this fiber we can build a unique horizontal lift of γ which canonically maps back on γ. When we move to a different point on γ the intersection between the fiber over this new point and the horizontal lift of γ through p_0 is a unique point of this new fiber. Doing this transport now for various $p_0 \in \pi^{-1}(x_0)$ it is like we map all points p_0 of a fiber into all the points of another fiber, and this map is parameterized by the base curve. This mapping is actually a fiber isomorphisms, and it is called the *parallel displacement* of the fibers along the given curve.

One of the most important results of differential geometry is that to each connection we can associate a 1-form on P, $\mathfrak{g}$-valued as follows

Definition 36. A connection form ω of a given connection Γ is a differentiable 1-form on P with values in $\mathfrak{g}$ such that for each $V \in T_pP$ we have $(\omega; V) = \{A \in \mathfrak{g} \mid A^* = vX\}$.

In other words, a connection form maps a vector field V on P to a Lie algebra vector whose fundamental vector field is exactly the vertical component of V. In a physicist language a connection form is a vector field defined on a bundle of frames such that its directional derivatives in any directions provide one-dimensional Lie algebras of symmetry (flows) in the vertical component of those directions.

Definition 37. Let ϕ be a differentiable r-form on P. We call the $(r+1)$-form $D\phi$ exterior covariant derivative with action on vector fields of P given by $D\phi = (d\phi)\mathrm{Pr}_H$, where D is the exterior derivative (see definition 23) and Pr_H is the projection on the horizontal space of the vector fields.

The exterior covariant derivative of the connection form is called *curvature form* $D\omega = \Omega$, and we have the *structure equation*

$$d\omega = -\frac{1}{2}[\omega, \omega] + \Omega, \tag{4.34}$$

acting on any pair of vector fields on P. The proof is immediate and it is based on (4.25), and on the vertical/horizontal direct sum properties. A connection is flat if and only if its curvature form is null. In a similar manner we define the *torsion form* $\Theta = D\theta$ and we have another structure equation [158, 306]

$$d\theta = -\frac{1}{2}[\omega, \theta] + \Theta. \tag{4.35}$$

The canonical form θ, the connection form ω, the curvature form Ω, the torsion form Θ, and their exterior covariant derivatives fulfil two important relations

$$D\Theta = \Omega \wedge \theta, \quad D\Omega = 0, \tag{4.36}$$

which are nothing but the Bianchi identities, (4.55), (4.56), expressed in the differential forms language.

A connection defined in the bundle of linear frames is a linear connection, and if it is defined in a bundle of affine frames it is an affine connection. On any manifold of positive dimension there are infinitely many affine connections. The choice of an affine connection is equivalent to prescribing a way of differentiating vector fields which satisfies several reasonable properties (linearity and the Leibniz rule). This yields a possible definition of an affine connection as a covariant derivative or (linear) connection on the tangent bundle. A choice of affine connection is also equivalent to a notion of parallel transport, which is a method for transporting tangent vectors along curves. This also defines a parallel transport on the frame bundle. In the bundle of orthonormal frames we have a metric induced by the action of the orthogonal group. So, we define a *Riemannian connection* (or *Levi-Civita connection*) a linear connection with zero torsion.

In order to build the *covariant derivative* of a cross section $\varphi : X \to TX$ in the $\boldsymbol{X} \in TX$ direction we have to lift this last vector to its horizontal component $\boldsymbol{X}^* \in H \subset TFX$. Following the projections we have $FX \ni u \to x = \pi(u) \to \varphi(x)$ which actually defines a cross section in FX. So, we can apply the directional

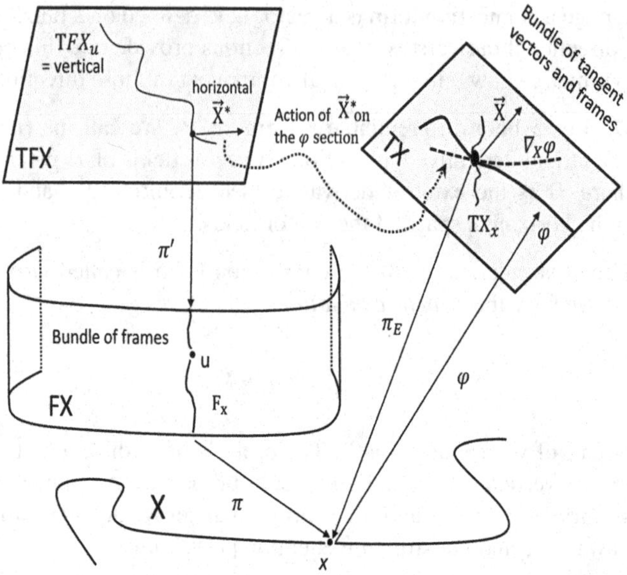

Fig. 4.10 Pictorial interpretation of the covariant derivative. We have the principal bundle of frames FX and its projection π on top of the manifold X, and the tangent bundles to each of these: TX, and TFX, respectively, with their projections π_E, π'. We also represented the local fibres. At $TX_x = \pi_E^{-1}(x) \in TX$ we have two vectors: the arbitrary direction $\boldsymbol{X}$, and the vector cross section φ. The first one is horizontally lifted in TFX as $\boldsymbol{X}^*$ and then acts upon φ generating its covariant derivative $\nabla_X \phi$, as a new cross section (dashed line) in TX

derivative $\boldsymbol{X}^*(\varphi(x(u))) = \nabla_X \varphi$, and this is the requested covariant derivative, see Fig. 4.10. Basically, it is the horizontal component of the directional derivative.

In order to express the connection form ω and consequently its covariant derivative in components we first need to define a canonical basis $\{\boldsymbol{e}_i\}_{i=1,\dots,n}$ in the standard fiber $F \sim \mathbb{R}^n$, and a canonical basis $\{\boldsymbol{E}_{ij}\}_{i,j=1,\dots,n}$ for the Lie algebra $\mathfrak{g}(n, \mathbb{R})$. Since the canonical form θ is $\mathbb{R}^n$-valued, and the connection form ω is $\mathfrak{g}(n, \mathbb{R})$-valued we have

$$\theta = \theta^i \boldsymbol{e}_i; \quad \omega = \omega^{ij} \boldsymbol{E}_{ij},$$

while the two structure (4.35), (4.34) can be written now

$$\begin{aligned} d\theta^i &= -\omega^{ij} \wedge \theta^j + \Theta^i \\ d\omega^{ij} &= -\omega^{ik} \wedge \omega^{kj} + \Omega^{ij}. \end{aligned} \tag{4.37}$$

Obviously, for Riemannian connections on manifolds imbedded in flat spaces the structure equations reduce to

$$d\theta = -\omega \wedge \theta, \; d\omega = -\omega \wedge \omega, \tag{4.38}$$

with the simple interpretation, [299], that the canonical form accounts for the position changes at a change of frame, and the connection form accounts for the twisting of the frames when we move the point

$$\begin{aligned} d\boldsymbol{r} &= \theta^i \boldsymbol{e}_i \text{ change of position,} \\ d\boldsymbol{e}_i &= \omega^{ij} \boldsymbol{e}_j \text{ change of frame.} \end{aligned} \tag{4.39}$$

Let us assign local coordinates in the n-dimensional space X in the form $x \leftrightarrow (x^i)$. The coordinates in the tangent bundle are covariant vectors $\partial/\partial x^i$, a frame in FX is described by the vector fields $X^i_j(x)\partial/\partial x^i$, and the local coordinates in the bundle of frames are (x^i, X^i_j), namely a point and a basis of n-vector fields. Consequently, a frame u ∈FX is represented by the components of the basis fields $u \leftrightarrow X^i_j$ which is exactly the $n \times n$ linear isomorphism u from F onto $T_x X$. The canonical 1-form and the connection 1-form can be written

$$\theta = (X^{-1})^i_j dx^j \boldsymbol{e}_i, \tag{4.40}$$

$$\omega = [(X^{-1})^i_k dX^k_j + (X^{-1})^i_k \Gamma^k_{ml} X^l_j dx^m] \boldsymbol{E}^j_i, \tag{4.41}$$

where the connection coefficients Γ are the Christoffel's symbols. The basis vectors $\partial/\partial x^j$ in TX can be horizontally lifted to

$$\left(\frac{\partial}{\partial x^j}\right)^* = \frac{\partial}{\partial x^j} - \Gamma^i_{jk} X^k_l \frac{\partial}{\partial X^i_l},$$

that is we subtract from the tangent vector its vertical component, which is represented by its connection part (ω or Γ). The covariant derivative acts on the basis (covariant) vectors as follows

$$\nabla_{\partial/\partial x^j} \frac{\partial}{\partial x^i} = \Gamma^k_{ji} \frac{\partial}{\partial x^k}. \tag{4.42}$$

Equation (4.42) and the linearity of the covariant derivative direct us to the coordinate expression of the covariant derivative of a vector field $V = V_i \partial/\partial x^i$ defined on X with respect to the directions of the local frame

$$\nabla_j V_i = \frac{\partial V_i}{\partial x^j} - \Gamma^k_{ij} V_k. \tag{4.43}$$

We illustrate these constructions with an example. Let us have a unit radius spherical surface $X = S_2$ embedded in $\mathbb{R}^3$ with coordinates $x^1 = \theta \in [0, \pi], x^2 = \phi \in [0, 2\pi)$. The tangent space is TS_2 generated by the basis vectors $\{\boldsymbol{e}_\theta, \boldsymbol{e}_\phi\}$. The bundle of the orthonormal frames OS_2 has coordinates $(\theta, \phi, \hat{R}(\alpha))$ where the last one represents an element of the Lie structure group $O(2, \mathbb{R})$, i.e. a rotation of angle

α of the tangent frame around the normal to the sphere. The covariant derivatives have the form

$$\nabla_{e_\theta} e_\theta = 0,\ \nabla_{e_\phi} e_\theta = e_\phi \cot\theta,\ \nabla_{e_\phi} e_\phi = e_\theta \sin\theta \cos\theta,$$

and the horizontal lift of the basis vectors is

$$e^*_\theta = e_\theta - n\cos\theta,\ e^*_\phi = e_\phi - n\sin\theta\cos\theta.$$

We can check this by noticing that at $\theta = \pi/2$ the covariant derivatives cancel, as well as the vertical projections, which is correct since this equatorial circle is actually a geodesic and performs a parallel transport for the tangent vectors. If we want to find, for example (see [299] pp. 66), how is parallel-transported a tangent vector field we can choose a vector which is e_ϕ at an initial point, and we transport it along a parallel to the sphere at $\theta = \theta_0$, parameterized by $t \in [0, 2\pi)$. The resulting parallel-translating vector is

$$V(t) = \sin(\theta_0)\sin(t\cos\theta_0)e_\theta + \cos(t\cos\theta_0)e_\phi,\ \nabla_{e_\phi} V = 0.$$

4.10 Tensor Analysis

In the following we present in more details the covariant derivative concept in an Euclidean manifold. Let $\mathbf{E}_n$ be an n-dimensional Euclidean space (loosely speaking this is an inner product space that has forgotten which point is its origin) and two orthonormal bases $\{e_i\}_{i=1,\dots,n}$ and $\{f_j\}_{j=1,\dots,n}$, fulfilling the linear transformation $f_j = a_{jk}e_k, a_{jk}a^t_{ki} = \delta_{ji}$, where superscript t represents the transpose. To any point $p \in E_n$ we associate the position vector and components $r = x^i e_i = \bar{x}^j f_j$ with the corresponding transformation $\bar{x}^j = a_{jk}x^k$. In an Euclidean space, for linear transformations, there is no difference between contravariant and covariant components, so there is no specific rule for placing the indices in lower or higher position. Any n-uple of numbers A^i that transforms under the same law $\bar{A}^i = a_{ij}A^j$ represents an affine vector. The definition can be extended by direct linear product to affine tensors of any rank in E_n.

Fluids and curved surfaces in motion require for spaces differential manifolds instead of just Euclidean spaces. When generalizing the above definitions to differentiable manifolds, and to nonlinear coordinate transformations

$$\bar{x}^i = \bar{x}^i(x^1, x^2, \dots, x^n), \tag{4.44}$$

more refinements should be introduced to assure that the differentials and derivatives of vectors and tensors are still vectors and tensors. However, for intuitive description we use some times the background of an Euclidean space.

We defined in Sect. 4.1 what is a transformation of coordinates in an n-dimensional differentiable manifold X. In the followings all the transformations will be considered differentiable functions. We define a *tensor field of type* (r, s) (i.e., r-times contravariant and s-times covariant) at a point $p \in X$ of coordinates x, a set of n^{r+s} differentiable functions $T^{(r,s)} = T(x)^{i_1,i_2,\ldots i_r}_{j_1,j_2,\ldots j_s}$ that transform under a change of coordinates (4.44) by the law

$$\bar{T}(\bar{x})^{i_1,\ldots i_r}_{j_1,\ldots j_s} = \frac{\partial \bar{x}^{i_1}}{\partial x^{k_1}} \cdots \frac{\partial \bar{x}^{i_r}}{\partial x^{k_r}} \frac{\partial x^{m_1}}{\partial \bar{x}^{j_1}} \cdots \frac{\partial x^{m_r}}{\partial \bar{x}^{j_s}} T(x)^{k_1,\ldots k_r}_{m_1,\ldots m_s}. \tag{4.45}$$

Like in the case of vector fields, a tensor field generically denoted $T(x)$ is singular at x if $T(x) = 0$. For tensors of order equal or lower than 2 we can use for the transformation $a(x) \to A(\bar{x})$ the matrix notation

$$\begin{aligned} &A^{(1,0)} = A = Ja, \;\; A^{(0,1)} = J^{-1}a, \\ &A^{(2,0)} = JaJ^t, \;\; A^{(0,2)} = (J^{-1})^t aJ^{-1}, \;\; A^{(1,1)} = JaJ^{-1}, \end{aligned} \tag{4.46}$$

where

$$J = J^i_{.j} = \frac{\partial \bar{x}^i}{\partial x^j}, \;\; J^{-1} = \frac{\partial x^i}{\partial \bar{x}^j}$$

is the transformation matrix, and superscript t means transposed. Note that J in components is nothing but the differential $d\bar{x}(x)$ of the coordinate transformation (4.44).

Definition 38. An n-dimensional differential manifold X endowed with a (0,2) type nonsingular tensor field is called Riemannian manifold.

This differentiable tensor field is called *Riemannian structure* or *Riemannian metric* on X, and it is usually denoted by g.

The problem is that the differential of a tensor (vector) field is not anymore a tensor field. One needs to introduce a specific type of differentiation which preserves the tensor character, and this happens through the *affine connection* introduced in Sect. 4.9.2. The differential manifold X endowed with such a structure is now an affine-connected space and the new differential is called the *covariant differential* (or the *absolute differential*) of a vector field $\boldsymbol{A}$

$$\blacktriangledown A^j = dA^j + \Gamma^j_{ik} A^i dx^k, \tag{4.47}$$

where Γ^j_{ik} are the components of the affine connection (or simply connection coefficients), see (4.42), (4.43). The connection coefficients transform at a change of coordinates such that the quantities

$$\Gamma^j_{ik} - \Gamma^j_{ki}$$

form the component of a tensor of type $(1, 2)$ called *torsion tensor.* Obviously, a symmetric connection has zero torsion. Equivalently, a covariant vector has the covariant differential

$$\nabla A_j = dA_j - \Gamma^i_{kj} A_i dx^k. \tag{4.48}$$

As we derived in Sect. 4.9 the connection allows us to introduce the covariant derivative with the action on vector fields

$$\nabla_k A^i = \frac{\partial A^i}{\partial x^k} + \Gamma^i_{jk} A^j \tag{4.49}$$

$$\nabla_k A_i = \frac{\partial A_i}{\partial x^k} - \Gamma^j_{ik} A_j, \tag{4.50}$$

such that transforms a tensor of type (r, s) into a tensor of type $(r, s + 1)$.

A direct way to build a connection in a Riemannian manifold, that is on a manifold (X, g_{jk}) endowed with a $(0, 2)$ type of symmetric nonsingular tensor field $g_{ij}(x)$ of class at least $C_1(X)$, is to obtain the Christoffel's symbols of the first kind from the metric

$$\Gamma^{(g)}_{ijk} = \frac{1}{2}\left(\frac{\partial g_{kj}}{\partial x^i} + \frac{\partial g_{ji}}{\partial x^k} - \frac{\partial g_{ik}}{\partial x^j}\right), \tag{4.51}$$

and then calculate the Christoffel's symbols of the second kind

$$\Gamma^{(g)i}_{jk} = g^{li} \Gamma^{(g)}_{jlk}. \tag{4.52}$$

The Christoffel symbols of the second kind form the components of an affine connection in X, with respect to the $(0, 2)$ type of symmetric tensor field g. As a direct consequence we have

$$\nabla_k g_{ij} = \nabla_k g^{ij} = 0, \tag{4.53}$$

which is nothing but *Ricci's Lemma* for Riemannian manifolds. The Christoffel symbols and their derivatives (second-order covariant derivatives) generate two new tensors, namely the *curvature tensor* K^j_{lhk} (sometimes $K_{jlhk} = g_{js} K^s_{lhk}$ is called curvature tensor) and the *torsion tensor* S^l_{hk}, from the relations

$$\nabla_k \nabla_h A^j - \nabla_h \nabla_k A^j = K^j_{lhk} A^l - S^l_{hk} \nabla_l A^j \tag{4.54}$$

and

$$K^j_{lhk} \equiv \frac{\partial \Gamma^j_{lh}}{\partial x^k} - \frac{\partial \Gamma^j_{lk}}{\partial x^h} + \Gamma^j_{mk} \Gamma^m_{lh} - \Gamma^j_{mh} \Gamma^m_{lk},$$

$$S^l_{hk} \equiv \Gamma^l_{hk} - \Gamma^l_{kh}.$$

The sum of the two terms in (4.54), considered as a linear operator acting on A, is the so-called *Riemann–Christoffel tensor*. When an affine connection is generated from Christoffel symbols the torsion is zero, yet the second-order covariant derivatives still does not commute. These commuting relations are called Ricci identities [10, 181]. In Riemannian manifolds (Definition 38), where the affine connection and the Christoffel symbols are the same quantities, the (1,3) type curvature tensor K^j_{lhk} is denoted R^j_{lhk}, and the (0,4) curvature tensor K_{jlhk} is denoted R_{jlhk}, respectively. In this case one can write two important relations called the Bianchi identities

$$R_{jlhk} + R_{jklh} + R_{jhkl} = 0, \tag{4.55}$$

and

$$\nabla_m R^j_{lhk} + \nabla_k R^j_{lmh} + \nabla_h R^j_{lkm} = 0. \tag{4.56}$$

The most important application of the covariant derivative is the generalization of the "parallel" transport of vectors along curves. If A is vector field, and C a curve of equation $x^i(t)$, both of class $C_1(X)$, the field is *parallel* transported along C if ∇A^j if

$$\frac{dA^j}{dt} + \Gamma^j_{hk} A^j \frac{dx^k}{dt} = 0. \tag{4.57}$$

By construction, the parallel transport of a given vector field is not unique unless $\Gamma^j_{hk} A^h \dot{x}^k dt$ is an exact differential. Consequently one can say that there is no absolute parallelism in general. For a complete study of these differential tools we recommend any book on differential geometry which presents the calculations also in coordinate form. For example a good selection of monographs that complete one another could be given by [10, 19, 33, 68, 181, 235, 242].

4.11 The Mixed Covariant Derivative

This section is in direct relation with the Sects. 6.3 and 6.5. In this section we treat the general n-dimensional case, without going into much details. In Sect. 6.3 we investigate specifically two-dimensional surfaces embedded in $\mathbb{R}^3$ and we go deeper in consequences for the surface differential operators, which themselves are analyzed in detail in Sect. 6.5.

The covariant derivative introduced in (4.49) and (4.50) does not work in the case of mappings between manifold of different dimensions, like in the case of a two-dimensional surface embedded in a three-dimensional space. For example, $M \subset N$ is a submanifold of dimension m of the manifold N of dimension $n > m$. The submanifold M is an embedding, defined by the equations $x^i = x^i(u^\alpha)$, where u^α are the local coordinates in N, and x^i are the local coordinates in M. If we have $m = n - 1$, then M is a hypersurface in N. We introduce the Jacobian matrix associated to the mapping $M \to N$ by

$$B^i_\alpha = \frac{\partial x^i}{\partial u^\alpha}, \ i = 1, \ldots, n, \ \alpha = 1, \ldots, m. \tag{4.58}$$

In both these manifolds we can introduce transformations of coordinates independently, namely $\bar{x}^i = \bar{x}^i(x^j)$ and $\bar{u}^\alpha = \bar{u}^\alpha(u^\beta)$. The B matrix is a contravariant tensor relative to the change of coordinates in N, and it is a covariant tensor relative to the change of coordinates in M. In addition, we can always define a tensor field $Y^i_\alpha(u)$ on M which is also of $(1,0)$ type of tensor with respect to N, and $(0,1)$ type with respect to M. However, neither the derivatives nor the covariant derivatives of Y^i_α are tensors. To construct a tensor quantity by differentiation from such a mixed object, we need to introduce the *mixed covariant derivative*. That is

$$\widetilde{\nabla}_\beta Y^j_\alpha \equiv \frac{\partial Y^j_\alpha}{\partial u^\beta} - \Gamma_{\alpha\ \beta}^{\ \lambda} Y^j_\lambda + \Gamma_{h\ k}^{\ j} Y^h_\alpha B^k_\beta, \tag{4.59}$$

where Γ are the Christoffel symbols of the corresponding manifolds (4.52). The mixed derivative (4.59) is tensor field of type $(1,0)$ with respect to the transformation of coordinates in N, and tensor field of type $(0,2)$ with respect to the transformation of coordinates in M.

We can apply the mixed covariant derivative to the B matrix, and the resulting tensor is of some importance in the geometry of the embedded surface. We define the mixed tensor of $(1,0) - x^i$ type and $(0,2) - u^\alpha$ type as

$$H_{\alpha\ \beta}^{\ j} = \widetilde{\nabla}_\beta B^j_\alpha \equiv \nabla_\beta B^j_\alpha - \Gamma_{\alpha\ \beta}^{\ \lambda} B^j_\lambda + \Gamma_{h\ k}^{\ j} B^h_\alpha B^k_\beta, \tag{4.60}$$

where the two Γ are Christoffel symbols, each defined in another manifold (being Riemannian, in the two manifolds the Christoffel symbols coincide with the affine connection). With the help of these tensor one can enunciate the famous *Equation of Gauss*

$$K_{\alpha\beta\gamma\epsilon} = K_{ljhk} B^l_\alpha B^j_\beta B^h_\gamma B^k_\epsilon + a_{jh}(H_{\alpha\ \gamma}^{\ j} H_{\beta\ \epsilon}^{\ h} - H_{\alpha\ \epsilon}^{\ j} H_{\beta\ \gamma}^{\ h}). \tag{4.61}$$

This theorem expresses the curvature tensor of the subspace M in terms of the curvature tensor of the embedding space N, and the mixed covariant derivatives of the Jacobian matrix. For hypersurfaces $m = n - 1$, there is a great simplification of (4.61), because one can define a unique normal at each point of M. For such a situation one can also define a 2-form called the generalized *second fundamental form* on M. For $n = 3$ case, see Chap. 6. This form $\Pi_{\alpha\beta} du^\alpha du^\beta$ is defined from

$$\widetilde{\nabla}_\beta B^j_\alpha = \pm N^j \Pi_{\alpha\beta}, \tag{4.62}$$

where N^j is the normal to M in x^j coordinates. Consequently, for the Riemannian hypersurfaces case Equation of Gauss becomes

$$K_{\alpha\beta\gamma\epsilon} = K_{ljhk} B^l_\alpha B^j_\beta B^h_\gamma B^k_\epsilon + \Pi_{\alpha\gamma}\Pi_{\beta\epsilon} - \Pi_{\alpha\epsilon}\Pi_{\beta\gamma}. \tag{4.63}$$

The second fundamental form is responsible for the principal directions in M, i.e., its eigenvectors. The coefficients of the characteristic polynomial associated to this eigenvector–eigenvalue problem are related to the curvatures of M. For example, the coefficient of the free term in the characteristic polynomial $\det(\Pi_\alpha^\beta - \lambda\delta_\beta^\alpha) = 0$, denoted $H_{(1)}$ is the *mean curvature* and the coefficient of the highest power, denoted $H_{(n-1)}$ is the *Gaussian curvature*. We also mention the relations

$$H_{(n-1)} = (-1)^{n-1} \det \Pi_\beta^\alpha = (-1)^{n-1} \frac{\det \Pi_{\alpha\beta}}{\det a_{\alpha\beta}}. \tag{4.64}$$

For the dynamics of fluid surfaces case, $n = 3$, we have

$$H_{(2)} \equiv K = \frac{\det \Pi_{\alpha\beta}}{\det a_{\alpha\beta}}, \tag{4.65}$$

and since the only nonzero component of $K_{\alpha\beta\delta\epsilon}$ is K_{1212}, we have

$$K = \frac{K_{1212}}{\det a}. \tag{4.66}$$

4.12 Curvilinear Orthogonal Coordinates

The expression of the differential operators in arbitrary curvilinear coordinates is the best illustration of how the covariant derivative works. A curvilinear coordinate system is defined by three regular (differentiable and locally invertible) transformation functions of the Cartesian coordinates of a three-dimensional Euclidean space $x^i(q^\alpha) : D \subset \mathbb{R}^3 \to C \subset \mathbb{R}^3$, $i, \alpha = 1, 2, 3$. We define as Lamme coefficients the functions

$$H(q)_\alpha = \left| \frac{\partial \boldsymbol{r}}{\partial q^\alpha} \right| = \sqrt{\sum_{i=1}^{3} \left(\frac{\partial x^i}{\partial q^\alpha} \right)^2}, \tag{4.67}$$

and the metrics coefficients

$$g_{\alpha,\beta} = \frac{\partial \boldsymbol{r}}{\partial q^\alpha} \cdot \frac{\partial \boldsymbol{r}}{\partial q^\beta}, \tag{4.68}$$

and we note that $g = \det(g_{\alpha,\beta}) = \prod_{\alpha=1}^{3} H_\alpha^2$ and $H_\alpha = \sqrt{g_{\alpha\alpha}}$ without summation. The unit tangent vectors to each of the three coordinate curves $\boldsymbol{r}(q_\alpha)$ are

$$\boldsymbol{e}_\alpha = \frac{\partial \boldsymbol{r}}{\partial q^\alpha} \left| \frac{\partial \boldsymbol{r}}{\partial q^\alpha} \right|^{-1} = \frac{1}{H_\alpha} \frac{\partial \boldsymbol{r}}{\partial q^\alpha} \text{w.s.} \tag{4.69}$$

The curvilinear coordinates are *orthogonal* if at each point of space $g_{\alpha\beta} = 0$ for $\alpha \neq \beta$. If the curvilinear coordinates are orthogonal we have at each point of space two orthonormal frames: the Cartesian frame $\{\boldsymbol{e}_i\}$ and the curvilinear frame $\{\boldsymbol{e}_\alpha\}$, so any contravariant vector defined in the space $\boldsymbol{A}(\boldsymbol{r}) \in T\mathbb{R}^3$ has two sets of components

$$\boldsymbol{A} = A^i \boldsymbol{e}_i = A^\alpha \boldsymbol{e}_\alpha,$$

with the transformation law

$$A^i = \frac{\partial x^i}{\partial q^\alpha} A^\alpha.$$

The same definition occurs for covariant vectors $\boldsymbol{A} = (A_i)$. With the definition of the unit vectors, the Lamme coefficients can be understood as cosines of the angles between the Cartesian and new basis vectors [10].

We want to make a comment. In many works, when one changes the coordinates, especially in abstract spaces, it may happen that the new coordinates are not normalized, i.e., the new basis is not normalized like in (4.69). In this situation, in addition to the geometrical separation in contravariant and covariant vectors, $A_\alpha = g_{\alpha\beta} A^\beta$, we need to make distinction between "normalized" (or physical) components (components defined in a orthonormal frame) and "not normalized" components (defined in a frame which is just orthogonal). We have the relations $A^\alpha_{norm} = H_\alpha A^\alpha$ and $A_{norm,\alpha} = H^{-1}_\alpha A_\alpha$ without summation. It is interesting to note that the normalized components lose their contravariant/covariant identity. Indeed, $A^\alpha_{norm} = A_{\alpha,norm}$. There are reasons for using one or the other definition: normalized components are more physical from the point of view of units, but they do not form anymore the components of a contravariant/covariant vector. For example the gradient in curvilinear coordinates

$$(\nabla\Phi)_\alpha = \frac{\partial \Phi}{\partial q^\alpha},$$

is a covariant vector, while its "normalized" components

$$(\nabla\Phi)_{\alpha,norm} = \frac{1}{H_\alpha}\frac{\partial \Phi}{\partial q^\alpha},$$

do not form a covariant vector anymore. There is a certain deal of confusion from these conventions. For example, the divergence of a contravariant vector in nonnormalized components reads

$$\nabla \cdot \boldsymbol{A} = \frac{1}{\sqrt{g}}\frac{\partial}{\partial q^\alpha}(\sqrt{g}A^\alpha),$$

and it is a scalar field. The same divergence can be expressed in terms of the normalized coordinates (like it is defined for example in [10, 33])

$$(\nabla \cdot \boldsymbol{A})_{norm} = \frac{1}{\sqrt{g}} \frac{\partial}{\partial q^\alpha} \left(\frac{\sqrt{g}}{H_\alpha} A^\alpha_{norm} \right),$$

and it is not any more a scalar, and the same happens with the curl, etc. The explanation is that by normalization we apply the action of a dilation local group of transformations, which (being local) interferes with the contravariant/covariant character.

Gradient

The gradient of a scalar field $\Phi(\boldsymbol{r}(q))$ is defined as the covariant derivative, and it is a covariant vector

$$\nabla_\alpha \Phi = \frac{\partial \Phi}{\partial q^\alpha}. \tag{4.70}$$

Its contravariant components are

$$\nabla^\alpha \Phi = g^{\alpha\beta} \frac{\partial \Phi}{\partial q^\beta} = \frac{1}{H_\alpha^2} \frac{\partial \Phi}{\partial q^\alpha} \text{ w.s..} \tag{4.71}$$

The normalized components of both covariant and contravariant gradient coincide (though in this form they are not anymore the components of a vector), and these are the components usually provided in mathematical physics books [10, 33]

$$(\nabla^\alpha \Phi)_{norm} = (\nabla_\alpha \Phi)_{norm} = \frac{1}{H_\alpha} \frac{\partial \Phi}{\partial q^\alpha}, \tag{4.72}$$

or in explicit component notation

$$(\nabla \Phi)_{norm} = \sum_{q=1}^{3} \frac{1}{H_q} \frac{\partial \Phi}{\partial q^q} \boldsymbol{e}_q, \tag{4.73}$$

where $\boldsymbol{e}_j$ are the local basis unit vectors.

Divergence

For any contravariant vector field $\boldsymbol{A}(\boldsymbol{r}(q))$ the divergence is obtained by applying the covariant derivative and contracting over the indices

$$\nabla \cdot \boldsymbol{A} = \nabla_\alpha A^\alpha = \frac{1}{\sqrt{g}} \frac{\partial \sqrt{g} A^\alpha}{\partial q^\alpha}. \tag{4.74}$$

In terms of the local curvilinear frame divergence reads

$$\nabla \cdot \boldsymbol{A} = \frac{1}{H_1 H_2 H_3} \sum_{\alpha=1}^{3} \frac{\partial (V^\alpha H_\beta H_\gamma)}{\partial q^\alpha}, \quad \{\alpha, \beta, \gamma\} \in \mathcal{P}_3, \tag{4.75}$$

where $\mathcal{P}_3$ is the set of permutation of 3. Equally, for a covariant vector we have

$$\nabla \cdot \boldsymbol{A} = g^{\alpha\beta}\nabla_\beta A^\alpha = \frac{1}{\sqrt{g}}\frac{\partial \sqrt{g} H_\alpha^2 A^\alpha}{\partial q^\alpha}. \tag{4.76}$$

Both normalized and unnormalized components provide the same expression.

Curl

The curl is an absolute contravariant vector and it is defined as the skew-symmetric linear combination of the components of the covariant derivative

$$(\nabla \times \boldsymbol{A})^\alpha = \epsilon^{\alpha\beta\gamma}\nabla_\beta A^\gamma = \epsilon^{\alpha\beta\gamma} g_{\gamma\delta}\nabla_\beta A^\delta, \tag{4.77}$$

where $\epsilon^{\alpha\beta\gamma}$ are the Levi–Civita symbols. In terms of the local curvilinear frame curl reads

$$\nabla \times \boldsymbol{A} = \sum_{\alpha=1}^{3} \frac{1}{H_\beta H_\gamma}\left(\frac{\partial(A^\gamma H_\gamma)}{\partial q^\beta} - \frac{\partial(A^\beta H_\beta)}{\partial q^\gamma}\right)\boldsymbol{e}_\alpha, \quad \{\beta, \alpha, \gamma\} \in \mathcal{P}_3. \tag{4.78}$$

Laplacian

The Laplacian (also called Laplace–Beltrami operator) in curvilinear coordinates is the contraction of the double covariant differentiated scalar

$$\Delta\Phi = g^{\alpha\beta}\nabla_\alpha\nabla_\beta = \frac{1}{\sqrt{g}}\frac{\partial}{\partial q^\beta}(\sqrt{g} g^{\alpha\beta}\frac{\partial \Phi}{\partial q^\alpha}). \tag{4.79}$$

For example, in spherical coordinates we have [33]

$$\begin{aligned} \boldsymbol{A} &= A^r \boldsymbol{e}_r + A^\theta \boldsymbol{e}_\theta + A^\phi \boldsymbol{e}_\phi; \;\; \boldsymbol{e}_r = (\sin\theta\cos\phi, \sin\theta\sin\phi, \cos\theta); \\ \boldsymbol{e}_\theta &= (\cos\theta\cos\phi, \cos\theta\sin\phi, -\sin\theta); \;\; \boldsymbol{e}_\phi = (-\sin\phi, \cos\phi, 0). \end{aligned} \tag{4.80}$$

The operators are

$$\nabla\Phi = \frac{\partial\Phi}{\partial r}\boldsymbol{e}_r + \frac{1}{r}\frac{\partial\Phi}{\partial\theta}\boldsymbol{e}_\theta + \frac{1}{r\sin\theta}\frac{\partial\Phi}{\partial\phi}\boldsymbol{e}_\phi$$

$$\nabla \cdot \boldsymbol{A} = \frac{1}{r^2}\nabla \cdot \boldsymbol{A} = \frac{1}{r^2}\frac{\partial(r^2 A^r)}{\partial r} + \frac{1}{r\sin\theta}\frac{\partial(\sin\theta A^\theta)}{\partial\theta} + \frac{1}{r\sin\theta}\frac{\partial A^\phi}{\partial\phi}$$

$$\nabla \times \boldsymbol{A} = \frac{1}{r\sin\theta}\left(\frac{\partial(\sin\theta A^\phi)}{\partial\theta} - \frac{\partial(A^\theta)}{\partial\phi}\right)\boldsymbol{e}_r + \left(\frac{1}{r\sin\theta}\frac{\partial(A^r)}{\partial\phi} - \frac{1}{r}\frac{\partial(rA^\phi)}{\partial r}\right)\boldsymbol{e}_\theta$$

$$+\left(\frac{1}{r}\frac{\partial(rA^{\theta})}{\partial r}-\frac{1}{r}\frac{\partial(A^{r})}{\partial\theta}\right)\boldsymbol{e}_{\phi}$$

$$\Delta\Phi=\frac{1}{r^2}\frac{\partial}{\partial r}\left(r^2\frac{\partial\Phi}{\partial r}\right)+\frac{1}{r^2\sin\theta}\frac{\partial}{\partial\theta}\left(\sin\theta\frac{\partial\Phi}{\partial\theta}\right)+\frac{1}{r^2\sin^2\theta}\frac{\partial^2\Phi}{\partial\phi^2}. \tag{4.81}$$

4.13 Special Two-Dimensional Nonlinear Orthogonal Coordinates

For some practical applications one needs to build some special orthogonal coordinates which provide the differential operators, or at least the solutions, to look simpler. This chapter is a mathematical one, but we make an exception and give here a physical, even experimental motivation: in a surface wave tank, or water soliton tank the experimentalist faces the problem to generate a wave of a given initial profile, some times it may be required to have even an initial soliton profile. In principle this could be done by using conducting liquids (salted water, mercury) and try to shape the initial surface by applying an electric field upon the liquid, then turn it off and release the wave. To provide such help, we introduce the so-called plane *soliton coordinates* (Fig. 4.11). They are defined implicitly by their coordinate curves in the Euclidean plane (x, y). For topological soliton shapes (tanh), we have

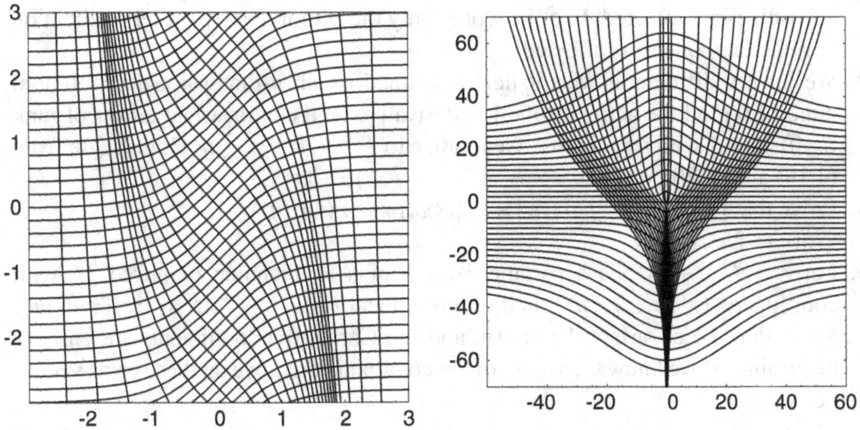

Fig. 4.11 Soliton coordinates in 2-dimensions. The coordinate curves match soliton shapes. *Left*: topological solitons represented by the tanh function. *Right*: Non-topological solitons represented by the square of the function sech

$(x, y) \to (\alpha, \beta)$

$$y_\alpha(x, \alpha) = \tanh(x) + \alpha,$$
$$y_\beta(x, \beta) = -\frac{x}{2} - \frac{\sinh(2x)}{4},$$

and for nontopological soliton shapes (sech), we have $(x, y) \to (\xi, \eta)$

$$y_\xi(x, \xi) = \xi\left(1 + \operatorname{sech}^2 \frac{x}{\xi}\right),$$
$$y_\eta(x, \eta) = \frac{\eta}{2}\left(\frac{1}{2}\cosh^2 \frac{2x}{\eta} + \ln\left[\sinh\left(\left|\frac{x}{\eta}\right|\right)\right]\right.$$

4.14 Problems

1. Find what is the difference between the contravariant and covariant components of a vector at an infinitesimal transformation of coordinates. Use for example a model of an infinitesimal transformation in the form
$$x = q^1, \quad y = q^2, \quad z = q^3 + \epsilon h(q^1, q^2),$$
with $0 < \epsilon \ll 1$ and h is a bounded differentiable function. Prove that $A^1 \to A_1 = A^1 + \epsilon h_x A^3 + O_2(\epsilon)$, etc. Find the $g_{\alpha\beta}$ matrix, the determinant g, and prove that the Christoffel symbols are in $O_2(\epsilon)$. Prove that with respect to the covariant derivative only the z component changes, and only the horizontal derivatives are affected. In other words only the action of the parallel gradient on normal components is affected.
2. We have a differential vector field $\boldsymbol{v}$ defined on an n-dimensional differential manifold. Find the action of the Lie derivative on a contravariant tensor of rank $(k, 0), k > 1$ with respect to $\boldsymbol{v}$. Generalize to $T^{(k,p)}, k + p \geq n$. Hint: Begin from (4.19) and (4.20), and use $T^{(2,0)} = T^{ij}(\partial/\partial x^i)(\partial/\partial x^j)$.
3. Prove that $[f\boldsymbol{v}, g\boldsymbol{w}] = fg[\boldsymbol{v}, \boldsymbol{w}] + f(\boldsymbol{v}(g))\boldsymbol{w} - g(\boldsymbol{w}(f))\boldsymbol{v}$.
4. Prove (4.23), (4.24).
5. Let $\boldsymbol{r} : U \to M$ be a parametrization of a Riemannian manifold M with coordinates $(u^1, u^2, \ldots, u^n)$, and define a local *basis* in T^*M by $x_i = \partial/\partial u_i$. Show that the covariant derivative, and the Christoffel coefficients are entirely determined if we know the values of the covariant derivative on these basis vector, i.e. $\nabla_{x_j} x_i$.

6. Starting from the spherical coordinates (r, θ, φ) in $\mathbb{R}^3$ we introduce the so called dipole coordinates (a, b, φ) by the relations

$$a = \frac{r}{\sin^2\theta}, \quad b = \frac{r^2}{\cos\theta}, \quad \varphi = \varphi.$$

Prove they are non-singular and orthogonal coordinates (in what range of parameters). Find the expression of the Laplace operator in dipole coordinates. Find, as a physical application, the expression of the magnetic potential $\boldsymbol{A}$ generated by a point-like magnetic dipole placed at the origin of the axes.

7. For the dipole coordinates defined above show that in the approximation $r \sim a$ (which is equivalent to the approximation $r^2/b << 1$) the Laplace operator becomes linear and has the form

$$\Delta \sim \frac{\partial^2}{\partial a^2} + \frac{4}{a}\frac{\partial}{\partial a} + \frac{b^2}{a^6}\frac{\partial}{\partial b}\left(b^2\frac{\partial}{\partial b}\right) + \frac{1}{a^2}\frac{\partial^2}{\partial \varphi^2}.$$

Show that the Schrödinger equation for a free particle in dipole coordinates, in the above approximation, can be reduced by a simple conformal transform to the Heun differential equation [199, 275].

8. *Gaetano Vilasi Lemma*. For two differential vector fields $\boldsymbol{X}, \boldsymbol{Y}$ on a Riemannian manifold find a linear relation between the covariant derivative, the Lie derivative and the exterior derivative of a 1-form.

Chapter 5
Geometry of Curves

In this chapter we introduce elements of the differential geometry of curves in an Euclidean space with three dimensions. We begin with the Serret-Frenet equations and their consequences and we devote a section on results related to closed curves.

5.1 Elements of Differential Geometry of Curves

In the following we use the traditional definition of a parameterized curve from [46, 162, 299].

Definition 39. A parameterized differentiable curve is a differentiable (class C_k) map $\boldsymbol{r}(u)$ from the open real interval $u \in I = (a, b) \subset \mathbb{R}$ into $\mathbb{R}^3$. If $k = \infty$ the parameterized curve is called smooth.

In the next chapters we will simply call the parameterized differential curves just parameterized. Occasionally, we will use for the domain of a parameterized curve $I = [0, l], l > 0$, without any loss of generality. The set $\boldsymbol{r}(I) \subset \mathbb{R}^3$ is the *trace* of the curve.

Definition 40. The point $P = \boldsymbol{r}(u)$, $u \in I$ is a *regular point* if $\boldsymbol{r}'(u) \neq 0$. If all the points of a curve are regular the curve is *regular*. A curve is *simple* if the map $\boldsymbol{r} : I \to \boldsymbol{r}(I)$ is an injection.

For example, a one-turn helix is a simple curve, but if it winds more than one turn it is not simple anymore.

We consider a continuous family of parameterized differential curves denoted $\boldsymbol{r}(u, \beta)$ where each curve of the family is parameterized by $u \in (0, u_{max}) \subset \mathbb{R}$, and different curves in the family are assigned different values of the parameter $\beta \in \mathbb{R}$. Later on, this β parameter will be associated with time, and the mappings of curves for different values of β will be associated with their deformation and motion in time. Among different possible parameterizations of a curve, there is always

A. Ludu, *Nonlinear Waves and Solitons on Contours and Closed Surfaces*,
Springer Series in Synergetics, DOI 10.1007/978-3-642-22895-7_5,

one unique representative parametrization with important geometrical significance called the natural parametrization along the curve. In Euclidean geometry, the natural parametrization is accomplished by referring a curve to its arc-length as a parameter. In order to obtain this special parametrization we define the metric on the curve

$$g(u,\beta) = \sum_{i=1}^{3} \frac{\partial x^i}{\partial u}\frac{\partial x^i}{\partial u} = \boldsymbol{r}_u \cdot \boldsymbol{r}_u, \tag{5.1}$$

where $\cdot$ represents the scalar, or dot product, and the subscript represents partial derivative. The arc-length s of a curve is defined by

$$s(u,\beta) = \int_0^u \sqrt{g(u',\beta)}du', \tag{5.2}$$

and now we can use either (u,β) or (s,β) as coordinates for a point on the curve. At every point of the curve we can define an orthonormal frame constructed by the vectors: $\boldsymbol{t}, \boldsymbol{n}, \boldsymbol{b}$ called the unit tangent, the principal normal and the binormal, respectively. This local orthonormal frame is called the Serret-Frenet trihedron (or frame). These vectors are defined by the corresponding Serret-Frenet formulas

$$\begin{aligned} \boldsymbol{t} &= \frac{\partial \boldsymbol{r}}{\partial s} = g^{-\frac{1}{2}}\frac{\partial \boldsymbol{r}}{\partial u} \\ \frac{\partial \boldsymbol{t}}{\partial s} &= \kappa \boldsymbol{n} \\ \frac{\partial \boldsymbol{n}}{\partial s} &= -\kappa \boldsymbol{t} + \tau \boldsymbol{b} \\ \frac{\partial \boldsymbol{b}}{\partial s} &= -\tau \boldsymbol{n}. \end{aligned} \tag{5.3}$$

In all these expressions, we mean by differentiation with respect to s, the partial differentiation, i.e., $\partial(\ )/\partial s \equiv [\partial(\ )/\partial s]_{\beta=\text{conts.}}$. We note one important feature of the parametrization by the arc-length: the derivative of the position vector with respect to s is automatically normalized, $||\boldsymbol{t}|| = 1$, while the normalization of the other two Serret-Frenet vectors is by construction. The two functions

$$\kappa(s,\beta) = \left|\left|\frac{\partial \boldsymbol{t}}{\partial s}\right|\right|, \quad \tau(s,\beta) = \left|\left|\frac{\partial \boldsymbol{b}}{\partial s}\right|\right|$$

are the curvature and the torsion, respectively. The Serret-Frenet local frame together with the curvature and torsion of a curve form the intrinsic geometry of the curve. The three vectors fulfill the relations

$$\boldsymbol{n} = \boldsymbol{b} \times \boldsymbol{t}, \quad \boldsymbol{b} = \boldsymbol{t} \times \boldsymbol{n}, \quad \boldsymbol{t} = \boldsymbol{n} \times \boldsymbol{b}. \tag{5.4}$$

The relations fulfilled by the three vectors can also be written in matrix form

$$\begin{pmatrix} \boldsymbol{t}_s \\ \boldsymbol{n}_s \\ \boldsymbol{b}_s \end{pmatrix} = \begin{pmatrix} 0 & \kappa & 0 \\ -\kappa & 0 & \tau \\ 0 & -\tau & 0 \end{pmatrix} \begin{pmatrix} \boldsymbol{t} \\ \boldsymbol{n} \\ \boldsymbol{b} \end{pmatrix} \tag{5.5}$$

Each of the three mutually orthogonal coordinate planes determined by the Serret–Frenet trihedron has a name: the plane generated by $\boldsymbol{b}$ and $\boldsymbol{n}$ is the *normal plane*, the plane generated by $\boldsymbol{b}$ and $\boldsymbol{t}$ is the *rectifying plane*, and the plane generated by $\boldsymbol{t}$ and $\boldsymbol{n}$ is the *osculating plane*. If we reduce the family β of curves to only one curve, and we know its curvature and torsion as function of s, we understand (5.3) as a linear ODE system. This expresses the fact that if $\kappa \neq 0, \tau \neq 0$ the three unit vectors form a local orthonormal basis in $\mathbb{R}^3$. If we write (5.3) in components with respect to a fixed orthogonal coordinate system $\{\boldsymbol{e}_i\}_{i=1,\ldots,3}$, this ODE system of nine equations reads

$$\begin{aligned} t^i_s &= \kappa n^i \\ n^i_s &= -\kappa t^i + \tau b^i \\ b^i_s &= -\tau n^i, i = 1, \ldots 3 \end{aligned} \tag{5.6}$$

The Serret–Frenet system has three first integrals in the form

$$(t^i)^2 + (n^i)^2 + (b^i)^2 = \text{const.}, \quad i = 1, \ldots 3. \tag{5.7}$$

To prove (5.7) we multiply, for a fixed i, each of the three equations in (5.6) with the same component of the vector in the LHS, nondifferentiated, like $t^i t^i_s = \kappa t^i n^i$, etc. By adding these relations we obtain $t^i t^i_s + n^i n^i_s + b^i b^i_s = 0$ (without summation over i), and if we integrate once the resulting relation is (5.7). The three constants of integration are 1 if we choose a particular coordinate system having its axes parallel to the Serret–Frenet unit vectors for some particular s.

A curve is plane (two-dimensional) if and only if its torsion is zero. In such a situation, the Serret-Frenet relations reduce to

$$\boldsymbol{t} = \frac{\partial \boldsymbol{r}}{\partial s}, \quad \frac{\partial \boldsymbol{t}}{\partial s} = \kappa \boldsymbol{n}. \quad \frac{\partial \boldsymbol{n}}{\partial s} = -\kappa \boldsymbol{t}, \quad \tau = 0. \tag{5.8}$$

In the following we give some examples.

Definition 41. A curve γ is called a *helix* if its tangent lines make a constant angle with a fixed direction.

Proposition 1. *A curve γ is a helix if and only if κ/τ =const.*

The general form of a helix is

$$\boldsymbol{r}(s) = \left(\frac{\sqrt{b^2+c^2}}{c} \int \sin\theta(s)ds, \frac{\sqrt{b^2+c^2}}{c} \int \cos\theta(s)ds, \frac{b}{c}s \right),$$

where $\kappa/\tau = b/\sqrt{b^2+c^2}$. In a similar way we have

Proposition 2. *The parameterized curve γ is a helix if and only if the normal lines (i.e., the lines containing $\boldsymbol{n}(s)$ and passing through γ) are parallel to a fixed plane,*

and

Proposition 3. *The parameterized curve γ is a helix if and only if the binormal lines (i.e., the lines containing $\boldsymbol{b}(s)$ and passing through γ) make a constant angle with a fixed direction.*

In particular, if both the curvature and the torsion are constant we call γ a *cylindric helix*. The cylindric helix depends on two parameters: the radius R of the base circle and the "pitch" b . By denoting the parameter along the curve $u = \phi$, the circular helix is defined by the equation

$$\boldsymbol{r} = (R\cos(\phi), R\sin\phi, b\phi), \tag{5.9}$$

has the metrics $g = R^2 + b^2$ and the arc-length $ds = \sqrt{R^2+b^2}d\phi$. The Serret–Frenet frame is

$$\begin{aligned}
\boldsymbol{t} &= \frac{1}{\sqrt{R^2+b^2}}(-R\sin\phi, R\cos\phi, b), \\
\boldsymbol{n} &= (-\cos\phi, -\sin\phi, 0) \\
\boldsymbol{b} &= \frac{1}{\sqrt{R^2+b^2}}(b\sin\phi, -b\cos\phi, R).
\end{aligned} \tag{5.10}$$

The curvature and torsion are $\kappa = \frac{R}{R^2+b^2}$ and $\tau = \frac{b}{R^2+b^2}$, and we note that the circular helix is the only three-dimensional curve with constant curvature and constant torsion. For $b = 0$ it reduces to a circle, and for $R = 0$ it reduces to a line.

Another way to express the curvature (the torsion) of a parameterized curve as a function of the first two (three) derivatives is

$$\begin{aligned}
\kappa(u,\beta) &= \frac{|\boldsymbol{r}_u \times \boldsymbol{r}_{uu}|}{|\boldsymbol{r}|^3} \\
\tau(u,\beta) &= \frac{(\boldsymbol{r}_u \times \boldsymbol{r}_{uu}) \cdot \boldsymbol{r}_{uuu}}{|\boldsymbol{r}_u \times \boldsymbol{r}_{uu}|^2}.
\end{aligned} \tag{5.11}$$

In the case of plane curves, (5.8) there are interesting integral properties. We have

$$\boldsymbol{r}(L,\beta)-\boldsymbol{r}(0,\beta)=\int_0^L \boldsymbol{t}(s',\beta)ds' \tag{5.12}$$

and this expression is zero for a closed curve. Another important quantity for a closed plane curve is its rotation index $\theta(s,\beta)$, defined as $\kappa=\partial\theta/\partial s$. We have

$$\theta(L,\beta)-\theta(0,\beta)=\int_0^L \kappa(s',\beta)ds'=2\pi N, \tag{5.13}$$

where N is the Euler–Poincaré characteristic of the domain having the curve as its boundary. For a convex plane domain $N=1$ (Santalo theorem). It is also easy to check that the area of a plane closed curve is

$$A=\frac{1}{2}\int_0^L \boldsymbol{r}\times\boldsymbol{t}ds=\oint g^{\frac{1}{2}}\boldsymbol{r}\times\boldsymbol{t}du. \tag{5.14}$$

Indeed, for a differentiable curve, the area of an infinitesimal sector of curve subtended by $\boldsymbol{r}(s+ds,\beta)-\boldsymbol{r}(s,\beta)=\boldsymbol{t}(s,\beta)ds$ is $\boldsymbol{r}\times\boldsymbol{t}ds/2$. The origin of the coordinates does not matter since a translation of a fixed vector does not contribute to the closed integral.

In the two-dimensional case, the curvature given as a function of the arc-length $\kappa=\kappa(s)$, and the initial condition $\boldsymbol{r}(0)$ defines a curve completely. One can integrate the parametric equations of the curve and obtain the solutions in terms of the Fresnel integrals

$$x(s)=\int_0^s \cos\left(\int_0^{s'}\kappa(s'')ds''\right)ds'+x_0,$$

$$y(s)=\int_0^s \sin\left(\int_0^{s'}\kappa(s'')ds''\right)ds'+y_0. \tag{5.15}$$

In the following we elaborate on parameterized curves of class $k\geq 3$ with nonzero vanishing curvature. Actually, the set of points that describes the curve is an equivalence class in the set of allowable parameterizations, i.e., the set of $C_3(I)$ functions from the interval I into the real space $\mathbb{R}^3$.

Every intrinsic or geometric property of a curve should not depend on its parameterizations. However, there is a specific representation of a given curve that has different structures for different parameterizations, and it is called the spherical images of the curve. We can assume that the three unit vectors of the moving Serret-Frenet trihedron can undergo a parallel displacement toward the origin O of the Cartesian coordinate system in $\mathbb{R}^3$. While bonded to the origin, when the chosen parameter along the curve describes it, the three ending points of these unit vectors lie on the surface of the S_2 sphere, and describe themselves three curves

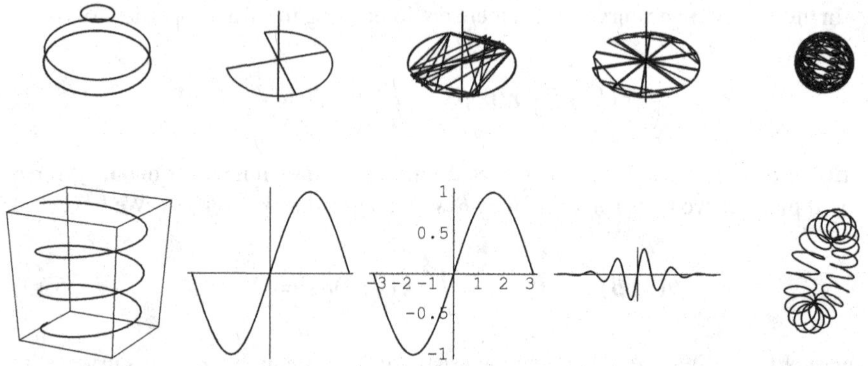

Fig. 5.1 Spherical images (*upper row*) of some curves (*lower row*): circular helix $(x, \sin(x))$, plane curve $(x^2, \sin(x^2))$, same sine, and function, which is the same set of points, but a different parametrization, modulated sech function (breathing NLS soliton), and torus

named *indicatrices*, or spherical images of the curve. These are the tangent, principal normal, and binormal indicatrix. In Fig. 5.1 we have chosen some traditional curves (upper row) and in the lower row we show the corresponding indicatrices. For a circular helix the spherical images are three parallel circles. We have always closed curves for the spherical images, when the original curve is periodic. For plane curves, one of the indicatrix is a vertical axis, and the other two indicatrices form two horizontal plane curves. For example in the case of the graphics of a sine function, i.e., the curve given by the parameterizations $(x, \sin(x))$, the spherical images look like in the second column of Fig. 5.1 However, the same curve, parameterized differently, e.g., $(f(x), \sin(f(x)))$ for $f(x)$ an arbitrary real homeomorphism, the spherical images look totaly different.

The linear arc elements of the indicatrices fulfill the Lancret formula

$$ds_n = \sqrt{ds_t^2 + ds_b^2}, \tag{5.16}$$

where $ds_t^2 = \boldsymbol{t}_s \cdot \boldsymbol{t}_s ds^2$, etc.

The most important result for the differential geometry of curves is the possibility of characterizing the curve in a manner independent of the coordinates, except for the position of the curve in space, i.e., to find representations of the curve invariant to all possible congruent transformations. So we are looking to construct representations of the curve with quantities and parameters independent of the choice of coordinates, but depending only on the geometric shape of the curve. Any set of two independent functional relations between s, κ, and τ are called the *natural* or *intrinsic* equations of the curve. Actually, if the curvature and torsion are continuous functions of s on a given interval, they generate an arc of curve, uniquely determined modulo its position in space. Consequently, any invariant with respect to congruent transformation of the space is expressible in terms of the curvature and torsion.

Theorem 10. *Let $\kappa(s)$ and $\tau(s)$ be $C_0[0,a]$ functions of the real variable s. Then there is one and only one arc $\mathbf{r}(s)$ of a curve, determined up to a direct congruent transformation.*

The proof of the theorem is based on the theorem of existence and uniqueness of a linear system of differential equations. There is a simple and intuitive presentation of Theorem 10, beyond its traditional analysis in terms of the existence and uniqueness of the solution of linear ODE system of equations with variable coefficients and given initial data. This intuition is based on the so-called *canonical representation of a curve*, which is just the description of the shape of the curve in the neighborhood of any of its points. Let us assume we have curves of class $k \geq 3$ and we expand the equation of the points of this curve in Taylor series, up to the third order

$$\boldsymbol{r}(s) = \boldsymbol{r}(0) + \sum_{k=1}^{3} \frac{s^k}{k!} \frac{d^k \boldsymbol{r}}{ds^k} + \mathcal{O}(4). \tag{5.17}$$

From the Serret–Frenet equations we have

$$\boldsymbol{r}_s = \boldsymbol{t}, \quad \boldsymbol{r}_{ss} = \boldsymbol{t}_s = \kappa \boldsymbol{n}, \quad \boldsymbol{r}_{sss} = \kappa_s \boldsymbol{n} - \kappa^2 \boldsymbol{t} + \kappa\tau \boldsymbol{b}. \tag{5.18}$$

Since we can always choose the Cartesian frame such that its origin coincides with the beginning point $s = 0$ of the curve, and such that $\boldsymbol{t}(0) = (1,0,0), \boldsymbol{n}(0) = (0,1,0), \boldsymbol{b}(0) = (0,0,1)$, we can write (5.18) in the form

$$\boldsymbol{r}(s) = \left(s - \frac{\kappa_0^2 s^3}{6}, \frac{\kappa_0^2 s^2}{2} + \frac{\dot{\kappa}_0 s^3}{6}, \frac{\kappa_0 \tau_0 s^3}{6} \right) + \mathcal{O}(4), \tag{5.19}$$

where $\kappa_0 = \kappa(0)$, etc. Equation (5.19) proves that on an infinitesimal interval any curve of class $k \geq 3$ can be sufficiently well approximated with polynomials in s with coefficient uniquely determined by the curvature and torsion around that point. It is interesting to mention that in a neighborhood of any of its points, a $k \geq 3$ class curve can be represented with approximation as a parabola in the osculating plane, a cubical function in the rectifying plane, and a semicubical parabola in the normal plane.

In the following we find useful to introduce elements of the theory of contacts between curves and curves and surfaces.

Definition 42. A curve $\boldsymbol{r}_\Gamma(s)$ of class $m+1$ has a contact of order m with another curve $\boldsymbol{r}_{\Gamma^*}(s^*)$ of the same class, at a point P_0, if $\boldsymbol{r}_\Gamma^{(k)}(s_{P_0}) = \boldsymbol{r}_{\Gamma^*}^{(k)}(s^*_{P_0})$ for $k = 1, 2 \ldots m$ and $\boldsymbol{r}_\Gamma^{(m+1)}(s_{P_0}) \neq \boldsymbol{r}_{\Gamma^*}^{(m+1)}(s^*_{P_0})$.

This definition can be further extended for the contact between a curve and a surface:

Definition 43. A curve Γ of class $m+1$ has a contact of order m with a surface Σ at a point P_0 if there exists at least one curve Γ^* on Σ that has contact of order

m with Γ at P_0, and there does not exist a curve on Σ that has higher-order contact than m with Γ at P_0.

It is interesting to mention that a curve has at least contact of second order with its corresponding osculating plane. The proof is immediate since the osculating plane at s_0 is spanned by the vectors $\boldsymbol{r}_s(s_0)$ and $\boldsymbol{r}_{ss}(s_0)$, so up to the second-order derivative, the Taylor approximation of the curve lies in this plane. These definitions provide a very interesting geometrical characterization of the contact of a curve with a surface:

Theorem 11. *Let be Γ a curve of class $k \geq m+1$ that has contact of order m with a surface Σ of class k at point P_0. If m is even then Γ punctures Σ at P_0, if m is odd, there is always a neighborhood of P_0 such that Γ lies on one side of Σ in this neighborhood.*

In order to prove this theorem we notice that the surface is of class greater or equal than $m+1$, we can choose a neighborhood $\mathcal{V}(P_0)$ in $\mathbb{R}^3$ such that on $\mathcal{V}(P_0)$ the surface can be represented by a function $F(\boldsymbol{r}) = 0$, and such that F is positive on one side of Σ and negative on the other side of Σ. Since the contact is of order m, the first nonzero Taylor term of $F(\boldsymbol{r}(s)) = F(\boldsymbol{r}_\Gamma(s))$ on $\mathcal{V}(P_0)$ is proportional to s^{m+1} and to the $m+1$ derivative of F with respect to s. Because this $m+1$ derivative is continuous, and nonzero in P_0, the sign of $F(s)$ is uniquely determined by s^{m+1}, which proves the theorem.

5.2 Closed Curves

A regular parameterized curve $\boldsymbol{r}(u)$, $u \in [0, l]$ is considered closed if as many as possible of the derivatives of its equation with respect to the parameter agree at 0 and l. For curves in the Euclidean space we have a more rigorous definition: a *closed curve* is a regular parameterized curve $\boldsymbol{r}(u) : [0, l] \subset \mathbb{R} \to \mathbb{R}^3$ with the property that it has a smooth intersection at $t = 0$ and $t = l$, namely $\boldsymbol{r}^{(k)}(0) = \boldsymbol{r}^{(k)}(l)$, $\forall k = 0, 1, \ldots$. In practice, and especially for numerical calculations, we can relax this definition and use instead the following criteria

$$\begin{aligned} \boldsymbol{r}(0) &= \boldsymbol{r}(l) \\ \kappa(0) &= \kappa(l) \\ \theta(l) - \theta(0) &= 2\pi N, \quad N = \text{integer}. \end{aligned} \tag{5.20}$$

A closed curve $\gamma \subset \mathbb{R}^3$ is a differential immersion $\gamma : S_1 \to \mathbb{R}^3$. The points of $\mathbb{R}^3$ where the curve has *self-intersections* (that is for $x \neq y$ we have $\gamma(x) = \gamma(y)$) are also called *double points* or *crossing points*. If there are no three distinct points of $[0, l]$ having the same image the closed curve is called *self-transverse*, that is all self-intersection points. The curve in Fig. 5.3 is self-transverse and it has 4 self-intersections.

A closed plane curve ($\gamma : [0, l] \to \mathbb{R}^2$) is *simple* if its self-intersections are at 0 and l, only. Obviously, a curve without any intersection is not necessarily simple (like in the case of a helix with more than one turn). For a closed regular plane curve γ we can define the following special elements:

Definition 44. • A loop of γ is any restriction of γ to a closed sub-interval $[l_1, l_2] \subsetneqq [0, l]$ such that $\gamma|_{[l_1, l_2)}$ is injective and $\gamma(l_1) = \gamma(l_2)$.
• A vertex is a point $x \in [0, l]$ where $\kappa(x) = 0$.

A simple closed convex curve has at least four vertices (the four-vertex theorem) [46]. A closed plane curve with n self-intersections has maximum $2n$ loops case in which is called maximally looped [47].

In the following, we present some global theorems for closed plane curves whose proofs can be found in [46, 162, 299] for example. An important global invariant for a closed plane curve is its *winding number* (also called *index*) relative to a point. Basically, this number describes the number of turns performed by a vector originating at a fixed point p_0 while its end covers the curve. The best approach is to use the concept of covering map $cov : C \to X$ between two topological spaces, see Definition 1. For any closed parameterized curve $\gamma(t) : [0, l] \to X$, $\gamma(0) = \gamma(l) = p_0$ in the topological space X we define a *lifting* of γ any closed curve $\tilde{\gamma} \subset C$ with the property $\gamma = cov \circ \tilde{\gamma}$. It can be easily proved that there is a unique lifting of γ which contains a given point $\tilde{p}_0 \in C$ and $cov(\tilde{p}_0) = p_0$. Moreover, if the base space X is arcwise-connected (which we know it is stronger than just connected) there is a one-to-one correspondence between the sets $\{cov^{-1}(p)|p \in X\}$ and $\{cov^{-1}(q)|q \in X\}$, $p \neq q$ which actually means that $card\{cov^{-1}(p)|p \in X\}$ is the number of sheets of the covering map, and this number is independent of the choice of the point p. The lifting procedure preserves homotopy, meaning that any two lifted curves are homotopic.

In the following we choose a covering map from the real axis to the unit circle defined by

$$cov : \mathbb{R} \to S_1, \; cov(x) = (\cos x, \sin x) = p \in S_1, \qquad (5.21)$$

presented in Fig. 5.2. Let $\phi : [0, l] \to S_1$ and $\phi(0) = \phi(l) = p \in S_1$ a closed arc γ in the unit circle. Since S_1 is arcwise-connected there is a unique lifting $\tilde{\gamma}$ of γ defined by $\tilde{\phi} : [0, l] \to \mathbb{R}$ such that if $x \in \mathbb{R}$ then $cov(x) = p$, that is $cov(\tilde{\phi}) = \phi$. From here it results that $cov(\tilde{\phi}(0)) = cov(\tilde{\phi}(l))$ which means that 2π divides $\tilde{\phi}(0) - \tilde{\phi}(l)$. In this way we have

Definition 45. The degree of the map ϕ defined as above is the number

$$deg\, \phi = \frac{\tilde{\phi}(l) - \tilde{\phi}(0)}{2\pi}.$$

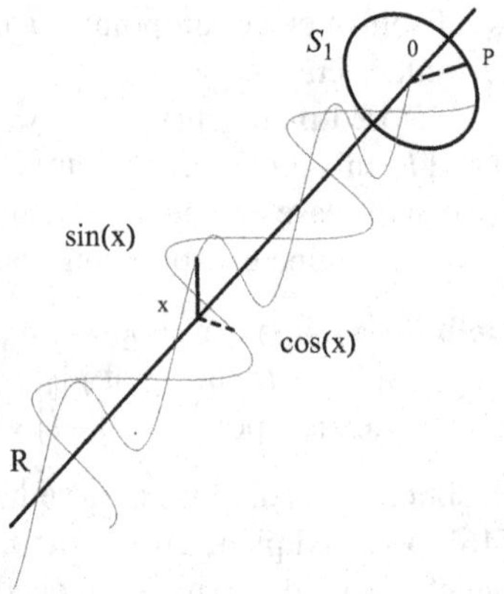

Fig. 5.2 The covering map from (5.21). For each $x \in \mathbb{R}$ we construct the two-dimensional vector of components $(\cos x, \sin x)$ and project it onto the unit circle

The degree of the map ϕ is independent of the choice of the points x, p. Based on Definition 45 we can generalize the index characterization to arbitrary closed plane curves γ of equation $\boldsymbol{r}(t) : [0, l] \to \mathbb{R}^2$ and we can introduce

Definition 46. The winding number (or index) of a plane close curve $\gamma \subset \mathbb{R}^2$ relative to a point $p_0 \in \mathbb{R}^2$ is the degree of the map

$$ind\,\gamma = deg\,\frac{\boldsymbol{r}(t) - p_0}{|\boldsymbol{r}(t) - p_0|}. \tag{5.22}$$

As an example we choose a fivefold circle in polar coordinates of the form $r = 1 + \sin(\phi/5)$ and three reference points. For each such point we calculated the winding number by evaluating the arg function for the fraction in the RHS of (5.22). The number of singularities of this arg provides exactly the winding number. This example is presented in Fig. 5.3 where the three reference points are chose at the center of each little frame. Inside each frame we plot the arg function, and we can count the number of singularities (spikes in the frames) associated to the winding number relative to that point.

An important result of the theory of closed plane curves is the *Jordan curve theorem* , which basically says that such a curve divides the plane in two disjoint regions.

Theorem 12. *If $\boldsymbol{r}(u) : [0, l] \to \mathbb{R}^2$ is a plane, regular, closed, and simple curve, then the region obtained by eliminating the curve from the plane (i.e., $\mathbb{R}^2 - \boldsymbol{r}([0, l])$) has exactly two connected components, and $\boldsymbol{r}([0, l])$ is their common boundary.*

In other words every regular closed plane curve without self-intersections separates its plane in two disjoint regions. The crucial point of the proof is to show that the difference between winding numbers of the curve relative to two points placed on different sides of the curve is not zero. We choose the two points close enough to this curve such that we can approximate the curve with a polygonal line. Then, by

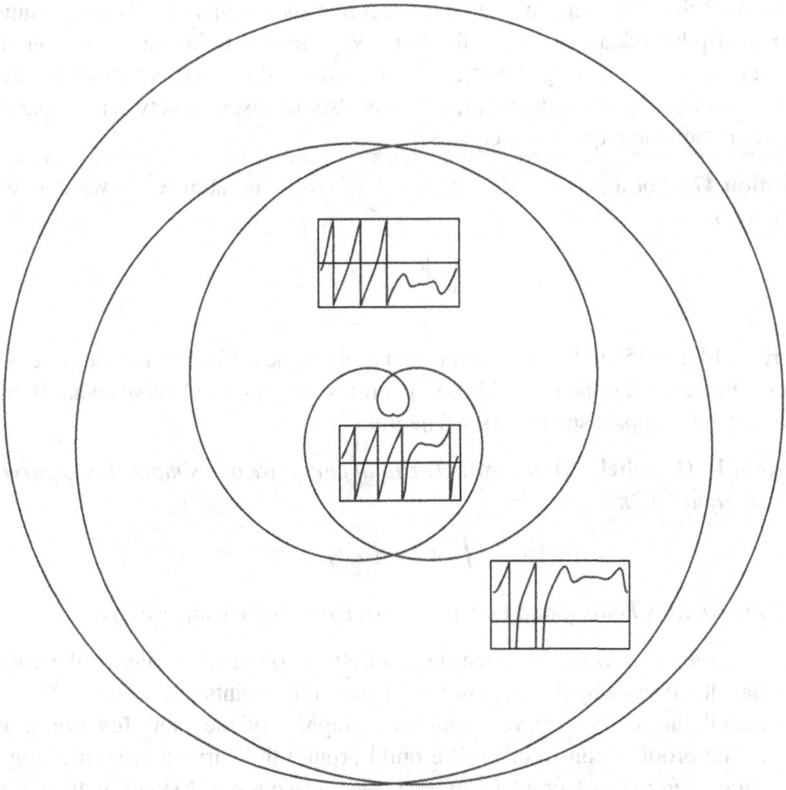

Fig. 5.3 A plane closed curve and three little frames whose centers are located at three reference points. In each frame we plot the complex argument of the winding number function from (5.22). The number of singularities recorded in each of the three framed graphics is exactly the winding number of the curve relative to the point at the center of the corresponding frame

using homotopy (smooth deformations), and by knowing that the winding number is constant in each connected component of a set, we prove that we have two disjoint components.

The region of the plane bounded by γ is called *interior* of the curve and it is homeomorphic with the open unit disc in $\mathbb{R}^2$. We have a simple closed curve *positively oriented* if we choose its parameter such that when we move along the curve and the parameter is increasing, we have the interior of the curve to the left.

Theorem 13. *For any simple closed plane curve γ of length L and area of the region bounded by γ, A we have*

$$L^2 \leq 4\pi A,$$

where equality holds if γ is a circle.

The proof of this theorem is based on the fact that such a curve is always contained within a strip bounded by two parallel lines. We can also fit between these parallel lines a circle. By comparing A and L with the area and the perimeter of this circle, last ones depending on the diameter, which is also the distance between the parallel lines, we obtain the requested inequality.

Definition 47. For a curve $\boldsymbol{r}(u) : [0, l] \to \mathbb{R}^3$ with curvature $\kappa(s)$, we define its *total curvature*

$$\int_0^l |\kappa(s)| ds.$$

Theorems 14 and 15 are the most important tools in the differential theory of closed curves. The Fenchel's and Fary–Milnor theorems are for closed curves what Bonnet theorem is for compact surfaces (see Theorem 20).

Theorem 14 (Fenchel's Theorem). *The total curvature of a simple closed curve is larger or equal to* 2π

$$\int_0^l \kappa ds \geq 2\pi,$$

where the equality holds if and only if the curve is a plane convex curve.

A plane curve is *convex* if its trace lies entirely on one side of any of the closed half-plane determined by the tangent line at any of the points of the curve. A circle, or a parabola are convex curves, while the graphics of the "sin" function is not convex. The proof is constructive. We build around the curve a *tube* of radius ρ, i.e., a parameterized surface $\boldsymbol{r}_\Sigma(s, v) = \boldsymbol{r}(s) + \rho(\boldsymbol{n} \cos v + \boldsymbol{b} \sin v)$, with $\boldsymbol{n}, \boldsymbol{b}$ the normal and binormal vectors of α. We choose ρ small enough such that the tube does not self-intersect. If the curve $\boldsymbol{r}(s)$ is a simple closed curve, then the tube is homeomorphic to a torus. We notice that the Gaussian curvature K and the area element dA of the tube surface have the property

$$\iint_{[0,l]\times(v_1,v_2)} K dA = -\int_0^l \int_{v_1}^{v_2} \frac{\kappa \cos v}{\rho(1 - \rho\kappa \sin v)} \sqrt{EG - F^2} ds dv$$
$$= \int_0^l \kappa(s) ds (\sin v_1 - \sin v_2).$$

We can choose the angles $v_{1,2}$ such that the unit normal of the tube in this range of v covers the entire unit sphere S_2, and also $K > 0$ in this range. Indeed, this is possible because we can approach the tube with a plane coming from infinity, from any direction in $\mathbb{R}^3$. The first point of the tube encountered by this plane has positive Gaussian curvature (is an elliptic point). So, if we apply the Gauss–Bonnet Theorem 20 for K we obtain the requested inequality for the total curvature of $\boldsymbol{r}(s)$.

If the curve $\boldsymbol{r}(s)$ is closed but knotted we have the following.

Theorem 15 (Fary–Milnor Theorem). *The total curvature of a knotted simple closed curve is greater or equal to* 4π.

By taking profit of the integral formulas for surface differential operators we can obtain two useful relations for closed curves. Let Σ be a regular parameterized surface and $D \subset \Sigma, \Gamma = \partial D$, and ∇_Σ be the surface gradient operator defined in Sect. 6.5. From the surface divergence integral theorem (6.61), by using the fact that $\nabla_\Sigma \times \boldsymbol{r} = 0$, we obtain

$$\oint_\Gamma \boldsymbol{r} \cdot \boldsymbol{t} ds = 0, \tag{5.23}$$

$$\oint_\Gamma \boldsymbol{t}^\perp \times \boldsymbol{r} ds = 2 \iint_D H(\boldsymbol{N} \times \boldsymbol{r}) dA. \tag{5.24}$$

In these equations $\boldsymbol{r}$ is the position vector, $\boldsymbol{N}$ is the unit normal to Σ, and $\boldsymbol{t}^\perp = \boldsymbol{N} \times \boldsymbol{t}$ belongs to the Darboux frame associated to Σ, Γ.

The theory of closed curves is not closed. There are still open questions, and a simple theorem to provide an analytic differential criterium for closeness in terms of curvature and torsion does not exist in general. Some more information on the topics can be found in [47, 62, 221, 225, 339].

5.3 Curves Lying on a Surface

We will see in Chap. 6, where we study the geometry of surfaces, that in a way a surface is a reunion of curves, so it is natural to study the geometry of surfaces starting with, and through the methods of the geometry of curves. However, in this section we consider the (non-flat) surface as given and fixed, and we look at the behavior of curves lying on this surface. In particular closed curves on surfaces are the most important since they are related to the homotopy and the homology of the surface. The first interesting observation is that two important theorems holding true for closed plane curves do not hold in the case of non-flat surfaces: the Jordan curve theorem (Theorem 12) and the *four-vertex theorem.*

The Jordan theorem is not valid anymore if $\mathbb{R}^2$ is substituted with an arbitrary surface [47]. Indeed, a circle laying on a torus T_1 divides its surface in either $S_1 \bigcup \{p\}$ (where p is a point on the torus) like in the plane case, or just puts it into homotopy with S_1, depending on where the curve is placed. For example, we present a double self-intersecting closed curve lying on a torus in Fig. 5.4; it is impossible to map the same figure in the plane because the order of increasing value of the parameter at the self-intersections is not the same.

Likewise,the four-vertex theorem for curves in $\mathbb{R}^3$ shows, [62], that simple closed curves lying on strictly convex surfaces in space exist only if their torsion does not vanish. Points of null torsion are actually third order singularities for the equations of curves in space. This observation brings an understanding of the relation between

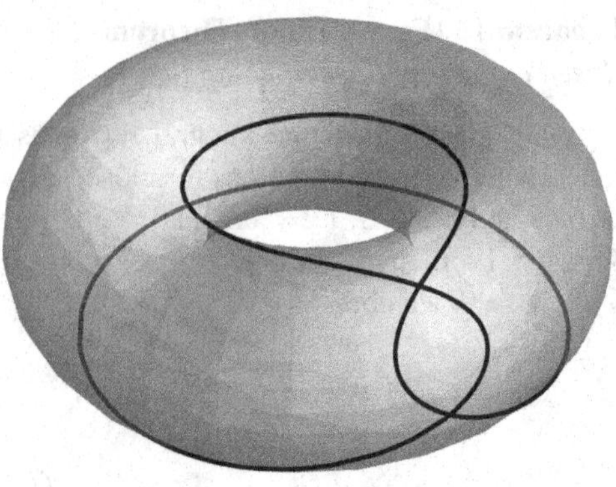

Fig. 5.4 A closed curve on a torus. The order of the values of the parameter at the two self-intersecting points is not the same as the order of the same points for the same curve on $\mathbb{R}^2$

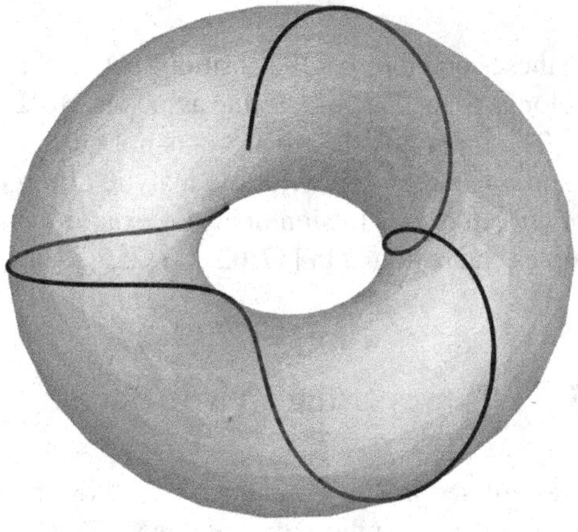

Fig. 5.5 A twisted (1, 3)-closed curve of nonvanishing torsion on a torus

singularities of order $m \leq 3$ (the derivatives of order m of the equation of the curve are linear dependent) of closed curves lying on surfaces, and the convexity of that surface. The four-vertex theorem can be generalized in $\mathbb{R}^3$ as follows:

Theorem 16. *A simple closed curve lying on a strictly-convex surface has at least four points of vanishing torsion (vertices).*

A good example of a non-strictly convex surface is provided by the torus T_1. We define a (q, p)-curve winding on the surface of the torus as a regular closed curve that "spirals" around the torus p times in the vertical sense, and q times in the horizontal sense. There is an exact relation between the large radius of the torus (the small radius is kept 1) and the ratio p/q which guarantees a curve a twisted character. Basically, for a given ratio a twisted (q, p)-curve exists on a torus only for a finite interval of the large radius. In Fig. 5.5 we present a twisted $(1, 3)$-curve of equation

$$x + iy = [2 + \cos(9 \sin t)]e^{3i \sin t}, \; z = \sin(9 \sin t),$$

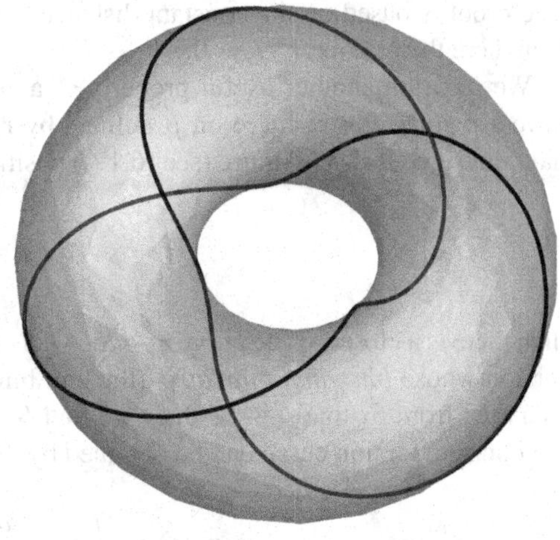

Fig. 5.6 A non-twisted (2, 3)-closed curve on a torus has always points of null torsion

and in Fig. 5.6 we present a trefoil (2, 3)-curve on the same torus, of equation

$$x = (2 + \cos(3t))\cos(2t), \ y = x = (2 + \cos(3t))\sin(2t), \ z = \sin(3t).$$

A regular closed curve γ defined by $\boldsymbol{r}(u) : [0, l] \to S_2$ lies on the unit sphere $S_2 \subset \mathbb{R}^3$ if and only if

$$\int_0^L \tau(s)ds = 0, \tag{5.25}$$

where τ is the torsion of γ and s is the arc-length. The proof consists in noticing that, if θ is the angle between the principal normal of γ, $\boldsymbol{n}$, and the unit normal to the sphere, $\boldsymbol{N}$, i.e., $\cos\theta = \boldsymbol{N} \cdot \boldsymbol{n}$, we have

$$\frac{d\theta}{ds} = \tau - \tau_g,$$

where τ_g is the geodesic torsion (see Definition 59). Because the curve is closed and regular we have

$$\int_0^L \tau ds - \int_0^L \tau_g ds = 2\pi n,$$

where n is integer. However, all curves lying on a sphere are lines of curvature, so their geodesic torsion is zero. Knowing that n is a topological invariant, and since all closed curves lying on a sphere are homotopic to a point, we have $n = 0$, which proves the affirmation.

Also, a curve lying on a sphere fulfills the following relation

$$\left(\frac{1}{\kappa}\right)^2 + \left(\frac{\partial}{\partial s}\frac{1}{\kappa}\right)^2 \left(\frac{1}{\tau}\right)^2 = \text{const.}$$

The proof is based on the constant distance between the center and the curve, and from here the relation $\boldsymbol{r} \cdot \boldsymbol{t} = 0$.

We mention another useful property of a curve lying on a sphere [339]. If we have a regular closed curve on β defined by $\boldsymbol{r}_\beta(t) : S_1 \to S_2$ with $\boldsymbol{r} \in C_2(S_1)$, having its geodesic curvature (see (6.19)) positive, and fulfilling the property

$$\int_{S_1} \boldsymbol{r}_\beta \times d\boldsymbol{r}_\beta = 0,$$

then there is a closed space curve $\boldsymbol{r}_\gamma : S_1 \to S_2$ with positive curvature, and constant torsion whose *binormal indicatrix* (the unit binormal vector of a curve understood as a map from S_1 to S_2) is β. Once we find β with the properties requested above, the constant torsion curve can be obtained by the integration

$$\boldsymbol{r}_\gamma(t) = \int \boldsymbol{r}_\beta \times \frac{d\boldsymbol{r}_\beta}{dt} dt.$$

Curves of constant torsion are important for biology in the study of stiff polymer chains, elastic properties of DNA, structure and dynamics of axonemal cells, and for motile cells swimming by the use of flagella or cilia. For example, the axoneme is a quasi-rigid, quasi-flexible structure of almost parallel microtubules that can bend, and hence can generate motions and swimming by their relative sliding and shearing. The torsion of an axoneme can reduce or cancel the relative slide between the elements and then can annihilate the bending. In this way twisting in an axonemal system can work like a system of shifting gears. For the same reason braided steel cables are used in constructions and bridges since they have very soft bending rigidity.

5.4 Problems

1. Show that there exists at least one closed regular curve in $\mathbb{R}^3$ with positive curvature and constant torsion.
2. Show that simple closed curves in $\mathbb{R}^3$ with nonzero torsion everywhere, lying on strictly convex surfaces (i.e., $K > 0$, $H > 0$ everywhere) do not exist. A counter example would be a spiral on a torus.
3. Prove that for any closed knotted curve in $\mathbb{R}^3$ there is a plane that intersects the curve in at least six points.
4. The Euler equations for "elastica" (that is curves in $\mathbb{R}^3$ describing ideal elastic rods of constant length with clamped or hinged boundary conditions at their ends) have the form

$$2\kappa_{ss} + \kappa^3 - 2\kappa\tau^2 = 0, \quad \kappa^2\tau = \text{constant},$$

in terms of the curvature and torsion in the arc-length parametrization. Find particular elastica solutions in terms of Jacobi elliptic functions. Show that planar elastica solutions belong to only three classes: one class with non-vanishing curvature, another whose curvature has alternating signs, and last class with constant curvature.

5. Let us have a fixed closed plane curve and a moving segment of constant length which keeps its ends on this curve. Prove that the area between this curve and any curve described by a point rigidly attached to this segment is constant, and depends only on the product of the distances from this point to the ends of the segment. Also prove that this second curve is also closed (A. D. Risteen, *Annals of Mathematics* 1887).
6. A convex curve is a plane curve with the property that any straight line which intersects this curve, intersects it at most in two points. Prove that a convex curve has no self-intersections.
7. Prove that if a parameterized curve γ, which is not a helix, has its curvature and torsion fulfilling the relation $C_1\kappa(s) + C_2\tau(s) = C_3$ for all s and with C_i constants, then there is one and only one other curve $\tilde{\gamma}$ having the same normal lines as γ (if γ has the above property is called a Bertrand curve, and $\tilde{\gamma}$ is its Bertrand mate).
8. Show that for a plane closed curve $\boldsymbol{r} : S_1 \to \mathbb{R}^2$ with n self-intersections and more than n loops, n must be odd.
9. In Sect. 5.3 a (q, p)–closed curve in $\mathbb{R}^3$ is defined as a curve that spins p times in the vertical sense, and q times in the horizontal sense. Find a rigorous mathematical definition for such (q, p)-curves.

Chapter 6
Geometry of Surfaces

There are two main differences between the theory of regular curves and regular surfaces in three-dimensional Euclidean spaces. On one hand, smooth curves are mappings (i.e., $\gamma(s) : I \to \mathbb{R}^3$), while regular surfaces are submanifolds. On the other hand, all curves have the natural arc-length parameter, while surfaces do not have a natural parametrization. Moreover, curves are uniquely defined (up to a rigid motion) by two real functions (curvature and torsion), while surfaces are defined uniquely up to rigid motions by six real functions (E, F, G, e, f, and g).

Definition 48. $S \subset \mathbb{R}^3$ is a *regular surface* if for any of its points $p \in S$ we can define locally (in a neighborhood of p) a regular, differentiable homeomorphism between an open set $U \in \mathbb{R}^2$ and S. That is, $\forall p \in S, \forall V(p) \in \mathcal{V}(p, \mathbb{R}^3)$ we have $\exists \boldsymbol{r} : U \to V(p) \in S$ such that:

1. $\boldsymbol{r}$ is differentiable.
2. $\boldsymbol{r}$ is a homeomorphism.
3. $d\boldsymbol{r}_q : \mathbb{R}^2 \to \mathbb{R}^3, q \in U$ has maximal rank.

See also Fig. 6.1. The last requirement is equivalent that the tangent map $d\boldsymbol{r}_q$ at q is one-to-one, or, equivalently, its Jacobian has rank 2.

Definition 49. A *parameterized surface* is a differential map $\boldsymbol{r} \in \text{Diff}\,(U, \mathbb{R}^3)$, with $U \in \mathbb{R}^2$.

The map $\boldsymbol{r}_q$ is regular if $d\boldsymbol{r}_q$ is one-to-one at any point $q \in U$. For a regular parameterized surface S, the curves $u = u_0, v \in \mathbb{R}$ and $v = v_0, u \in \mathbb{R}^2$ in U are mapped by $\boldsymbol{r}$ into the *coordinate* curves $\boldsymbol{r}(u_0, v)$ and $\boldsymbol{r}(u, v_0)$, respectively. The tangent plane to S at $p = (u_0, v_0)$ is defined as the subspace of $\mathbb{R}^2$ generated by $\boldsymbol{r}_u, \boldsymbol{r}_v$, evaluated at (u_0, v_0). Here subscripts mean differentiation with respect to the parameters. The tangent map of $\boldsymbol{r}(u, v)$ takes values in the tangent plane, $d\boldsymbol{r}_q \in T_{p=\boldsymbol{r}(q)}S$. Actually, as we underlined in Definition 48, according to the second interpretation of the tangent map, $d\boldsymbol{r}$ maps the canonical basis (u, v) from U into a local basis in S, $\{\boldsymbol{r}_u, \boldsymbol{r}_v\}$. Let us have a curve in the arc-length parametrization. Then, in the local basis $\boldsymbol{r}_u, \boldsymbol{r}_v$ of $T_{\boldsymbol{r}(u,v)}S$, the values of the tangent

A. Ludu, *Nonlinear Waves and Solitons on Contours and Closed Surfaces*,
Springer Series in Synergetics, DOI 10.1007/978-3-642-22895-7_6,

Fig. 6.1 Regular surface

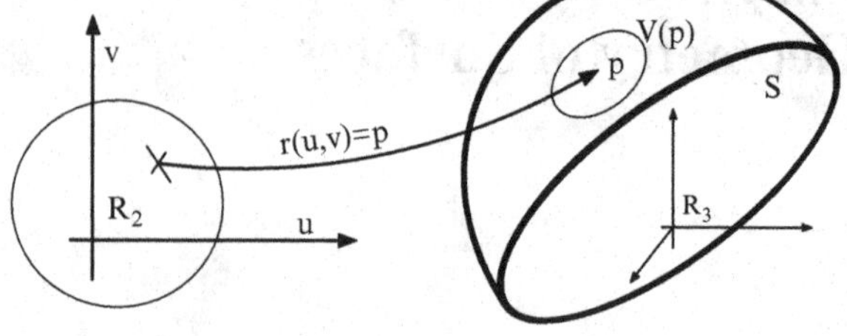

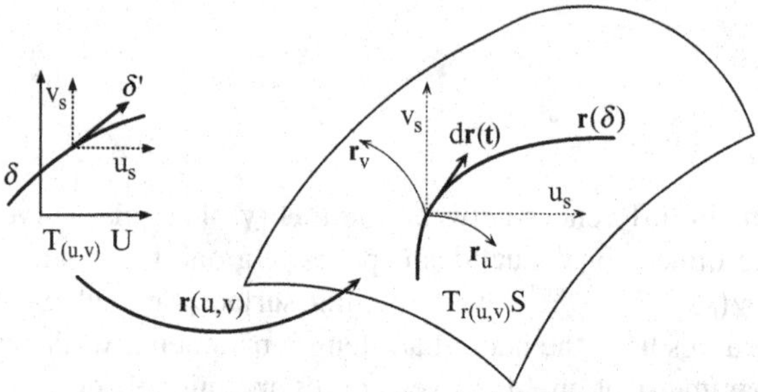

Fig. 6.2 The tangent map $d\boldsymbol{r}(\boldsymbol{t})$. Let $\boldsymbol{r} : U \subset \mathbb{R}^2 \to S \subset \mathbb{R}^3$ be a differential map representing a regular surface. Let $\gamma(s) \subset U$ be a regular parameterized curve with tangent $\boldsymbol{t}(s)$, where s is γ's arc-length parameter. In a local basis of the tangent space $T_{\boldsymbol{r}(u,v)}S$, the tangent map $d\boldsymbol{r}(\boldsymbol{t}(s))$ has the same components (u_s, v_s) as the unit tangent $\boldsymbol{t}(s)$ has in $T_{(u,v)}U$, i.e., $d\boldsymbol{r}(\boldsymbol{t}) = \boldsymbol{r}_u u_s + \boldsymbol{r}_v v_s$. In terms of $\mathbb{R}^3$ coordinates $(x^i, i = 1, 2, 3)$, $d\boldsymbol{r}(\boldsymbol{t}) = \frac{\partial x^i}{\partial \eta^j}\frac{d\eta^j}{ds}$ with $\eta^j = (u, v)$

map $d\boldsymbol{r}(\boldsymbol{t}(s))$ have the same components (u_s, v_s) as the $\boldsymbol{t}(s)$ has in $T_{(u,v)}U$, i.e., $d\boldsymbol{r}(\boldsymbol{t}) = \boldsymbol{r}_u u_s + \boldsymbol{r}_v v_s$ (see Fig. 6.2).

We mention that different parameterizations around p span the same tangent plane (for a proof see [299]). We denote the components of any vector lying in the tangent space of S at p, $\boldsymbol{w} \in T_pS$, as $\boldsymbol{w} = (a, b) = a\boldsymbol{r}_u + b\boldsymbol{r}_v$. For example, the unit tangent vector to a regular parameterized curve α on S looks like

$$\boldsymbol{t} = \frac{d\gamma}{ds} == \frac{d\boldsymbol{r}}{ds}(u(s), v(s)) = \boldsymbol{r}_u u_s + \boldsymbol{r}_v v_s,$$

where s is the arc-length parameter along γ.

Definition 50. S is *orientable* if it admits a differential vector field of unit normal vectors defined on the whole surface. By choosing such a field one chooses an *orientation* for S.

The traditional choice for the *unit normal vector field* is

$$\boldsymbol{N}(u, v) = \frac{\boldsymbol{r}_u \times \boldsymbol{r}_v}{|\boldsymbol{r}_u \times \boldsymbol{r}_v|}.$$

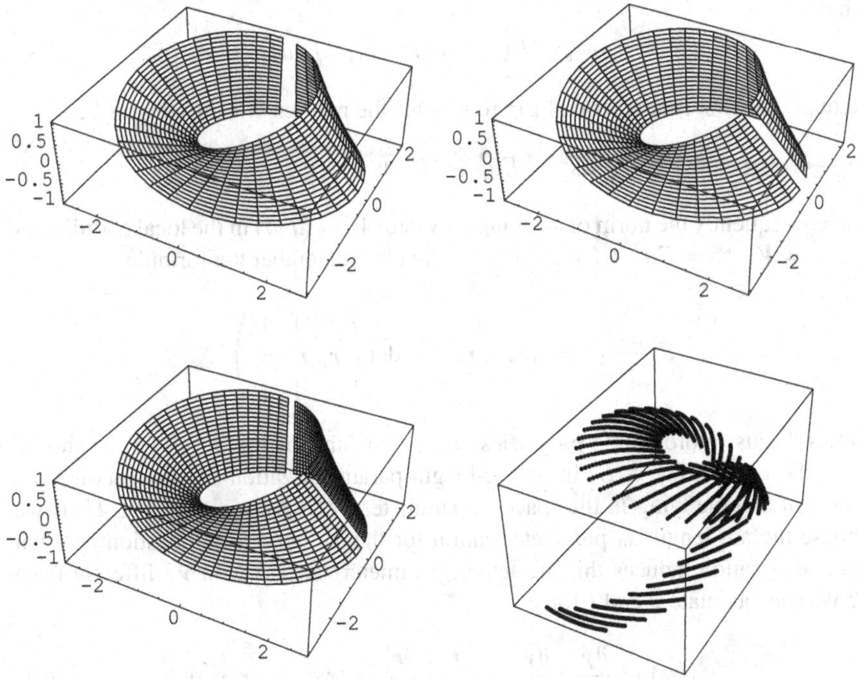

Fig. 6.3 *Upper row*: the two coordinate charts that form the atlas for the Möbius strip. *Lower row left*: the intersection of the two coordinate charts is not connected, but it has two connected components. *Lower row right*: the unit normal is not well defined; it has two possible orientations in the same point

Such a field does not exist on not-orientable surfaces, like in the case of a Möbius strip (for example, see Fig. 6.3).

6.1 Elements of Differential Geometry of Surfaces

The *first fundamental form* on a parameterized surface is the equivalent of the metrics in the case of a curve.

Definition 51. At every point $p = \boldsymbol{r}(u, v)$ of a regular surface, we can define a symmetric second-order tensor field, the first fundamental form on S, $g_p : U \to$ *Diff* $(S, \mathbb{R})$ whose action on tangent vectors $\boldsymbol{V} = (a, b), \boldsymbol{U} = (c, d) \in T_pS$ is defined as

$$g_p(\boldsymbol{V}, \boldsymbol{U}) = Eac + F(ad + bc) + Gbd = \begin{pmatrix} E & F \\ F & G \end{pmatrix} \begin{pmatrix} V \\ U \end{pmatrix},$$

where

$$E(u,v) = \boldsymbol{r}_u \cdot \boldsymbol{r}_u; \quad F(u,v) = \boldsymbol{r}_u \cdot \boldsymbol{r}_v; \quad E(u,v) = \boldsymbol{r}_v \cdot \boldsymbol{r}_v.$$

Actually the first fundamental form represents the metrics of a curve γ on S,

$$ds^2 = Eu_t^2 + 2Fu_t v_t + Gv_t^2,$$

and consequently the norm of any tangent vector $V = (a,b)$ in the local coordinates $||V||^2 = V \cdot V = Ea^2 + 2Fab + Gb^2$. We also remember the formula

$$EF - G^2 = |\boldsymbol{r}_u \times \boldsymbol{r}_v|^2 = \det \begin{pmatrix} | & | & | \\ \boldsymbol{r}_u & \boldsymbol{r}_v & N \\ | & | & | \end{pmatrix}^2 .$$

Let us discuss more on the properties of the first fundamental form. Let us choose a regular curve $\boldsymbol{\gamma}(s) \in S$, in its arc-length parameterization s. This curve is the image of a curve lying in the space of parameters, namely $\boldsymbol{\gamma}_0 = \boldsymbol{\gamma}^{-1} \subset U$. If we choose the arc-length s_0 parameterization for the $\boldsymbol{\gamma}_0$ curve, the question is: what parameterization induces this arc-length parameterization s_0, on $\boldsymbol{\gamma}$, different from s? We can calculate g and $\boldsymbol{t}$ for $\boldsymbol{\gamma}$

$$g_{\gamma}(s_0) = \frac{\partial \boldsymbol{\gamma}}{\partial s_0} \cdot \frac{\partial \boldsymbol{\gamma}}{\partial s_0} = \frac{\partial \boldsymbol{r}}{\partial s_0} \cdot \frac{\partial \boldsymbol{r}}{\partial s_0} = (\boldsymbol{r}_u u_{s0} + \boldsymbol{r}_v v_{s0})^2, \tag{6.1}$$

which is a quadratic form $Eu_{s0}^2 + 2FEu_{s0}v_{s0} + Gv_{s0}^2 = g(u_{s0}, v_{s0})$. This last relation should be the definition of the second fundamental form, which acts $g : TS \to R$, while the expression defined in this relation acts on vectors $(u_{s0}, v_{s0}) \in TU$. The salvation comes from the fact that (u_{s0}, v_{s0}) are also the components the vector $\frac{\partial r}{\partial s_0}$ in the local basis $\{\boldsymbol{r}_u, \boldsymbol{r}_v\}$. Consequently, we can introduce a canonical isomorphism iso : $T_{(u_{s0},v_{s0})}U \to T_{\boldsymbol{r}(u_{s0},v_{s0})}S$ by using the local basis defined by the parameterization of S in TS. The unit tangent of the curve $\boldsymbol{\gamma} \in S$ is

$$\boldsymbol{t}_{\alpha} = \frac{\boldsymbol{r}_u u_{s0} + \boldsymbol{r}_v v_{s0}}{\sqrt{g(u_{s0}, v_{s0})(s_0)}} = \boldsymbol{\gamma}',$$

and we can define the action of the directional derivative upon a differential function defined on S

$$D_{\gamma'} f = (\boldsymbol{t}_{\gamma} \cdot \nabla_{u,v}) f(u,v) = \frac{1}{\sqrt{g}} (u_{s0} f_u + v_{s0} f_v).$$

We also have $g(\boldsymbol{r}_u) = g(1,0) = E$ and $g(\boldsymbol{r}_v) = g(0,1) = G$. So far $g = g[E,F,G]$ is a quadratic form defined on TS. It depends on three functions, while a metric depends only on one function. In conclusion, the interpretation of the first fundamental form in terms of curve properties is the arc-length $ds_{\alpha} = \sqrt{g(u(s_0), v(s_0))} ds_0$. The interpretation in terms of a quadratic form defined on the tangent space $\boldsymbol{v} \in TS$ is $g(\boldsymbol{v}) = |\boldsymbol{v}|^2$ in the basis $\{\boldsymbol{r}_u, \boldsymbol{r}_v\}$.

For any quadratic form $Q : \mathbb{V} \to \mathbb{R}^1$, there is an associate symmetric bilinear form $B : \mathbb{V} \to \mathbb{R}^1$ defined by

$$B(\boldsymbol{u},\boldsymbol{v}) = \frac{1}{4}(Q(\boldsymbol{u}+\boldsymbol{v}) - Q(\boldsymbol{u}+\boldsymbol{v})).$$

Consequently, we can extend g to a symmetric bilinear form

$$g(\boldsymbol{u},\boldsymbol{v}) = \frac{1}{4}(E(u_1+v_1)^2 + 2F(u_1+v_1)(u_2+v_2) + G(u_2+v_2)^2 \\ -E(u_1-v_1)^2 - 2F(u_1-v_1)(u_2-v_2) - G(u_2-v_2)^2),$$

such that the first fundamental form $g : TS \times TS \to R$ (or more precisely $\forall p \in S, g_p : T_pS \times T_pS \to R$) is defined as

$$g(\boldsymbol{u},\boldsymbol{v}) = Eu_1v_1 + 2F(u_1v_2 + u_2v_1) + Gu_2v_2.$$

The geometric significance of the form is in terms of the scalar product in any T_pS. In the $\{\boldsymbol{r}_u, \boldsymbol{r}_v\}$ basis, we have $\boldsymbol{u} \cdot_p \boldsymbol{v} = g_p(\boldsymbol{u},\boldsymbol{v})$ with matrix representation

$$g_p = \begin{pmatrix} E & F \\ F & G \end{pmatrix}, \tag{6.2}$$

so $\boldsymbol{v} = v_\alpha \boldsymbol{e}_\alpha = g_p(\boldsymbol{v}, \boldsymbol{e}_\alpha)\boldsymbol{e}_\alpha$. We have $g(\boldsymbol{r}_u, \boldsymbol{r}_u) = E$, $g(\boldsymbol{r}_v, \boldsymbol{r}_v) = G$, and $g(\boldsymbol{r}_u, \boldsymbol{r}_v) = F$. Also, $\sqrt{\det g_p} = \sqrt{EG - F^2} = ||\boldsymbol{r}_u \times \boldsymbol{r}_v||$.

Definition 52. For a regular parameterized surface $\boldsymbol{r} : U \to S$, we define the *area* of a bounded region of $\mathcal{R} = \boldsymbol{r}(Q) \subset S$, with $Q \in U$ by the expression

$$A(\mathcal{R}) = \iint_Q |\boldsymbol{r}_u \times \boldsymbol{r}_v| du dv = \iint_Q \sqrt{EG - F^2}\, du\, dv.$$

The first fundamental form is also called the *metric of the surface*. If the surface is deformable, the surface equation depends smoothly on a parameter λ that could be the time (moving surfaces) or just the label for a family of smooth surfaces, $\boldsymbol{r} = \boldsymbol{r}(u, v, \lambda)$. In this case, it could be interesting to calculate how the first fundamental form g_p does change with this parameter. That will provide information on how the elementary area and the arc-length change when we change λ. We consider (6.2) as defining the covariant components of the rank 2 tensor $g_p = g_{\alpha\beta}$ in a two-dimensional Euclidean space, $\alpha, \beta = 1, 2$. The associated contravariant tensor (the dual) will be

$$g^{\alpha\beta} = \frac{1}{EG - F^2}\begin{pmatrix} G & -F \\ -F & E \end{pmatrix}. \tag{6.3}$$

From here we have $g^{\alpha\beta} g_{\beta\gamma} = \delta^{\alpha}_{\gamma}$, and if we differentiate this identity with respect to λ we have

$$\frac{dg^{\alpha\delta}}{d\lambda} = -g^{\alpha\beta} g^{\delta\gamma} \frac{dg_{\beta\gamma}}{d\lambda}. \tag{6.4}$$

Moreover, if $g = \det g_{\alpha,\beta}$, we can obtain by straightforward calculations the interesting relation

$$\frac{dg}{d\lambda} = g g^{\alpha\beta} \frac{dg_{\alpha\beta}}{d\lambda}. \tag{6.5}$$

Definition 53. The map

$$N : S \to S_2 \subset \mathbb{R}^3,$$

is called *the Gauss map*.

The tangent map of the Gauss map

$$dN : T_p S \to T_{N(p)} S_2,$$

is a linear self-adjoint operator. We have:

$$\begin{array}{ccccc}
U \subset \mathbb{R}^2 & \xrightarrow[r]{} & S \subset \mathbb{R}^3 & \xrightarrow[= \frac{r_u \times r_v}{|r_u \times r_v|}]{N(p) = N(r(u,v))} & S_2 \subset \mathbb{R}^3 \\
\downarrow & & \downarrow & & \downarrow \\
T(U) \simeq \mathbb{R}^2 & \xrightarrow[dr]{} & T_p(S) \simeq \mathbb{R}^2 & \xrightarrow[dN]{} & T_{N(p)} S_2 \simeq \mathbb{R}^2
\end{array} \tag{6.6}$$

The expression of the tangent map of the unit normal in components is

$$dN_{(r)}(\xi) = \frac{\partial N}{\partial x^i} \xi^i = \frac{\partial N}{\partial u} \xi_u + \frac{\partial N}{\partial v} \xi_v,$$

and the same expression is obtained if we use the tangent map. The tangent map of the unit normal has an interesting property. Let γ be a parameterized curve on S. The action of dN on an arbitrary vector $\boldsymbol{\xi} \in T_p S$ is given by the action of the $\boldsymbol{\xi}$ vector field:

$$dN(\boldsymbol{\xi}) = \boldsymbol{\xi}[N] = D_{\boldsymbol{\xi}} N.$$

In some loose sense, a smooth parameterized surface is a continuous collection of smooth curves, so it is natural to understand the properties of the surface by looking at the curves that can lie on it. In that, let us take a plane generated by a certain tangent vector $\boldsymbol{\xi}$ and by the unit normal $\boldsymbol{N}$, and choose a curve $\boldsymbol{\gamma}$ lying in the intersection of this plane with the surface. Of course we have $\boldsymbol{\xi} = \boldsymbol{\gamma}'$. By using (4.5), we can compute

$$\boldsymbol{\gamma}' \cdot D_{\boldsymbol{\gamma}'} \boldsymbol{N} = \boldsymbol{\gamma}' \cdot (\boldsymbol{N} \circ \boldsymbol{\gamma})'.$$

On the other hand, the tangent to the curve is in the tangent plane to the surface, hence it is perpendicular on the unit normal, so we can write $\boldsymbol{\gamma}' \cdot \boldsymbol{N} = -\boldsymbol{\gamma} \cdot \boldsymbol{N}' = -\boldsymbol{\gamma} \cdot (\boldsymbol{N} \circ \boldsymbol{\gamma})'$. So, from this equation and the equation above we find

$$\boldsymbol{\gamma}' \cdot (\boldsymbol{N} \circ \boldsymbol{\gamma})' = -\boldsymbol{\gamma}'' \cdot \boldsymbol{N} = \kappa \boldsymbol{n} \cdot \boldsymbol{N} = \pm\kappa,$$

where κ is the curvature of the curve γ lying in the *normal* plane to S. Of course, the scalar product $\cdot$ is taken in the sense of the first fundamental form. This is an interesting relation between the curvature of such a "normal" curve, its tangent and the unit normal to S (again, we speak here about a curve lying in the intersection of the surface with the plane generated by the tangent $\boldsymbol{\gamma}'$ to the curve and the unit normal $\boldsymbol{N}$). In this situation the rate of change of the unit normal of S in the direction of the tangent to the curve (directional derivative), projected upon the tangent, is nothing but plus or minus the curvature of the curve:

$$\boldsymbol{\gamma}' \cdot D_{\boldsymbol{\gamma}'} \boldsymbol{N} = \pm\kappa. \tag{6.7}$$

We can generalize this quadratic form to a bilinear form $\boldsymbol{w} \cdot D_{\boldsymbol{v}} \boldsymbol{N}$, defined on the tangent plane.

Definition 54. The symmetric bilinear form $\Pi_p : T_pS \times T_pS \to \mathbb{R}$ defined in any point p of the surface S by

$$\Pi_p(\boldsymbol{w}, \boldsymbol{v}) = \boldsymbol{w} \cdot d\boldsymbol{N}_p(\boldsymbol{v}) = \boldsymbol{v} \cdot d\boldsymbol{N}_p(\boldsymbol{w})$$

is called *second fundamental form* of the surface.

The explicit form of the second fundamental form can be derived from its action on tangent vectors to curves $\boldsymbol{v} = \boldsymbol{\gamma}' = (u_s, v_s)$:

$$\Pi(\boldsymbol{\gamma}', \boldsymbol{\gamma}') = e u_s^2 + 2 f u_s v_s + g v_s^2,$$

where

$$e(u, v) = \boldsymbol{N} \cdot \boldsymbol{r}_{uu} = -\boldsymbol{N}_u \cdot \boldsymbol{r}_u,$$

$$g(u, v) = \boldsymbol{N} \cdot \boldsymbol{r}_{vv} = -\boldsymbol{N}_v \cdot \boldsymbol{r}_v,$$

$$f(u, v) = \boldsymbol{N} \cdot \boldsymbol{r}_{uv} = -\boldsymbol{N}_u \cdot \boldsymbol{r}_v = -\boldsymbol{N}_v \cdot \boldsymbol{r}_u,$$

with the properties

$$\boldsymbol{N} \cdot \boldsymbol{r}_u = \boldsymbol{N} \cdot \boldsymbol{r}_v = 0.$$

Definition 55. For a regular curve $\gamma \subset S$, we define the *normal curvature* at p, the number $\kappa_n(p) = \kappa(p) \cos\theta$, evaluated at p, where κ and $\boldsymbol{n}$ are the curvature and the principal normal of γ at $p \in S$, and $\cos\theta = \boldsymbol{n}(p) \cdot \boldsymbol{N}(p)$.

In other words, the normal curvature of a curve γ is the projection of the vector $\kappa\boldsymbol{n}$ over $\boldsymbol{N}$ at p. All regular curves that intersect S at p and have their tangent vectors in the tangent plane of S at p have the same normal curvature. We also have $\Pi_p(v, v) = \kappa_n(p)$ if $v \in T_pS$ and $|v| = 1$.

Definition 56. Being a self-adjoint linear operator, $D_v(N) = dN(v)$ (also called the *shape operator*) has two real eigenvalues, traditionally denoted $-\kappa_{1,2}(p)$, called *principal curvatures* of S at p. The corresponding eigenvectors are called *principal directions*, and they are orthogonal.

A curve is called *line of curvature* if its tangent vector at each point is a principal direction. The principal curvatures at p are actually the minimum and maximum values of the normal curvature at p and $\kappa_1(p) < \Pi_p(S_1 \subset T_pS) < \kappa_2(p)$. We have

Theorem 17. *If $\gamma \subset S$ is a regular connected curve, and if it is a line of curvature on S, then*

$$\frac{dN}{ds} = -\kappa_n(s)\frac{d\gamma}{ds},$$

where s is the arc-length parameter along γ.

The tangent map of the Gauss map has three important properties:

1. $$dN = \begin{pmatrix} -\kappa_1 & 0 \\ 0 & -\kappa_2 \end{pmatrix}$$

   in the basis of the principal directions.

2. $$\det(dN_p) = \kappa_1\kappa_2 = K, \tag{6.8}$$

   where K is called the *Gaussian curvature*.

3. $$-\frac{\mathrm{Tr}(dN_p)}{2} = \frac{\kappa_1 + \kappa_2}{2} = H, \tag{6.9}$$

   where Tr is the trace operator, and H is called the *mean curvature*.

We can also express the Gaussian (6.8) and mean curvatures (6.9) in terms of the coefficients of the first and second fundamental forms:

$$H = \frac{1}{2}(\kappa_1 + \kappa_2) = \frac{1}{2}\frac{eG - 2fF + gE}{EG - F^2}, \tag{6.10}$$

$$K = \kappa_1\kappa_2 = \frac{eg - f^2}{EG - F^2}. \tag{6.11}$$

A simple interpretation of the two curvatures is the following. When Gaussian curvature in a point is positive, the point is called *elliptical*, and all curves on S passing through such a point are locally contained in one side only of the tangent plane through this point (or the principal normals to these curves all point toward one side of the tangent plane). A "point-like" particle would have a stable position of equilibrium in such a point, so elliptical points are good "confiners." The sign of H does not matter for such elliptical points, though H measures the *degree of asymmetry* in stability of such a confinement, between the principal directions. For example if one of the principal curvatures k_1 is very small, and the other one is very large $k_2 >> k_1$, the Gaussian curvature is small, showing a weak confinement (the particle can escape along the principal direction associated with k_1); but $H = k_1 + k_2 \simeq k_2$ is still large, showing a high asymmetry in the two directions. If both principal curvatures are small, and the surface is almost planar, hence not at all confining in any direction, both K and H are small, showing little confinement, and also little asymmetry in the two directions.

Points with negative K are hyperbolic and they describe somehow unstable equilibrium points in a potential energy picture. This point is also called a *saddle* point. The particle is confined along one direction, but it is highly unstable along the perpendicular direction. In this case, the sign of H decides if the point is more like stable or unstable (which of $k_{1,2}$ is larger). *Parabolic* points have one direction of indifferent equilibrium, and one of stable equilibrium. A regular surface S where $H = 0$ is a *minimal surface*.

The Gaussian curvature K represents the factor by which the Gaussian map N distorts a *principal* infinitesimal area on S, as it maps on the sphere S_2. Indeed, if we have an infinitesimal curvilinear rectangle on S at p of area a, with sides along the principal directions, then the image of this rectangle under the Gauss map is also a rectangle of area $K = \kappa_1 \kappa_2 a$ (interpretation due to Gauss).

Theorem 18. *The Gaussian curvature is determined by only the first fundamental form. That is K can be computed from just E, F, G and their partial derivatives up to order* 2.

For a sphere of radius R in $\mathbb{R}^3$ parameterized by $(\theta, \phi) = (u, v)$ the above coefficients are:

$$\begin{aligned}
\boldsymbol{r}(u,v) &= R(\sin u \cos v, \sin u \sin v, \cos u),\\
\boldsymbol{r}_u &= R(\cos u \cos v, \cos u \sin v, -\sin u),\\
\boldsymbol{r}_v &= R(-\sin u \sin v, \sin u \cos v, 0),\\
E &= R^2, F = 0, G = R^2 \sin^2 u,\\
e &= -R, f = 0, g = -R \sin^2 u,
\end{aligned}$$

and

$$K = \frac{1}{R^2}, \quad H = -\frac{1}{R}.$$

We notice that we choose the unit normal of the sphere to be directed outside, along the radius vector. For this reason the principal curvatures are negative, and so is H. The Gaussian curvature does not depend on the orientation of the surface. For a torus, for example, we have $\boldsymbol{r}(u,v) = ((a + R\cos u)\cos v, (a + R\cos u)\sin v, R\sin u)$:

$$E = R^2,\ F = 0,\ G = (a + r\cos u)^2,\ e = -R,\ f = 0,\ g = -\cos u(a + R\cos u)$$

and

$$K = \frac{\cos u}{R(a + R\cos u)}, \quad H = -\frac{a + 2R\cos u}{2R(a + R\cos u)}.$$

From the differential geometry of surface point of view, we can relate the six functions E, F, G, e, f, g, and K with the components of the derivatives $\boldsymbol{r}_{i,j}$ expressed in the $\boldsymbol{r}_u, \boldsymbol{r}_v$ basis. These are the famous Gauss and Codazzi equations. More general, from the general differential geometry point of view, these relations introduce the Christoffel symbols, and further relate the second-order derivatives of the surface equation to the covariant derivative. For fluid surface dynamics these relations are very important because they help mapping Euclidean three-dimensional vectors of the embedding space $\mathbb{R}^3$ to two-dimensional vectors in the tangent plane of the surface, hence facilitating the construction of momentum conservation theorems for fluid surfaces (see for example Sects. 8.3 and 8.4).

In surface theory the *Christoffel symbols* are introduced simply by calculating the second-order derivatives of the equation of the surface, namely

$$\begin{aligned} \boldsymbol{r}_{uu} &= \Gamma^u_{uu}\boldsymbol{r}_u + \Gamma^v_{uu}\boldsymbol{r}_v + e\boldsymbol{N}, \\ \boldsymbol{r}_{uv} &= \Gamma^u_{uv}\boldsymbol{r}_u + \Gamma^v_{uv}\boldsymbol{r}_v + f\boldsymbol{N}, \\ \boldsymbol{r}_{vv} &= \Gamma^u_{vv}\boldsymbol{r}_u + \Gamma^v_{vv}\boldsymbol{r}_v + g\boldsymbol{N}. \end{aligned} \tag{6.12}$$

Example of Christoffel symbols for common surfaces can be seen in [46, 299]. The Christoffel symbols fulfill two sets of important equations: the Codazzi equations

$$\begin{aligned} e_v - f_u &= e\Gamma^u_{uv} + f(\Gamma^v_{uv} - \Gamma^u_{uu}) - g\Gamma^v_{uu}, \\ f_v - g_u &= e\Gamma^u_{vv} + f(\Gamma^v_{vv} - \Gamma^u_{uv}) - g\Gamma^v_{uv}, \end{aligned} \tag{6.13}$$

and the Gauss equations

$$\begin{aligned} EK &= (\Gamma^v_{uu})_v - (\Gamma^v_{uv})_u + \Gamma^u_{uu}\Gamma^v_{uv} + \Gamma^v_{uu}\Gamma^v_{vv} - \Gamma^u_{uv}\Gamma^v_{uu} - (\Gamma^v_{uv})^2, \\ FK &= (\Gamma^u_{uv})_u - (\Gamma^u_{uu})_v + \Gamma^v_{uv}\Gamma^u_{uv} - \Gamma^v_{uu}\Gamma^u_{uv}, \\ FK &= (\Gamma^v_{uv})_v - (\Gamma^v_{vv})_u + \Gamma^u_{uv}\Gamma^v_{uv} - \Gamma^u_{vv}\Gamma^v_{uu}, \\ GK &= (\Gamma^u_{vv})_u - (\Gamma^u_{uv})_v + \Gamma^u_{vv}\Gamma^u_{uu} + \Gamma^v_{vv}\Gamma^u_{uv} - (\Gamma^u_{uv})^2 - \Gamma^v_{uv}\Gamma^u_{vv} \end{aligned} \tag{6.14}$$

Proofs of these equations can be found in [46, 299]. Based on the Codazzi–Gauss (6.13) and (6.14), we can use the *fundamental theorem of surfaces*.

Theorem 19. *Two parameterized surfaces $\boldsymbol{r}_1, \boldsymbol{r}_2 : U \to \mathbb{R}^3$ are* congruent *(i.e., differ by a rigid motion) if and only if $g_1 = g_2$ and $\Pi_1 = \pm\Pi_2$.*

This is the equivalent of the fundamental theorem of curve geometry (Theorem 10) introduced in Sect. 5.1. There is an *existence* version of the fundamental theorem (for example [299, Chap. 2.3]). Given the six differentiable functions

$E, F, G, e, f, g : U \to \mathbb{R}$ with $E > 0$ and $EG - F^2 > 0$, and satisfying (6.13) and (6.14), there exists a (locally) parameterized surface $\boldsymbol{r}(u, v)$ with the respective g and Π.

6.2 Covariant Derivative and Connections

The following calculations on two-dimensional surfaces embedded in $\mathbb{R}^3$ are based on the concepts of covariant derivative, Christoffel symbols, and connections that have been introduced for general differential manifolds in Sect. 4.10. An useful operator acting on a surface S is the *covariant derivative* of a vector field $\boldsymbol{Y}$ along another vector field $\boldsymbol{X}$, namely

$$\nabla_X \boldsymbol{Y} = D_X \boldsymbol{Y} - \boldsymbol{N}(\boldsymbol{N} \cdot D_X \boldsymbol{Y}). \tag{6.15}$$

The covariant derivative of $\boldsymbol{Y}$ with respect to $\boldsymbol{X}$ at $p \in S$ represents the directional derivative of $\boldsymbol{Y}$ with respect to $\boldsymbol{X}$ $(D_X \boldsymbol{Y})$, projected onto $T_p S$. The covariant derivative becomes more important if the field $\boldsymbol{X}$ is the unit tangent to a curve $\gamma \subset S$. For a parameterized curve along S, $\boldsymbol{X} = \boldsymbol{t}_\gamma = \gamma'$, the covariant derivative along γ of the unit normal to the surface $\boldsymbol{N}$ is nothing but its directional derivative along γ, $\nabla_{\gamma'} \boldsymbol{N} = D_{\gamma'} \boldsymbol{N} \in TS$. The covariant (or directional) derivative of the unit normal along γ can be decomposed in terms of the local basis $\{\boldsymbol{r}_u, \boldsymbol{r}_v\}$. More interestingly, we can decompose this derivative along the unit tangent $\gamma' = \boldsymbol{t}$, and along the perpendicular $\boldsymbol{t}^\perp$ to the unit tangent, defined by $\boldsymbol{t}^\perp \cdot \boldsymbol{t} = 0, \boldsymbol{t}^\perp \in T_p S$.

$$d\boldsymbol{N}(\gamma') = D_{\gamma'} \boldsymbol{N} = \nabla_{\gamma'} \boldsymbol{N} = \kappa_n \boldsymbol{t} + \tau_g \boldsymbol{t}^\perp, \tag{6.16}$$

where τ_g is the *geodesic torsion* of the curve γ, defined as

$$\tau_g = \frac{d\boldsymbol{N}}{ds}(0) \cdot \boldsymbol{t}_p^\perp = (D_{\gamma'} \boldsymbol{N}) \cdot \boldsymbol{t}_p^\perp. \tag{6.17}$$

So, the covariant derivative of the unit normal along a curve is the sum of the normal curvature $(\kappa_n(\gamma') = \Pi(\gamma'))$ times the unit tangent, and the geodesic torsion times the direction orthogonal to the unit tangent, into the tangent plane. This property of the unit normal is called *parallel transport* along γ.

In general if the covariant derivative of a vector field is zero along a curve, we say that this field is parallel transported along that curve. In general, the covariant derivative of a tangent vector field also contains a component along the unit normal of the surface. We also mention another property: if a curve belonging to a surface has its geodesic torsion zero, then its unit tangent is always along the local principal direction, and conversely. We call such curves *lines of curvature*. For example, the intersecting curves between a system of (triple) orthogonal curvilinear coordinates are lines of curvature.

Definition 57. A parameterized curve γ in a surface S is a *geodesic* if its tangent vector is parallel along the curve.

For any point $p \in S$ and any direction $\boldsymbol{v} \in T_pS$, there is $\epsilon > 0$ and a unique geodesic $\gamma(s) : (-\epsilon, \epsilon) \to S$ such that $\gamma(0) = p$ and $\boldsymbol{\gamma}'(0) = \boldsymbol{v}$. The most important property is that geodesics are locally *distance minimizing*. This property is valid in general only locally. This happens because for an arbitrary surface, even regular and connected, either the existence of a geodesic through any point, or its property to be the minimum distance between two given points, are not mandatory. Parameterized geodesic curves could be distance-minimizing curves in a global sense (over the whole surface) depending on the surface. If a geodesic passing through an arbitrary point of a regular surface $p \in S$ can be indefinitely extended on S, in any direction of T_pS, S is called a *complete surface*. On a complete surface a geodesic defined locally can be extended "for all time" (this is the famous Hopf–Rinow theorem, see [46,299]). Imagine a punctured sphere without North pole $S_2\,\{N\}$ and a great circle (i.e., a geodesic curve on the sphere) that passes through this point. This geodesic curve also passes through the South pole. Points very close to N can be joined by smaller arcs than the geodesic curve joining them through South. This is an example of a not complete surface.

Definition 58. A regular connected surface S is *extendable* if it is a proper subset of another regular connected surface $\tilde{S}$, $S \subsetneq \tilde{S}$. A regular connected surface S is *complete* if $\forall p \in S, \forall \gamma : (0, \epsilon) \to S$ parameterized geodesic with $\gamma(0) = p$, there is an extended parameterized geodesic $\tilde{\gamma} : \mathbb{R} \to S$, $\tilde{\gamma}|_{(0,\epsilon)} = \gamma$.

A complete surface is nonextendable. A closed surface is complete, and a compact surface, being closed, is also complete. A complete surface which is not closed is for example an asymptotic convergent cylindric spiral (see Fig. 6.4). A parameterized minimal surface, in an isothermal parameterization, is nonextensible surface, without being complete. In general, given any oriented regular surface S and arc-length parameterized curve $\gamma(s)$ lying on S, we can build at any point $p = \gamma(s) \in S$ a local trihedron, called the *Darboux trihedron* (or frame). This right-handed orthonormal frame is more natural when working with curves lying on surfaces, than the Serret–Frenet frame.

Definition 59. The Darboux frame (Fig. 6.5) is defined by the unit tangent of γ $\boldsymbol{t}(s) = \gamma'(s)$, $\boldsymbol{t}^{\perp}(s) = \boldsymbol{N}(s) \times \boldsymbol{t}(s)$, and the unit normal to the surface, $\boldsymbol{N}(s)$ by

$$\frac{\partial}{\partial s}\begin{pmatrix} \boldsymbol{t} \\ \boldsymbol{t}^{\perp} \\ \boldsymbol{N} \end{pmatrix} = \begin{pmatrix} 0 & \kappa_g & \kappa_n \\ -\kappa_g & 0 & \tau_g \\ -\kappa_n & -\tau_g & 0 \end{pmatrix}\begin{pmatrix} \boldsymbol{t} \\ \boldsymbol{t}^{\perp} \\ \boldsymbol{N} \end{pmatrix},$$

where $\kappa_n(s)$ is the *normal curvature*, $\tau_g(s)$ is the *geodesic torsion*, and $\kappa_g(s)$ is the *geodesic curvature*.

The normal curvature was introduced in Definition 55, and the two geodesic coefficients were involved in (6.16). The geodesic curvature can be understood even

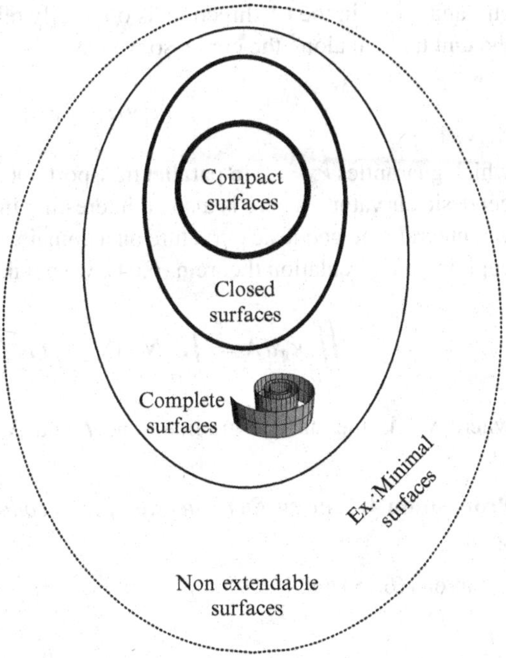

Fig. 6.4 Relations between classes of surfaces

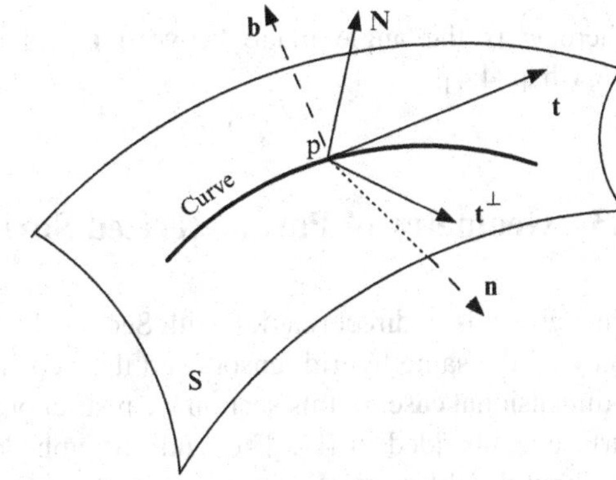

Fig. 6.5 The Serret–Frenet $(\boldsymbol{t}, \boldsymbol{n}, \boldsymbol{b})$ and Darboux $(\boldsymbol{t}, \boldsymbol{N}, \boldsymbol{t}^{\perp})$ frames

better if we decompose the *curvature vector* (i.e., the rate of change of the tangent along the curve) $\kappa \boldsymbol{n}$ along the two orthogonal directions in the tangent plane to S

$$\frac{d\boldsymbol{t}}{ds} = \kappa \boldsymbol{n} = \underbrace{(\kappa \boldsymbol{n} \cdot \boldsymbol{t}^{\perp})}_{\kappa_g}\boldsymbol{t}^{\perp} + \underbrace{(\kappa \boldsymbol{n} \cdot \boldsymbol{N})}_{\kappa_n}\boldsymbol{N}. \tag{6.18}$$

Again, the coefficient of the normal component is the normal curvature from Definition 55 and Theorem 17. The tangent component which defines the geodesic curvature, i.e., in the $\boldsymbol{t}^{\perp}$ direction, is obviously related to the covariant derivative of the unit tangent along the curve, so we have

$$|\nabla_t \boldsymbol{t}| = |\kappa_g|, \tag{6.19}$$

which guaranties $\kappa_g = 0$ in parallel transport. Obviously geodesic curves have zero geodesic curvature. There is also an interesting integral consequence of this fact. If we integrate the geodesic curvature on a domain of the surface (from (6.60)) and by applying the circulation theorem (6.64), we obtain

$$\iint_D \kappa_g dA = \iint_D \boldsymbol{N} \cdot (\nabla_\Sigma \times \boldsymbol{t}) dA = \oint_{\partial A} \boldsymbol{t} d\boldsymbol{r} = 0,$$

where ∇_Σ is the surface gradient, and $\boldsymbol{t}$ and $\boldsymbol{N}$ have their usual interpretations. That is

Proposition 4. *The surface integral of the geodesic curvature over any domain is zero.*

Equations (6.18) and (6.19) imply $\kappa^2 = \kappa_n^2 + \kappa_g^2$ and we have

$$\kappa_g = \frac{d\varphi}{ds},$$

where φ is the angle made between $\boldsymbol{t}$ and a parallel direction to the curve [46, Chap. 4.4].

6.3 Geometry of Parameterized Surfaces Embedded in $\mathbb{R}^3$

This section is in direct relation with Sects. 4.11 and 6.5. Section 4.11, for example, analyzes the same hybrid tensors and their covariant derivative, but in the general n-dimensional case. In this section we restrict our analysis only to two-dimensional surfaces embedded in $\mathbb{R}^3$. The study of embedded surfaces in Euclidean spaces, and how the differential operators map from one space to the other, is necessary for setting correct balance equations and boundary conditions for fluid surfaces. Let us have a parameterized surface Σ defined by the regular change of coordinate functions $\boldsymbol{r}(u, v) = (x^i(u^\alpha), \alpha = 1, 2, i = 1, 2, 3$. We introduce the mixed Jacobian matrix

$$B = B^i_\alpha = \frac{\partial x^i}{\partial u^\alpha}, \tag{6.20}$$

which is a *hybrid tensor*. This tensor is nothing but the $T\Sigma$ basis $\{\boldsymbol{r}_u, \boldsymbol{r}_v\}$ introduced earlier, written in a consistent covariant form. A contravariant surface vector A^α is a vector field defined on $T\Sigma$ that changes its components at a coordinate change $u^\alpha \to \tilde{u}^\alpha$ like

$$\tilde{A}^\alpha = \frac{\partial \tilde{u}^\alpha}{\partial u^\beta} A^\beta. \tag{6.21}$$

Examples of contravariant vectors are the tangent vectors to curves lying in Σ. The first fundamental form (the *metric tensor*) on Σ is represented by $(0,2)$-type of tensor defined on $\Omega^2(T\Sigma)$ (Sect. 4.2, definition 38), and it has the expression

$$g_{\alpha\beta} = \begin{pmatrix} \boldsymbol{r}_u \cdot \boldsymbol{r}_u & \boldsymbol{r}_u \cdot \boldsymbol{r}_v \\ \boldsymbol{r}_v \cdot \boldsymbol{r}_u & \boldsymbol{r}_v \cdot \boldsymbol{r}_v \end{pmatrix} = \begin{pmatrix} E & F \\ F & G \end{pmatrix}, \tag{6.22}$$

and we have

$$ds^2 = B^i_\alpha B^i_\beta du^\alpha du^\beta = g_{\alpha\beta} du^\alpha du^\beta = Edu^2 + 2Fdudv + Gdv^2$$

and also $g_{\alpha\beta} g^{\beta\gamma} = \delta^\gamma_\alpha$ where δ is the Kronecker symbol. Another useful equation is

$$g_{\alpha\beta} = B^i_\alpha B^i_\beta. \tag{6.23}$$

The contravariant components of the metric tensor are

$$g^{\alpha\beta} = \frac{1}{EG - F^2} \begin{pmatrix} G & -F \\ -F & E \end{pmatrix}, \tag{6.24}$$

and both covariant and contravariant metric tensors are used to lift or lower indices of various tensors.

An example of covariant vector field is the surface gradient of a function $f : \Sigma \to \mathbb{R}$

$$\nabla_\Sigma f = \left(\frac{\partial f}{\partial u^\alpha} \right) = (f_u, f_v). \tag{6.25}$$

The contravariant components of the surface gradient are

$$\nabla f^\alpha = g^{\alpha\beta} \nabla f_\beta = \left(\frac{Gf_u - Ff_v}{EG - F^2}, \frac{Ef_v - Ff_u}{EG - F^2} \right), \tag{6.26}$$

see also [338, Chap. XII] or [46, Sect. 2.5]. Sometimes in literature this operator is also denoted $\nabla_{\|}$, and it can be also written in the form

$$\nabla_\Sigma f = \nabla_{\|} f = \frac{G\boldsymbol{r}_u - F\boldsymbol{r}_v}{EG - F^2} f_u + \frac{E\boldsymbol{r}_v - F\boldsymbol{r}_u}{EG - F^2} f_v. \tag{6.27}$$

If the surface is isothermal ($F = 0$), (6.27) reduces to the well-known gradient in some orthogonal curvilinear coordinate system

$$\nabla_\Sigma f = \left(\frac{1}{H_u} f_u, \frac{1}{H_v} f_v \right), \tag{6.28}$$

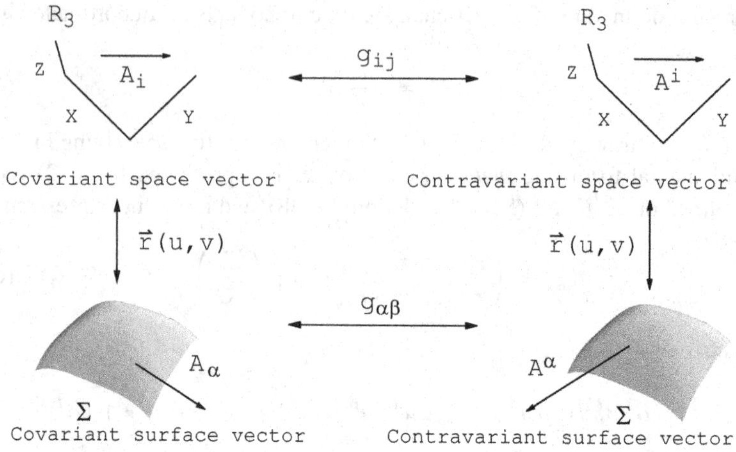

Fig. 6.6 Mappings between three-dimensional vectors and surface vectors

where $H_{u,v} = \boldsymbol{r}_{u,v}/|\boldsymbol{r}_{u,v}|^2$ are the Lamme coefficients defined in Sect. 4.12. The surface gradient fulfills $< \nabla_\Sigma f(x), \boldsymbol{v} >_x = df_x(\boldsymbol{v}) = D_{\boldsymbol{v}} f(x)$ for any $\boldsymbol{v} \in T\Sigma$, where df_x is the differential of the mapping f taken at x, $<,>$ is the Euclidean scalar product on $T\Sigma$ taken at x, and $D_{\boldsymbol{v}}$ is the directional derivative. A more detailed analysis of this operator is done in Sect. 6.5.1.

If $\boldsymbol{A}(x) = (A^i) \in (T\mathbb{R}^3)_x$ and $\boldsymbol{a}(u) = (a^\alpha) \in T\Sigma_u$ are an Euclidean and a surface vector, respectively, we can map their contravariant components by

$$\boldsymbol{A} = \boldsymbol{r}_u a_u + \boldsymbol{r}_v a_v = (B^i_\alpha a^\alpha), \tag{6.29}$$

and conversely

$$\boldsymbol{a} = (\boldsymbol{r}_u \cdot \boldsymbol{A}, \boldsymbol{r}_v \cdot \boldsymbol{A}) = (a_u, a_v).$$

If the embedding space is just Riemannian manifold (and not Euclidean), the above equations change and we have to use the metric on this space, too. For example, we would have $a_\alpha = B^i_\alpha A_i = B^i_\alpha g_{ij} A^j$, and so on. The algebraic relations between three-dimensional vectors and surface vectors are represented in Fig. 6.6. A traditional example is the normal to the surface which is a covariant vector

$$N_i = \frac{1}{2g} \epsilon^{\alpha\beta} \epsilon_{ijk} B^j_\alpha B^k_\beta, \tag{6.30}$$

where the two ϵ are the Levi–Civita symbols in the two spaces, and $g = \det(g_{\alpha\beta})$.

6.3.1 Christoffel Symbols and Covariant Differentiation for Hybrid Tensors

We investigated already such hybrid tensors in Sect. 4.11 in the general case of m-dimensional Riemannian submanifold embedded into an n-dimensional Riemannian manifold, both being nonflat. In this section, we continue along the same line given in Sect. 6.3, specifically studying differential hybrid operators on two-dimensional regular parameterized surfaces Σ embedded in $\mathbb{R}^3$. Consequently, $g_{ij} = \delta_{ij}$ and $\Gamma_{ij}^k = 0$. Also, since Σ is Riemannian (has a metric defined) we know that the affine connection on Σ comes from Christoffel symbols, and consequently its torsion is zero, $S_{\alpha\beta}^{\gamma} = 0$ (4.54). The Christoffel symbols on Σ are defined as

$$\Gamma_{\alpha\beta}^{\delta} = \frac{1}{2} g^{\gamma\delta} \left(\frac{\partial g_{\gamma\alpha}}{\partial u^\beta} + \frac{\partial g_{\beta\gamma}}{\partial u^\alpha} - \frac{\partial g_{\alpha\beta}}{\partial u^\gamma} \right), \tag{6.31}$$

see also (4.51). Christoffel symbols are introduced on a manifold in a variety of ways [19, 46, 119, 158, 162, 299]. One simple way to look at them is to consider a change of coordinates in Σ from an arbitrary system of coordinates to an isothermal system of coordinates, i.e., $u^\alpha \to \tilde{u}^\alpha$, such that $ds^2 = g_{\alpha\beta} du^\alpha du^\beta = (d\tilde{u}^1)^2 + (d\tilde{u}^2)^2$ with Jacobian

$$J_\alpha^\beta = \frac{\partial \tilde{u}^\beta}{\partial u^\alpha}.$$

Then, the Christoffel symbols are nothing but the *law of derivation* of the Jacobian matrix

$$\frac{\partial J_\beta^\alpha}{\partial u^\gamma} = \Gamma_{\beta\gamma}^{\delta} J_\delta^\alpha.$$

Also they fulfill the relation

$$\frac{1}{2g} \frac{\partial g}{\partial u^\alpha} = \Gamma_{\beta\alpha}^{\beta} = \Gamma_{\alpha\beta}^{\beta}.$$

For example, for a surface parameterized by $\boldsymbol{r}(x, y) = (x, y, f(x, y))$ we have $g = 1 + f_x^2 + f_y^2$ and

$$\Gamma_{\alpha\beta}^{\delta} = \frac{f_{\alpha\beta} f_\delta}{1 + f_x^2 + f_y^2}.$$

The covariant derivative was repeatedly introduced in this text in either (4.49) and (6.15), or even the hybrid one in general (4.59). In the case $\Sigma \subset \mathbb{R}^3$, for a hybrid tensor A_α^i we define a *hybrid surface covariant derivative* as

$$\nabla_\beta A_\alpha^i = \frac{\partial A_\alpha^i}{\partial u^\beta} - \Gamma_{\alpha\beta}^{\gamma} A_\gamma^i. \tag{6.32}$$

It has the properties

$$\nabla_\gamma g_{\alpha\beta} = \nabla_\gamma g^{\alpha\beta} = 0. \tag{6.33}$$

Let

$$\Pi_{\alpha\beta} = \frac{\partial^2 x^i}{\partial u^\alpha \partial u^\beta} N_i = \begin{pmatrix} e & f \\ f & g \end{pmatrix},$$

be the tensor associated with the second fundamental form on Σ (from Definition 54). We can express the second-order derivatives in (6.33) in another way. From (6.23) we have

$$\nabla_\gamma g_{\alpha\beta} = 0 = (\nabla_\gamma B^i_\alpha) B^i_\beta + B^i_\alpha (\nabla_\gamma B^i_\beta),$$

from where, by symmetry, we obtain

$$(\nabla_\gamma B^i_\alpha) B^i_\beta = 0.$$

Since B^i_α are actually the basis vectors of the tangent space for Σ, from the above relation it results that the hybrid surface covariant derivatives of the hybrid tensor B^i_α are orthogonal to the tangent space. So, they are proportional to the normal

$$\nabla_\beta B^i_\alpha \sim (\text{some tensor})_{\alpha\beta} N^i.$$

By using (6.32) in the LHS term of the above relation, we obtain

$$\frac{\partial^2 x^i}{\partial u^\alpha \partial u^\beta} = \Gamma^\gamma_{\alpha\beta} B^i_\gamma + (\text{some tensor})_{\alpha\beta} N^i, \tag{6.34}$$

but this is just the definition of Christoffel symbols given previously (6.12). So we infer

$$\nabla_\beta B^i_\alpha = \Pi_{\alpha\beta} N^i, \tag{6.35}$$

or, in an equivalent form

$$\Pi_{\alpha\beta} = \nabla_\beta B^i_\alpha N_i. \tag{6.36}$$

We mention a useful relation that can be obtained from (6.36)

$$\Pi_{\alpha\beta} = \frac{1}{2\sqrt{g}} \epsilon^{\rho\sigma} \epsilon_{ijk} (\nabla_\beta B^i_\alpha) B^j_\rho B^k_\sigma. \tag{6.37}$$

We can rewrite the important results from Sect. 6.1 in this covariant formalism. For example, from (6.10) we have a very compact way of calculating the mean curvature

$$2H = g^{\alpha\beta} \Pi_{\alpha\beta}, \tag{6.38}$$

and the Gaussian curvature

$$\Pi_{\alpha\beta} \Pi_{\gamma\delta} g^{\beta\gamma} g^{\delta\alpha} = 4H^2 - 2K. \tag{6.39}$$

6.4 Compact Surfaces

The most important result in the differential geometry of surfaces is the Gauss–Bonnet theorem. In the following we present only a corollary of the global version of this theorem. For the complete differential and global versions on surfaces with boundaries, we suggest [46, 119, 162, 299].

Theorem 20 (Gauss–Bonnet Theorem). *If S is an orientable compact surface, then*

$$\iint_S K dA = 2\pi\chi(S),$$

where K is the Gaussian curvature, and $\chi(S)$ is the Euler–Poincaré characteristic of the surface S.

In other words, the total curvature of a compact surface (i.e., a finite closed surface without boundaries) can only be $-4\pi(n-1)$, where the positive integer n is the number of "handles" (or holes) of the surface. The *Euler – Poincaré characteristic* χ of a manifold can be calculated from the ranks of the homology groups of the surface (Sect. 2.2) by using triangulation procedures. For details we recommend [112, 235], and for the proof we recommend [46, 119, 162]. The χ characteristic is a topological (homotopy) invariant. It can also be expressed in the form $\chi = 2-2g$, where g is the *genus* of the surface, and it is equal to n defined above. Any surface homeomorphic with a sphere has $\chi = 2$, the torus has $\chi = 0$, etc. In Fig. 6.7 we present an example of a closed surface of genus $g = 6$. The genus can be calculated as the largest number of nonintersecting simple closed curves on a surface that still do not separate

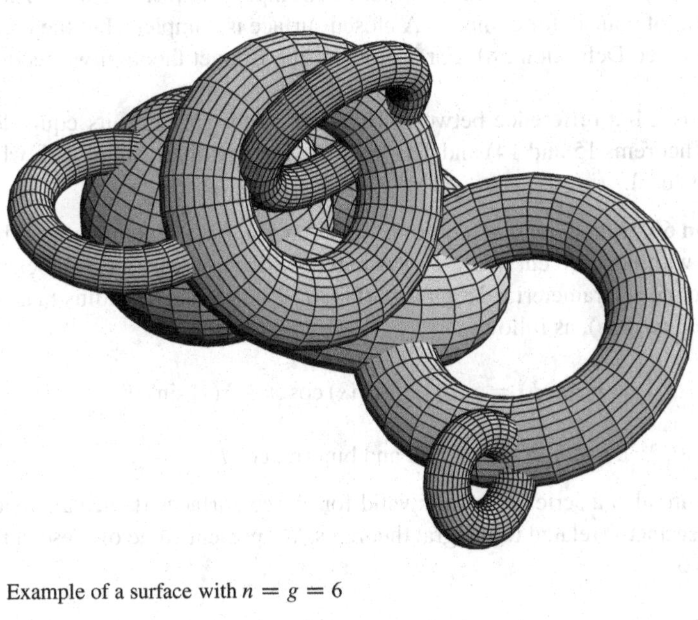

Fig. 6.7 Example of a surface with $n = g = 6$

it into disconnected sets. The spectacular fact about the Gauss–Bonnet theorem is that no matter how we smoothly (homeomorphic) deform a surface, its curvature distributes itself in such a way that the total curvature does not change. For example, for the unit sphere we have $\iint_{S_2} = 4\pi$. If we deform the sphere such that half of it becomes flat, we still have the same total curvature, in spite of the fact that half of the surface reduced its curvature to zero. This is because we have big accumulation of curvature along the sharp diameter, i.e., a region of area zero times infinite curvature. Theorem 20 is related with Theorems 15 and 14 for curves. All these theorems provide necessary criteria for a curve or surface to be bounded.

The question is what do we have for the converse affirmation: what criterium should the curvature fulfill to assure compactness for the surface? The answer is provided by another very powerful theorem. However, this theorem is valid only for complete surfaces.

Theorem 21 (Bonnet Theorem). *If the Gaussian curvature K of a* complete surface *S satisfies the condition*

$$K \geq \delta^2 > 0,$$

then S is compact and the diameter *ρ of S satisfies the inequality*

$$\rho \leq \frac{\pi}{\delta}.$$

This theorem holds if the surface is closed in the topological sense. That is, if the surface contains all its accumulation points. Complete is just a generalization for closed, and of course, for compact. A closed surface is complete, but the reciprocal is not true (see Definition 58). For a proof of the Bonnet theorem we recommend [46, Sect. 5-4].

There is a big difference between Theorem 20 for surfaces, its equivalent for curves (Theorems 15 and 14) and Theorem 21. The first three are global, while the last one is local.

Definition 60. For a regular curve Γ of equation $\boldsymbol{r}(s)$, parameterized by arc-length s, with nonzero curvature everywhere, and for any positive number $r_0 > 0$, we can define a parameterized regular surface T_Γ, called *tube* of radius r_0 around Γ (or *tubular surface*), as follows

$$\boldsymbol{r}_T(s, \phi) = \boldsymbol{r}(s) + r_0(\boldsymbol{n}(s) \cos\varphi + \boldsymbol{b}(s) \sin\varphi),$$

with $\varphi \in [0, 2\pi]$, and $\boldsymbol{n}, \boldsymbol{b}$ the normal and binormal of Γ.

There are also a series of results valid for closed surfaces (hence also valid for compact surfaces) related to integral theorems. We present some of these at the end of Sect. 6.5.

6.5 Surface Differential Operators

This section is in direct relation with Sects. 6.3 and 4.11. In this section we introduce some of the properties and applications of differential operators defined on a surface $\Sigma \subset \mathbb{R}^3$. The reason for such a construction is the following. When working with fluids with free surfaces, like so many examples in this book, a necessary condition is to match the conserving quantities at the fluid boundaries, which are free surfaces. For this reason we have to handle sometimes only the tangent components of the conserving quantities. These tangent, or parallel, components fulfill a different type of differential geometry than those in $\mathbb{R}^3$, yet a surface geometry induced by the $\mathbb{R}^3$ geometry. The action of differential operators on surfaces was first described in terms of *differential invariants* (or historically called *differential parameters*) by Beltrami and Darboux, and later on developed by Weatherburn [337, 338], Oldroyd [241], and Scriven [292]. Useful reviews of the matter can be found in [10, Chaps. 9, 10] and [210, Chap. 1].

In the following, we are interested in expressing differential operators that can "see" only the dependence on the point of the surface, and factorize upon the dependence in normal direction. It is interesting to reformulate the well-known vector analysis formulas that require zero value for the curl(grad), and div(curl) ($\nabla \times \nabla$, $\nabla \cdot (\nabla \times)$). Since the normal direction plays somehow the role of a kernel, we expect these formulas to be still valid modulo some no-zero components along the normal direction to the surface.

We consider a regular parameterized surface $\boldsymbol{r}(u, v) : D \to \Sigma \subset \mathbb{R}^3$ with its first fundamental form coefficients E, F, and G, unit normal $\boldsymbol{N}$, and mean curvature H. We define a scalar differential function $\tilde{\Phi} : \Sigma \to \mathbb{R}$ and $\Phi(u, v) = \tilde{\Phi}(\boldsymbol{r}(u, v))$.

6.5.1 Surface Gradient

The surface gradient was already introduced in coordinates in Sect. 6.3. We introduce the surface gradient of Φ to be the vector field with values in the tangent bundle $\nabla_\Sigma \Phi \in T\Sigma$ defined by

$$\nabla_\Sigma \Phi = \frac{1}{EG - F^2}(G\Phi_u - F\Phi_v)\boldsymbol{r}_u + \frac{1}{EG - F^2}(E\Phi_v - F\Phi_u)\boldsymbol{r}_v, \tag{6.40}$$

where subscript means differentiation, and $\boldsymbol{r}_{u,v}(u, v)$ form a basis in the tangent plane $T_{(u,v)}\Sigma$. Equation (6.40) is independent of the parameterization of the surfaces, and in that it is a *differential invariant*. The function $\nabla_\Sigma \Phi$ defines a tangent vector field perpendicular on the $\Phi = \text{const.}$ lines on Σ. Indeed, if $\nabla_\Sigma \Phi|_\Sigma = 0$, it results (through $G\Phi_u = F\Phi_v$, $E\Phi_v = F\Phi_u$) $\Phi = \text{const.}$, like in the case of the full gradient operator on $\mathbb{R}^3$. Otherwise, curves with $\nabla_\Sigma \Phi = 0$ are called *level curves*. Actually, we can define only the surface-gradient operator by

$$\nabla_\Sigma = \frac{1}{EG - F^2}\left[\left(G\frac{\partial}{\partial u} - F\frac{\partial}{\partial v}\right)\boldsymbol{r}_u + \left(E\frac{\partial}{\partial v} - F\frac{\partial}{\partial u}\right)\boldsymbol{r}_v\right] = \nabla_1 \boldsymbol{r}_u + \nabla_2 \boldsymbol{r}_v, \tag{6.41}$$

or simply (∇_1, ∇_2) in the $\{\boldsymbol{r}_u, \boldsymbol{r}_v\}$ basis. For orthogonal parametric curves ($F = 0$) we have

$$\nabla_\Sigma = \frac{\boldsymbol{r}_u}{E}\partial_u + \frac{\boldsymbol{r}_v}{G}\partial_v,$$

where $\partial_u = \partial/\partial u$, etc. Since

$$\oint_\Gamma \nabla_\Sigma \cdot d\boldsymbol{r} = 0,$$

for any closed curve $\Gamma \subset \Sigma$, the condition for a tangent field $\boldsymbol{a} : \Sigma \to T\Sigma$ to be a gradient field is

$$\oint_{\forall \Gamma} \boldsymbol{a} \cdot d\boldsymbol{r} = 0.$$

In Fig. 6.8 we present some examples of surface-gradient fields $\boldsymbol{a} = \nabla_\Sigma Y_{lm}(\theta, \varphi)$ defined on a sphere (Y_{lm} are the spherical harmonics). It is interesting to relate these fields with the *hairy ball theorem*, see problems at the end of this chapter.

6.5.2 Surface Divergence

Let $\boldsymbol{a}(u, v) = a_1(u, v)\boldsymbol{r}_u + a_2(u, v)\boldsymbol{r}_v$ be a vector field in the tangent space. We define the surface divergence acting on a vector field $\boldsymbol{a}$

$$\begin{aligned} \nabla_\Sigma \boldsymbol{a} &= (\nabla_1, \nabla_2)\boldsymbol{a} = (\boldsymbol{r}_u \nabla_1 + \boldsymbol{r}_v \nabla_2) \cdot \boldsymbol{a} = \boldsymbol{r}_u \cdot \nabla_1 \boldsymbol{a} + \boldsymbol{r}_v \cdot \nabla_2 \boldsymbol{a} \\ &= \frac{1}{EG - F^2}\left[\left(G\frac{\partial \boldsymbol{a}}{\partial u} - F\frac{\partial \boldsymbol{a}}{\partial v}\right)\boldsymbol{r}_u + \left(E\frac{\partial \boldsymbol{a}}{\partial v} - F\frac{\partial \boldsymbol{a}}{\partial u}\right)\boldsymbol{r}_v\right]. \end{aligned} \tag{6.42}$$

We have a remarkable property.

Proposition 5.

$$\nabla_\Sigma \cdot \boldsymbol{N} = -2H.$$

Proof. From (6.42) we have

$$\begin{aligned} \nabla_\Sigma \boldsymbol{N} &= \frac{1}{EG - F^2}(G\boldsymbol{r}_u \cdot \boldsymbol{N}_u - F\boldsymbol{r}_u \cdot \boldsymbol{N}_v + E\boldsymbol{r}_v \cdot \boldsymbol{N}_v - F\boldsymbol{r}_v \cdot \boldsymbol{N}_u) \\ &= -\frac{eG + Eg - 2Ff}{EG - F^2} = -2H, \end{aligned}$$

according to (6.11), where e, g, f are from Definition 54. □

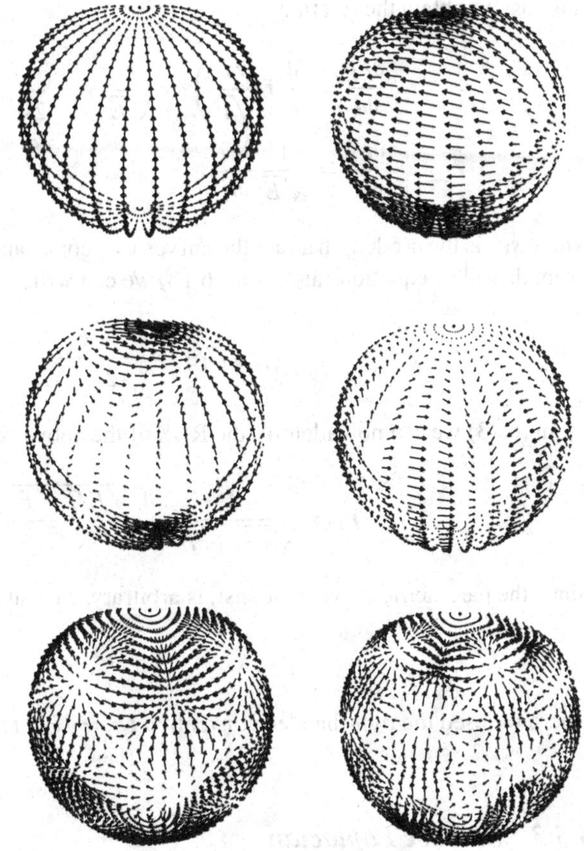

Fig. 6.8 Surface-gradient fields on sphere $\nabla_\Sigma Y_{lm}(\theta, \varphi)$. From *upper left* to *lower right* $l = 1, m = 0$; $l = 3, m = -1$; $l = 5, m = 3$; $l = 1, m = 1$; $l = 3, m = 3$; $l = 9, m = 4$

We can generalize the action of the surface divergence (6.42) to arbitrary vector fields $\boldsymbol{A} = A_1\boldsymbol{r}_u + A_2\boldsymbol{r}_v + A_n\boldsymbol{N}$ in $\mathbb{R}^3$

$$\nabla_\Sigma \cdot \boldsymbol{A} = -2HA_n + \frac{1}{\sqrt{EG - F^2}}\left[(\sqrt{EG - F^2}A_1)_u + (\sqrt{EG - F^2}A_2)_v\right], \quad (6.43)$$

where subscripts represent differentiation. For an application see Exercise 2 at the end of the chapter.

Surface divergence is intimately related to the geodesic curvature. To verify this we choose an orthogonal parameterization $\{\boldsymbol{r}_u, \boldsymbol{r}_v\}$ with $F = 0$ on Σ and normalize it to

$$\boldsymbol{r}_1 \equiv \frac{\boldsymbol{r}_u}{\sqrt{E}}, \quad \boldsymbol{r}_2 \equiv \frac{\boldsymbol{r}_v}{\sqrt{G}}.$$

It is easy to obtain the relations

$$\frac{\partial \boldsymbol{r}_1}{\partial u} = \frac{\partial}{\partial u}\left(\frac{\boldsymbol{r}_u}{\sqrt{E}}\right) = \frac{e}{\sqrt{E}}\boldsymbol{N} - \frac{E_v}{2\sqrt{EG}}\boldsymbol{r}_2.$$

$$\frac{\partial \boldsymbol{r}_1}{\partial s_1} = \frac{1}{\sqrt{E}}\frac{\partial \boldsymbol{r}_1}{\partial u},$$

where $s_{1,2}$ is the arc-length along the curves $v =$ const. and $u =$ const., respectively. From these last equations and from (6.18) we can write

$$\kappa_g|_{v=\text{const.}} = -\frac{E_v}{2E\sqrt{G}}.$$

From (6.43) we can now identify the RHS of the above equation with the relation

$$\nabla_\Sigma \cdot \boldsymbol{r}_2 = \frac{1}{\sqrt{EG-F^2}}\frac{\partial}{\partial v}\frac{\sqrt{EG-F^2}}{\sqrt{G}} = -\kappa_g.$$

Since the parametric curve $v =$const. is arbitrary, we can enounce [338]

$$\kappa_g = -\nabla_\Sigma \cdot \boldsymbol{t}^\perp, \tag{6.44}$$

where we used the right-handed convention $\boldsymbol{t}^\perp = \boldsymbol{N} \times \boldsymbol{t}$.

6.5.3 *Surface Laplacian*

We define the *surface Laplacian* of a scalar function in the usual way

$$\Delta_\Sigma \Phi = \nabla_\Sigma \cdot \nabla_\Sigma \Phi = \frac{1}{\sqrt{EG-F^2}}\left[\left(\frac{G\Phi_u - F\Phi_v}{\sqrt{EG-F^2}}\right)_u + \left(\frac{E\Phi_v - F\Phi_u}{\sqrt{EG-F^2}}\right)_v\right].$$

When studying the motion of a free surfaces $\boldsymbol{r}(u, v, t)$, it is useful to have a simpler relation for the surface Laplacian of the position vector

$$\Delta_\Sigma \boldsymbol{r} = \frac{1}{\sqrt{EG-F^2}}\left[\left(\frac{G\boldsymbol{r}_u - F\boldsymbol{r}_v}{\sqrt{EG-F^2}}\right)_u + \left(\frac{E\boldsymbol{r}_v - F\boldsymbol{r}_u}{\sqrt{EG-F^2}}\right)_v\right]. \tag{6.45}$$

By using the Christoffel symbols Γ^c_{ab} (4.51) and (4.52), we obtain the following expression

$$\Delta_\Sigma \boldsymbol{r} = \frac{F}{\sqrt{EG-F^2}}\Big[(G^2\Gamma^v_{uu} - FG\Gamma^v_{uv} + 2F^2\Gamma^u_{uv} - FE\Gamma^u_{vv} - GE\Gamma^u_{uv})\boldsymbol{r}_u$$
$$+ (E^2\Gamma^u_{vv} - FG\Gamma^v_{uu} + 2F^2\Gamma^v_{uv} - FE\Gamma^u_{uv} - GE\Gamma^v_{uv})\boldsymbol{r}_v$$
$$+ \frac{2fF(F^2-EG) + E^2gG - eF^2G + EeG^2 - gF^2E}{F}\boldsymbol{N}\Big], \qquad (6.46)$$

decomposed along the tangent $\{\boldsymbol{r}_u, \boldsymbol{r}_v\}$ basis and the unit normal $\boldsymbol{N}$ to the surface Σ. Equation (6.46) is used to provide relations between the Laplacian of the position vector $\boldsymbol{r} = (x^i) \in \mathbb{R}^3$ and the mean (H) and Gaussian (K) curvatures of the surface Σ. For example, from (6.11) and (6.46), the Laplacian of the normal component of the position vector is

$$(\Delta_\Sigma \boldsymbol{r})_n = 2H\boldsymbol{N}. \qquad (6.47)$$

It is interesting to compare this result with (10.56) $\Delta \boldsymbol{r} = 2EH\boldsymbol{N}$ from Theorem 28. In the full three-dimensional case, for isothermal parameterization the relation between the Laplacian and mean curvature contains an additional factor of E.

In the case of orthogonal parameterization (u, v) on the surface, we have $F = 0$ and consequently

$$\Delta_\Sigma \boldsymbol{r} = 2H\boldsymbol{N}. \qquad (6.48)$$

Also we can write [338]

$$(\Delta_\Sigma \boldsymbol{r})^2 = 2K + \sum_{i=1}^{3}(\nabla_\Sigma \times \nabla_\Sigma x^i)^2, \qquad (6.49)$$

and for the normal component

$$\Delta_\Sigma(\boldsymbol{r}\cdot\boldsymbol{N}) = (\boldsymbol{r}\cdot\boldsymbol{N})(2K - 4H^2) - 2H + 2\nabla_\Sigma\cdot(H\boldsymbol{r}). \qquad (6.50)$$

Another useful relation occurs if we apply (6.49) to $\boldsymbol{N}$

$$\boldsymbol{N}\cdot\Delta_\Sigma\boldsymbol{N} + (\nabla_\Sigma\cdot\boldsymbol{N})^2 = 2K.$$

In the case of minimal surfaces ($H = 0$) we have (from (6.50)) the special relation $\Delta_\Sigma \boldsymbol{r} = 0$, and also the relation

$$\Delta_\Sigma(\boldsymbol{r}\cdot\boldsymbol{N}) = 2(\boldsymbol{r}\cdot\boldsymbol{N})K. \qquad (6.51)$$

6.5.4 Surface Curl

For a three-dimensional differential vector field $\boldsymbol{A}$ we introduce the surface curl by

$$\nabla\times\boldsymbol{A} = \frac{1}{EG-F^2}\left[\boldsymbol{r}_u\times\left(G\frac{\partial \boldsymbol{A}}{\partial u} - F\frac{\partial \boldsymbol{A}}{\partial v}\right) + \boldsymbol{r}_v\times\left(E\frac{\partial \boldsymbol{A}}{\partial v} - F\frac{\partial \boldsymbol{A}}{\partial u}\right)\right]. \qquad (6.52)$$

If $A = A_1 r_u + A_2 r_v + A_n N$ we have a very useful relation

$$\nabla_\Sigma \times A = \frac{1}{\sqrt{EG-F^2}}\Big[(FA_1+GA_2)_u - (EA_1+FA_2)_v\Big]N + \frac{1}{\sqrt{EG-F^2}}\Big[(fA_1+gA_2)r_u - (eA_1+fA_2)r_v\Big] + \nabla_\Sigma A_n \times N. \tag{6.53}$$

The terms in the second line of (6.53) represent the tangent components of the surface curl. There are some interesting properties

$$\nabla_\Sigma \times N = 0 \tag{6.54}$$

$$\nabla_\Sigma \times r(u,v) = 0 \tag{6.55}$$

$$\nabla_\Sigma \times (\Phi N) = \nabla_\Sigma \Phi \times N. \tag{6.56}$$

Equation (6.54) raises the question: according to the Helmholtz theorem (Theorem 29) of representation in three dimensions, we know that a curl-free vector field is the gradient of some scalar field. What happens in the case of surface curl? Does it mean that the normal is a surface-gradient field? The answer is of course no, and it will be proved so in Lemma 5. Basically, to be a surface gradient, the vector field has to be tangent, in addition of being curl-free, which is not the case of the normal field.

In the following we are interested to verify if the well-known three-dimensional relation $\nabla \times (\nabla\Phi) = 0$ has an equivalent in terms of surface operators. The answer is given by:

Proposition 6. *If $a = \nabla_\Sigma \Phi$ then $\nabla_\Sigma \times a \in T\Sigma$. A necessary condition for*

$$\nabla_\Sigma \times \nabla_\Sigma \Phi = 0, \tag{6.57}$$

is $K = 0$, i.e., the surface curl of a surface gradient is zero only on surfaces with zero Gaussian curvature. On such surfaces (6.57) *is satisfied if*

$$\frac{E\Phi_v - F\Phi_u}{G\Phi_u - F\Phi_v} = -\frac{f}{g}. \tag{6.58}$$

Very interesting, and contrary to the $\mathbb{R}^3$ case, the surface curl of a surface gradient is not necessarily zero, but belongs to the tangent bundle. It can be zero but only on special types of surfaces, and for specific scalar fields only. For the proof we use (6.53). Obviously $(\nabla_\Sigma \Phi)_n = 0$. The first part of Proposition 6 is immediate by checking that the normal part of the curl is zero

$$\nabla_\Sigma \Phi = \frac{1}{EG-F^2}(G\Phi_u - F\Phi_v)r_u + \frac{1}{EG-F^2}(E\Phi_v - F\Phi_u)r_v,$$

then

$$(F(\nabla_\Sigma \Phi)_1 + G(\nabla_\Sigma \Phi)_2)_u - (E(\nabla_\Sigma \Phi)_1 + F(\nabla_\Sigma \Phi)_2)_v = 0.$$

The second part of the proposition results also from (6.53) and the compatibility of the linear system

$$A_1 f + A_2 g = 0$$

$$A_1 e + A_2 f = 0.$$

Basically, (6.58) tells that the surface curl of the surface gradient of a scalar field Φ is zero if for any displacement (du, dv) orthogonal to the level lines of Φ in Σ we have

$$\frac{du}{dv} = -\frac{g}{f}.$$

In other words, the tangent vector surface gradient of Φ makes at every point a certain prescribed angle with the local frame $\{\boldsymbol{r}_u, \boldsymbol{r}_v\}$.

The next question addresses the problem of the surface divergence of a surface curl. For any space vector $\boldsymbol{A} = A_1 \boldsymbol{r}_u + A_2 \boldsymbol{r}_v + A_n \boldsymbol{N}$ we calculate

$$\nabla_\Sigma \cdot (\nabla_\Sigma \times \boldsymbol{A}) = \frac{2H}{\sqrt{EG - F^2}}[(EA_1 + FA_2)_v - (FA_1 + GA_2)_u] + \frac{1}{\sqrt{EG - F^2}}[(fA_1 + gA_2)_u - (eA_1 + fA_2)_v], \quad (6.59)$$

and we notice it is independent of A_n. We have

Proposition 7. *If $\boldsymbol{A} \perp T\Sigma$, i.e., $\boldsymbol{A} = A_n \boldsymbol{N}$, then*

$$\nabla_\Sigma \cdot (\nabla_\Sigma \times \boldsymbol{A}) = 0.$$

In other words the surface divergence of the surface curl is zero if the vector field is normal, but not in general. For an arbitrary vector field the equation $\nabla_\Sigma \cdot (\nabla_\Sigma \times \boldsymbol{A}) = 0$ is a complicated PDE, involving Christoffel symbols and second-order derivatives of $\boldsymbol{N}$.

We leave the proof of this Proposition as an exercise to the reader (Hint: use (7.55)).

Like in the case of the surface divergence, there is a relation between the surface curl and the geodesic curvature. From (6.44), (6.69), and (6.54) we obtain

$$\kappa_g = \boldsymbol{N} \cdot \nabla_\Sigma \times \boldsymbol{t}, \quad (6.60)$$

that is the geodesic curvature of a curve lying on Σ is the normal component of the curl of the unit tangent to the curve.

When the partial derivative is substituted with the covariant derivative, in all the above surface differential operators, some of the relations between operators change. This happens because of the noncommutativity property of the second-order covariant derivative (4.54).

6.5.5 Integral Relations for Surface Differential Operators

There are equivalent forms for the integral theorems of Stokes, Gauss, and Green in terms of surface differential operators, relating integrals on domains of the surface and line integrals around the boundaries of such domains. Like previously we denote by $\boldsymbol{A} = A_1(u,v)\boldsymbol{r}_u + A_2(u,v)\boldsymbol{r}_v + A_n(u,v)\boldsymbol{N} \in \mathbb{R}^3$ a three-dimensional differential vector field.

We consider a domain $D \subset \Sigma$ with smooth boundary given by the arc-length parameterized curve $\partial D = \Gamma \subset \Sigma$. At any point of Γ we have the Serret–Frenet trihedron $\{\boldsymbol{t}, \boldsymbol{n}, \boldsymbol{b}\}_{u(s),v(s)}$, and the unit surface normal $\boldsymbol{N}(u,v)$. We define the unit vector tangent to the surface and normal to the curve $\boldsymbol{t}^{\perp} \in T\Sigma, \boldsymbol{t} \cdot \boldsymbol{t}^{\perp} = 0$, see Example 7 in Sect. 6.6. A possible way to define it is $\boldsymbol{t}^{\perp} = \boldsymbol{N} \times \boldsymbol{t}$. The direction of $\boldsymbol{t}^{\perp}$ is chosen outward from the region D enclosed by Γ. We have another trihedron composed by $\{\boldsymbol{t}, \boldsymbol{t}^{\perp}, \boldsymbol{N}\}$. The equivalent of the Gauss divergence theorem is given by

$$\iint_D \nabla_\Sigma \cdot \boldsymbol{A} dA = \oint_\Gamma \boldsymbol{A} \cdot \boldsymbol{t}^{\perp} ds - 2 \iint_D H \boldsymbol{A} \cdot \boldsymbol{N} dA, \tag{6.61}$$

where dA is the infinitesimal area element. This equation is the Gauss divergence theorem analog for surfaces. The LHS is the integral over the domain of the (surface) divergence of a vector field $\boldsymbol{A}$. Contrary to the three-dimensional case, where this term is balanced only by an integral over the boundary of the domain, in the surface case we have two terms. The first term in the RHS is indeed the "flux" of the vector field (Γ curve) through the boundary, in this case in the direction $\boldsymbol{t}^{\perp}$. The second term in the RHS is additional, depends on the surface geometry, and represents the transfer of flux of $\boldsymbol{A}$ in the normal direction through the domain D. This term cancels if the surface is minimal (case when the Gauss theorem for three-dimensional domains and (6.61) are identical) fact which can be used as equilibrium criterium for the energy balance. If, for example, we examine an incompressible flow $\nabla \boldsymbol{v} = 0$ and we consider Σ a free fluid surface (so we have no normal flow across the surface), by substituting $\boldsymbol{A} = \boldsymbol{v}$ in (6.61), we obtain a zero circulation theorem

$$\oint_\Gamma \boldsymbol{v} \cdot \boldsymbol{t}^{\perp} ds = \oint_\Gamma v_{\perp} ds = 0, \tag{6.62}$$

for any closed curve lying on the free surface. This conservation law is true even for an arbitrary surface when we have fluid flow across it. When we assume an orthogonal parameterization for simplicity, and from (6.43) we notice that in this case $0 = \nabla \boldsymbol{v} = \nabla_\Sigma \boldsymbol{v} + 2H v_n$, so the LHS in (6.43) cancels the second term in the RHS, and we have again (6.62).

Another consequence of (6.43), useful in some applications, is obtained if we choose A =const.

$$\oint_{\Gamma} t^{\perp} ds = 2 \iint_D H N dA = 2 \iint_D \boldsymbol{H} dA. \tag{6.63}$$

The Green and Stokes integral theorems for surface differential operators have the same form as in the full three-dimensional case. For more details the reader can find details in the book of Weatherburn [338, Articles 120–130]. We write here only the (geometrical) *circulation theorem*, also known under the name of Stokes theorem

$$\iint_D N \cdot (\nabla_{\Sigma} \times A) dA = \oint_{\partial D} A \cdot t ds, \tag{6.64}$$

where the RHS is called the *circulation of the field* A *around the loop* $\Gamma = \partial A$. An immediate consequence of the circulation theorem is the following [338].

Lemma 5. *If a tangent vector field* $\boldsymbol{a} \in T\Sigma$ *has its surface curl tangent to the surface, too,* $\nabla_{\Sigma} \times \boldsymbol{a} \in T\Sigma$, *this vector is the surface gradient of a scalar function defined on the surface.*

Proof. The LHS in (6.64) is zero and by the circulation theorem the RHS is zero, for any arbitrary loop. According to Sect. 6.5.1 the tangent field A is the gradient of some scalar function $\Phi : \Sigma \to \mathbb{R}$. □

Consequently, contrary to the three-dimensional case where the necessary condition for a vector field to be the gradient of some scalar field was to have the curl zero, in the surface case the field also needs to be tangent (see also Exercises 8 and 9 of this chapter).

6.5.6 Applications

In the following we illustrate the above propositions with examples from cylindrical, spherical, and toroidal surfaces.

6.5.6.1 Cylindrical Surfaces

We choose an infinite right cylinder of radius R with parameterization $u = \varphi$ (the polar angle in the xOy base plane), and $v = z$, and we have $G = 1, E = R^2, F = g = f = 0, e = R$. The normal is $N = (\cos\varphi, \sin\varphi, 0)$, the Gaussian curvature is obviously 0 and $H = 1/(2R)$. The surface differential operators are

$$\nabla_{Cyl} \Phi = \left(-\frac{\sin\varphi}{R} \frac{\partial \Phi}{\partial \varphi}, \frac{\cos\varphi}{R} \frac{\partial \Phi}{\partial \varphi}, \frac{\partial \Phi}{\partial z} \right),$$

$$\nabla_{Cyl}\cdot\boldsymbol{A} = \frac{\partial A_1}{\partial\varphi} + \frac{\partial A_2}{\partial z} + \frac{A_n}{R},$$

$$\nabla_{Cyl}\times\boldsymbol{A} = \left(\frac{1}{R}\frac{\partial A_2}{\partial\varphi} - R\frac{\partial A_1}{\partial z}\right)\boldsymbol{N} + \frac{1}{R}\frac{\partial A_n}{\partial z}\boldsymbol{r}_u + \left(A_1 - \frac{1}{R}\frac{\partial A_n}{\partial\varphi}\right)\boldsymbol{r}_v,$$

$$\Delta_{Cyl}\Phi = \frac{1}{R^2}\frac{\partial^2\Phi}{\partial\varphi}^2 + \frac{\partial^2\Phi}{\partial z^2}.$$

We also check by direct calculation that $\nabla_{Cyl}\times(\nabla_{Cyl}\Phi) = 0$ if $\Phi = \Phi(z)$, i.e., the curl of the gradient is zero on scalar fields with cylindrical symmetry only. Also, $\nabla_{Cyl}\cdot(\nabla_{Cyl}\times\boldsymbol{A}) = 0$ only if $A_2 = A_2(z)$.

6.5.6.2 Spherical Surfaces

We have a sphere of radius R with parameterization $u=\theta$ and $v=\varphi$, and we have $G = R^2\sin^2\theta$, $E = R^2$, $F = f = 0$, $e = -R$, $g = -R\sin^2\theta$. The normal is $\boldsymbol{N} = (\sin\theta\cos\varphi, \sin\theta\sin\varphi, \sin\theta\cot\varphi)$, the Gaussian curvature is $K = 1/R^2$ and $H = -1/R$. The surface differential operators are

$$\nabla_{Sph}\Phi = \frac{1}{R}\left(\cos\theta\cos\varphi\frac{\partial\Phi}{\partial\theta} - \frac{\sin\varphi}{\sin\theta}\frac{\partial\Phi}{\partial\varphi}, \cos\theta\sin\varphi\frac{\partial\Phi}{\partial\theta}\right.$$
$$\left. + \frac{\cos\varphi}{\sin\theta}\frac{\partial\Phi}{\partial\varphi}, -\sin\theta\frac{\partial\Phi}{\partial\theta}\right),$$

$$\nabla_{Sph}\cdot\boldsymbol{A} = \frac{1}{\sin\theta}\frac{\partial}{\partial\theta}(\sin\theta A_1) + \frac{\partial A_2}{\partial\varphi} + \frac{2}{R}A_n,$$

$$\nabla_{Sph}\times\boldsymbol{A} = \frac{1}{2}\left(-2A_1\sin\varphi + A_2\sin(2\theta)\cos\varphi - 2\cos\varphi\frac{\partial A_1}{\partial\varphi}\right.$$
$$+2\cos\varphi\sin^2\theta\frac{\partial A_2}{\partial\theta} + \frac{2}{R}\sin\varphi\frac{\partial A_n}{\partial\theta} + \frac{2\cot\theta\cos\varphi}{R}\frac{\partial A_n}{\partial\varphi}$$
$$+\frac{2}{R}\sin\varphi\frac{\partial A_n}{\partial\theta}, 2A_1\cos\varphi + A_2\sin(2\theta)\sin\varphi$$
$$-2\sin\varphi\frac{\partial A_1}{\partial\varphi} + 2\sin^2\theta\sin\varphi\frac{\partial A_2}{\partial\theta}$$
$$+\frac{2}{R}\cot\theta\sin\varphi\frac{\partial A_n}{\partial\varphi} - \frac{2}{R}\cos\varphi\frac{\partial A_n}{\partial\theta}, A_2(3+2\cos(2\theta))$$
$$\left.-2\cot\theta\frac{\partial A_1}{\partial\varphi} + \sin(2\theta)\frac{\partial A_2}{\partial\theta} - \frac{1}{R}\frac{\partial A_n}{\partial\varphi}\right),$$

$$\triangle_{Sph}\Phi = \frac{1}{R^2}\left(\cot\theta\frac{\partial\Phi}{\partial\varphi} + \frac{\partial^2\Phi}{\partial\theta^2} + \frac{1}{\sin^2\theta}\frac{\partial^2\Phi}{\partial\varphi^2}\right).$$

As an example let us find the condition for a vector field $\boldsymbol{a} = a_1\boldsymbol{r}_u + a_2\boldsymbol{r}_v$ tangent to a sphere to fulfill the property in Proposition 7. We have

$$\nabla_{Sph}\cdot(\nabla_{Sph}\times\boldsymbol{a}) = 2a_2\cos\theta - \frac{1}{\sin\theta}\frac{\partial a_1}{\partial\varphi} + \sin\theta\frac{\partial a_2}{\partial\theta} = 0,$$

and this equation results in the following condition for the components of the field

$$\frac{\partial}{\partial\theta}(a_2\sin^2\theta) = \frac{\partial a_1}{\partial\varphi}.$$

For example, if we choose

$$a_2 = P_{3,1}(\cos\theta)\cos(4\varphi)\sin(2\varphi),$$

from the above condition we obtain the expression

$$a_1 = \frac{d}{d\theta}\left(P_{3,1}(\cos\theta)\sin^2\theta\right)\int^{\varphi}\cos(4\varphi')\sin(2\varphi')d\varphi'.$$

The field fulfilling this conditions is presented in Fig. 6.9.

6.5.6.3 Toroidal Surfaces

We set toroidal coordinates (u, v) in the form

$$\boldsymbol{r}(u,v) = ((a + R\cos u)\cos v, (a + R\sin u)\sin v, R\sin u),$$

where a, R are the small and large radii of a torus. We have $E = R^2, G = (a + R\cos u)^2, F = f = 0, g = \cos u(a + R\cos u), e = R$. The surface differential operators are

$$\nabla_{tor}\Phi = \frac{1}{R^2}\frac{\partial\Phi}{\partial u}\boldsymbol{r}_u + \frac{1}{(a + R\cos u)^2}\frac{\partial\Phi}{\partial v}\boldsymbol{r}_v,$$

$$\begin{aligned}\nabla_{tor}\cdot\boldsymbol{A} &= -\frac{R\sin u}{a + R\cos u}A_1 + \frac{\partial A_1}{\partial u}\\ &+ \frac{\partial A_2}{\partial v} - \frac{a^2 + R^2 + 3aR\cos u + R^2\cos^2 u}{R(a + R\cos u)^2}A_n,\end{aligned}$$

$$(\nabla_{tor}\times\boldsymbol{A})_n = -\frac{R}{a + R\cos u}\frac{\partial A_1}{\partial v} + \frac{a + R\cos u}{R}\frac{\partial A_2}{\partial u} - 2\sin u A_2,$$

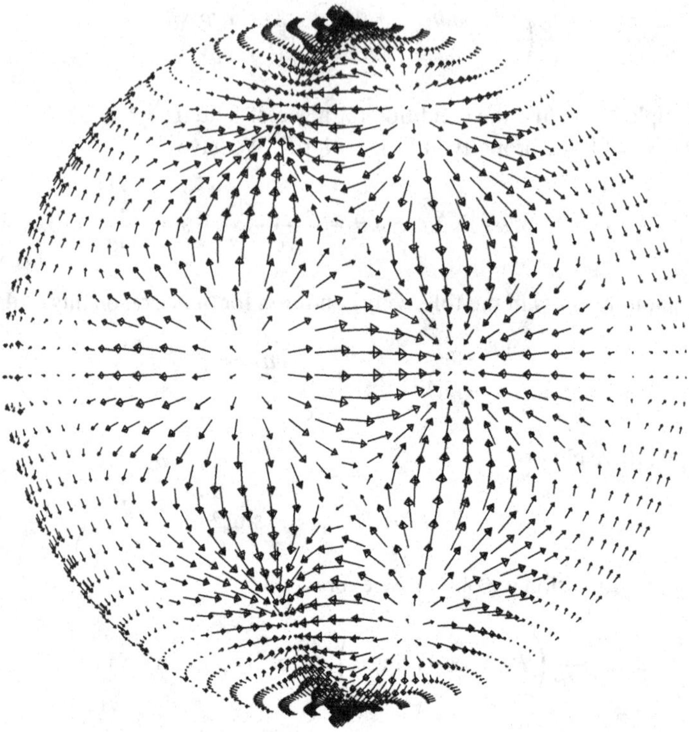

Fig. 6.9 An example of a tangent vector field on the surface of a sphere fulfilling the condition of zero divergence of the curl discussed in Proposition 7

$$(\nabla_{tor} \times \boldsymbol{A})_1 = \frac{\cos u}{R} A_2 + \frac{1}{R(a + R\cos u)} \frac{\partial A_n}{\partial v},$$

$$(\nabla_{tor} \times \boldsymbol{A})_2 = -\frac{1}{a + R\cos u} A_1 - \frac{1}{R(a + R\cos u)} \frac{\partial A_n}{\partial u},$$

and the Laplacian

$$\Delta_{tor} \Phi = -\frac{\sin u}{R(a + R\cos u)} \frac{\partial \Phi}{\partial u} + \frac{1}{R^2} \frac{\partial^2 \Phi}{\partial u^2} + \frac{1}{(a + R\cos u)^2} \frac{\partial^2 \Phi}{\partial v^2}.$$

6.5.6.4 Closed Surfaces

There are some interesting consequences of the integral equations for the surface operators. For example, a consequence of the divergence integral formula (6.61) is that on a closed surface Σ we have the LHS of (6.63) approaching zero. It results

Proposition 8. *The average value of the mean curvature vector is zero on any closed surface*

$$\iint \boldsymbol{H} dA = \iint H\boldsymbol{N} dA = 0. \tag{6.65}$$

Other interesting relations holding on closed surfaces are

$$\iint_{\Sigma} \boldsymbol{N}\cdot(\nabla_{\Sigma}\times \boldsymbol{A})dA = 0 \text{ and } \iint_{\Sigma} \boldsymbol{N}\times\nabla_{\Sigma}\Phi dA = 0, \tag{6.66}$$

for any vector or scalar field $\boldsymbol{A}$ and Φ, respectively.

6.6 Problems

1. Find a proof for Proposition 5 by using (6.7).
2. There are some ambiguities in the notation of vector components in different orthogonal bases. For example let us have on a sphere $S_2 \subset \mathbb{R}^3$ parameterized by $(u,v) = (\theta,\varphi)$ the orthonormal basis $\{\boldsymbol{e}_\theta, \boldsymbol{e}_\varphi\}$. We have $\boldsymbol{r}_u = (\cos\theta\cos\varphi, \cos\theta\sin\varphi, -\sin\theta) = \boldsymbol{e}_\varphi$ because its norm is 1. However, $\boldsymbol{r}_v = \sin\theta(-\sin\varphi,\cos\varphi,0) = \sin\theta\boldsymbol{e}_\varphi$. Now, a tangent field can be expressed in either way $\boldsymbol{a} = a_\theta\boldsymbol{e}_\theta + a_\varphi\boldsymbol{e}_\varphi$ or $\boldsymbol{a} = a_u\boldsymbol{r}_u + a_v\boldsymbol{r}_v$, and we have the relations $a_\theta = a_u$, $a_\varphi = \sin\theta a_v$. Show that
$$\nabla_{\Sigma}\boldsymbol{a} = \frac{1}{\sin\theta}\frac{\partial}{\partial\theta}(\sin\theta a_\theta) + \frac{1}{\sin\theta}\frac{\partial a_\varphi}{\partial\varphi}.$$
3. A parameterization of a surface Σ is called *isometric* if $E = G$ and $F = 0$. The name comes from the resulting arc-length relation $ds^2 = \lambda(du^2 + dv^2)$. Show that we have an isometric system of coordinates (u,v) on Σ defined by curves $u =$ const. and their orthogonal complements, if and only if
$$\frac{\Delta_\Sigma u}{|\nabla_\Sigma u|}$$
is a function of u only. Hint: check [338].
4. Prove (use [338, Article 120]) that the following usual algebraic relations fulfilled by differential operators in $\mathbb{R}^3$ are also valid for surface differential operators

$$\nabla_\Sigma\cdot(\Phi\boldsymbol{A}) = \nabla_\Sigma\Phi\cdot\boldsymbol{A} + \Phi\nabla_\Sigma\cdot\boldsymbol{A}, \tag{6.67}$$

$$\nabla_\Sigma\times(\Phi\boldsymbol{A}) = \nabla_\Sigma\Phi\times\boldsymbol{A} + \Phi\nabla_\Sigma\times\boldsymbol{A}, \tag{6.68}$$

$$\nabla_\Sigma\cdot(\boldsymbol{A}\times\boldsymbol{B}) = \boldsymbol{B}\cdot\nabla_\Sigma\times\boldsymbol{A} - \boldsymbol{A}\cdot\nabla_\Sigma\times\boldsymbol{B}, \tag{6.69}$$

$$\Delta_\Sigma(\Phi\boldsymbol{A}) = \Phi\Delta_\Sigma\boldsymbol{A} + 2\nabla_\Sigma\Phi\nabla_\Sigma\cdot\boldsymbol{A} + \boldsymbol{A}\Delta_\Sigma\Phi. \tag{6.70}$$

5. Let $\Phi(\theta, \varphi)$ be a scalar differentiable field defined on a sphere S_2. Show by direct calculation that $\nabla_{Sph} \times (\nabla_{Sph}\Phi) = 0$ only if $\Phi = \text{const.}$, and compare this result with Proposition 6 (i.e., $K_{Sph} \neq 0$).
6. Prove that $\nabla_\Sigma \cdot \Delta_\Sigma \boldsymbol{r} = -4H^2$.
7. A curve C lies on a surface $\boldsymbol{r}(u, v) \in \Sigma$. Prove that the unit perpendicular $\boldsymbol{t}^\perp$ to the tangent $\boldsymbol{t}$ of the curve, contained in the tangent plane, has the expression

$$\boldsymbol{t}^\perp = \frac{(Fu_s + Gv_s)\boldsymbol{r}_u - (Eu_s + Fv_s)\boldsymbol{r}_v}{E(G - F)u_s + G(F - E)v_s}.$$

8. For a minimal surface $H = 0$ so the surface divergence of the normal is zero. Does it result from here that in the case of minimal surfaces the unit normal can be expressed as a surface curl (like in the three-dimensional case)?
9. Find out: is there a surface equivalent (in terms of surface differential operators) of the Helmholtz representation theorem?
10. For a $(1, 1)$-type of tensor defined on $\Sigma \subset \mathbb{R}^3$ A_α^β, prove that

$$\nabla_\alpha^\alpha = \frac{1}{\sqrt{g}} \frac{\partial}{\partial u^\alpha} (\sqrt{g} A^\alpha).$$

11. Find properties of the surface differential operators arising for the *Hairy ball theorem*, i.e., there is no zero everywhere tangent vector field on the 2-sphere.

Chapter 7
Motion of Curves and Solitons

A large class of physical, chemical, and biological systems can be modeled in terms of their *contour dynamics*, namely the kinematics and dynamics of their boundaries [248, 249, 329]. In many situations (e.g., when the inside bulk has the property of being "incompressible") such contour representations are the most natural, and are simpler ones. Basically, the contour dynamics approach reduces the problem to the study of motion of curves and surfaces, especially the closed ones. In this chapter, we focus on the analysis of the motion of curves in the three-dimensional Euclidean space.

The study of two-dimensional contour dynamics models are important for flat liquid droplets [137,138,270,332], quantum Hall electron droplets in high magnetic field [340, 341], growth of dendritic crystals in a plane [32], planar motion of interfaces (like for example oil spots surrounded by water) [147, 295], dynamics of polymers [66,290,345], vortex structures in geophysical fluid dynamics and plasma [110], motile cells immobilized in vitro [29], etc. Two-dimensional contours can be plane curves or curves lying on surfaces. In the three-dimensional case, in addition to the above mentioned fields, interesting applications can be found in the dynamics of vortex filaments in fluid dynamics [123, 177], KdV flows on star-shaped curves [39], DNA models [325], long and stiff polymer chains, flagellar swimming for motile cells [30, 126, 176, 193, 324], level set method [296], and solitons in the Euler elastica equation [227, 228].

All these applications have in common the properties of preserving global geometric quantities like area and perimeter. Imposing global geometrical constrains on contour dynamics leads to the occurrence of nonlinearities in the dynamical equations. This is because, on one hand, the global constraints involve the fundamental forms of surfaces (or at least metrics of curves), and these forms contain quadratic or higher-order terms as combinations of the metrics, the Serret–Frenet and Darboux vectors and their derivatives. On the other hand, global constraints involve strong nonlocality and long-range interactions in the system, like for example in hydrodynamics [20, 108, 233, 248, 249, 329]. An example of global constraint interaction from biology is the swimming of a flagellated cell. A local

A. Ludu, *Nonlinear Waves and Solitons on Contours and Closed Surfaces*, Springer Series in Synergetics, DOI 10.1007/978-3-642-22895-7_7,

constraint applied to the free end of the flagellum, which is a bundle of filaments attached to the cell membrane, could prevent the existence of relative shear between the filaments in the bundle, which is the very cause of bending, twisting and hence swimming. Since the local shear is related to the curvature, the local condition at the end generates a global constraint: the total curvature of the bundles should be zero, i.e., allowable shapes have to have zero total curvature.

The occurrence of nonlinearities in the contour dynamics problems involves the connection between this dynamics and the integrable evolution equations. Indeed, the motion of curves is intimately related to the Korteweg–de Vries (KdV), modified Korteweg–de Vries (MKdV), and nonlinear Schrödinger equations (NLS) [2, 169]. This leads to the existence of soliton-like solutions in the motion of curves, as well as the existence of infinite number of conservation laws that can be put into relation with global geometric quantities. The purpose of the next sections is to describe these relations, for the two-dimensional and three-dimensional case.

The problem of the dynamics of moving curves is not completely solved. There are systems, especially in the world of microorganisms with very complicated shapes, where the interaction between the two-dimensional contours (like the cell membrane) and one-dimensional attachments (like flagella, cilia, etc.) cannot be neglected, to understand the physics of their exquisite motility. A general model for such type of interaction should lie somewhere between the geometry of curves and surfaces, like for example the geometry of a $(1 + \epsilon)$-dimensional manifold. Such situations occur for example while investigating the propagation of waves created in a one-dimensional system into a two-dimensional surface, or conversely, the motions induced in a bundle of cilia by membrane oscillations.

7.1 Kinematics of Two-Dimensional Curves

In this section we study the dynamics of two-dimensional contours from the perspective of differential geometry of closed curves and the hierarchy of integrable systems like KdV and MKdV systems. The association of the Serret–Frenet equations with nonlinear integrable systems (like the cubic Schrödinger equation for example) is somehow natural, because the Serret–Frenet equations are known to be equivalent to a Riccati equation (see Sect. 18.2). Moreover, through an exponential integral transformation of curvature and torsion into a complex function Hasimoto has shown in [123] that the Serret–Frenet equations can be directly mapped into the cubic Schrödinger equation. We mention, however, that there are possible many other two-dimensional curve motions that are not integrable. A comprehensive discussion about this reduction can be found in [88, 310].

We begin our studies of purely *local* surface dynamics with a simple model, i.e., the motion of plane curves. Later on we will generalize the result to three-dimensional curves. We need to use the concepts developed in Sect. 5. We consider a differentiable (class $k \geq 3$) two-dimensional curve parameterized by u at any moment of time t. The evolution of the shape of the curve in time is describable by the geometry of a family of curves, each curve parameterized by u, and labeled in the family by t. Basically, we need to use the formalism in Sect. 18.4 and substitute the β parameter with the time t. We mention that there should be no notation confusion between t as time parameter and $\boldsymbol{t}$ as tangent unit vector. The points of the curve at a certain moment of time t are described by $\boldsymbol{r}(u,t)$ or $\boldsymbol{r}(s(t),t)$ where s is the natural arc-length parameter along the curve, which itself depends on time through the metric. The metric on the curve is $g(u,t)$ and we associate the Serret–Frenet trihedron, also at any moment of time t.

In this book we take into account only curve motions produced by local interaction. Consequently, the kinematics of the curves depends only on the local intrinsic geometrical variable of the two-dimensional curve, i.e., $\kappa(s)$. This further means that the kinematics of the curve depends on s only through $\kappa^{(k)}(s,t)$, $k = 0, 1, \dots$, the derivatives of the curvature. The kinematics is described in terms of the velocity $\boldsymbol{V}$ of the points on the curve

$$\frac{d\boldsymbol{r}}{dt} = \dot{\boldsymbol{r}}(u,t) = U(\kappa, \kappa_s, \dots)\boldsymbol{n}(u,t) + W(\kappa, \kappa_s, \dots)\boldsymbol{t}(u,t), \tag{7.1}$$

where $\boldsymbol{n}, \boldsymbol{t}$ are the unit principal normal and tangent to the curve, and (U, W) are the normal and tangent components of the curve velocity at the point described by the (s,t) coordinates. These velocities are purely locally defined quantities, as stated above. *In general we will denote partial derivative with respect to t by using the subscript t and the total derivative with a dot.* In the case of (u,t) parametrization these two coincide, which is not the case of the (s,t) parametrization.

Usually in literature it is vaguely mentioned that the W term in (7.1) is irrelevant, because it is only related to a reparameterization of the curve. We provide in the following the Epstein–Gage theorem which brings clarifications and limitations to this observation [55].

Theorem 22. *Let us consider (7.1) with the particular dependence* $U = U(\kappa,\theta)$, $W = W(\kappa,\theta)$, *where* θ *is the tangent angle*

$$\theta(s,t) = \int^{s} \kappa ds.$$

If U, W *are* C_3 *differentiable and periodic of period* 2π *in* θ, *then for any solution of* $\boldsymbol{r}(s,t)$ *of (7.1) there is a reparameterization* $s' = s'(s,t)$ *of the curve* $\boldsymbol{r}(s,t)$, *such that*

$$\frac{ds'}{ds} > 0, \quad s'(s,0) = s,$$

and $\boldsymbol{r}(s'(s,t),t)$ *is a solution of the equation*

$$\frac{d\boldsymbol{r}}{dt}(s',t) = U(\kappa,\theta)\boldsymbol{n}(s',t).$$

The reparameterization function fulfills the equation

$$\frac{ds'}{dt} = -|\boldsymbol{r}(s',t)|W(\kappa(s',t),\theta(s',t)).$$

That is, the motion of such a curve depends only on its normal velocity. However, there are cases when the tangent speed matters. We give such an example at the end of Sect. 7.3.

If we have a parameterized "rigid" (that is $g_t = 0$) closed curve $\boldsymbol{r}(u,t)$ in uniform translation and uniform rotation, so $U = \dot{\boldsymbol{r}}\cdot\boldsymbol{n}$, $W = \dot{\boldsymbol{r}}\cdot\boldsymbol{t}$. Both components have time variation because the motion of the points of the curve is accelerated. However, if we eliminate the translation both components become constant. The reason for this is that the local frame $\{\boldsymbol{t},\boldsymbol{n}\}$ moves together with the rigid curve. In Fig. 7.1, we show a uniform rotated figure-8 shape.

We will now investigate the equation of motion of a two-dimensional parameterized curve. The Serret–Frenet relation plus the expression (7.1) for the velocity of the points on the curve allow us to obtain the time evolution of each quantity.

First set of relations is obtained on behalf of the commuting relations between derivations. The position vector is at least third derivative continuous function, so each of its partial derivative of order 2 or less commute, if we take them with respect to the independent coordinates, i.e., u and t. Consequently, we can write

$$\begin{aligned}\boldsymbol{r}_{ut} &= (g^{\frac{1}{2}}\boldsymbol{r}_s)_t = \frac{1}{2}g^{-\frac{1}{2}}g_t\boldsymbol{r}_s + g^{\frac{1}{2}}(\boldsymbol{r}_s)_t\\ &= \frac{1}{2}g^{-\frac{1}{2}}g_t\boldsymbol{t} + g^{\frac{1}{2}}\boldsymbol{t}_t = \frac{1}{2}g^{-\frac{1}{2}}g_t\boldsymbol{t} + g^{\frac{1}{2}}\dot{\boldsymbol{t}},\end{aligned} \tag{7.2}$$

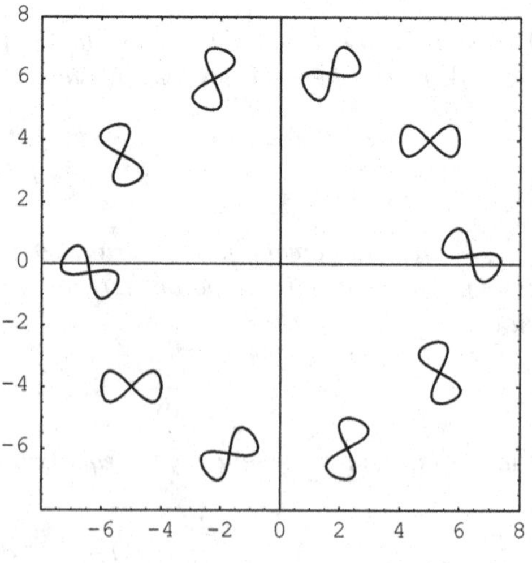

Fig. 7.1 A figure-8 shape in uniform rotation. The velocity (U, W) in the local Serret–Frenet frame is constant at all times

where we used $\partial_u = g^{1/2}\partial_s$, the Serret–Frenet equation in the plane (5.8). On the other side, we have

$$\boldsymbol{r}_{tu} = g^{\frac{1}{2}}(U\boldsymbol{n} + W\boldsymbol{t})_s = g^{\frac{1}{2}}(U_s\boldsymbol{n} + U\boldsymbol{n}_s + W_s\boldsymbol{t} + W\boldsymbol{t}_s)$$

$$= g^{\frac{1}{2}}\left[(U_s + \kappa W)\boldsymbol{n} + (W_s - \kappa U)\boldsymbol{t}\right], \tag{7.3}$$

where the subscript U_s means total derivative of U with respect to s, that is $U_s = U_\kappa \kappa_s + U_{\kappa_s}\kappa_{ss}$, etc. Since $||\boldsymbol{t}|| = 1 \rightarrow \dot{\boldsymbol{t}} \cdot \boldsymbol{t} = 0$, and since in the plane $\boldsymbol{t} \perp \boldsymbol{n}$ it results that $\dot{\boldsymbol{t}} \parallel \boldsymbol{n}$ and $\dot{\boldsymbol{n}} \parallel \boldsymbol{t}$. Consequently, if we equate (7.2) and (7.3), we have to equate the coefficients of $\dot{\boldsymbol{t}}$ and $\boldsymbol{t}$. It results

$$\boldsymbol{t}_t = (U_s + \kappa W)\boldsymbol{n} \tag{7.4}$$

$$g_t = 2g(W_s - \kappa U). \tag{7.5}$$

When the curve moves it changes its shape, but also its intrinsic geometry since in general $s = s(t)$. Indeed, if we neglect the time dependence of s, and we use commuting of derivatives in the form $\partial_t\partial_s = \partial_s\partial_t$ instead of that one in u and t, we would obtained instead of (7.5), $W_s = \kappa U$, which is something else. This general approach does not conserve arc-length locally, but we can always introduce this conservation law if we request that the LHS of (7.5) to be zero.

The second set of relations is obtained from the second Serret–Frenet relations, namely $\boldsymbol{t}_s = \kappa\boldsymbol{n}$. When we differentiate this equation with respect to time, and transform the s-derivatives into u-derivatives through $\partial_u = g^{1/2}\partial_s$ we obtain

$$-\frac{1}{2}g^{-\frac{3}{2}}g_t\boldsymbol{t}_u + g^{-\frac{1}{2}}\boldsymbol{t}_{ut} = \kappa_t\boldsymbol{n} + \kappa\boldsymbol{n}_t. \tag{7.6}$$

By using again the commutativity between derivatives, by substituting $\boldsymbol{t}_t$ from (7.4), and by using again Serret–Frenet equations we have

$$-\frac{1}{2}g^{-1}g_t\kappa\boldsymbol{n} + (U_s + \kappa W)_s\boldsymbol{n} - (U_s + \kappa W)\kappa\boldsymbol{t} = \kappa_t\boldsymbol{n} + \kappa\boldsymbol{n}_t.$$

We know that $\boldsymbol{n}_t \parallel \boldsymbol{t}$, so, by identifying the coefficients of the two orthogonal directions we have

$$\boldsymbol{n}_t = -(U_s + \kappa W)\boldsymbol{t} \tag{7.7}$$

$$\kappa_t = U_{ss} + \kappa^2 U + \kappa_s W. \tag{7.8}$$

Equation (7.6) can be obtained directly form (7.4) if we differentiate with respect to time the identity $\boldsymbol{t} \cdot \boldsymbol{n}$ and use the fact that in the plane $\boldsymbol{n}_t$ is parallel to $\boldsymbol{t}$. The time evolution of the arc-length $s(u,t)$ can be obtained directly if we differentiate with respect to time its integral definition (5.2), and use (7.5). We obtain

$$s_t(u,t) = W(u,t) - W(0,t) - \int_0^u \kappa U ds'. \tag{7.9}$$

Because s depends implicitly on time, we can write $\dot{\kappa} = \frac{d\kappa}{dt} = \kappa_s \dot{s} + \kappa_t$. By using (6.8) we have

$$\frac{\partial \kappa}{\partial t} = \frac{\partial^2 U}{\partial s^2} + \kappa^2 U + \frac{\partial \kappa}{\partial s} W(0,t) + \frac{\partial \kappa}{\partial s} \int_0^s \kappa U ds'. \tag{7.10}$$

This partial differential equations shows that in the two-dimensional case the curvature of the moving curve is determined only by the normal component of the velocity and the initial value of the tangent velocity. It results that the tangent velocity W only determines how the points parameterized by u move along the curve, without affecting the shape, namely the tangent velocity introduces just a reparameterization. In the diagram below we present the flowchart of procedures to determine the two-dimensional curve motion:

$$\begin{array}{ccccc}
\Big(U(u,t), W(u,t)\Big) & \xrightarrow[(7.10)]{\int \ldots du} & \kappa(u,t) & \xrightarrow[(7.5)]{\int \ldots du} & g(u,t) \\
 & & \downarrow & & \downarrow \\
 & & & & s(u,t) \\
 & & \downarrow & & \downarrow \\
 & & & \xrightarrow[(7.7),(7.8)]{\int \ldots du} & \{\boldsymbol{t}(s,t), \boldsymbol{n}(s,t)\}
\end{array}$$

7.2 Mapping Two-Dimensional Curve Motion into Nonlinear Integrable Systems

The theory of plane curves motion represents first of all a warming-up study for the motion of the real three-dimensional curves, it has some direct applications in fluid dynamics and in biophysics by itself, but most importantly it can be put in relation with the theory of nonlinear integrable systems.

In the work of Nakayama et al. [233], the authors show that the Serret–Frenet relations for plane curves, in the form (5.6), form a set of integrable evolution equations compatible with the MKdV hierarchy [2, 169]. By compatible one understands that both Serret–Frenet and MKdV hierarchy systems of nonlinear PDE can be described by the same type of scattering problem, i.e., the matrix of the two-component linear system associated with the nonlinear equation has the same form. The explicit form of the resulting nonlinear equation in the curvature is obtained from (7.10) by additional choices for U. Different choices for the normal velocity of the curve will provide different types of nonlinear equations in the curvature. Only for some special classes of motion of curves, like for example

$$U(s,t) = -\kappa(s,t)_s, \tag{7.11}$$

the dynamical equation for the curvature (7.10) becomes

$$\kappa_t = \frac{3}{2}\kappa^2\kappa_s + \kappa_{sss}, \tag{7.12}$$

which is precisely the MKdV integrable system [169] with stable solitons. These types of curves (7.11) belong to a general class sometimes called curvature-driven curves, since they move faster in the normal direction where the curvature has larger tangent gradient, see for example a recent mathematical study on this topics in [85]. Since solutions of (7.12) are known from the inverse scattering methods, we can use these solutions for the curvature, integrate the corresponding Fresnel relations (5.15) to find the shape of the curve, and implement the curve equation in (7.1) to find the velocity of the curve. A consequence of the choice (7.11) is the existence of the conservation law

$$(\ln\sqrt{g})_t = \left(W + \frac{\kappa^2}{2}\right)_s.$$

If the solution generates a loop, we also have conservation of perimeter and area in time, by the periodicity conditions. Indeed, from (7.33) and (7.40) we have

$$\dot{L} = -\int_0^L \kappa U ds = \int_0^L \kappa\kappa_s ds = 0,$$

$$\dot{A} = -\int_0^L U ds = \int_0^L \kappa_s ds = 0. \tag{7.13}$$

As an example, for the MKdV one-soliton solution of (7.12)

$$\kappa = \kappa_0 \text{sech}\left[\frac{\kappa_0}{2}\left(s - \frac{t\kappa_0^2}{4}\right)\right] \tag{7.14}$$

we obtain the curve velocities

$$U = \frac{\kappa_0^2}{2}\frac{\sinh\frac{\kappa_0}{8}(4s-\kappa_0^2 t)}{\cosh^2\frac{\kappa_0}{8}(4s-\kappa_0^2 t)}$$

$$W = -\frac{5\kappa_0^2}{2}\text{sech}^2\frac{\kappa_0}{8}(4s-\kappa_0^2 t). \tag{7.15}$$

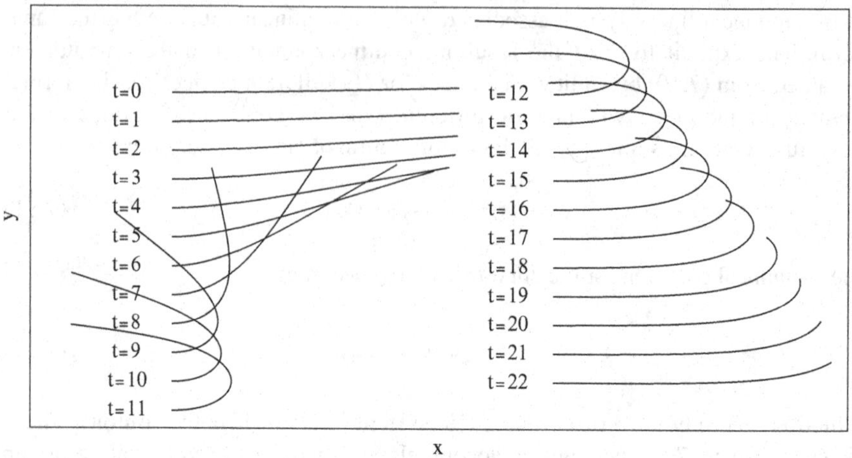

Fig. 7.2 Motion of a plane curve under an MKdV one-soliton solution in curvature. Different paths represent the curve at different moments $t = 0, \dots, 22$. At $t = 0$ the curve has zero curvature everywhere, and then MKdV curvature-soliton propagates from the right to the left end of the curve, and bends it locally. For the first 11.5 units of time the curve bents to the left and upwards, and for later moments of time it bends back executing *swimming or beats patterns*

The resulting curves are always open, because asymptotically the soliton is zero. In Fig. 7.2, we present a straight line run by an MKdV one-soliton in curvature. Such moving shapes occur in the beats, oscillations and swimming of flagella and cilia for microscopic organisms [30, 126, 176, 193, 195]. A richer traveling solution for the MKdV equation is obtained by the substitution $(s,t) \to \xi, \partial/\partial t \to V$, and by integrating two times (7.12) until we obtain the generic form

$$(\kappa_s)^2 = -\frac{1}{4}\kappa^4 + V\kappa^2 + C_1\kappa + C_2 = -\frac{1}{4}(\kappa-\kappa_1)(\kappa-\kappa_2)(\kappa-\kappa_3)(\kappa-\kappa_4), \tag{7.16}$$

where $C_{1,2}$ are constants of integration, and the roots $f_{1,\dots,4}$ can be determined by identification. This equation has the general solution

$$F\left(\arcsin\sqrt{\frac{\kappa-\kappa_1}{\kappa-\kappa_2}\cdot\frac{\kappa_2-\kappa_3}{\kappa_1-\kappa_1}}\,\middle|\,\frac{(\kappa_1-\kappa_3)(\kappa_2-\kappa_4)}{(\kappa_2-\kappa_3)(\kappa_1-\kappa_4)}\right)$$

$$= \pm\frac{\xi}{4}\sqrt{(\kappa_1-\kappa_4)(\kappa_3-\kappa_2)} + C_3, \tag{7.17}$$

where F is the incomplete elliptic integral of the first kind (Sect. 18.3), and the vertical bar represents the usual notation for the separation between the argument and the parameter [5]. The explicit traveling solution reads

$$\kappa(\xi) = \frac{A + B\,\mathrm{cn}(\frac{\xi}{\Lambda}|m)}{D + F\,\mathrm{cn}(\frac{\xi}{\Lambda}|m)}, \tag{7.18}$$

with the modulus of the Jacobi functions

$$m = -\frac{F(2BF^2 - BD^2 - ADF)}{2(AD - BF)(D^2 - F^2)},$$

and the width of the solitary wave

$$\Lambda = \sqrt{\frac{2DF(D^2 - F^2)}{[DF(A^2 + B^2) - AB(D^2 + F^2)]}},$$

and the traveling speed

$$V = \left[\frac{AB}{3DF} + 2\frac{A^2D^2 + B^2F^2 - 2ABDF}{3(D^2 - F^2)^2}\right]. \tag{7.19}$$

This solution is periodic of period $4\Lambda K(m)$ (Sect. 18.3). The curve γ is a loop if the tangent $\boldsymbol{t}$ is periodic modulo 2π at 0 and L

$$\int_0^L \kappa(s,t)ds = 2\pi I,$$

where I is the rotation index of γ.

For different choices of the parameters A, B, D, F, we can have different types of loops, usually self-intersecting ones. For example, for very small values of F in (7.18) the curvature represents a circle of radius D/A plus a traveling perturbation. In Fig. 7.3, we present a numerical integrated shape of a curve with such a curvature soliton fulfilling the condition of closure between A, B, and D with $F = 0$. In Fig. 7.4, we present the result of a numerical integration of the Fresnel integrals for the curvature given in (7.18) with $F = 0$, but the curve is not closed, and repeats itself with an angular shift at every turn toward a chaotical shape. In the

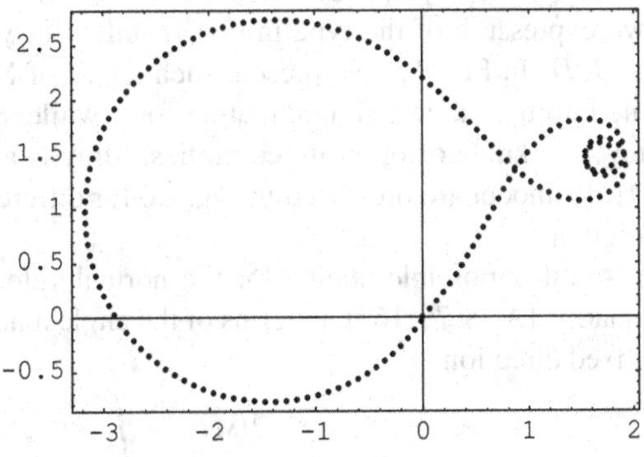

Fig. 7.3 An MKdV soliton solution in curvature generating a closed loop. The frame and ticks represent the $x - y$ system of coordinates

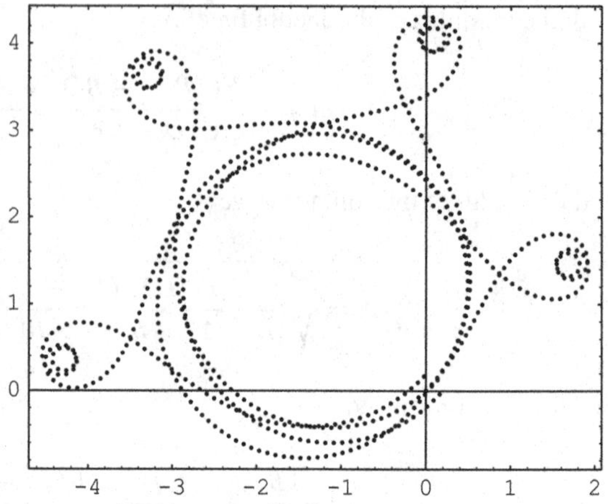

Fig. 7.4 An MKdV soliton solution in curvature where the periodicity (closure) condition is not fulfilled. It generates an open curve. The frame and ticks represent the $x - y$ system of coordinates

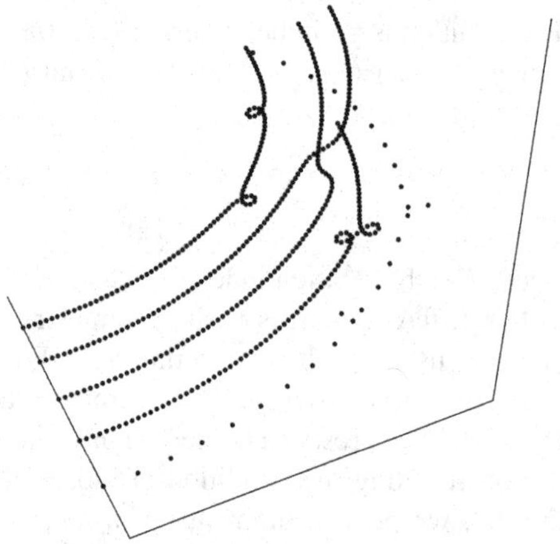

Fig. 7.5 A 3-D view of a curve whose curvature contains a MKdV soliton–antisoliton pair running one against the other

case of a soliton–antisoliton solution [169], when the two bumps far separated in s, i.e., the asymptotic zone, we can approximate the solution with a sum of two expression of the type in (7.18) shifted in s, and having different parameters A, B, D. In Fig. 7.5, we present such a pair of MKdV soliton–antisoliton running one into the other and annihilating for a while. Such a pair is represented by two traveling knots of opposite chiralities. More examples of curves generated by the MKdV model are presented in Fig. 7.6 for different values of the parameters in the solution.

Another possible choice for the normal velocity can lead to the sine–Gordon equation [2,78,79,169], in terms of the angle made by the tangent of the curve with a fixed direction

$$\theta(s,t) = \int^{s} \kappa(s',t)ds'. \tag{7.20}$$

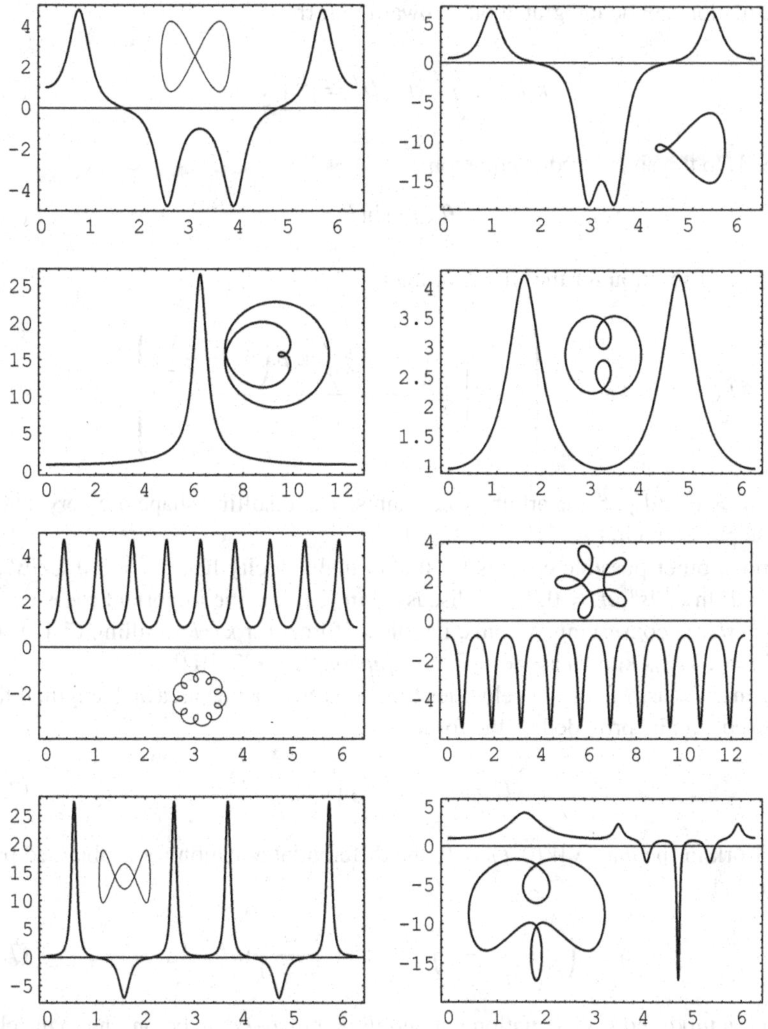

Fig. 7.6 Shapes generated by the solution (7.18) for the MKdV model for curvature, plotted together with the graphics of their curvature vs. s. From left to right and toward downward the curves are called: the first row are lemniscate (or figure-8), then hypocycloid or ratio 1:2, the next three are hypotrochoids of different ratios, then it is a "pretzel" knot, and the last one is a combination between a hypotrochoid and a epicycloid

To obtain the sine–Gordon equation we choose to work in the "gauge" $W(0,t)=0$. The expression of U is given by solving an integrodifferential equation, hence the system models a nonlocal interaction. The condition can be written in the operatorial form

$$\kappa_s = \left(\int^s ds' \left[\left(\frac{\partial^2}{\partial s'^2} + \kappa^2(s') \right) \delta(s-s') + \frac{\partial \kappa}{\partial s}(s')\kappa(s') \right] \right)^2 U(s').$$

This equation can be integrated once toward to form

$$\kappa_{st} + \kappa \int^{s} \kappa \kappa_{s'} ds' = \kappa + C, \tag{7.21}$$

and leads to the sine–Gordon equation

$$\theta_{st} = \sin \theta. \tag{7.22}$$

A typical solution for the sine–Gordon is

$$\theta(s,t) = 4 \arctan \left\{ \gamma \exp\left[\pm \frac{\left(as + \frac{t}{a}\right) \pm \beta \left(as - \frac{t}{a}\right)}{\sqrt{1-\beta^2}} \right] \right\}, \tag{7.23}$$

where $a > 0$ and γ, β are arbitrary constants. The resulting shape are very similar to those presented in Fig. 7.2.

Among other possible choices for the normal velocity like $U = \pm \partial^n \kappa / \partial s^n$, or $U = \pm \partial^n \ln \kappa / \partial s^n$, $n = 0, 1, \ldots$ discussed in [233], some important cases are the perimeter/area conserving systems, i.e., those forms for κ, U fulfilling (7.13). The case $U = -\kappa_{ss}$ is known as the *surface diffusion* flow [37, 102].

A general class of normal velocities functions conserving area and length, if they are closed, can be provided by the forms

$$U = \kappa^p \kappa_s, \quad p \text{ integer}. \tag{7.24}$$

If we work in the gauge $W(0,t) = 0$, the differential equation for κ obtained from (7.10) reads

$$\kappa_t = \left(1 + \frac{1}{p+2}\right) \kappa^{p+2} \kappa_s + \frac{1}{p+1} (\kappa^{p+1})_{sss} \tag{7.25}$$

which is a modified KdV equation with *nonlinear dispersion*, belonging to the class denoted $K(p+2, p+1)$ class of nonlinear PDE with *compacton solutions*. For $p = 0$ we recover the MKdV equation. If we look for standing traveling solutions in $\xi = s - Vt$, (7.25) can be integrated into its *potential picture* form

$$(\kappa_\xi)^2 = -\frac{2(C_3 + V)}{p+2} \kappa^{2-p} - \frac{1}{(p+1)(p+2)} \kappa^2 + C_1 \kappa^{1-p} + C_2 \kappa^{-2p}, \tag{7.26}$$

with C_i being arbitrary constants of integration. The RHS term of this equation can be plotted as a functional of variable κ like in a phase space (Fig. 7.7). This potential picture shows the existence of two valleys, which according to the analysis performed in [86], leads to the existence of two solitary wave solutions in curvature. In terms of the shape of the curve, this can lead to something similar with a double

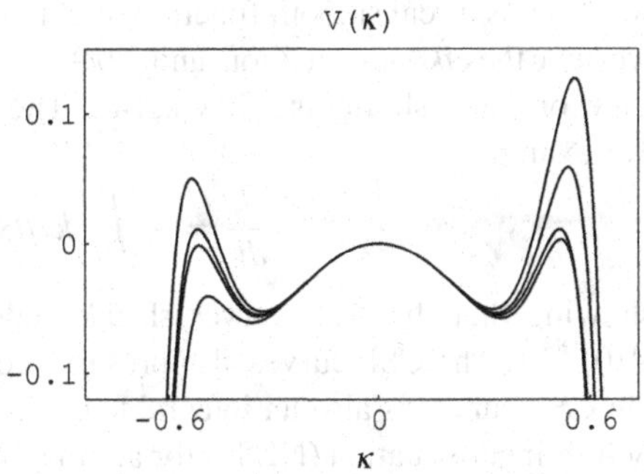

Fig. 7.7 Potential picture for the PDE for κ associated with area and length conservation, $U = \kappa^p \kappa_s$, for several values of p

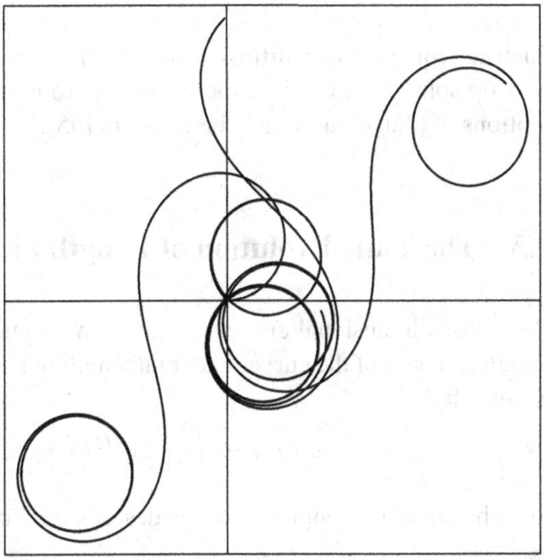

Fig. 7.8 Time evolution of the curve generated by $U = \kappa\kappa_s$ presented in the $x - y$ plane. The two attractors with asymptotically constant curvature correspond to the two valleys in the potential picture in Fig. 7.7. The curve is not closed

spiral. Equation (7.26) is not integrable in general, but for some particular values of p we can find some exact solutions. For example for $p = 4$, $C_{1,2} = 0$ we have

$$\xi = \frac{\sqrt{18\kappa^4 + 5V}\ln[6\kappa^2 + \sqrt{2(18\kappa^4 + 5V)}]}{6\sqrt{2}\kappa\left(-\frac{6\kappa^2}{5} - \frac{V}{3\kappa^2}\right)}, \tag{7.27}$$

see Fig. 7.8. In Fig. 7.8, we present the evolution in time of the curve determined by (7.27) for $p = 1$, which is again an integrable case.

Another interesting example of curves from the MKdV hierarchy is provided by the so-called *curve-shortening* equations [55], i.e., when the normal speed has the form

$$U = f(\kappa), \tag{7.28}$$

where f is a real smooth function of curvature. Examples are f = constant, when we have the *eikonal* equation, and $f(\kappa) = \kappa$ we have the so-called curvature-eikonal flow, or curve-shortening flow (CSF). The CSF curves have interesting properties. For example

$$\frac{dL}{dt} = -\int_0^L k^2 ds, \quad \frac{dA}{dt} = -2\pi, \tag{7.29}$$

showing that the CSF curves shrink under the flow and cease to exist beyond $A(0)/2\pi$. The CSF curves also preserve convexity. The nonlinear PDE fulfilled by the CSF curves is also an integrable evolution system, namely the cubic nonlinear Schrödinger equation (NLS3) for an imaginary time

$$\kappa_t = \kappa_{ss} + \kappa^3. \tag{7.30}$$

Such an equation is a diffusion nonlinear equation with superlinear growth with blowup solutions in finite time. More rigorous results and numerous examples of motions of planar curves can be found in [55].

7.3 The Time Evolution of Length and Area

For a two-dimensional curve we have two geometric global quantities of interest: length and area of the curve. The total length of a moving curve Γ of metrics $g(u,t)$ is given by

$$L(t) = \int_0^{u_{max}} g^{1/2}(u,t)du = \int_0^L ds. \tag{7.31}$$

The change of the length in time is described by its time derivative

$$\frac{dL}{dt} = W(u_{max}, t) - W(0,t) - \int_0^L kUds, \tag{7.32}$$

where W and U are the tangent and normal velocities of the curve, respectively, and k is the curvature. For a loop the time variation becomes simply

$$\frac{dL}{dt} = -\int_0^L kUds = -\oint_0^{\theta_{max}} Ud\theta, \tag{7.33}$$

where $kds = d\theta$ is the turning angle of the tangent. Equations (7.32) and (7.33) represent a conservation law, and the normal velocity is the "flow of length" in the turning angle representation.

For example, an interesting application is the case of "shortening" closed curves [55], i.e., curves where the normal velocity is proportional with some positive power σ of the magnitude of the curvature $U = U(k) = U_0|k|^{\sigma+1}$. Equation (7.32) becomes

$$\frac{dL}{dt} = -\oint_{\Gamma} |k|^{\sigma+1} ds,$$

such that the length of the curve is strictly decreasing in time. Since

$$\int_{\Gamma} |k| ds \geq \oint_{\Gamma} k ds = \oint_{\Gamma} d\theta = 2\pi,$$

by using the Hölder inequality we obtain an upper bound for the negative derivative

$$\frac{dL}{dt} \leq -\frac{(2\pi)^{\sigma+1}}{L^{\sigma}},$$

and by integrating once, we find that there is always $t_0 > 0$ such that

$$t \leq \frac{L^{\sigma+1}(0) - L^{\sigma+1}(t_0)}{(\sigma+1)(2\pi)^{\sigma+1}},$$

meaning we have an upper bound of the life time of the loop, which depends only on the initial length of the curve. Examples of such shortening curves evolution equations are provided by the elastic energy, for example, where $\sigma = 1$. Equation (7.31) is useful for finding the expression of the change of the infinitesimal arc-length $dL = ds = g^{1/2} du$ for a moving curve. During an infinitesimal amount of time δt we have from (7.5)

$$\delta dL = \frac{\partial (dL)}{\partial t} \delta t = \frac{\partial}{\partial t} g^{\frac{1}{2}} \delta t du = \left(\frac{\partial W}{\partial s} - kU \right) \delta t ds, \tag{7.34}$$

which reads

$$\delta dL = -kU ds \delta t + \frac{\partial W}{\partial s} ds \delta t. \tag{7.35}$$

Equations (7.34) and (7.35) provide the variation in time of the infinitesimal arc-length of a moving curve, function of the local velocity of the curve. The same equation can be written just in terms of variations

$$\delta dL = -k \delta u ds + \delta dw,$$

where δu and δw are the normal and tangent displacements of a point on the curve, during its infinitesimal motion. The second term on the RHS of the above equation and (7.35) represent the contribution to the variation of the infinitesimal arc-length due to the stretch or compression of the curve along its local tangent. This term is a total differential, hence for loops this term is zero

$$\delta dL_{loop} = -kU \delta t. \tag{7.36}$$

Although the tangent shift term δdw can be always canceled by using a convenient reparameterization of the curve (see Theorem 22), there are applications where this term plays some role. For example if we choose a finite line segment with a fixed, and having the other end moving in an arbitrary direction with uniform motion, we have a nonuniform extension of the length of the segment, which is described by this tangent term. The first term in the RHS of (7.35) is usually known in hydrodynamics literature in the approximated form

$$-kUds = dL' - dL \simeq \frac{\delta\xi}{R} dL,$$

where $\delta\xi$ is the usual notation (in hydrodynamics books) for the infinitesimal displacement along the normal to the curve, and R is the radius of curvature.

The area associated with a curve is defined as

$$A(t) = -\frac{1}{2}\int_0^L \boldsymbol{r}\cdot\boldsymbol{n}ds = \frac{1}{2}\int_0^L |\boldsymbol{r}\times\boldsymbol{t}|ds, \tag{7.37}$$

where the equality between the two forms is guaranteed by $\boldsymbol{r}\cdot\boldsymbol{n} = \pm|\boldsymbol{r}\times\boldsymbol{n}|$. This equation emerges from the integration of the area $dA = \boldsymbol{r}\cdot\boldsymbol{n}ds/2$ of an elementary triangle generated by the infinitesimal arc-length ds and the two position vectors from the origin of the coordinate system toward the ends of this infinitesimal arc. The signs in front of the area expressions are related to a certain convention of running the curve. For an arc covered CCW the area is considered positive. In the case of a loop, (7.37) provides the area inside the loop. For an open curve, expression (7.37) provides the sum of the areas of the surfaces bounded by the curve from 0 to L, and the two lines drawn from the origin of the coordinate system to these two ends. These two lines may cross the curve many times, and the corresponding areas are taken with plus or minus accordingly to the resulting sign according to the sign convention stated above. In the case of an open curve, if we change the origin of the coordinate system $O \to O'$ by a translation $OO' = \boldsymbol{R}$, the area in (7.37) changes in an additive way. If we have a curve lying from O to some point L, its area measured from O' reads

$$A'(OL) = -\frac{1}{2}\int_0^L \boldsymbol{r}'\cdot\boldsymbol{n}ds = A(OL) + \frac{\boldsymbol{R}}{2}\int_0^L \boldsymbol{n}ds,$$

where $\boldsymbol{r}' = \boldsymbol{r} - \boldsymbol{R}$. From the definition of the turning angle of the tangent $d\theta = kds$, and the theorem of derivation of implicit functions, we have

$$\boldsymbol{n}ds = \frac{\frac{d\boldsymbol{t}}{ds}}{k}ds = \frac{\frac{d\boldsymbol{t}}{ds}}{\frac{d\theta}{ds}}ds = \frac{d\boldsymbol{t}}{d\theta}ds = -\boldsymbol{r}ds,$$

and consequently

$$A'(OL) = A(OL) + A_{\Delta OLO'}.$$

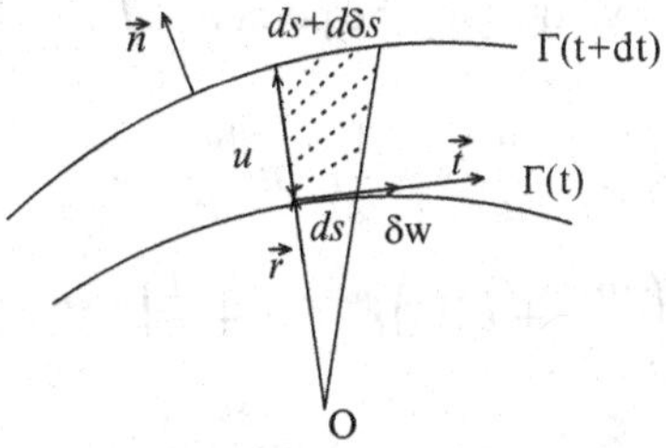

Fig. 7.9 The infinitesimal arc-length ds on the curve Γ in $\boldsymbol{r}$ at moment t, transforms into the new infinitesimal arc-length $ds + \delta ds$ on the moved curved at $t + dt$. The infinitesimal displacement $d\boldsymbol{r}$ can be projected onto the normal and tangent to the curve $d\boldsymbol{r} = \delta u\boldsymbol{n} + \delta w\boldsymbol{t}$. The *dashed area* represents δdA

Fig. 7.10 The infinitesimal displacement of the curve Γ with $\delta\boldsymbol{r}$

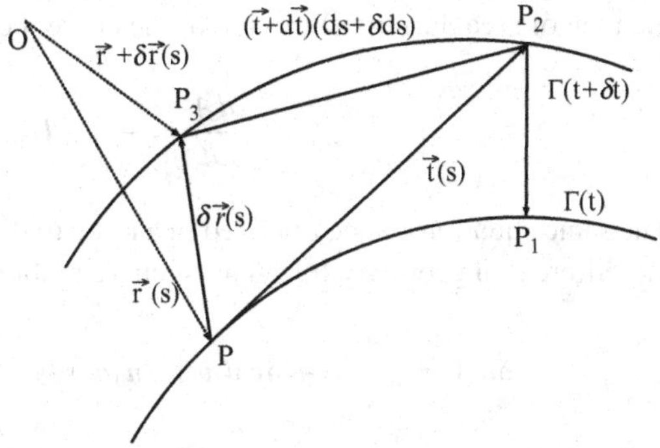

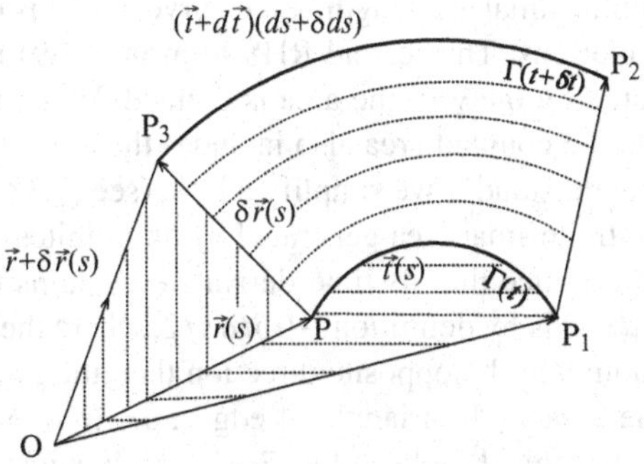

Fig. 7.11 The infinitesimal variation of the elementary swept area of a moving curve Γ. *Horizontal dashed area* is the initial elementary area of Γ (OP_1P), and the total area of the figure $(OP_1P_2P_3)$ is the elementary area of the *shifted curve* $\Gamma(t + \delta t)$. The swept area $(PP_1P_2P_3)$ is *dashed with curved lines*, and the residual area (OPP_3) is *dashed with vertical lines*

Now we provide the equation for the variation of the infinitesimal area δdA for a curve in motion. In other words, this is the infinitesimal area swaped by an infinitesimal arc-length during an infinitesimal interval of time of curve motion (Figs. 7.9–7.11).

If V represents the total velocity of the point $\boldsymbol{r} \in \Gamma$, we have from (6.37)

$$\begin{aligned}\frac{\partial dA}{\partial t} &= -\frac{1}{2}\boldsymbol{V}\cdot\boldsymbol{n}ds - \frac{1}{2}\boldsymbol{r}\cdot\frac{\partial \boldsymbol{n}}{\partial t}ds - \frac{1}{2}\boldsymbol{r}\cdot\boldsymbol{n}\frac{\partial g^{\frac{1}{2}}}{\partial t}du \\ &= -\frac{1}{2}Uds + \frac{1}{2}\left(\frac{\partial U}{\partial s} + kW\right)\boldsymbol{r}\cdot\boldsymbol{t}ds + \frac{1}{2}\left(kU - \frac{\partial W}{\partial s}\right)\boldsymbol{r}\cdot\boldsymbol{n}ds. \qquad (7.38)\end{aligned}$$

After regrouping the terms we can write (7.38) in the form

$$\delta dA = -Uds\delta t + \frac{1}{2}\frac{\partial}{\partial s}\Big(\boldsymbol{r}\cdot(U\boldsymbol{t} - W\boldsymbol{n})\Big)ds\delta t, \qquad (7.39)$$

which represents the infinitesimal variation in time (δt) of the infinitesimal (d) element of area during the motion of the curve. For example, for a loop we have

$$\frac{dA}{dt} = -\int_0^L Uds. \qquad (7.40)$$

The same equation can be obtained in a more traditional way (without the help from the differential geometry formulas of curve motion)

$$\delta dA = -\frac{1}{2}(\boldsymbol{r} + \delta\boldsymbol{r})(\boldsymbol{n} + \delta\boldsymbol{n})d(s + \delta s) + \frac{1}{2}\boldsymbol{r}\cdot\boldsymbol{n}ds + \cdots.$$

Let us discuss the two terms on the RHS of (7.39). The first term is the real swept area of the moving infinitesimal arc-length of the curve, and this is the term we need in the following calculations. The second RHS term in (7.39) is just a "residual" area, and it is originated by the way the area is defined. When the position vector sweeps the arc-length, the counted area also includes the area swept by this vector itself. This is easy to understand if we simplify a little (see (7.38) and (7.39)). From the definition of the infinitesimal area generated by an infinitesimal arc-length, the integrand in (7.37), we notice that its time derivative contains three terms (7.38). The first one, $-\delta\boldsymbol{r}\cdot\boldsymbol{n}ds/2$ is by definition $-U\delta t ds/2$, where the minus sign occurs because the normal points in the opposite direction than the normal motion of the curve. This part is the area of the triangle of edges: $\delta\boldsymbol{r}, (\boldsymbol{t} + \delta\boldsymbol{t})(ds + \delta ds)$, and $\delta\boldsymbol{r} + \boldsymbol{t}ds$. That is the triangle PP_2P_3 in Fig. 7.10, which is just half of the needed swept area:

$$\delta dA_{PP_2P_3} = -\frac{1}{2}\delta\boldsymbol{r}(s)\times(\boldsymbol{t} + \delta\boldsymbol{t})(ds + \delta ds).$$

Because of the recursion-like relations (7.7) and (7.8), which describe the time variation of the tangent and the normal, the last two terms in the RHS of (7.38) mix together and produce the other half of the swept area (PP_1P_2 in Fig. 7.10)

$$\delta dA_{PP_1P_2} = -\frac{1}{2}\delta\boldsymbol{r}(s + \delta s)\times(\boldsymbol{t}(s)ds$$

and a total differential

$$\delta dA_{residual} = \frac{1}{2}\frac{\partial}{\partial t} A_{OPP_3} ds \delta t.$$

We can illustrate this even better in Fig. 7.11. In this figure, the initial infinitesimal area $dA(t) = dA_{OP_1P}$ is presented dashed with horizontal lines. The new infinitesimal area $DA(t+\delta t) = dA_{OP_1P_2P_3}$ is the total area presented in this figure. The variation of the infinitesimal area, $\delta dA = dA(t+\delta t) - dA(t)$ is of course the area dashed with vertical and curved lines, $A_{OPP_1P_2P_3}$. The portion dashed with curved lines is the correct one, the swept area in the curve motion, given by the first term in the RHS of (7.38) and (7.39). The area dashed with vertical lines is the so-called "residual" one.

Let us test (7.39), and this last comment, with two examples. If we choose a straight segment along Ox, moving upward along the Oy axis, $x \in [0, L], y = Ut$, we can figure out that the swept area has to be $U\delta t ds$. On the other hand, (7.39) provides $\delta dA = -U\delta t ds + (1/2)t \cdot U t \delta t ds = (1/2)U\delta t ds$ just half of the swept area. This happens because it took into account the "residual" term. If we take into account only the first term in its RHS, we obtain the correct result. Another example can be given by a unit segment rotating with its origin fixed in O, with angular speed ω. Equation (7.39) provides

$$\delta dA = -\omega s \delta t ds + \frac{1}{2}\frac{\partial}{\partial s}[s(\cos\omega t, \sin\omega t)\cdot(\cos\omega t, \sin\omega t)\omega s - 0]\delta t ds = 0.$$

Again wrong, since the real swept area by this rotating segment is actually $\omega s \delta t ds$. If we retain just the first term in the equation, we obtain the correct result. The zero result of the total infinitesimal area is produced by the fact that at any moment of time, the area of this curve is actually zero (it is a straight segment). So its area is constant, so its total time derivative is of course zero.

For the sake of completeness we present here another approach the infinitesimal variation of the area swept by a moving curve, namely the variational approach. We have

$$L(t+\delta t) = \int_0^{u_{max}} g^{\frac{1}{2}} du = \int_0^{u_{max}} \sqrt{\frac{\partial \mathbf{r} + \delta \mathbf{r}}{\partial u} \cdot \frac{\partial \mathbf{r} + \delta \mathbf{r}}{\partial u}} du$$
$$= \left(\frac{\partial(x_i + \delta x_i)}{\partial u}\frac{\partial(x_i + \delta x_i)}{\partial u}\right)^{\frac{1}{2}}. \tag{7.41}$$

By variational calculation we obtain

$$\delta L = -\int_0^L k \delta x_{normal} ds + \delta W \Big|_0^L, \tag{7.42}$$

which is in agreement with (7.32).

7.4 Cartan Theory of Three-Dimensional Curve Motion

A moving parameterized curve $\gamma(t) \subset \mathbb{R}^3$, which can be described at any moment of time by the Serret-Frenet frames, generates a set of points Σ_γ. Any parameterized surface $\Sigma \subset \mathbb{R}^3$ can be described by its tangent bundle $T\Sigma$, but we need a more sophisticated vector bundle to describe the hypothetical surface obtained through the curve motion than the available tangent bundle $T\Sigma_\gamma$. Moreover, in order to approach a moving curve as a regular surface some restrictions should apply to this motion. The curve should not self-intersect during the motion in order to have fulfilled the immersion condition for a regular surface. The time dependence of the position of any point on the curve should be a differentiable function, which requests some extra structure relations (or compatibility equations) between the mixed time and arc-length second order derivatives. In conclusion, the surface obtained by the motion of the curve has to fulfil some extra constraints.

In order to define the differentiable motion of a curve in arbitrary direction, like for example along $\{\boldsymbol{t}(t), \boldsymbol{n}(t), \boldsymbol{b}(t)\}$, we have to define vector fields along the curve that do not belong only to the tangent space of the curve $T\gamma$. However, it would be simpler if we could describe such vector fields in the moving Serret-Frenet frames. For that we have to immerse the local Serret-Frenet frames in the frame bundle for the affine space $\mathbb{R}^3$.

The immersion can be obtained by mapping different vector bundles over orthogonal groups $O(n, \mathbb{R})$ into vector sub-bundles over orthogonal subgroups, correspondingly. Then, the homomorphisms between different orthogonal groups provide the requested mappings between the frame bundles. If such mappings are constructed, by using their pull-backs, the covariant derivative in $\mathbb{R}^3$ induces a covariant derivative in the curve. This allows us to define vertical and horizontal vector spaces for the vector bundle of the frames along the curve. Consequently we can identify "orthogonal" spaces to the curve, and the vectors in these spaces will provide the local directions of motion of the curve.

The imbedded parameterized curve γ is a Riemannian sub-manifold of $\mathbb{R}^3$, and it has a natural Riemannian connection defined on it. Let $x \in \gamma$ and we have the vector subspace relation $T_x\gamma \lhd T_x\mathbb{R}^3$. We denote by $(T_x\mathbb{R}^3)^\perp$ the orthogonal complement of $T_x\gamma$ in $T_x\mathbb{R}^3$ which is called the normal space to the immersion γ at x. We can build the following two orthogonal frame bundles, and when we denote them we skip from the notation the structure groups, which obviously are the corresponding orthogonal groups. We have $\mathrm{OF}(\gamma)$ over γ with canonical projection π', and $\mathrm{OF}(\mathbb{R}^3)$ over $\mathbb{R}^3$ with canonical projection π. Also, we can factorize $\mathrm{OF}(\mathbb{R}^3)/\gamma = \{v \in \mathrm{OF}(\mathbb{R}^3) | \pi(v) \in \gamma\}$ which is a principal bundle of orthonormal frames over γ with symmetry group the orthogonal real Lie group $O(3, \mathbb{R})$.

We define the bundle of *adapted* frames $\mathrm{OF}(\mathbb{R}^3, \gamma)$ over γ with symmetry group $O(2, \mathbb{R}) \times O(1, \mathbb{R})$. This is actually a sub-bundle of $\mathrm{OF}(\mathbb{R}^3)/\gamma$ obtained through the map i (see the diagram in (7.43)) in a natural way: it contains the frames over $\mathbb{R}^3$ which are also frames over the curve, and have one axis along the tangent to the curve. The $O(2, \mathbb{R})$ part in the symmetry group takes care of the possible rotations of this frames around the curve tangent, while the $O(1,\mathbb{R}) = \{1,-1\}$ part describes the two possible chiralities along the curve. Mapping 3-dimensional vectors along the curve, and in the normal plane induces two orthogonal Lie groups natural homomorphisms $h' : O(2,\mathbb{R}) \times O(1,\mathbb{R}) \to O(1,\mathbb{R})$ and $h'' : O(1,\mathbb{R}) \times O(2,\mathbb{R}) \to O(2,\mathbb{R})$, which induce on their own two corresponding fiber bundles homomorphisms which we denoted with same letters, see Fig. 7.43.

$$\begin{array}{ccccc}
 & & \mathrm{OF}(\mathbb{R}^3) & \xrightarrow[O(3,\mathbb{R})]{\pi} & \mathbb{R}^3 \\
 & & \uparrow j & & \\
 & & \mathrm{OF}(\mathbb{R}^3)/\gamma & \xrightarrow[O(3,\mathbb{R})]{\pi} & \gamma \\
 & & \uparrow i & & \\
\mathrm{OF}(\gamma) = \mathrm{OF}(\mathbb{R}^3,\gamma)/O(2,\mathbb{R}) & \xleftarrow[h']{} & \mathrm{OF}(\mathbb{R}^3,\gamma) & \xrightarrow[h'']{} & \mathrm{OF}(\mathbb{R}^3,\gamma)/O(1,\mathbb{R}) \\
O(1,\mathbb{R}) \Big\downarrow \pi' & & O(1,\mathbb{R})\times O(2,\mathbb{R}) \Big\downarrow \pi & & O(2,\mathbb{R}) \Big\downarrow \pi'' \\
\gamma & & \gamma & & \gamma
\end{array} \tag{7.43}$$

Finally, we become even more abstract and construct the vector normal bundle of γ as $T(\gamma)^{\perp} = \bigcup_{x\in\gamma}(T_x\gamma)^{\perp}$ associated to the bundle of normal frames, with standard fibre $\mathbb{R}^2$ and group $O(2,\mathbb{R})$. If we denote by Γ_3 the Riemannian connection form on $\mathrm{OF}(\mathbb{R}^3)$ then the composite pull-back $i^* j^* \Gamma_3$ is the connection form in $\mathrm{OF}(\mathbb{R}^3,\gamma)$. Geometrically this connection form defines parallel displacement of the normal space $T_x\gamma^{\perp}$ onto the normal space $T_y\gamma^{\perp}$ along the curve γ.

In the following we express the covariant derivative for the curve. We denote the directional and covariant derivatives in $\mathbb{R}^3$ along $\boldsymbol{v} \in T\mathbb{R}^3$ by $D_{\boldsymbol{v}} = \nabla_{\boldsymbol{v}}$, and we assign a basis $\{\boldsymbol{e}_i\}$ in $T\mathbb{R}^3$. We need the expression of the covariant derivative $\nabla_i = \nabla_{\boldsymbol{e}_i}$ from (4.43). For imbedded manifolds the connection Γ simply becomes the second fundamental form define on the submanifold ([158] Chap. VII, [299] pp. 64, or [46] Sect. 4-4) and the result is called Gauss' formula if $\boldsymbol{V}$ belongs to the tangent space, or Weingarten's formula if V belongs to the normal space, respectively

$$\nabla_{\boldsymbol{e}_i}\boldsymbol{V} = D_{\boldsymbol{e}_i}\boldsymbol{V} - \boldsymbol{\Pi}(\boldsymbol{e}_i, \boldsymbol{V}).$$

The vector $\boldsymbol{\Pi}$ is the vertical component of the directional derivative, usually called the *second fundamental* form. It is defined on X with values in the vertical space. We remember that if X is a surface with unit normal $\boldsymbol{n}$ we have $\boldsymbol{\Pi} = \Pi\boldsymbol{n}$, definition 54. For any two vector fields $\boldsymbol{v}, \boldsymbol{w} \in T\gamma$ we define the covariant derivative associated to the (natural) Riemannian connection of γ at a point $x \in \gamma$, (4.43)

$$(\nabla_{\boldsymbol{v}}\boldsymbol{w})_x = (D_{\boldsymbol{v}}\boldsymbol{w})_x - \boldsymbol{\Pi}_x(\boldsymbol{v}, \boldsymbol{w}) \in T_x\gamma. \tag{7.44}$$

Here $\boldsymbol{\Pi}_x(\boldsymbol{v},\boldsymbol{w}) \in T_x\gamma^{\perp}$ is the *second fundamental form* (see definition 54) of γ at x, i.e. a symmetric bilinear differential form with values in the normal space to γ. The vector second fundamental form $\boldsymbol{\Pi}$ allows us to define directional derivatives along the normal space to γ at points on γ.

In the following we give a practical example, in coordinates. We know we can always choose two differential orthonormal fields of vectors ξ_1, ξ_2 (i.e. two sections) of the normal bundle $T\gamma^{\perp}$. Let us also choose $x_0 \in \gamma$ and note that it is always possible to choose an adapted orthogonal frame with a system of normal coordinates $\{y^1, y^2, y^3\}$ with origin in x_0 such that $(\partial/\partial y^1)_{x_0}$ spans $T_{x_0}\gamma$ and $\{\xi_1 = (\partial/\partial y^2)_{x_0}, \xi_2 = (\partial/\partial y^3)_{x_0}\}$ spans $T_{x_0}\gamma^{\perp}$. Let s be the arc-length in a neighborhood $U(x_0) \subset \gamma$ and let $y^i = y^i(s)$ be the equations describing the imbedding of U into $\mathbb{R}^3$. We have the action of the second fundamental form Π on tangent vectors of γ given by

$$\Pi\left(\frac{\partial}{\partial s}\bigg|_{x_0}, \frac{\partial}{\partial s}\bigg|_{x_0}\right) = \left(\frac{\partial^2 y^1}{\partial s^2}\right)_{x_0}\frac{\partial}{\partial y^1} + \left(\frac{\partial^2 y^2}{\partial s^2}\right)_{x_0}\frac{\partial}{\partial y^2}. \tag{7.45}$$

The proof is simple and it is based on direct calculation of the Hessian of transformation from x to y coordinates, and on the fact that the Christoffel symbols for the Riemannian connection in $\mathbb{R}^3$ are zero (see for example second volume of [158], Chap. VII). It is easy to check that (7.45) includes the Serret-Frenet relations (5.3), namely (7.45) represents $\boldsymbol{\Pi}(\boldsymbol{t},\boldsymbol{t}) = \kappa\boldsymbol{n}$. Let us choose $y^1 = s, y^2 = -\boldsymbol{r}(s_0)\cdot\boldsymbol{n}(s_0)$, and $y^3 = \boldsymbol{r}(s_0)\cdot\boldsymbol{b}(s_0)$. We have

$$\frac{\partial^2 y^2}{\partial s^2}\bigg|_{s_0} = -\frac{\partial}{\partial s}(\boldsymbol{r}_s\cdot\boldsymbol{n}^{\cdot} + \boldsymbol{r}\cdot\boldsymbol{n}_s)_{s_0} = (\tau_s y^3 + \tau^2 y^2 - \kappa_s y^1 + \kappa + \kappa^2 y^2)_{s_0} = \kappa,$$

and in the same way $\partial^2 y^3/\partial s^2 = 0$ at s_0 which proves the affirmation.

7.5 Kinematics of Three-Dimensional Curves

In the following we relate the general frame bundle formalism developed in Sect. 7.4 to three-dimensional curve motions in space. On each point of arc-length coordinate s along the parameterized curve γ we define the adapted (orthonormal) Serret-Frenet frame $\{\boldsymbol{e}_i\}_{i=1,2,3} = \{\boldsymbol{t},\boldsymbol{n},\boldsymbol{b}\}$ of vectors in the principal bundle $\mathrm{OF}(\mathbb{R}^3,\gamma)$ over γ, (7.43). Let be (s,n,b) the local coordinates in this frames, and $(s,n,b,\alpha_1,\alpha_2,\alpha_3)$ local coordinates in the principal bundle, where α_i represent the three angles of frame rotations in $O(3,\mathbb{R})$. The canonical 1-form has the generic expression

$$\theta = \boldsymbol{\theta}_1 ds + \boldsymbol{\theta}_2 dn + \boldsymbol{\theta}_3 db + \sum_{i=1}^{3}\boldsymbol{\theta}_i d\alpha_i,$$

and its action on tangent vectors from the principal bundle is given by (4.33), (4.39) in the form

$$d\boldsymbol{r} = (\theta^i; \boldsymbol{X})\boldsymbol{e}_i = W\boldsymbol{t} + U\boldsymbol{n} + B\boldsymbol{b}, \tag{7.46}$$

with W, U, B arbitrary 1-form coefficients. When we consider the time motion of the curve these coefficients become the pull-back 1-forms of a cross-section in the principal bundle determined by γ. Namely, they are the coefficients of the velocity of the curve in the local Serret-Frenet frames

$$d\boldsymbol{r} = \boldsymbol{V}(s,t)dt = \frac{\partial \boldsymbol{r}}{\partial t}dt = (W\,dt)\boldsymbol{t} + (U\,dt)\boldsymbol{n} + (B\,dt)\boldsymbol{b},$$

according to the definition of curve velocity introduced, for example, in [109, 168, 172, 233, 289, 289], which is basically the same definition used in Sect. 7.1. Let us denote by Γ_{ij}^k the Christoffel symbols associated with the connection defined on this principal bundle. We determine them by using (7.44) and (4.42)

$$D_t\boldsymbol{t} = \kappa\boldsymbol{n} \to \nabla_1\boldsymbol{e}_1 = D_1\boldsymbol{e}_1 - \boldsymbol{\Pi}(\boldsymbol{e}_1,\boldsymbol{e}_1) = 0, \text{ so } \Gamma_{11}^1 = 0,$$
$$D_t\boldsymbol{n} = -\kappa\boldsymbol{t} + \tau\boldsymbol{b} \to \nabla_1\boldsymbol{e}_2 = D_1\boldsymbol{e}_2 - \boldsymbol{\Pi}(\boldsymbol{e}_1,\boldsymbol{e}_2) = -\kappa\boldsymbol{e}_1, \text{ so } \Gamma_{12}^1 = -\kappa,$$
$$\ldots$$
$$\begin{aligned} D_b\boldsymbol{b} &= -\boldsymbol{b}\cdot\frac{\partial \boldsymbol{t}}{\partial b}\boldsymbol{t} - \boldsymbol{b}\cdot\frac{\partial \boldsymbol{n}}{\partial b}\boldsymbol{n} \to \nabla_3\boldsymbol{e}_3 = D_3\boldsymbol{e}_3 - \boldsymbol{\Pi}(\boldsymbol{e}_3,\boldsymbol{e}_3) \\ &= -\boldsymbol{b}\cdot\frac{\partial \boldsymbol{t}}{\partial b}\boldsymbol{e}_1, \text{so } \Gamma_{33}^1 = -\boldsymbol{b}\cdot\frac{\partial \boldsymbol{t}}{\partial b}, \end{aligned} \tag{7.47}$$

and so on. In order to obtain the connection form, in addition to the Christoffel symbols, we need the transformations of the orthonormal adapted frames in the bundle of frames, (4.41). in the form of three 2×2 rotation matrices $\hat{R}$ as 1-parameter Lie subgroups of $O(2,\mathbb{R})$

$$\frac{\partial \boldsymbol{e}_i}{\partial x^q} = \hat{R}_q^{ij}\boldsymbol{e}_j$$

with $i = 2, 3$, $q = 1, 2, 3$ and $x^1 = s, x^2 = n, x^3 = b$. For $q = 1$ we have obviously

$$\hat{R}_1 = \begin{pmatrix} 0 & \tau \\ -\tau & 0 \end{pmatrix}.$$

By applying the structure conditions (4.37) in the (4.38) form, we obtain the relations describing the change of frames along the local frame directions, that is the Gauss-Weingarten (4.39), in the form

$$d\boldsymbol{e}_i = \omega_q^{ij}\,dx^q\boldsymbol{e}_j. \tag{7.48}$$

There is a simple curvilinear coordinates-like language in which the connection form coefficients have an intuitive form [289]

$$\frac{\partial}{\partial n}\begin{pmatrix} \boldsymbol{t} \\ \boldsymbol{n} \\ \boldsymbol{b} \end{pmatrix} = \begin{pmatrix} 0 & -\Gamma_{22}^1 & -\Gamma_{23}^1 \\ \Gamma_{22}^1 & 0 & \boldsymbol{b}\cdot\frac{\partial \boldsymbol{n}}{\partial n} \\ \Gamma_{23}^1 & -\boldsymbol{b}\cdot\frac{\partial \boldsymbol{n}}{\partial n} & 0 \end{pmatrix}\begin{pmatrix} \boldsymbol{t} \\ \boldsymbol{n} \\ \boldsymbol{b} \end{pmatrix}, \tag{7.49}$$

$$\frac{\partial}{\partial b}\begin{pmatrix} \boldsymbol{t} \\ \boldsymbol{n} \\ \boldsymbol{b} \end{pmatrix} = \begin{pmatrix} 0 & -\Gamma_{32}^2 & -\Gamma_{33}^1 \\ \Gamma_{32}^2 & 0 & \boldsymbol{b}\cdot\frac{\partial \boldsymbol{n}}{\partial b} \\ \Gamma_{33}^1 & -\boldsymbol{b}\cdot\frac{\partial \boldsymbol{n}}{\partial b} & 0 \end{pmatrix}\begin{pmatrix} \boldsymbol{t} \\ \boldsymbol{n} \\ \boldsymbol{b} \end{pmatrix}, \tag{7.50}$$

and of course the derivatives with respect of s are the Serret-Frenet relations (5.5). Moreover, by defining the vector field

$$\boldsymbol{X} = \boldsymbol{t}\frac{\partial}{\partial s} + \boldsymbol{n}\frac{\partial}{\partial n} + \boldsymbol{b}\frac{\partial}{\partial b} \in T\mathrm{OF}(\mathbb{R}^3, \gamma), \tag{7.51}$$

we can construct the other curvilinear differential operators like the curvilinear divergence of the tangent

$$\nabla\cdot\boldsymbol{t} = \boldsymbol{n}\cdot\frac{\partial \boldsymbol{t}}{\partial n} + \boldsymbol{b}\cdot\frac{\partial \boldsymbol{t}}{\partial b}, \tag{7.52}$$

where we used $\boldsymbol{t}\cdot\partial\boldsymbol{t}/\partial s = 0$

$$\nabla\cdot\boldsymbol{n} = -\kappa + \boldsymbol{b}\cdot\frac{\partial \boldsymbol{n}}{\partial b}, \quad \nabla\cdot\boldsymbol{b} = -\boldsymbol{b}\cdot\frac{\partial \boldsymbol{n}}{\partial n}.$$

The curvilinear curl has the form

$$\begin{aligned} \nabla\times\boldsymbol{t} &= \boldsymbol{t}\times\frac{\partial \boldsymbol{t}}{\partial s} + \boldsymbol{n}\times\frac{\partial \boldsymbol{t}}{\partial n} + \boldsymbol{b}\times\frac{\partial \boldsymbol{t}}{\partial b} \\ &= \kappa\boldsymbol{b} + \boldsymbol{n}\times\left(\frac{\partial \boldsymbol{t}}{\partial n}\cdot\boldsymbol{b}\right)\boldsymbol{b} + \boldsymbol{b}\times\left(\frac{\partial \boldsymbol{t}}{\partial b}\cdot\boldsymbol{n}\right)\boldsymbol{n} = \kappa\boldsymbol{b} + \Omega_s\boldsymbol{t}, \end{aligned}$$

where $\Omega_s = \boldsymbol{t}\cdot(\nabla\times\boldsymbol{t})$ is called the total moment of the $\boldsymbol{t}$ field or abnormality. Similarly we have

$$\begin{aligned} \nabla\times\boldsymbol{n} &= -(\nabla\cdot\boldsymbol{b})\boldsymbol{t} + \Omega_n\boldsymbol{n} - \Gamma_{22}^1\boldsymbol{b}, \\ \nabla\times\boldsymbol{b} &= (\kappa + \nabla\cdot\boldsymbol{n})\boldsymbol{t} + \Gamma_{33}^1\boldsymbol{n} + \Omega_b\boldsymbol{b}, \end{aligned}$$

with $\Omega_n = \Gamma_{32}^2 - \tau$, $\Omega_b = -\Gamma_{23}^1 - \tau$ being the other two abnormalities.

It is interesting to mention a relation between the three rotational abnormalities

$$\Omega_s - \tau = \frac{1}{2}(\Omega_s + \Omega_n + \Omega_b).$$

According to [289] this relation is a consequence of the Dupin's theorem (i.e. the intersections of surfaces of orthogonal curvilinear coordinates are lines of curvature). Expressing the motion of three-dimensional curves through the abnormalities forms has the advantage of classification of motions in three categories, function of which abnormality we choose to keep zero. For example, the very well known *binormal motion* happens when the normal abnormality vanishes, $\Omega_n = 0$. This is typical vortex filament motion, and it will be studied in more detail in Chap. 15. In the binormal motion the s–lines and b–lines are contained in a one-parameter surface U =constant, perpendicular on $\boldsymbol{n} = \nabla U/|\nabla U|$. Consequently, the normal field is quasi-potential (is derived as the product between a scalar function and a gradient). All equations and forms of the surface generated by a binormal motion can be easy calculated. For example, following the Weatherburn theorem ([338] XII, 121) $K = \boldsymbol{N} \cdot \mathrm{curl}_U \boldsymbol{t} \times \mathrm{curl}_U \boldsymbol{b}$, we have the Gaussian and mean curvature in the form

$$K = -\kappa(\kappa + \nabla \cdot \boldsymbol{n}) - \tau^2, \; H = \nabla \cdot \boldsymbol{n},$$

respectively, while the Gauss-Codazzi equations and Gauss' Theorema Egregium are encapsulated in a very simple expression

$$K = \frac{\partial \Gamma^1_{33}}{\partial s} + (\Gamma^1_{33})^2.$$

In the case when the b parameter can be considered time (the so-called pure binormal motions) it results that $\boldsymbol{r}_b = \boldsymbol{r}_t = g^{1/2}\boldsymbol{b}$ and, most importantly, $s_t = 0$ which draws a spectacular conclusion: *pure binormal motions are possible only for inextensible curves.* This could be the geometrical insight of the strong stability of vortex filaments having this type of motion.

From the structure equations for the connection form $d\omega = -\omega \wedge \omega + \Omega$ we obtain the expression of the curve motion in time, as function of the velocity. It is easy to note that $\partial b/\partial t = B, \partial n/\partial t = U$ and we have

$$\frac{\partial}{\partial s}\begin{pmatrix} W \\ U \\ B \end{pmatrix} = \begin{pmatrix} 0 & \kappa & 0 \\ -\kappa & -\Gamma^1_{22} & -\Gamma^2_{32} + \tau \\ 0 & -\Gamma^1_{23} - \tau & -\Gamma^1_{33} \end{pmatrix}\begin{pmatrix} W \\ U \\ B \end{pmatrix} + \begin{pmatrix} \frac{\dot{g}}{2g} \\ 0 \\ 0 \end{pmatrix},$$

where we note that the change in time of the arc-length accounts for a non-zero curvature of the connection. We can re-write (7.49–7.50) in terms of the components of the velocity

$$\frac{d\boldsymbol{t}}{dt} = \left(\frac{\partial U}{\partial s} - \tau B + \kappa W\right)\boldsymbol{n} + \left(\frac{\partial B}{\partial s} + \tau U\right)\boldsymbol{b},$$

$$\frac{d\boldsymbol{n}}{dt} = -\left(\frac{\partial U}{\partial s} - \tau B + \kappa W\right)\boldsymbol{t} + \left[\frac{1}{\kappa}\frac{\partial}{\partial s}\left(\frac{\partial B}{\partial s} + \tau U\right) + \frac{\tau}{\kappa}\left(\frac{\partial U}{\partial s} - \tau B + \kappa W\right)\right]\boldsymbol{b},$$

$$\frac{d\boldsymbol{b}}{dt} = -\left(\frac{\partial B}{\partial s} + \tau U\right)\boldsymbol{t} - \left[\frac{1}{\kappa}\frac{\partial}{\partial s}\left(\frac{\partial B}{\partial s} + \tau U\right) + \frac{\tau}{\kappa}\left(\frac{\partial U}{\partial s} - \tau B + \kappa W\right)\right]\boldsymbol{n},$$

$$\frac{dg}{dt} = 2g\left(\frac{\partial W}{\partial s} - \kappa U\right). \tag{7.53}$$

The total (material) time derivative can be broken into the partial derivative and an extra term

$$\frac{d}{dt} = \frac{\partial}{\partial t} + \left(W - \int^s \kappa U ds'\right)\frac{\partial}{\partial s}.$$

From the above relations we can derive the dynamical connections between the velocity components and curvature and torsion of γ

$$\frac{\partial \kappa}{\partial t} = \frac{\partial^2 U}{\partial s^2} + (\kappa^2 - \tau^2)U + \frac{\partial \kappa}{\partial s}\int^s \kappa U ds' - 2\tau\frac{\partial B}{\partial s} - B\frac{\partial \tau}{\partial s}, \tag{7.54}$$

$$\frac{\partial \tau}{\partial t} = \frac{\partial}{\partial s}\left[\frac{1}{\kappa}\frac{\partial}{\partial s}\left(\frac{\partial B}{\partial s} + \tau U\right) + \frac{\tau}{\kappa}\left(\frac{\partial U}{\partial s} - \tau B\right) + \tau\int^s \kappa U ds'\right] + \kappa\tau U + \kappa\frac{\partial B}{\partial s}. \tag{7.55}$$

On behalf of the fundamental theorem of curves (Theorem 10), once we integrate (7.54) and (7.55) and find κ, τ the curve is uniquely determined in the arc-length parametrization, up to rigid motions in space. Obviously, as a check, if we cancel the torsion we obtain the equations of motion for the two-dimensional curves.

7.6 Mapping Three-Dimensional Curve Motion into Nonlinear Integrable Systems

Like in the case of motion of two-dimensional curves, there are integrable three-dimensional motions in direct relation with integrable evolution equations (in this case it will be the cubic nonlinear Schrödinger (NLS) hierarchy), and also nonintegrable motions.

In order to map the three-dimensional curve motion into a nonlinear integrable system we follow [123, 168], as well as an older suggestion of Darboux, and we introduce the complex curvature–torsion function by the *Hasimoto transformation*

$$\Phi(s,t) = \kappa(s,t)e^{i\int^s \tau(s',t)ds'}. \tag{7.56}$$

By introducing (7.56) in (7.54) and (7.55), we obtain a complex equation in the form

$$\frac{\partial \Phi}{\partial t} = \left[\frac{\partial^2}{\partial s^2} + |\Phi|^2 + i\Phi \int^s \tau \Phi^* ds' + \frac{\partial \Phi}{\partial s} \int^* \Phi^* ds'\right] U e^{i\int^s \tau(s',t)ds'}$$
$$+ \left[i\frac{\partial^2}{\partial s^2} + i|\Phi|^2 + \Phi \int^s \tau \Phi^* ds' - i\Phi \int^s \frac{\partial \Phi^*}{\partial s'} ds'\right] B e^{i\int^s \tau(s',t)ds'}, \tag{7.57}$$

where $*$ is complex conjugation, and the square parentheses are operators acting to the right. A simple example is immediate: if we choose a binormal type of motion with $B = \kappa$, and zero normal velocity $U = 0$, (7.57) reduces to the (focusing) version of the nonlinear Schrödinger equation

$$i\frac{\partial \Phi}{\partial t} + \frac{\partial^2 \Phi}{\partial s^2} + \frac{3}{2}|\Phi|^2 \frac{\partial \Phi}{\partial s} = 0. \tag{7.58}$$

If we consider a more complex type of motion with $U = -\kappa_s$, and $B = -\kappa\tau$ we obtain instead the equation

$$\frac{\partial \Phi}{\partial t} + \frac{\partial^3 \Phi}{\partial s^3} + \frac{3}{2}|\Phi|^2 \frac{\partial \Phi}{\partial s} = 0, \tag{7.59}$$

which is an MKdV equation for a complex function. Of course (7.57–7.59) reduce to the previously studied two-dimensional case if $\tau = 0$, i.e., the imaginary part of all equations vanishes.

Another example of mapping is provided by the binormal motion of curves with constant curvature, i.e. $\Omega_n = 0$ (or $\partial \mathbf{r}/\partial b = g^{1/2}\mathbf{b}$) and κ =const. The resulting equation for torsion can be mapped, after a scaling, into either the Dym nonlinear equation, or the Camassa-Holm equation from hydrodynamics. If the initial curve is a helix, a binormal motion with constant curvature generates the so-called *soliton surfaces*, [289], which are periodic surfaces of revolution representing the motion of a soliton along a circular helix.

7.7 Problems

1. Find the PDE equation fulfilled by the curvature of a moving curve on the surface of a unit sphere S_2. Find criteria for this curve to be closed.
2. Show that (7.26) is integrable for $p = 1$ and for $p = -4$, and find the solutions for κ. Study the integrability of (7.26) function of p.
3. Find a more compact form for (7.4), by introducing a complex vector $\Xi = \mathbf{t} + i\mathbf{n}$. Hint: use (7.6).
4. Prove that a rigid unit circle in uniform rotation around its venter has indeed $U = W = 0$.

5. Show that the most general Euclidean motion of a rigid curve fulfills the equations $W_s = \kappa U$, $U_s = -\kappa W + C_0 e^{\pm i\varphi(t)}$, where $\varphi(t)$ is an arbitrary rotation angle and C_0 is an arbitrary constant. Show that in the tangent angle representation these equations read $W_\theta = U, U_\theta = -W + C_0 e^{\pm i\varphi(t)}$ or simply $W_{\theta\theta} + W = C_0 e^{\pm i\varphi(t)}, U_{\theta\theta} + U = 0$.
6. Re-obtain the results for the motion of plane curves by using differential forms, including a tangent motion with velocity W. Hint: $\theta_1 = ds + Wdt$. Show that in this case we can obtain the complete dynamical equation for curvature $\kappa_t = U_{ss} + \kappa^2 U + \kappa_s W$.
7. Show that in the case of 2-dimensional curves, the Cartan frame method provides $\theta = ds\boldsymbol{t} + Udt\boldsymbol{n}$, and $d\boldsymbol{t} = \omega_{12}\boldsymbol{n}, d\boldsymbol{n} = -\omega_{21}\boldsymbol{t}$. From the structure conditions we have $d\theta_1 = \omega_{12} \wedge \theta_2, d\theta_2 = \theta_1 \wedge \omega_{12}$. which provide the nontrivial solutions $\omega_{12} = \kappa ds, \omega_{12} = U_s dt$, that is the Serret-Frenet relations for the $W = 0$ case: $d\boldsymbol{t} = (\kappa ds + U_s dt)\boldsymbol{n}$, $d\boldsymbol{n} = -(\kappa ds + U_s dt)\boldsymbol{t}$.

Chapter 8
Theory of Motion of Surfaces

In this chapter we focus on the kinematics and dynamics of moving surfaces, in the same way we did in Chap. 7 for curves. The boundary conditions obtained from this geometrical approach will be used in the next chapters for the study of nonlinear oscillations and waves of liquid drops. In this chapter we assume that all transformations of coordinates are continuous at least of class C_2, and they have nonvanishing Jacobian functions.

8.1 Differential Geometry of Surface Motion

In the following we consider a time parameterized family of regular surfaces defined by the immersions $\boldsymbol{r}(t, u^\alpha) : [0, \infty] \times U \subset \mathbb{R} \times \mathbb{R}^2 \rightarrow \Sigma(t) \subset \mathbb{R}^3$. We assume it is possible to define at any moment of time t an orthonormal basis $\{\boldsymbol{e}_\alpha, \boldsymbol{N}\}_{\alpha=1,2}$ in $\mathbb{R}^3$ where $\boldsymbol{e}_\alpha = (\partial \boldsymbol{r}/\partial u^\alpha)/||\partial \boldsymbol{r}/\partial u^\alpha||$ and $g_{\mu\nu} = \boldsymbol{r}_u \cdot \boldsymbol{r}_v$ is the first fundamental form. We apply the Cartan frame formalism described in Sect. 7.4 for the principal bundle of adapted frames $\mathrm{OF}(\mathbb{R}^3, \Sigma(t))$ over $\Sigma(t)$ which are actually the Darboux frames (Definition 59), and we can write the canonical form

$$
\begin{aligned}
(\theta; \boldsymbol{X}) = d\boldsymbol{r} &= \boldsymbol{r}_\mu du^\mu + W^\mu \boldsymbol{e}_\mu dt + U\boldsymbol{N} dt \\
&= \sum_{\alpha=1}^{2} \underbrace{(\sqrt{g_{\alpha\alpha}} du^\alpha + W^\alpha dt)\boldsymbol{e}_\alpha}_{\theta_\alpha} + \underbrace{U\boldsymbol{N} dt}_{\theta_3},
\end{aligned}
\tag{8.1}
$$

where W^α, U are the tangent and normal components of surface velocity, respectively. By using the Gauss and Weingarten equations, with the notations from Chap. 6, namely $\boldsymbol{r}_{\mu\nu} = \Gamma^\lambda_{\mu\nu} \boldsymbol{r}_\lambda + \boldsymbol{\Pi}_{\mu\nu}$, $\boldsymbol{\Pi} = \Pi \boldsymbol{N}$, and $\boldsymbol{N}_\mu = -g^{\lambda\nu} \boldsymbol{r}_\lambda \Pi_{\nu\mu}$, and such that the Christoffel symbols are derived from the Riemannian metric on $\Sigma(t)$, we can write the connection form

A. Ludu, *Nonlinear Waves and Solitons on Contours and Closed Surfaces*, Springer Series in Synergetics, DOI 10.1007/978-3-642-22895-7_8,

$$(\omega;\boldsymbol{X})|_{T\Sigma} = d\boldsymbol{r}_\mu = \Gamma^\lambda_{\mu\nu}\boldsymbol{r}_\lambda du^\nu + \boldsymbol{N}\Pi_{\mu\nu}du^\nu + \Upsilon_{\mu\nu}\boldsymbol{r}_\nu dt + \Xi_\mu \boldsymbol{N}dt,$$

$$d\boldsymbol{N}|_{T\Sigma^\perp} = -g^{\lambda\mu}\Pi_{\mu\nu}\boldsymbol{r}_\lambda du^\nu + \Upsilon_\mu \boldsymbol{r}_\mu dt + \Xi \boldsymbol{N}dt, \tag{8.2}$$

where the 1-forms $Udt, W^\mu dt, \Upsilon_{\mu\nu}dt, \Upsilon_\mu dt, \Xi_\mu dt, \Xi dt$ are responsible for the motion (tangent and normal) of the surface. By applying the structure conditions (4.37–4.38), we obtain a PDE system with eight equations for these nine unknown functions [217,234]. The indeterminacy is related to the fact that there is no natural parametrization on the surface. Also, from the structure equation (i.e. $d^2\boldsymbol{r} = 0$) we obtain six equations for the time dependence of the surface metric and of the second fundamental form

$$g_{\mu\nu,t} = g_{\mu\alpha}W^\alpha_\nu + g_{\nu\beta}W^\beta_\mu - 2\Gamma^\lambda_{\mu\nu}W^\alpha g_{\alpha\lambda} - 2\Pi_{\mu\nu}U,$$

$$\Pi_{\mu\nu,t} = U_{,\mu\nu} + \Pi_{\mu\lambda}W^\lambda_{,\nu} + \Pi_{\nu\lambda}W^\lambda_{,\mu} + (\Pi_{\mu\lambda}\Gamma^\lambda_{\rho\nu} + \Pi_{\nu\lambda}\Gamma^\lambda_{\rho\mu})W^\rho$$
$$+\Gamma^\lambda_{\mu\nu}U_{,\lambda} - g^{\rho\lambda}\Pi_{\rho\nu}\Pi_{\mu\lambda}U. \tag{8.3}$$

The comma subscript represents differentiation with respect to the variables after this comma. (8.3) represent the intrinsic formulation of surface motion, which (as opposed to the local formulation $\boldsymbol{r}(u^1, u^2, t)$) is not redundant and does not have the "z-axis" type of singularities. If we are given the surface velocity components, by integration of equations above we obtain the evolution of the surface at any moment of time, through the knowledge of its fundamental forms. Similar to the curve motion case, the W^α tangent velocity components are not essential: they just re-parameterize the surface, or "pushing" particles along the surface. We can note this by asking $U = 0$ for example and noticing that the resulting equations are linear in W components.

In order to verify if (8.3) describe the motion of the surface for real, we perform a limiting procedure reducing the surface to one of its curves of coordinates, and expecting to re-obtain the equations of motions for curves. However, like in any limiting process, we first have to write these equations in covariant form

$$g_{\mu\nu,t} = \nabla_\mu W_\nu + \nabla_\nu W_\mu - 2\Pi_{\mu\nu}U, \tag{8.4}$$

$$\Pi_{\mu\nu,t} = \nabla_\mu(\nabla_\nu U) + (\Pi_{\mu\lambda}\nabla_\nu + \Pi_{\nu\lambda}\nabla_\mu)W^\lambda - g^{\rho\lambda}\Pi_{\rho\mu}\Pi_{\lambda\nu}U. \tag{8.5}$$

In this form (8.4) plus the ten Gauss-Codazzi conditions ($d^2\boldsymbol{e}_\mu = d^2\boldsymbol{N} = 0$) provide sixteen equations for nine functions describing the surface and its motion: $E, F, G, e, f, g, W^1, W^2, U$. We apply the following limiting verification procedure: if we make $\partial\boldsymbol{r}/\partial u^2 = 0$, and consequently the surface shrinks to a moving plane curve $\Sigma(t) \to \gamma(t), \boldsymbol{N} \to \boldsymbol{n}$ we expect $g_{\mu\nu}(u^1, u^2, t) \to g(s,t)$, $\boldsymbol{W} \to W$, while U keeps having the same interpretation. Also, since the principal curvatures will approach $\kappa_1 \to \kappa, \kappa_2 \to 0$ we have

$$H = \frac{\kappa_1 + \kappa_2}{2} = \frac{eG - 2fF + gE}{2(EG - F^2)} \rightarrow \frac{e}{2E},$$

that is $\Pi_{\mu\nu} \rightarrow g\kappa$. In this limit (8.4) reduces to the regular time variation of the curve metric, (7.5), $g_t = 2g(W_s - \kappa U)$. In the same plane curve limit we have (8.5) approaching (7.10), namely $\kappa_t = U_{ss} + \kappa^2 U + \int^s \kappa U ds'$.

In literature there are basically three simplification approaches of the surface motion equation. The first one uses a sort of "diagonal philosophy" by using orthogonal particle-frozen coordinates in the surface that push back the particles in their original position when the surfaces changes. The other two approaches investigate particular cases of surfaces like developable surfaces ($K = 0$) or K-surfaces ($K < 0$ and constant).

In the first approach we use surface coordinates along the principal directions (the surface should have no umbilical points, though!) in $\Sigma(t)$ such that

$$g_{\mu\nu} = \begin{pmatrix} e^{a_1} & 0 \\ 0 & e^{a_2} \end{pmatrix}, \ \Pi_{\mu\nu} = \begin{pmatrix} \kappa_1 e^{a_1} & 0 \\ 0 & \kappa_2 e^{a_2} \end{pmatrix}, \tag{8.6}$$

with $a_\mu, \kappa_\mu \in C_2(\mathbb{R}^2)$. The "frozen particles" rigidity constraints $g_{12,t} = \Pi_{12,t} = 0$ reduce the equation of motion (8.4–8.5) to a system of total differentials with respect to time for the unknown functions a_μ, κ_μ

$$\left(\frac{\partial}{\partial t} - W^\mu \frac{\partial}{\partial u^\mu}\right) a_\nu = 2W^\nu_{,\nu} - 2\kappa_\nu U_{,\nu} \ (\nu \text{ w.s.}) \tag{8.7}$$

$$\left(\frac{\partial}{\partial t} - W^\mu \frac{\partial}{\partial u^\mu}\right) \kappa_\nu = \kappa_\nu^2 U + U_{,\sigma_\nu \sigma_\nu} + \frac{1}{2} e^{-a_{\nu'}} a_{\nu,\sigma_{\nu'}} U_{,\sigma_{\nu'}}, \ (\nu \text{ w.s.}), \tag{8.8}$$

with $\nu = 1, \nu' = 2$ or viceversa, without summations and we need to introduce the following coordinate transformation [217]

$$\sigma_1 = \int^{u^1} e^{\frac{1}{2} a_1 (u'^1, u^2)} du^{1'},$$

and a similar expression for σ_2. The moving surface is then described by the following Gauss-Weingarten relations

$$\frac{\partial}{\partial \sigma^1} \begin{pmatrix} \boldsymbol{r}_{\sigma^1} \\ \boldsymbol{r}_{\sigma^2} \\ \boldsymbol{N} \end{pmatrix} = \begin{pmatrix} \frac{1}{2} a_{1,1} & -\frac{1}{2} a_{1,2} e^{a_1 - a_2} & \kappa_1 e^{a_1} \\ \frac{1}{2} a_{1,2} & \frac{1}{2} a_{2,1} & 0 \\ -\kappa_1 & 0 & 0 \end{pmatrix} \begin{pmatrix} \boldsymbol{r}_{\sigma^1} \\ \boldsymbol{r}_{\sigma^2} \\ \boldsymbol{N} \end{pmatrix}, \tag{8.9}$$

$$\frac{\partial}{\partial \sigma^2}\begin{pmatrix} \boldsymbol{r}_{\sigma^1} \\ \boldsymbol{r}_{\sigma^2} \\ \boldsymbol{N} \end{pmatrix} = \begin{pmatrix} \frac{1}{2}a_{1,2} & \frac{1}{2}a_{2,1} & 0 \\ -\frac{1}{2}a_{2,1}e^{a_2-a_1} & \frac{1}{2}a_{2,2} & \kappa_2 e^{a_2} \\ 0 & -\kappa_2 & 0 \end{pmatrix} \begin{pmatrix} \boldsymbol{r}_{\sigma^1} \\ \boldsymbol{r}_{\sigma^2} \\ \boldsymbol{N} \end{pmatrix}. \tag{8.10}$$

When we confine to developable surfaces, the kinematic equations for the surface simplify considerable because the Gauss-Weingarten equations reduce to a vector form from a 2-tensor form. It is interesting that the motion of surfaces with constant non-positive Gauss curvature can be mapped into either the MKdV or sine-Gordon integrable systems [24].

8.2 Coordinates and Velocities on a Fluid Surface

In the case of moving fluid surfaces, it is more delicate to introduce Lagrangian, Eulerian, and convected coordinates. This is mainly because there is no natural differential mapping like in the case of the full three-dimensional space. To define such coordinates for fluid surface, we follow the geometric approach for shells given for example in [210, Sect. 1.5]. We define a *fluid surface* by a domain F in $\mathbb{R}^2$ and a general system of nonsingular curvilinear coordinates $(X^\alpha), \alpha = 1, 2$ for the points in this domain. Actually, these coordinates label the particles in the surface. Of course we can always endow $\mathbb{R}^2$ with a system of Euclidean coordinates (Z^α) for F. We have the coordinate transformations $Z^\alpha = Z^\alpha(X^1, X^2)$ and the inverse $X^\alpha = X^\alpha(Z^1, Z^2)$. The Euclidean coordinates have their unit Euclidean vectors as a basis, $\{\hat{I}_\alpha\}_{\alpha=1,2}$, while in the curvilinear coordinates we introduce the tangent vectors to the lines of coordinates, namely

$$\boldsymbol{E}_\alpha = \frac{\partial Z^\beta}{\partial X^\alpha}\hat{I}_\beta, \; \beta = 1, 2. \tag{8.11}$$

A configuration of F is a mapping $\boldsymbol{r} : F \to \mathbb{R}^3$, namely $\boldsymbol{r}(Z)$. We set the similar curvilinear (x^k) and Euclidean $(z^k), k = 1, \ldots, 3$ coordinates in $\mathbb{R}^3$ with their corresponding transformations of coordinates, and the basis

$$\boldsymbol{e}_k = \frac{\partial z^j}{\partial x^k}\hat{i}_j. \tag{8.12}$$

Let $C(F) = \bigcup_{\boldsymbol{r}:F\to\mathbb{R}^3} \boldsymbol{r}$ be the set of all configurations. A curve in $C(F)$ represents a motion of the fluid surface F. We can parameterize this curve with time, and we have the mapping from the curvilinear coordinates into the Euclidean three-dimensional space, as an embedding $t \to x \circ \boldsymbol{r}_t \circ Z = x \circ \boldsymbol{r} \circ Z(X, t) = (x^i(\boldsymbol{r}_t(X)))$. For the graphical intuition we present these systems in the left part of Fig. 8.1. We define the *material* (or Lagrangian) *velocity* of the fluid surface by the mapping $\boldsymbol{V}_L : F \to \mathbb{R}^3$

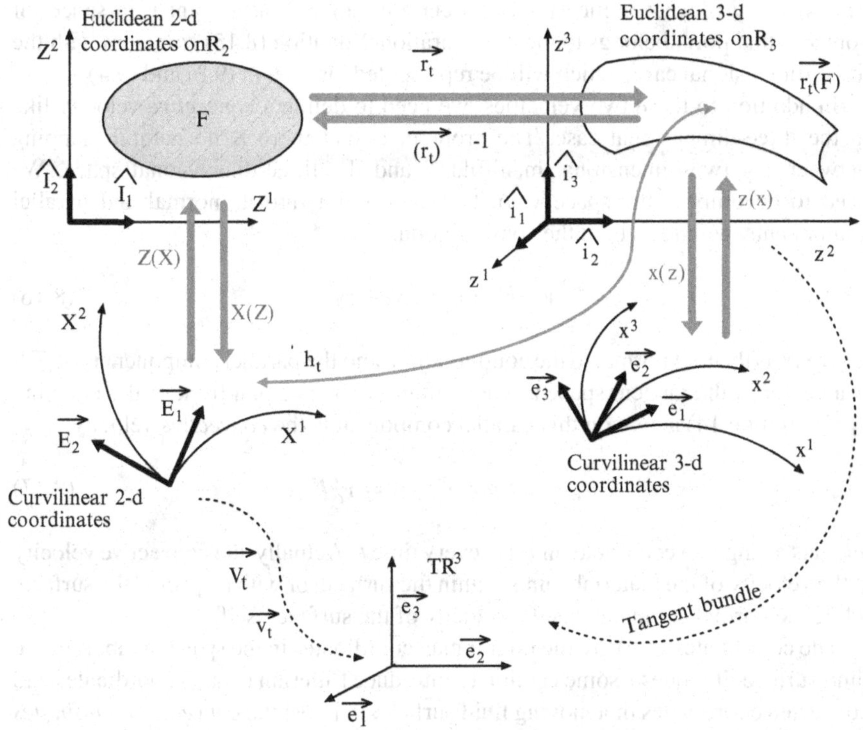

Fig. 8.1 The geometric description of a fluid moving surface. There are both Euclidean and curvilinear coordinate systems for both the surface F and $\mathbb{R}^3$, as well as basis vectors. The possible coordinate transformations, the configuration mapping (motion of the surface), and their inverses are drawn in *gray arrows*. The material and space velocities are also presented in the tangent bundle

$$\boldsymbol{V}_L = \boldsymbol{V}(X,t) = \frac{\partial x^i(X,t)}{\partial t}. \tag{8.13}$$

This vector is the three-dimensional velocity of the material particle labeled X and belonging to the configuration of the fluid surface. In components it reads $\boldsymbol{V}_L = V_L^i \boldsymbol{e}_i$.

A motion is *regular motion* if the mapping $\boldsymbol{r}_t$ is invertible with $\boldsymbol{r}_t^{-1} : \boldsymbol{r}_t(F) \to F$, and the mapping and its inverse are smooth functions. In this case we can define a *spatial* (or Eulerian) *velocity* by the composition of mappings

$$\boldsymbol{r}_t(F) \subset \mathbb{R}^3 \xrightarrow[\boldsymbol{r}_t^{-1}]{} F(Z^\alpha) \xrightarrow[X(Z)]{} \mathbb{R}^2(X^\alpha) \xrightarrow[\boldsymbol{V}_L]{} T\mathbb{R}^3. \tag{8.14}$$

In other words the space velocity reads

$$\boldsymbol{v}_E(\boldsymbol{r},t) = \boldsymbol{V}_L \circ X \circ \boldsymbol{r}_t^{-1}. \tag{8.15}$$

This velocity is what is measured at a certain moment, at a point $\boldsymbol{r}$ in space, of course if that point belongs to the configuration. Equation (8.15) coincides with the three-dimensional case, which will be represented later on in (9.3) and (9.4).

In addition to these two velocities, we need to define a *convective* velocity like in the three-dimensional case. The problem is that there is no natural mapping between the two-dimensional manifold F and the three-dimensional space. We need to decompose the space velocity at any point into its normal and parallel components, with respect to the configuration.

$$\boldsymbol{v}_E(\boldsymbol{r},t) = v_n \boldsymbol{N} + \boldsymbol{v}_{\parallel}, \tag{8.16}$$

where $\boldsymbol{N}$ is the unit normal to the configuration, and the parallel component $\boldsymbol{v}_{\parallel} \in TF$ is a vector in the tangent space to the configuration. The pull back of the mapping $\boldsymbol{r}_t$ (Definition 14) acting on this parallel component is the *convective* velocity

$$\boldsymbol{v}_c = (X \circ \boldsymbol{r}_t)^* \boldsymbol{v}_{E\,\parallel} = v_c^\alpha \boldsymbol{E}_\alpha, \tag{8.17}$$

and it is a tangent vector field on F for every time t. Actually, the convective velocity is the velocity of the material points within the surface, or with respect to the surface, while the normal component is the velocity of the surface itself.

The coordinates (X^α) are the Lagrangian coordinates in the space of labels of the fluid surface. It requests some caution to introduce Eulerian (space) coordinates and convected coordinates in a moving fluid surface. In [210] the *convected coordinates* for moving surfaces are introduced simply by the mapping

$$h_t : \boldsymbol{r}_t(F) \to \mathbb{R}^2, \;\; h_t^\alpha(\boldsymbol{r}) = X^\alpha(\boldsymbol{r}_t^{-1}), \tag{8.18}$$

so these coordinates label the points of the moving surface directly with the curvilinear (suitably chosen for this purpose) coordinates on F, and in that appear to be convected by the motion and move together with the surface. The convected coordinates defined like this have an interesting property.

Lemma 6. *The components of the convective velocity with respect to the coordinates (X^α) are exactly the same as the components of parallel projection of the space (or material) velocity on the surface with respect to the convected coordinates h_t^α.*

A proof of this lemma is to be found in [210]. We give here another proof. Since the basis vectors $\boldsymbol{E}_\alpha$ are pushed forward by the differential of h^{-1} into the basis vectors of the convected coordinates, $\boldsymbol{\xi}_\alpha = d(h_t^{-1})(\boldsymbol{E}_\alpha)$, we can write from (8.17)

$$\boldsymbol{v}_{E\,\parallel} = d(h_t^{-1})(\boldsymbol{v}_c) = d(h_t^{-1}) v_c^\alpha \boldsymbol{E}_\alpha = v_c^\alpha d(h_t^{-1}) \boldsymbol{E}_\alpha = v_c^\alpha \boldsymbol{\xi}_\alpha, \tag{8.19}$$

with $\alpha = 1, 2$, which proves the affirmation.

Now, if we want for a system of coordinates $X_E^\alpha = h_t^\alpha(\boldsymbol{r})$ to move together with the moving surface $\boldsymbol{r}_t(F)$, this coordinates should involve zero convected velocity

$\boldsymbol{v}_c = 0$. According to Lemma 6, from $\boldsymbol{v}_c = v_c^\alpha \boldsymbol{E}_\alpha$ we have $\boldsymbol{V}_{L\,\|} = \boldsymbol{v}_{E\,\|} = v^\alpha \boldsymbol{\xi}_\alpha$, and if $\boldsymbol{v}_c = 0$, by components the space and material velocities are normal to the surface in this points. Consequently they move together with it. In Aris' formalism [10], the convected coordinates are directly introduced by requesting that the points labeled by such coordinates move only in the normal direction to the surface. That is they "move" together with the surface. This definition works in many situations, but there are situations where this definition may request a special curvilinear coordinate system. In the following we give two examples.

Example 1. First example is in favor of using Eulerian coordinates. Let us introduce F as a half-plane of coordinates (X^α), X^1 being the distance from the point to the edge of this half-plane. The configuration will be this half-plane making a certain variable angle with a fixed system in $\mathbb{R}^3$, and the motion is the uniform rotation of this half-plane around the fixed edge with angular velocity ω. We can consider a thin layer of fluid adherent to this rotating half-plane and flowing away from the fixed axis, but in the half-plane, because, say, of the centrifugal force. A particle of Lagrangian label (X^α) is mapped into $\boldsymbol{r}_t = (\xi(t)\cos\omega t, \xi(t)\sin\omega t, 0)$ with $\xi(0) = X^1$. The Lagrangian velocity is in this case $\boldsymbol{V}_L = (-\xi\omega\sin\omega t, \xi\omega\cos\omega t, 0)$, the space velocity is $\boldsymbol{v}_E = (-\xi\omega\sin\omega t + \xi'\cos\omega t, \xi\omega\cos\omega t + \xi'\sin\omega t, 0)$, and the convective velocity is $(\xi', 0) \in TF$. In this case it is easy to associate Eulerian fixed coordinates: these are fixed points in the half-plane, describing concentric circles around the edge, because their velocities are normal.

Example 2. In our second example the Eulerian coordinates will not work so natural. It is the case of translation motion of closed surfaces. In such situation it is really difficult to construct a "fixed" coordinate system in a moving membrane (like an air bubble ascending to the surface, or the membrane of a motile cell while swimming). Let us consider $F_t, \boldsymbol{N}(X,t)$ being the configuration surface moving in time, and its normal, respectively, in the X parametrization. From any point $\boldsymbol{r}_t(X)$ of the configuration F_t, we can construct a flow box of curves which are always tangent to the instantaneous vector $\boldsymbol{N}$

$$\boldsymbol{r}_\epsilon(X,t) = \boldsymbol{r}(X,t) + \epsilon \boldsymbol{N}(X,t),$$

with arbitrary $\epsilon > 0$. This equation is just the normal variation of a surface, defined in (10.36) in Sects. 10.4.1 and 10.4.2. At $t+dt$ moment of time, the family of curves $\boldsymbol{r}_\epsilon(X,t)$ generated normally at t intersects the moved surface F_{t+dt} in some new points. These intersections represent the change from Lagrangian coordinates $\boldsymbol{r}_t$ to the Eulerian ones. If $\boldsymbol{r}_\epsilon(X,t)$ represent the Eulerian coordinates at moment t, the intersection between the normals at t and the moved surface at $t + dt$ are the new Eulerian coordinates. If the flow of the fluid surface is regular, and by using the flow box theorem (Theorem 6) in Sect. 4.4, we can integrate such positions for finite interval of time.

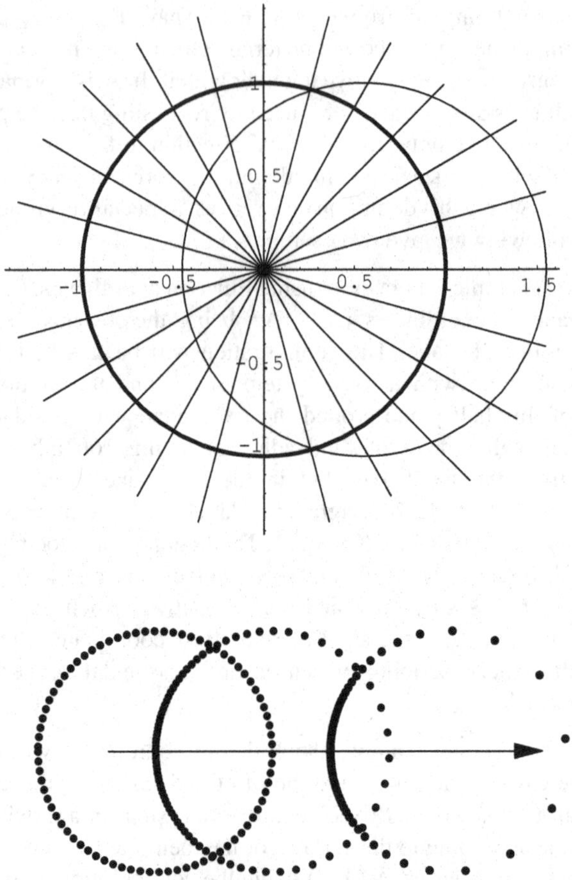

Fig. 8.2 Trying to assign "fixed" coordinates in a compact surface in translational motion. *Upper part*: the *thick circle* is the initial position of the surface, with the radii providing the normal directions to the surface. The intersections between these normals and the moved sphere (*thin circle*) provide the instantaneous Eulerian coordinates on the new sphere. *Bottom part*: transformation of the Eulerian coordinates in time, trying to keep moving only in the normal direction

Let us practice this definition by considering a sphere of radius R moving with constant translation velocity V along the z^1-axis, like it is represented in the upper part of Fig. 8.2.

The Lagrangian coordinates on the sphere move together with the sphere and keep for example the same polar and azimuthal angles. For example we can choose $B = (\varphi, \theta) \in [0, 2\pi] \times [0, \pi]$ and have spherical coordinates $\boldsymbol{r}_t = R(\sin\theta\cos\varphi + Vt, \sin\theta\sin\varphi, \cos\theta)$. In the following we focus on the big circle $\theta = \pi/2$. We have $\boldsymbol{V}_L = \boldsymbol{v}_E = (V, 0, 0)$, and

$$v_{E\,\parallel} = \frac{Vz^2(z^1 - Vt)}{(z^1 - Vt)^2 + (z^2)^2}\left(\frac{z^2}{z^1 - Vt}, -1\right).$$

From (4.6) and (8.17), we have $v_c = -V \sin\varphi$. The motion of the convected polar coordinates with this v_c can be noticed in the bottom frame of Fig. 8.2.

The Euler coordinates need to represent points that move only normal to the sphere (Fig. 8.2). Consequently the θ polar angle will transform according to the relation

$$\tan\theta_t = \frac{R\sin\theta_{t+dt}}{Vt + R\cos\theta_{t+dt}}, \tag{8.20}$$

and we present an example of this transformation in the bottom part of Fig. 8.2. For longer intervals of time transformation, (8.20) becomes singular. So, in this example, it is easier to work with Lagrangian coordinates.

To eliminate such nonconventional transformations of coordinates, we could introduce Eulerian coordinates in the moving configuration as follows. Begin with Lagrangian pair (X^α) at a certain moment of time. The transformation from this Lagrangian coordinates to the Eulerian coordinates is made by calculating $\boldsymbol{V}_L(X,t)$, and then by moving the $\boldsymbol{r}_t$ point along the surface with some tangent vector $\boldsymbol{w}$ such that the new Lagrangian velocity of this new point $\boldsymbol{V}_L + \boldsymbol{w}$ is normal to the surface. The relation between $\boldsymbol{r}_t$ and this new translated point provides the transformation from Lagrangian to Eulerian coordinates. To do this in the example with translating sphere, we have to expand (8.20) in Taylor series, take the first order of smallness and integrate the corresponding linear PDE

$$\theta(t) = \pm 2\arccos\frac{1}{\sqrt{1 + e^{\frac{2Vt}{R}}\tan^2\dfrac{\theta(0)}{2}}}.$$

For more elaborated discussions on the Eulerian, Lagrangian, convective coordinates or velocities, the reader can use any of the following sources [10, 210, 241, 292].

In the end, we make an observation regarding the mixed character of geometric objects in the kinematics of surfaces. For three-dimensional configurations, like theory of elasticity, it is more natural to define the convective velocity as a pull back, since all involved spaces are Riemannian manifolds of dimension 3. In the case of surfaces, we first have to project the space velocity on the tangent plane (8.16), then perform the pull back. However, there is a more general treatment, namely, to introduce a sort of *mixed covariant derivative* which assures the contravariant/covariant tensor character simultaneous in all spaces involved, no matter of the number of dimensions (2 or 3). We briefly introduce this mixed derivative with (4.59). A comprehensive treatment of the topic, for submanifolds and hypersurfaces in a general Riemannian space or dimension n, can be found in [181].

8.3 Kinematics of Moving Surfaces

Let (X^α) be the parametrization of the domain F and $\boldsymbol{r}_t = \boldsymbol{r}(X,t)$ be the corresponding moving regular configuration, i.e., a regular parameterized moving surface (Fig. 8.1). The convective velocity (8.17) $\boldsymbol{v}_c$ can be written in components in the form

$$\boldsymbol{v}_c(X,t) = \frac{dh_t^\alpha}{dt}\boldsymbol{\xi}_\alpha, \quad \alpha = 1,2, \tag{8.21}$$

and represents the velocity vector belonging to the tangent space to the surface, while its push forward by dh_t^{-1} is a velocity vector field tangent to the moving surface $X \circ \boldsymbol{r}_t(F)$, $\boldsymbol{v}(X,t) = v_{E\ \|}^{j}\hat{\boldsymbol{i}}_j = v_h^\alpha \boldsymbol{\xi}_\alpha$ (8.19). Also h_t^α are the convected coordinates in the surface (8.18). In the following the curvilinear coordinates X^α are time-independent coordinates, so they will be understood as Lagrangian coordinates on the surface, while the convected coordinates are time dependent by construction.

The area element is given by the first fundamental form of the surface g (or the metric tensor g in some books; Definitions 51 and 52): $dA(t) = \sqrt{g_L(t)}dX^1dX^2 = \sqrt{g_c(t)}dh_t^1dh_t^2$, where labels L and c refer to chosen system of coordinates. We need to mention that in the following, the symbol g is used in the sense of Sect. 6.1, and not in the sense of the unit basis vectors, like we did above (i.e., not the $\{\boldsymbol{g}_\alpha\}_{\alpha=1,2}$). We define by

$$\hat{J}(t) = \frac{\partial h_t^\alpha}{\partial X^\beta}, \tag{8.22}$$

the Jacobian matrix of transformation of coordinates. From (8.13) and (8.19), we can write

$$\frac{d\hat{J}}{dt} = \frac{d}{dt}\frac{\partial h_t^\alpha}{\partial X^\beta} = \frac{\partial v_E^\alpha}{\partial X^\beta} = \frac{\partial v_E^\alpha}{\partial h_t^\gamma}\frac{\partial h_t^\gamma}{\partial X^\beta} = \hat{\gamma}\hat{J}, \tag{8.23}$$

with $\hat{\gamma}$ defined as the *surface velocity-gradient matrix*. Consequently we write the time variation of the element of area in terms of the time-independent coordinates through the Jacobian matrix

$$\frac{d}{dt}dA = \left(\frac{1}{2g_c}\frac{dg_c}{dt}J + \frac{dJ}{dt}\right)\sqrt{g_c}dX^1dX^2, \tag{8.24}$$

with $J = \det\hat{J}$. We can formally integrate the matrix differential equation (8.23)

$$\hat{J}(t) = \hat{J}(0)e^{\int^t \hat{\gamma}(t')dt'}, \tag{8.25}$$

and take $\hat{J}(0)$ =gd. By using the matrix identity $\det e^A = e^{\mathrm{Tr}A}$, we have

$$J = \det\hat{J}(t) = e^{\mathrm{Tr}\int^t \hat{\gamma}(t')dt'},$$

and

$$\frac{dJ}{dt} = \mathrm{Tr}\hat{\gamma}(t)J(t), \tag{8.26}$$

since the trace operator is linear and hence commutes with the time derivative. We can express the trace of the surface velocity-gradient matrix by using the surface divergence operator (6.42) (Sect. 6.5.2), $\mathrm{Tr}\hat{\gamma} = \nabla_\Sigma \mathbf{v}_c = \boldsymbol{\xi}_\alpha \cdot \nabla_\alpha \mathbf{v}_c$, $\alpha = 1, 2$, where the surface Σ becomes here the time-dependent surface configuration F. We can write the equation for the rate of change in time of the element of moving area

$$\frac{d}{dt} dA = \left(\frac{1}{2g_c} \frac{dg_c}{dt} + \nabla_F \mathbf{v}_c \right) dA. \tag{8.27}$$

We mention that the area being a scalar, the time derivative coincides with the convective time derivative, which should actually be used.

To find the stretching of the surface along different directions, we need to find the equation for the rate of change in time of the arc-length in different coordinates

$$ds^2 = g_{L,\alpha,\beta} dX^\alpha dX^\beta = g_{c,\alpha,\beta} dh_t^\alpha dh_t^\beta.$$

We have

$$\frac{1}{ds} \frac{d_s}{dt} = \frac{\left(\frac{d}{dt} g_{L,\alpha\beta} \right) dX^\alpha dX^\beta}{2g_{L,\alpha\beta} dX^\alpha dX^\beta},$$

and we can define the Lagrangian *strain tensor* as

$$S_L = \frac{1}{2} \frac{dg_L}{dt}, \quad S_c = \frac{1}{2} \frac{dg_c}{dt}. \tag{8.28}$$

If we work in the convected coordinates, the appropriate approach is the use of the convective time derivative (see (9.16) in Sect. 9.2.6). Consequently, we can write a convected strain tensor in the form

$$S_{c,\alpha\beta} \to \frac{1}{2} \frac{d_c g_{c,\alpha\beta}}{dt} = \frac{1}{2} \frac{\partial g_{c,\alpha\beta}}{\partial t} + \frac{1}{2} (v_\gamma^c \nabla_\gamma g_{c,\alpha\beta} + g_{c,\alpha\gamma} \nabla_\beta v_c^\gamma + g_{c,\gamma\beta} \nabla_\alpha v_c^\gamma),$$

where ∇ represents here the covariant derivative. The first term in the RHS parenthesis is zero from (4.53), and for the remaining terms the action of the metric tensor g is just to lower the superscripts. So, the surface strain tensor reads

$$S_{c,\alpha\beta} = \frac{1}{2} \frac{\partial g_{c,\alpha\beta}}{\partial t} + \frac{1}{2} (\nabla_\beta v_{c,\alpha} + \nabla_\alpha v_{c,\beta}). \tag{8.29}$$

It is useful to compare this surface covariant result with the rate of strain tensor for the bulk flow in Euclidean three-dimensional space [111, 167, 171, 220, 224]

$$\frac{du_{ij}}{dt} = \frac{1}{2} \left(\frac{\partial v^i}{\partial x^k} + \frac{\partial v^k}{\partial x^i} \right). \tag{8.30}$$

8.4 Dynamics of Moving Surfaces

In the following, for simplicity, we denote the moving surface configuration $(X \circ \boldsymbol{r}_t)(F)$ by Σ, and since we calculate everything in the convected coordinates, we will drop the subscript c. Like in the three-dimensional hydrodynamics or elasticity [11, 170, 210], we can define a *surface stress*, i.e., the force acting on the unit of arc-length on the surface, as a contravariant vector field on the fluid surface $t^{\alpha}, \alpha = 1, 2$. Following the same similarity we define a two-dimensional *stress tensor* by the relations

$$\sigma^{\alpha\beta} n_{\alpha} = f^{\beta}, \tag{8.31}$$

where $n_{\alpha} = \boldsymbol{n} \cdot \boldsymbol{\xi}_{\alpha}$ are the projections of the principal normal of the arc-length (three-dimensional Euclidean vector) on the local basis vectors of the convective coordinate system. We can write the integral version of this stress equation to find the total stress on the surface in some arbitrary direction. Let $D \subset \Sigma$ and $\Gamma = \partial D$ be its boundary curve. For any arbitrary smooth covariant vector field $\boldsymbol{w} : \Sigma \to T\Sigma$ we can write

$$\oint_{\Gamma} \boldsymbol{f} \cdot \boldsymbol{w} ds = \oint_{\Gamma} \sigma^{\alpha,\beta} n_{\alpha} w_{\beta} ds = \iint_{D} \nabla_{\alpha} \sigma^{\alpha\beta} w_{\beta} dA, \tag{8.32}$$

where we used the regular Green theorem. For a stationary surface we assume that the surface stress is perpendicular to the surface, so from (8.31) it results that $\sigma^{\alpha,\beta} \sim g^{\alpha,\beta}$ with the proportionality constant being defined as *surface tension*, σ. Moreover, since the stationary surface is in equilibrium, we have zero total stress on any domain D, so by using (8.32) we obtain $\nabla_{\alpha} \sigma^{\alpha\beta} = 0$, and consequently $\nabla_{\alpha} \sigma = 0$, since the covariant derivative of the metric tensor g is always zero. It results that the surface tension σ must be constant over an equilibrium surface. There is a whole section devoted to the surface tension (namely Sect. 10.4), investigating it from the point of minimal surfaces.

Equations (8.31) and (8.32) can be used to obtain the three linear conservation laws of the fluid surfaces. If we denote by ς the surface mass density, and we assume that there is no exchange of matter between the surface and its surroundings, we have the conservation of mass

$$\frac{d}{dt} \iint_{D} \varsigma dA = 0.$$

By using (8.27) for the action of the time derivative we obtain the *surface continuity equation*

$$\frac{d\varsigma}{dt} + \varsigma \nabla_{\alpha} v_c^{\alpha} + \varsigma \frac{1}{2g} \frac{dg}{dt} = 0, \tag{8.33}$$

where ∇ is the covariant derivative, and $g = \det(g^{\alpha\beta})$. Similarly with the deduction of (8.32), we can obtain the momentum conservation equation. By using again an arbitrary vector field $\boldsymbol{w} : \Sigma \to T\Sigma$, and considering $\boldsymbol{F} : \Sigma \to T\Sigma, \boldsymbol{F} = (F^\alpha)$ some arbitrary external force tangent to the surface, we obtain

$$\varsigma A^\alpha = F^\alpha + \nabla_\beta \sigma^{\alpha\beta} = 0, \tag{8.34}$$

where

$$A^\alpha = \frac{dV^\alpha}{dt} = \frac{\partial V^\alpha}{\partial t} + V^\beta \nabla_\beta V^\alpha, \tag{8.35}$$

is the *material acceleration* on Σ. The time derivative in (8.35) is the material time derivative, and V is the convective velocity (actually it is the image of the convective velocity $\boldsymbol{v}_c$ through the differential of the map $Z \circ \boldsymbol{r}_t \circ x$).

In a similar way one can obtain an integral version for the angular momentum conservation equation for the fluid surface, by using (8.34) and (8.31) we have

$$\iint_D \epsilon_{\alpha\beta} \sqrt{g} \varsigma A^\alpha h_t^\beta \, dA = \iint_D \epsilon_{\alpha\beta} \sqrt{g} F^\alpha h_t^\beta \, dA + \oint_{\partial D} \epsilon_{\beta\gamma} \sqrt{g} f^\beta h_t^\gamma \, ds, \tag{8.36}$$

where $\epsilon_{\alpha\beta}$ is the Levi–Civita antisymmetric tensor in two dimensions, i.e., $\epsilon_{12} = 1, \epsilon_{21} = -1$, etc. Using the arbitrariness of D and the Green theorem on (8.36), we simply reduce it to $\sigma^{\alpha\beta} = \sigma^{\beta\alpha}$, i.e., the stress tensor is a symmetric $(2,0)$-type of tensor.

To write dynamical equations for the fluid surface, one needs to make *constitutive hypotheses* on the relations between stress and strain. The most usual hypothesis is the so-called *Newtonian fluid surface model*. A Newtonian fluid surface is described by a stress tensor that depends only on the strain tensor in a linear manner (if it is an arbitrary function of the strain, the fluid surface is called *Stokesian*), is an isotropic surface with respect to the stress, and in absence of the strain the stress is just the surface tension. Like in the case of a three-dimensional fluid, the only isotropic combination possible in two dimensions is provided by

$$\sigma^{\alpha\beta} = (\sigma + k g^{\lambda\mu} S_{\lambda\mu}) g^{\alpha\beta} + \epsilon (g^{\alpha\lambda} g^{\beta\mu} + g^{\alpha\mu} g^{\beta\lambda} - g^{\alpha\beta} g^{\lambda\mu}) S_{\lambda\mu}, \tag{8.37}$$

where $g^{\alpha\beta}$ is the matrix of the first fundamental form of the surface Σ (the metric tensor), and the constant coefficients k and ϵ are called *coefficient of interfacial dilatational viscosity* and *coefficient of interfacial shear viscosity*, respectively. For the rigorous deduction of (8.37), the reader can consult one of the following references [10, 11, 54, 167, 171, 210, 305]. It is useful to compare this covariant result with the stress tensor for the bulk flow in Euclidean three-dimensional space

$$\sigma^{ik} = -P\delta_i^k + \eta \left(\frac{\partial v^i}{\partial x^k} + \frac{\partial v^k}{\partial x^i} \right). \tag{8.38}$$

Indeed, the surface tension term σ in (8.37) becomes for the bulk fluid the pressure (modulo some convention of change of sign), and, because the metrics reduces to Kronecker symbol, the other terms in (8.37) group together in a symmetric tensor with combined coefficient $\epsilon + k$ which is just the bulk kinematic viscosity ν [111, 167, 171, 220, 224].

In the end, we can put together the dynamical equation for a Newtonian fluid surface, by using the constitutive equation (8.37) together with (8.31) and (8.34). Also, we can take profit of the symmetry of the stress tensor, and write

$$\begin{aligned} \varsigma A^{\alpha} &= F^{\alpha} + \nabla_{\beta}\sigma g^{\alpha\beta} + k g^{\alpha\beta}\nabla_{\beta}\cdot(g^{\lambda\mu}S_{\lambda\mu}) \\ &\quad + \epsilon\nabla_{\beta}[(g^{\alpha\lambda}g^{\beta\mu} + g^{\alpha\mu}g^{\beta\lambda} - g^{\alpha\beta}g^{\lambda\mu})S_{\lambda\mu}]. \end{aligned} \tag{8.39}$$

The different terms in (8.39) have different physical interpretations. The second term on the RHS is the surface gradient of the surface tension that introduces a force if this coefficient is not homogenous along the surface. The third term on the RHS can be processed by using (6.5), namely

$$\frac{dg}{dt} = g g^{\alpha\beta}\frac{dg_{\alpha\beta}}{dt}.$$

For the last term on the RHS, we use the noncommutativity property of the covariant derivative (see (4.54) and (4.66)). Because the surface Σ is two dimensional and it is embedded in a three-dimensional Euclidean space, (4.54) and (4.66) reduce to

$$(\nabla_{\beta}\nabla_{\gamma} - \nabla_{\gamma}\nabla_{\beta})V^{\beta} = K g_{\alpha\gamma}V^{\alpha}, \tag{8.40}$$

where K is the Gaussian curvature of Σ. After some tedious algebra on (8.39), and following Aris' suggestion [10] to artificially combine $k + \epsilon$ in the third term on the RHS of (8.39), we can express this equation in a two-dimensional covariant vector form

$$\varsigma \boldsymbol{A} = \boldsymbol{F} + \nabla\sigma + k\nabla(\nabla\cdot\boldsymbol{v}) + \epsilon\left[\Delta\boldsymbol{V} + K\boldsymbol{V} - \nabla\cdot\widehat{gg} - \nabla\left(\frac{1}{2g}\frac{dg}{dt}\right)\right] \tag{8.41}$$

where $\widehat{gg}_{\alpha}^{\beta} = g_{\alpha\gamma}dg^{\gamma\beta}/dt$, and all vectors in the equation are two dimensional, expressed in the covariant coordinates of $T\Sigma$, $\boldsymbol{A} = (A_{\alpha}), \nabla = (\nabla_{\alpha})$, etc. Equation (8.41) is the net force acting on the surface to be introduced in the equations for the balance of momentum across the surface, i.e., (10.22) in Sect. 10.2. We note that (8.41) contains terms which are nonzero only when the surface is time dependent. Also, there is one term proportional to the Gaussian curvature, which is responsible for a part of the shear surface viscosity. In different books there are different physical interpretations or definitions for each of the terms in (8.41), for example [10]. Here we limit ourselves to write the three-dimensional Navier–Stokes equation in comparison with (8.41)

$$\rho\boldsymbol{a} = \boldsymbol{F} - \nabla P + \eta\Delta\boldsymbol{v}. \tag{8.42}$$

8.5 Boundary Conditions for Moving Fluid Interfaces

In the following we obtain the most general dynamic equation of motion for a fluid surface that makes the separation between two bulk fluids. We follow the definitions and the geometric approach from Sects. 4.11 and 6.3, and for closer details we encourage the reader to check [10, Chap. 10]. Let $x^k(u^\alpha, t)$ be the equation of motion of the particle labeled u^α in Σ. The Lagrangian velocity of this particle belonging to the surface (defined in (8.13)) is an Euclidean vector

$$V_L^k = \frac{dx^k}{dt} = B_\alpha^k v_c^\alpha + \frac{\partial x^k}{\partial t},$$

where B is defined in (6.20). In general, next to the interface the fluid can flow past the interface so we can have sliding on both sides of the fluid surface. Consequently we need to define two more velocities $V_{i,e}$ as Lagrangian velocities of the bulk fluid next to the surface, interior and exterior, respectively. Each such Euclidean velocity induces a surface convective velocity $v_{c,\alpha}|_{int} = B_\alpha^k V_i^k, v_{c,\alpha}|_{ext} = B_\alpha^k V_e^k$. In general there is no kinematical constraint between these velocities, but if we request a *no slip* condition we need to equate their tangent components, i.e., the kinematical boundary condition at the interface reads

$$v_{c,\alpha} = B_\alpha^k \left(V_i^k - \frac{\partial x^k}{\partial t} \right) = B_\alpha^k \left(V_e^k - \frac{\partial x^k}{\partial t} \right). \tag{8.43}$$

If, in addition, there is no normal flow of fluid from or into the interface, we also have continuity of the normal components of the bulk and surface velocities

$$N_k (V_{i,e}^k - V_L^k) = 0, \tag{8.44}$$

and from here, with the help of (8.33), we can write the equation of continuity for the interface Σ. Namely, from the isolated fluid surface equation of continuity

$$\frac{d\varsigma}{dt} + \varsigma \nabla_\alpha v_c^\alpha + \frac{\varsigma}{2g}\frac{dg}{dt} = 0, \tag{8.45}$$

where ς is the surface mass density, we obtain

$$\frac{d\varsigma}{dt} + \varsigma \nabla_\alpha v_c^\alpha + \frac{\varsigma}{2g}\frac{dg}{dt} = [\rho_e V_e^k + (\rho_i - \rho_e) V_L^k - \rho_i V_i^k] N_k, \tag{8.46}$$

which reduces back to (8.45) if there is no interchange of matter through the interface. That is $V_i^k N_k = V_e^k N_k = V_L^k N_k = \partial x^k / \partial t N_k$. If, in addition, we have no slip the equation of continuity reduces even drastically to $V_L^k = V_e^k = V_i^k$. The equations obtained in this section may be related to the general Euclidean equations from Sect. 10.2.

8.6 Dynamics of the Fluid Interfaces

Let a domain $D \subset \Sigma$ and $\gamma = \partial D$ be its boundary. Let also n_α be the principal normal (Sect. 5.1) of the γ curve which is a surface vector lying in $T\Sigma$. The surface stress acting on an infinitesimal element of arc of γ in Σ is $\sigma^{\alpha\beta} n_\beta ds$, and its Euclidean components are

$$\sigma^i ds = B^i_\alpha \sigma^{\alpha\beta} n_\beta ds. \tag{8.47}$$

In a neighborhood of Σ, we have an Euclidean body force $\boldsymbol{F}$ in the bulk fluid which can be written in terms of its tangent components and the normal

$$F^k = B^k_\alpha F^\alpha + N^k N_j F^j. \tag{8.48}$$

Same equation applies to the material acceleration $\boldsymbol{A}$

$$A^k = B^k_\alpha A^\alpha + N^k N_j A^j. \tag{8.49}$$

In the following we follow the same procedure of using an additional vector field $\boldsymbol{\lambda}(x^i)$ in a region of space containing D. The momentum balance in the $\boldsymbol{\lambda}$ direction reads

$$\frac{d}{dt}\iint_D \varsigma V^k_L \lambda_k dA \equiv \iint \varsigma A^k \lambda_k dA = \iint_D F^k \lambda_k dA + \oint_{\partial D} \sigma^k \lambda_k ds, \tag{8.50}$$

from where, given the arbitrariness of $\boldsymbol{\lambda}$, D, and using Green theorem, we obtain

$$\varsigma A^k = \sigma^k + \nabla_\beta (B^k_\alpha \sigma^{\alpha\beta}). \tag{8.51}$$

Now it is the time to take profit of the formulas for the differential geometry of the surface from Chap. 6. Namely, from (6.35) we have

$$\nabla_\beta B^k_\alpha = \Pi_{\alpha\beta} N^k,$$

where Π is the $(0,2)$-type of tensor associated to the second fundamental form of the surface. From (8.47)–(8.49), (8.51), and the above relation, we can write the balance equations in the tangent plane (along the basis B^i_α) and along the normal, respectively

$$\varsigma A^\alpha = F^\alpha + \nabla_\beta \sigma^{\alpha\beta}, \quad \text{tangent} \tag{8.52}$$

$$\varsigma N_j A^j = N_j F^j + \Pi_{\alpha\beta}\sigma^{\alpha\beta}, \quad \text{normal.} \tag{8.53}$$

We can run a simple check of these relations by considering a fixed surface, i.e., $A^i = 0$. In this case we have $\sigma^{\alpha\beta} = \sigma g^{\alpha\beta}$, and by denoting $\boldsymbol{F} = \boldsymbol{F}_\parallel + F_\perp \boldsymbol{N}$, we obtain

$$F^{\alpha} = -g^{\alpha\beta}\nabla_{\beta}\sigma, \quad \text{tangent} \tag{8.54}$$

$$F_{\perp} = 2H\sigma, \quad \text{normal.} \tag{8.55}$$

So, for the stationary case, tangent forces occur if the coefficient of surface tension is not homogenous along the surface, but we always have a normal surface pressure (see for completion Sect. 10.4).

Equations (8.52) and (8.53) can be expressed even in more detail as functions of the velocity of the surface, the strain tensor, and the material coefficients k, ϵ defined in Sect. 8.4. We do not provide here the proof of the following balance equation, but the reader can find details in [10, Chap. 10]. The momentum balance equation for an interface reads in terms of Euclidean contravariant vectors

$$\begin{aligned} \varsigma\frac{d\boldsymbol{V}}{dt} &= \boldsymbol{F} + \hat{B}\nabla_{\Sigma}\sigma + (k+\epsilon)\hat{B}\nabla_{\Sigma}(\hat{B}\nabla_{\Sigma}\cdot\boldsymbol{V}) + 2\epsilon K\hat{B}\hat{B}\boldsymbol{V} - \frac{\epsilon}{g}\hat{B}\nabla_{\Sigma} \\ &\quad \times (\nabla_{\Sigma}\times\hat{B}\boldsymbol{V}) - \frac{2\epsilon}{g}\hat{B}\times\hat{\Pi}(\nabla_{\Sigma}\times\boldsymbol{V}_{\perp}) + 2\boldsymbol{N}H\sigma \\ &\quad + 2\boldsymbol{N}H(k+\epsilon)(\hat{B}\nabla_{\Sigma}\cdot\boldsymbol{V}) + \frac{2\epsilon}{g}\boldsymbol{N}\hat{B}\times\hat{\Pi}\nabla_{\Sigma}\times\boldsymbol{V}. \end{aligned} \tag{8.56}$$

In this equation we have $\boldsymbol{V} = \boldsymbol{V}_{i,e}$ and $\hat{B} = (B^{i}_{\alpha})$. Also we introduced $\boldsymbol{V}_{\perp} = (\boldsymbol{V}\cdot\boldsymbol{N})\boldsymbol{N}$ and $\hat{B}\times\hat{\Pi} = (B^{i}_{\alpha}\epsilon^{\alpha\lambda}\Pi_{\lambda\beta}\epsilon^{\beta\mu})$, etc. The first five terms on the RHS are tangent terms, and the last four terms are normal terms. The second term is the gradient of the surface pressure. This term occurs if the coefficient of surface tension is not uniform over the surface, or if it depends on the local curvature or velocity. The third term is dilatational force, and it is important, for example, for a surface that is highly contaminated with an insoluble surfactant. The fourth term has pure geometrical nature (proportional with velocity and Gaussian curvature). The fifth term is just the surface equivalent of a $\nabla\times\nabla\times$ curl–curl type of term, and the sixth term is responsible for surface vortexes. The seventh term is the normal surface tension term, and usually the dominant term in the dynamics of liquid drops, bubbles, and shells. The eighth term is also dilatational force (but normal) and the last one is the normal shear.

Equation (8.56) is the net force $\boldsymbol{F}_{\text{net}}$ acting on a material interface to be introduced in the equations for the balance of momentum across the surface, i.e., (10.22) in Sect. 10.2. To handle all the terms in (8.56) in different systems of curvilinear coordinates, we need to write them in components

$$\begin{aligned} \varsigma\frac{dV^{i}}{dt} &= F^{i} + B^{i}_{\alpha}g^{\alpha\beta}\nabla_{\Sigma,\beta}\sigma + (k+\epsilon)B^{i}_{\alpha}g^{\alpha\beta}\nabla_{\Sigma,\beta}(g^{\lambda\mu}B^{j}_{\lambda}\nabla_{\Sigma,\mu}V_{j}) \\ &\quad + 2\epsilon KB^{i}_{\alpha}g^{\alpha\beta}B^{j}_{\beta}V_{j} - \epsilon B^{i}_{\alpha}\epsilon^{\alpha\beta}\nabla_{\Sigma,\beta}[\epsilon^{\lambda\mu}\nabla_{\Sigma,\lambda}(B^{j}_{\mu}V_{j})] \\ &\quad - 2\epsilon B^{i}_{\alpha}\epsilon^{\alpha\lambda}g_{\lambda\beta}\epsilon^{\beta\mu}\nabla_{\Sigma,\mu}(N^{j}V_{j}) + 2\sigma HN^{i} \\ &\quad + 2H(k+\epsilon)N^{i}B^{j}_{\lambda}g^{\lambda\mu}\nabla_{\Sigma,\mu}V_{j} + 2\epsilon N^{i}B^{j}_{\lambda}\epsilon^{\lambda\alpha}g_{\alpha\beta}\epsilon^{\beta\mu}\nabla_{\Sigma,\mu}V_{j} \end{aligned} \tag{8.57}$$

where $\epsilon^{\alpha\beta}$ is the Levi–Civita symbol, and should not be mistaken for the dilatational viscosity coefficient ϵ which carries no labels.

8.7 Problems

1. Prove, by using properties of the normal to the surface, that points of constant convected coordinates on a moving surface have their space velocity normal to the surface. Hint: parameterize the surface F with (Z^1, Z^2) like in (8.11). Use the fact that we can define a direction normal to the surface $\boldsymbol{r}_u \times \boldsymbol{r}_v$ as a 2-form in $\mathbb{R}^3$ like

$$\omega_N = \frac{\partial z^i}{\partial Z^1}\frac{\partial z^j}{\partial Z^2} dx^i \wedge dx^j .$$

Then show that this 2-form is related to the area 2-form $A\, dZ^1 \wedge dZ^2$ defined in the two-dimensional manifold $\boldsymbol{r}_t(F)$ by the relation

$$\sqrt{\left(\omega_N ; \frac{\partial \boldsymbol{r}_t}{\partial Z^1}, \frac{\partial \boldsymbol{r}_t}{\partial Z^2}\right)}\, dZ^1 \wedge dZ^2 = A\, dZ^1 \wedge dZ^2,$$

where $(\omega; \boldsymbol{v}_1, \boldsymbol{v}_2)$ is the inner product between forms and vector fields defined in Sect. 4.6.

2. Generalize the definition of the convective velocity from Sect. 8.2 in terms of the action of the mixed covariant derivative (4.59) on a tensor field defined on F.

Part II
Solitons and Nonlinear Waves on Closed Curves and Surfaces

Many physical, chemical, and biological systems can be described to a satisfactory extent through the properties of their shapes. To model such systems, a description in terms of the dynamics of the boundaries is necessary, i.e., the evolution of shapes or contours, and their interactions with the inside and the exterior. The interaction with the inner part of the system is usually described through mathematical representation theorems, like the well-known Stokes, Gauss, Green, or Cauchy relations.

Examples of such systems can be given at any physical scale [32, 137, 138, 147, 270, 332]. In heavy and superheavy exotic nuclei, the potential energy of the nuclear shape is relevant for many phenomena including alpha decay, exotic radioactivity, existence of cold valleys, neutron-less fission, and ultra-heavy ions generation. Such topics are important subjects of fundamental physics, like the extension of the Periodic Table of Elements into the antimatter and strange-matter areas [113]. Other examples are provided by the flow of an incompressible fluid with free boundary, like droplets, bubbles, and liquid shells. Here the mechanics and thermodynamics of the free surface are related to the couplings between surface oscillation modes and waves, formation of necks, breakup process, etc. [9, 61, 91, 130, 157, 253, 319, 332]. Other examples are polymer chains, dynamics of vortex filaments in fluid dynamics, growth of dendritic crystals in a plane, or motion of interfaces.

A very important and recent field of research is represented by vortex patterns and dynamics in (mesoscopic) superconductors with direct applications in quantum dots, new generation of computers, spintronics and high temperature superconductivity [211, 229, 254]. At mesoscopic scale the characteristic length for the magnetic field diffusion in a superconductor of type II, and its size are larger than the average transverse size of a vortex (coherence length). This situation favors the penetration of stable quantized vortices in the material. The vortices can be detected through their induced currents, so they are an excellent candidate for fast and large capacity memory devices. The vortices are stable solutions of the nonlinear Ginzburg–Landau equation (GLE) and their dynamics resembles the solitons dynamics. Moreover, the geometry and topology of these vortex structures are strongly influenced by the boundaries (also a feature of the mesoscopic scale) which makes the subject interesting for the topics treated in this book.

The dynamics of the free surface is also important for biology, in swimming mechanisms, motile cell dynamics, pathogen agents spreading, and even the evolution of large populations of individuals like bird flocks or fish schools, etc. [222]. At a larger scale, the dynamics of the free surface of a neutron star is important in the study of the gravitational waves emitted by such tides or deformations [48,175,247].

In the description of systems with free boundaries, where the dynamics of the contours/shapes is important, there are interesting connections between local and global geometrical quantities. The local quantities are those intrinsic mathematical properties of the boundaries defined within neighborhood of points, like the fundamental forms, contact structures, curvature, torsion, etc. The global geometrical quantities are the integral quantities, like surface area, curve length or perimeter, geodesics structure, etc. The global quantities are related to geometrical and topological invariants, like homotopy and homology structure, and ultimately they are connected to the physical conservation laws of the system. In general, mathematical global constrains applied to free boundary systems result in long range, nonlocal interactions. Usually, such constraints are handled by considering them as Lagrange multipliers coupled to the general bulk plus surface conserved quantities, like mass, momentum, angular momentum, etc. In the case of contours or surfaces described by nonlinear integrable systems, with cnoidal waves, solitary waves, or soliton solutions for example, the Hamiltonian generates itself an infinite countable set of conservation laws, without introducing any other global constrains.

Chapter 9
Kinematics of Hydrodynamics

The goal of this chapter is to discuss the general frame of hydrodynamics, like particle trajectories (path lines), stream lines, streak lines, free surfaces, and fluid surfaces, and to compare their behavior in the Eulerian and Lagrangian frames. The following sections and chapters proceed on the assumption that the fluid is practically continuous and homogenous in structure. Of course, the concept of continuum is an abstraction that does not take into account the molecular and nuclear structure of matter. In that, we assume that the properties of the fluid do not change if we consider smaller and smaller amounts of matter [167]. May be the wisest point of view while we remain at the level of general laws of fluid dynamics (or fluid mechanics) is to keep the physical scales rather vague [220]. This aspect is in direct relation with the fact that these laws can be made dimensionless in a large variety of situations.

9.1 Lagrangian vs. Eulerian Frames

In fluid dynamics there are two possible approaches for the dynamical equations: the Lagrangian (also called material or convected) frame and the Eulerian (also called the spatial) frame. In the Lagrangian frame we identify and label individual particles of fluid, and we setup the frame such that particles retain their coordinate labels in time. In this approach, it is more likely to use topology and group continuous transformation tools. The Eulerian frame describes the fluid from a stationary lab frame. The motion of fluid is recorded at a fixed point vs. time. In this approach the mathematical tools are more related to geometry and field theory. In the following, we use the Eulerian approach, unless an explicit statement is made to the contrary. The fields that characterize the fluid are defined on some domains in the three-dimensional Euclidean space and they have a certain degree of mathematical smoothness. The degree of smoothness is chosen for a given fluid model such that the coarse grain structure of the infinitesimal fluid particles introduced above

A. Ludu, *Nonlinear Waves and Solitons on Contours and Closed Surfaces*, Springer Series in Synergetics, DOI 10.1007/978-3-642-22895-7_9,

is not seen by the differential equations (i.e., the molecular structure of the matter). In other words, the fluid particle is small enough to allow the existence of smooth space–time differentials, but large enough to average the molecular and quantum properties over its volume. The fields under consideration are the velocity field $\boldsymbol{v}(\boldsymbol{r},t)$, the nonnegative defined mass density $\rho(\boldsymbol{r},t)$, and the pressure field $P(\boldsymbol{r},t)$. Of course, function of necessity, we can add the distribution of energy, free energy, enthalpy, entropy, force density, or other fields of interest [167, 171] to these fields. We assume, unless otherwise specified, that these fields are smooth enough so that the standard calculations may be performed on them.

9.1.1 Introduction

In practice we consider $\boldsymbol{r} = (x, y, z) \in D$ a point in domain D filled with fluid, and consider the particles moving in space and time. In the Lagrangian approach, at every moment of time t we defined the spatial velocity of a certain particle of fluid as $\boldsymbol{V} = \frac{d\boldsymbol{r}}{dt}$.

The Eulerian velocity field (spatial velocity field) $\boldsymbol{V}(\boldsymbol{r},t)$, in principle not constant in time, is the velocity of a fluid particle that passes at moment t through the point $\boldsymbol{r}$. The Lagrangian frame is attached to that fluid particle, and it records the changes in velocity, density, etc., happening with this particle vs. its own local time, measured with a clock attached to it. In such a Lagrangian system, physical quantities have a complex time dependence. While traveling, the fluid particle has its physical quantities measured in the local frame, so they experience a global time variation (also called total or Lagrangian or material time derivative) denoted by $\frac{d}{dt}$, or identified by placing a dot on the top of the quantity (sometimes it is also denoted $\frac{D}{Dt}$). A part of this time variation happens because the particle travels through different domains of space, hence experiencing different constraintts. Such a partial variation is called Eulerian, or partial, and it is denoted $\frac{\partial}{\partial t}$ or simply by the subscript t. For example, we choose a fluid particle moving according to the law $\boldsymbol{r}_L(t)$, and we measure the scalar quantity $q(t) \equiv q(\boldsymbol{r}_L(t), t)$ associated to this particle, in this frame. The same quantity can be described in a fixed Eulerian frame, $Q(\boldsymbol{r},t)$. The relation between these two formalisms is given by

$$\dot{q} = \frac{dq}{dt}(\boldsymbol{r}_L(t), t) = \frac{\partial Q}{\partial t}(\boldsymbol{r},t) + \boldsymbol{V}(\boldsymbol{r},t) \cdot \nabla Q(\boldsymbol{r},t), \tag{9.1}$$

where ∇ is the gradient operator $(\frac{\partial}{\partial x}, \frac{\partial}{\partial y}, \frac{\partial}{\partial z})$, and $\cdot$ represents the usual Euclidean scalar product. Equation (9.1) is a well-known transformation law in hydrodynamic literature, yet is valid in a very restricted sense, namely only for scalar quantities and for the fluid velocity vector. If we try to apply the transformation (9.1) to a general vector field or to a covariant tensor field, the result fails, because the resulting quantity is not anymore a geometrical object of the same type. To keep the geometrical properties intact, we need a generalization of (9.1) for arbitrary covariant/contravariant geometrical objects ω. This is the *covariant time derivative* (also called convected or material time derivative) and it is defined by

$$\frac{d_c\omega}{dt} = \frac{\partial \omega}{\partial t} + \boldsymbol{v}(\omega), \tag{9.2}$$

where $\boldsymbol{v}(\omega)$ is the Lie derivative with respect to the flow $\boldsymbol{v}$. This generalization is introduced in Sect. 9.2.6.

9.1.2 Geometrical Picture for Lagrangian vs. Eulerian

We introduce the working space $(t, \boldsymbol{r}) \in \mathbb{R} \times \mathbb{R}^3$. From the Lagrangian point of view, the fluid particle motions are nonintersecting regular curves Γ_L in this base space, parametrized by time and described by equations $\boldsymbol{r}_L(t, \boldsymbol{r}_0)$. They are called *paths* or *material lines* [10] or *lines of motion* [167]. Since they do not intersect, each such curve is labeled by one of its points, $\boldsymbol{r}_0$, for example the position of the particle when $t = 0$. The tangent to this curve is

$$\boldsymbol{t}_L = \frac{(1, \boldsymbol{v}_L)}{\sqrt{1 + v_L^2}},$$

where $\boldsymbol{v}_L = \partial \boldsymbol{r}_L(t, \boldsymbol{r}_0)/\partial t$ is the Lagrangian velocity of the particle along the path. All these paths do not intersect and completely fill the base space when $\boldsymbol{r}_0 \in \mathbb{R}^3$.

If we choose a fixed point in space $\boldsymbol{r}$, some of the paths $\boldsymbol{r}_0$ will intersect this fixed point, $\boldsymbol{r}_L(t, \boldsymbol{r}_0) = \boldsymbol{r}$, so that we can write the "list" of these particles vs. time: $\boldsymbol{r}_0 = \boldsymbol{r}_0(t, \boldsymbol{r})$. Now, we can define the Eulerian velocity at $(t, \boldsymbol{r})$ by substituting this $\boldsymbol{r}_0(t, \boldsymbol{r})$ list in the velocity expression

$$\boldsymbol{v}_E(t, \boldsymbol{r}) = \boldsymbol{v}_L(t, \boldsymbol{r}_0(t, \boldsymbol{r})). \tag{9.3}$$

Example 1. We can illustrate the relation between Lagrangian and Eulerian velocities (9.3) with a simple one-dimensional example. Water is dripping downward from a hole in gravitational field, and different water molecules depart from the hole at different initial moments of time t_0. So the Γ_L curves are vertical parallel lines. Their laws of motion are

$$z(t) = \frac{g(t - t_0)^2}{2}.$$

In terms of some initial position z_0 their Lagrangian equations of motion read

$$z_L(t, z_0) = \frac{g}{2}\left(t - \sqrt{\frac{2z_0}{g}}\right)^2,$$

with

$$v_L(t, z_0) = g\left(t - \sqrt{\frac{2z_0}{g}}\right).$$

If we choose a reference level at z and equate $z = z_L$, we obtain

$$z_0 = \frac{g}{2}\left(t - \sqrt{\frac{2z}{g}}\right)^2$$

with the following signification: What is the initial position z_0 (at $t = 0$) of a particle to pass through the level z at the moment t? The resulting Eulerian velocity is, according to (9.3),

$$v_E(t, z) = v_L(t, z_0(z, t)) = \sqrt{2zg} = \text{const.},$$

as it should be from mechanics.

Now, we introduce a physical quantity Q defined for any fluid particle. For the particle labeled by $\boldsymbol{r}_0$ the Lagrangian value $Q_L(t, \boldsymbol{r}_0)$ is defined along Γ_L. Suppose this Γ_L intersects a fixed line $\boldsymbol{r}$ =const. at $\boldsymbol{r}_L(t, \boldsymbol{r}_0) = \boldsymbol{r}$. By solving this equation with respect to $\boldsymbol{r}_0$, we have $\boldsymbol{r}_0 = \boldsymbol{r}_0(t, \boldsymbol{r})$. We can define now the Eulerian value of Q by

$$Q_E(t, \boldsymbol{r}) = Q_L(t, \boldsymbol{r}_0(t, \boldsymbol{r})). \tag{9.4}$$

While following the particle in its motion, the quantity Q_L has a variation $dQ_L(t, \boldsymbol{r}_0) = (dQ_L/dt)dt$. At $\boldsymbol{r}$ =const., the quantity Q_E has another variation $dQ_E = (\partial Q_E/\partial t)dt$. By differentiation of (9.4) we have $dQ_L = dQ_E + (d\boldsymbol{r}_L \cdot \nabla Q_E)dt$. Since we follow the particle in its motion we have $d\boldsymbol{r}_L = \boldsymbol{v}_L dt$. Since all these relations are infinitesimal, and all are taken at $(t, \boldsymbol{r})$, we can use either $\boldsymbol{v}_E$ or $\boldsymbol{v}_L$ in them. In the end we obtain the classical relation between the Lagrangian and Eulerian variations of a physical quantity

$$\frac{dQ_L}{dt} = \left(\frac{\partial Q_E}{\partial t} + (\boldsymbol{v}_E \cdot \nabla)Q_E\right). \tag{9.5}$$

In local (Eulerian) coordinates $(t, \boldsymbol{r})$, this equation reads

$$(t, \boldsymbol{r}) \to \frac{dQ_L}{dt}(t, \boldsymbol{r}_0(t, \boldsymbol{r})) = \left(\frac{\partial Q_E}{\partial t} + (\boldsymbol{v}_E \cdot \nabla)Q_E\right)_{(t,\boldsymbol{r})}. \tag{9.6}$$

In the Lagrangian coordinates $(t, \boldsymbol{r}_0)$, same equation reads

$$(t, \boldsymbol{r}_0) \to \frac{dQ_L}{dt}(t, \boldsymbol{r}_0) = \left(\frac{\partial Q_E}{\partial t} + (\boldsymbol{v}_E \cdot \nabla)Q_E\right)_{(t,\boldsymbol{r}=\boldsymbol{r}_L(t,\boldsymbol{r}_0))}. \tag{9.7}$$

The Lagrangian motion of particles is represented by a family of curves Γ_L filling the base space, and the Lagrangian velocity is a vector field defined on this base space, parametrized by the flow lines. The Eulerian velocity is the same differential vector field, except is parametrized by local coordinates, like any regular field. Consequently, a Lagrangian physical quantity Q_L is represented by a family of curves Γ_Q lying in a base space $\mathbb{R} \times \mathbb{R}^3 \times \hat{Q}$, where $Q \in \hat{Q}$. The Eulerian value of the same quantity is a regular surface $Q_E(t, \boldsymbol{r})$ parametrized by the base space and immersed in $\mathbb{R} \times \mathbb{R}^3 \times \hat{Q}$. The Eulerian derivative is the partial derivative of Q_E. The particle paths Γ_L have tangents

$$\boldsymbol{t}_L = \frac{1}{\sqrt{1 + v_L^2}}(1, \boldsymbol{v}_L).$$

The curves for Q_L lying in the base space have tangents

$$\hat{t}_Q = \frac{1}{\sqrt{1 + v_L^2 + \dot{Q}_L^2}}(1, \boldsymbol{v}_L, \dot{Q}_L),$$

where the dot means time differentiation. In this geometrical context, the relation between Lagrangian and Eulerian variations (9.5) reads

$$\dot{Q}_L = D_{t_{\Gamma_L}} Q_E, \quad \text{or} \quad \dot{Q}_L(t) = (Q_E \circ \Gamma_L)'(t).$$

The Lagrangian derivative is just the directional derivative of the function Q_E along the particle path, see Fig. 9.1.

9.2 Fluid Fiber Bundle

Hydrodynamics studies the motion of fluid particles. The combination between the discrete labeling of the system of particles on one hand, and the smooth dependence of physical quantities on time on the other hand enhances the importance of families of curves for hydrodynamical systems. Somehow, this fact has a geometrical background arriving from the importance of compact submanifolds (closed curves, closed surfaces) for vector fields and flows (see Sect. 4.5) and [196].

9.2.1 Introduction

Curves of special interest, parametrized by time, are the path lines, stream lines, streak lines, and vorticity lines, studied from both Lagrangian and Eulerian points of view (Sect. 9.1.2). Moreover, there are the fluid particle lines (also called material

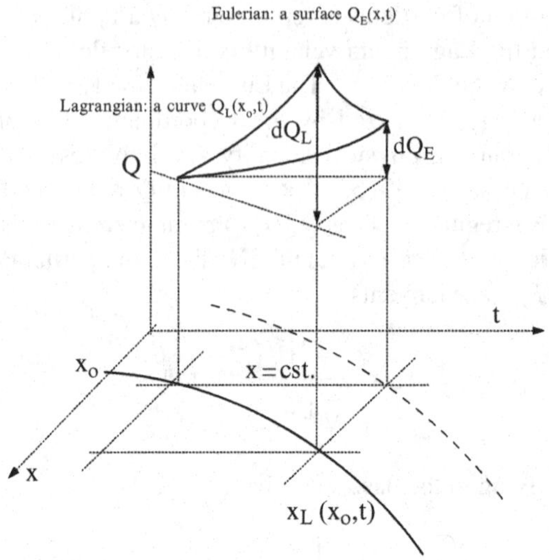

Fig. 9.1 The Lagrangian–Eulerian point of view for a one-dimensional flow. The path of a fluid particle is represented in the base horizontal plane by the curve $x_L(x_0, t)$; all such fluid paths are labeled by their x_0 initial points. The mapping of the fluid path into the base space of a physical observable Q is a curve $x_Q(x_0, t)$, i.e., the Lagrangian value of the physical quantity $Q_L(x_0, t)$. The Lagrangian variation along the fluid path is dQ_L in a certain dt. But, if we measure Q at a constant position x, we have its Eulerian value, and consequently its Eulerian variation dQ_E for the same time interval dt. The Eulerian value $Q_E(x, t)$ actually represents the Lagrangian value associated to another particle (*dashed line*) that actually moves through the same spot x at $t + dt$. When fluid particles fill up the space x and move, the Lagrangian values of the physical quantities associated to the particles of fluid generate curves, but the Eulerian values generate a surface

lines, particle contours, or circuit lines) and filaments especially important in conservation laws. We can raise the question if such particle contours are stable or they break at a certain point, or if they are invariant, etc. For example, to use the Kelvin or Ertel's theorems for closed contours (Theorem 10.3) related to invariants of the fluid dynamics, we need to have rigorous definition for the material lines of fluid particles than just intuition.

Example 2. To exemplify such a possible situation, when a particle contour can deform up to a breaking point (because of a stagnation point of the flow, for example) we choose an incompressible inviscid irrotational two-dimensional flow past a cylinder. To solve the flow we use a conformal mapping procedure. The velocity field is represented by $v(z) = \phi_x + i\phi_y$, $z = x + iy$, and it is tangent to the curves ϕ =const. because of the Riemann–Cauchy conditions. We build the holomorphic function $H(z) = \Phi(x, y) + i\Psi(x, y)$ where Φ is the potential function and Ψ is the stream function, i.e., the harmonic conjugate function to Φ. We have

$$v = \frac{dH^*}{dz},$$

and the cylinder contour Γ equation is $x^2 + y^2 = 1$. We perform the transformation $u + iv = \omega = f(z) = z + z^{-1}$. The cylinder contour transforms into $f(\Gamma) = \{z|v = 0\}$. A solution of the Laplace equation in the ω coordinates and for the boundary condition $\omega = 0$ on $f(\Gamma)$ is $G(\Phi) = \Phi_0\omega$. We have

$$H(z) = G \circ f(z) = A\left(z + \frac{1}{z}\right).$$

For example, in polar coordinates the stream lines (Ψ =const.) become

$$\Psi_0\left(r - \frac{1}{r}\right)\sin\phi = C = \text{const.}$$

The equation of the stream lines becomes

$$r(\phi) = \frac{r_0 + \sqrt{r_0^2 + 4\sin^2\phi}}{2\sin\phi}$$

and the Eulerian velocity is

$$v = \Psi_0\left(\frac{-y\cos\phi(x^2 + y^2 - 1) + x\sin\phi(x^2 + y^2 + 1)}{(x^2 + y^2)^{3/2}},\right.$$

$$\left.\times\frac{x\cos\phi(x^2 + y^2 - 1) + y\sin\phi(x^2 + y^2 + 1)}{(x^2 + y^2)^{3/2}}\right).$$

From the Euler equation the pressure becomes

$$P = \Psi_0^2\rho\frac{2(x^2 - y^2) - 1}{2(x^2 + y^2)^2},$$

where ρ is the density. In Fig. 9.2 we present the pressure distribution around the cylinder contour. The Lagrangian paths of fluid particles are obtained by numerical integration of the equations

$$\frac{\partial^2 x}{\partial t^2}\frac{\partial x}{\partial x_0} + \frac{\partial^2 y}{\partial t^2}\frac{\partial y}{\partial x_0} = -\frac{1}{\rho}\frac{\partial P}{\partial x_0}, \ldots \text{etc.}$$

In Fig. 9.3 we present the isobaric and stream lines, and the evolution of a particle contour line (thick line). Initially we choose all particles of this contour line to lie along a vertical segment. Then, we calculate their Lagrangian positions at a later moment of time. We notice the tendency of the contour line to spread and tear. In an extreme example this line may even be broken by possible abrupt changes in the Lagrangian velocities. This example shows that it makes sense to analyze the geometry and stability of particle contours for a general flow.

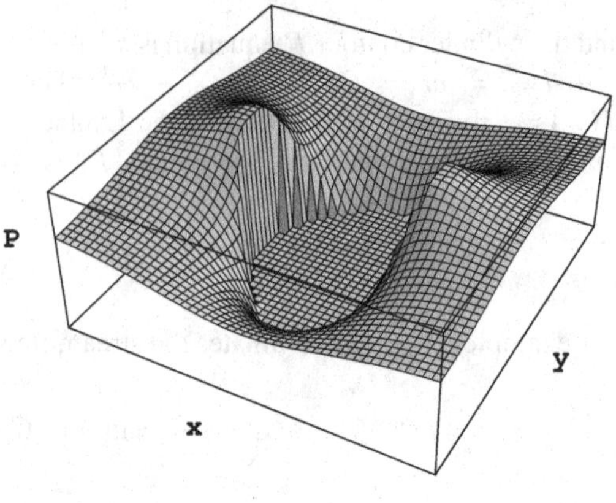

Fig. 9.2 Pressure distribution for a two-dimensional incompressible inviscid irrotational flow past a cylinder

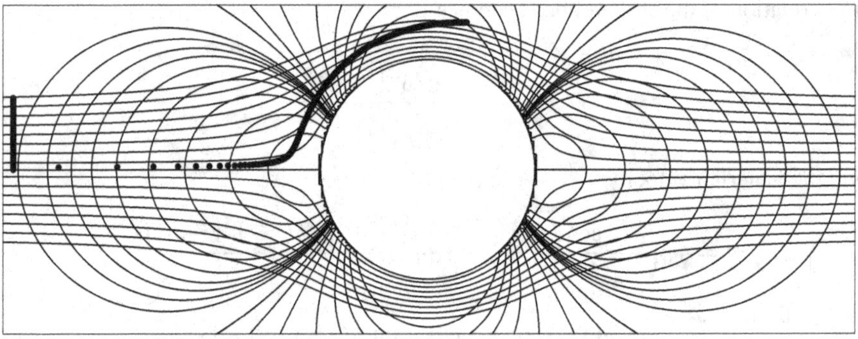

Fig. 9.3 Stream lines and isobaric lines (*thin lines*) for a two-dimensional incompressible inviscid irrotational flow past a cylinder. *Thick lines*: a finite particle contour at $t = 0$ (the vertical segment), and its Lagrangian flow at a later moment of time

9.2.2 Motivation for a Geometrical Approach

We can alwaysfs present a fluid using the following traditional picture of the flow, also introduced in Sect. 9.1.2. We introduce the available space for the fluid (the reference fluid container [212, 215]) as a domain D of $\mathbb{R}^3$, and add an extra dimension for time to form a base space $D \times \mathbb{R}$. The particle paths $\boldsymbol{r}_L(\boldsymbol{r}_0, t)$ are smooth time-parametrized curves in this base space. The projection on the horizontal planes (projections perpendicular on the time axis) of the tangent vectors to these curves represents the velocity fields of the particles. The two velocities, i.e., the Lagrangian (material) and Eulerian (spatial) velocities, have the same value at the same point of the base space. The only difference between these two types of velocities consists in the parametrization of the vector fields. The Lagrangian velocity field is defined along the particle paths in the base space, while the Eulerian velocity field is defined on the horizontal plane, in points where these paths intersect it, at a moment of time t. The integral curves of the Eulerian velocity field contained in any "horizontal" plane are the stream lines at that moment of time. However, the path lines do not identify with the lift of the stream lines in the base space. Namely,

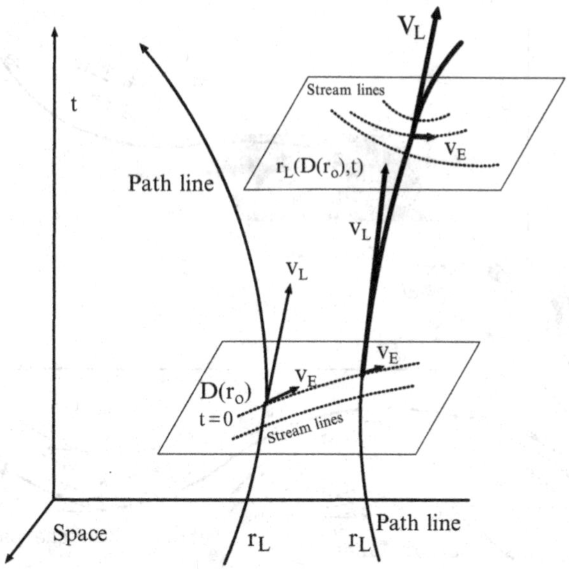

Fig. 9.4 A two-dimensional fluid domain $D(\boldsymbol{r}_0)$ shown at two moments of time 0, t, and two path lines $\boldsymbol{r}_L(t)$ whose tangents are the Lagrangian velocities $\boldsymbol{v}_L$. The projection of the Lagrangian velocity field on the tangent space of the fluid domain is the Eulerian velocity field $\boldsymbol{v}_{\mathcal{E}}$. The integral curves of the Eulerian vector field in the fluid domain, at a given moment of time t, are the stream lines at that moment (*dotted lines*). The projections of the path lines on the fluid domain do not coincide with path lines in general

if we choose a point $\boldsymbol{r}$ in some horizontal plane t and we compare the path line crossing through this point, and the vertical lift of the stream line crossing the same point, these two curves are different in general. An example is presented in Fig. 9.4.

In Fig. 9.5 we show another example of path lines and stream lines, when the particle moves along an open path, but locally the stream lines may appear to be closed.

For any given fixed point $\boldsymbol{r}_0$ in the initial plane, we can draw all paths crossing this at different moments of time (Fig. 9.6). The intersections of all these paths with a certain horizontal plane t generate a streak line initiated by a "nozzle" placed at $\boldsymbol{r}_0$. In traditional approaches, see for example [11, 54, 220, 305], the motion of the particles is described by a one-parameter (time) group of diffeomorphisms acting on the domain $D(\boldsymbol{r}_0)$. The Lagrange coordinate of a particle is the result of the action of this group on the corresponding element $\boldsymbol{r}_0$. If the motion is incompressible, the group of diffeomorphisms is volume preserving. In this formalism, the infinitesimal generator of the group is the Lagrangian field of velocities.

However, even practical, such a model is not quite perfect. That is because we tend to associate the same geometrical space to physical spaces with different signification, namely the material points (initial positions space), and the spatial points per se. Even if initially ($t = 0$) the positions $\boldsymbol{r}_0$ of all fluid particles, $\boldsymbol{r}_0 \in D$, belong to the position space during the motion, these vectors actually form

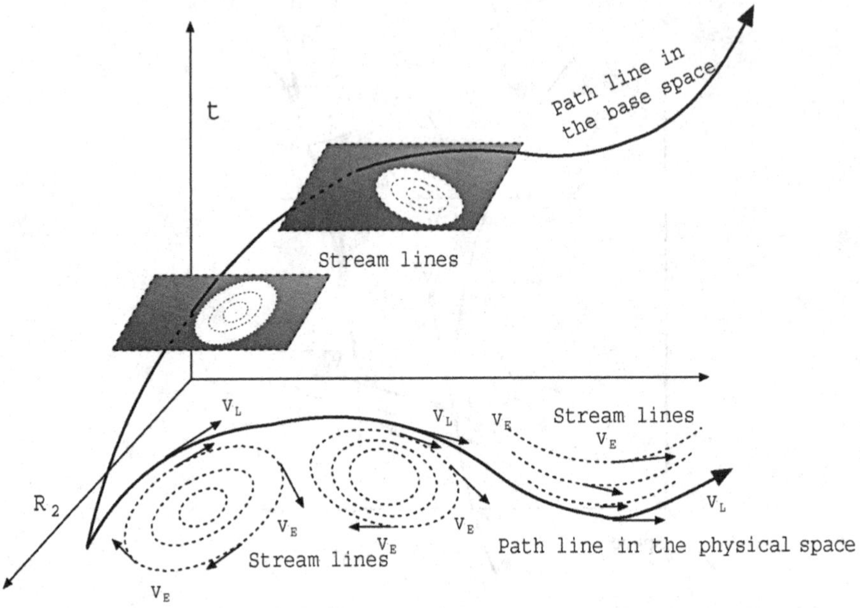

Fig. 9.5 A two-dimensional example. A path line in the physical space $\mathbb{R}^2$ (*horizontal solid curve*) and in the base space X (*lifted solid curve*), and associated stream lines at different moments of time (*dashed lines*)

Fig. 9.6 Same space as in Fig. 9.4, except we present several paths emerging from the "nozzle" point $\boldsymbol{r}_0$ (*dashed-dotted axis*) at different moments of time. The intersections of all such paths with a horizontal plane t provide a streak line (*dotted*) generated by the "nozzle" at t

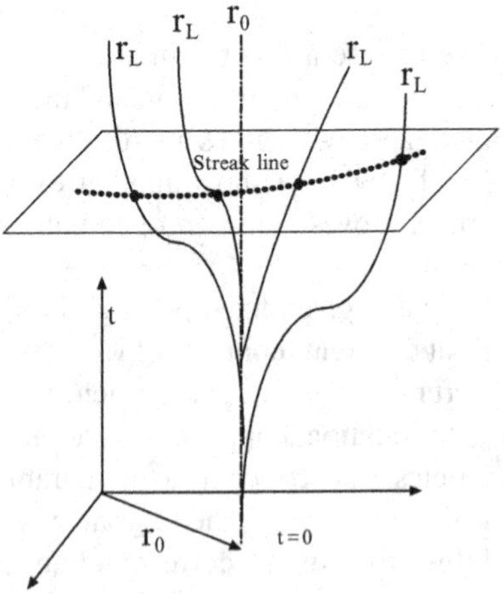

a space of parameters, labeling the particles. On the other hand, the positions of the particles at any arbitrary moment of time (given by the Lagrangian equations of motion $\boldsymbol{r}_L(\boldsymbol{r}_0, t)$) belong to a space of positions. The above picture does not make this difference a geometrical difference, and in that is incomplete and difficult to generalize for more complicated flows. For example, in Fig. 9.4, we can see that the stream lines at different moments of time belong to different planes. We need to make the distinction between the material space and the space of positions from a geometrical perspective. This is possible by using a fiber bundle structure instead of a common space.

9.2.3 The Fiber Bundle

We present a formalism in which a fluid is described using cross-sections σ in a fiber bundle $\mathcal{F}$ over some base manifold X. For the definitions and properties of a fiber bundle, the reader can check Sect. 4.9 and its references [101,212,215,306]. An intuitive picture of a fiber bundle consists in taking a certain manifold called fiber F, and assigns a homeomorphic transformation of F to any point of a base manifold X, constructing a sort of a local cartesian product. In the case of a fixed container for the fluid (even the case of the whole space), the traditional model is to consider the base as the space of particles (usually labeled by their initial positions) and the fiber is the space available for particle positions (see Fig. 9.7, left). On the contrary, a free surface introduces one more freedom in the problem. We cannot construct it using the same pattern (see Fig. 9.7, center) because we allow different particles to belong to different shapes simultaneously, which is impossible. A possible choice to build a fiber bundle is borrowed from the mechanics of deformable bodies (see Fig. 9.7, right). The base space is the manifold of all possible shapes, and the standard fiber is particle position space. The role of the particle labeling space is taken over by the nontrivial structure group.

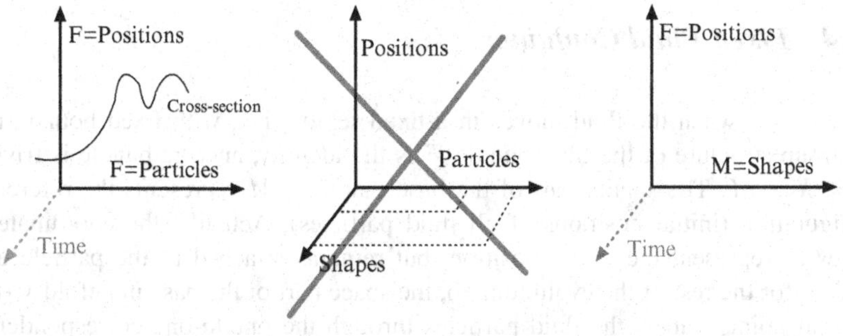

Fig. 9.7 Possible fiber bundle structures (M, F) for fluid dynamics problems. *Left*: In the case of no free surface the base space is the space of particles, and the fiber is the space available for the particles positions; *Center*: A free fluid surface introduces more freedom in the problem making the previous (Left) structure inoperable. It would allow different particles to belong to different shapes simultaneously, which is impossible; *Right*: Mechanics of deformable bodies model for the fiber bundle. The base space is the manifold of all possible shapes, and the standard fiber is particle position space. *Dotted line* means that time does not need necessarily to be included explicitly in the geometry picture

The base manifold (for the nonrelativistic case) is usually a space–time manifold built as a product between a smooth three-dimensional oriented Riemannian manifold (M, g), where g is the metric, and $\mathbb{R}$ for time, i.e., $X = M \times \mathbb{R}$. The coordinates in X are $x = (x^\mu) = (x^i, t) \in X$, with $i = 1, \ldots, 3, \mu = 1, \ldots, 4$. For fluid dynamics we can choose the fiber $F = M$ with coordinates $y \in F$ [215]. Consequently, the local coordinates in this $\mathcal{F}$ bundle over X are (x, t, y), and the projection is $\Pi : \mathcal{F} \to X, \quad (x, t, y) \to (x, t)$. Transformations and operations that affect only the base (spatial changes like rotations, etc.) are called fiber-preserving transformations. A lift of any geometrical object γ (a curve, surface, function, form, etc.) defined in the base space is a map of this object into the fiber bundle, $\gamma \to \gamma' \in \mathcal{F}$, such that it projects back down to the original object in M, $\Pi \circ \gamma' = \gamma$.

Cross-sections in this bundle $\sigma : X \to \mathcal{F}$ represent time-dependent *configurations*, i.e., particle position fields. The cross-section has the coordinates $\sigma(x) = (x^\mu, \sigma^i(x)) = (x^\mu, y^i)$. On the top of the configuration bundle $\mathcal{E}$, we can construct another fiber bundle $J^1\mathcal{F}$ over $\mathcal{F}$ called the first jet bundle [215,242], with the fiber above (x, y) consisting of linear maps from the tangent space of the base space to the tangent space of the bundle, $\gamma : T_x X \to T_{(x,y)}\mathcal{F}$, satisfying $d\pi \circ \gamma = Id_{T_x X}$.

For any cross-section σ in $\mathcal{F}$ over X, the differential $d\sigma_x$ at x (also called tangent map, see Sect. 4.1) is an element of the jet bundle $J^1\mathcal{F}_{\sigma(x)}$. Consequently, the map $x \to d\sigma_x$ is a cross-section of the jet bundle over X. This section, denoted $j^1\sigma$, is called the first jet extension of σ. In coordinates, it is given by $j^1\sigma(x) = (x^\mu, \sigma^i(x), \partial_\mu \sigma^i)$, where $\partial_\mu = (\partial_i, \partial_t)$. It is this triple which represents the fluid motion. The first three base coordinates space components x^i, originally coming from the initial positions of the fluid particles, now represent the particle labeling. The $\sigma^i(x)$ components identify the position of the x particle in space, and the $\partial_t \sigma^i$ components represent the velocity of the particle x.

9.2.4 Fixed Fluid Container

For the case when the fluid moves in a fixed region, i.e., with fixed boundaries, the group structure of the fiber bundle $\mathcal{F}$ is the identity, and the bundle is trivial, $\mathcal{F} = X \times M$. The spatial part of the base manifold M represents the reference configuration (initial positions of all fluid particles). Actually, the coordinate x ceases to represent the initial position, but remains attached to the particle and labels it for the rest of the evolution. So, the space part of the base manifold x (the material points) labels the fluid particles through the one-to-one correspondence between particles and their initial positions in the reference fluid container. The time base X corresponds to the time evolution. The fiber over any base point is the same manifold, meaning that the space available for any particle is the same at any moment of time. Its coordinates y are called spatial points. The fiber at any point $F_{(x,t)}$ represents the available space for particle x at the moment t, and it is diffeomorphic with M, i.e., the *reference fluid container* [212, 215]. In the case of $\mathcal{F}$, the requirement for the existence of a projection $\Pi : \mathcal{F} \to X$ from the definition of a fiber bundle (Sect. 4.9, Definition 30) guaranties that all points of the fiber, at any point of the base, are filled with fluid.

The fluid motion is described by a cross-section $\sigma(x,t)$ of the bundle $\mathcal{F}$ representing the particle placement field. Not any cross-section can represent a real motion of the fluid, and some *minimal constraintts* are needed. First, σ is not allowed to create or annihilate fluid particles, and second, two different particles cannot hold the same spatial point at the same moment of time. In the traditional approach presented above (the one not using geometry of a fiber bundle) these two constraintts are fulfilled by requesting that the Lagrangian paths of the fluid particles represent a diffeomorphism of the reference fluid container. In the fiber bundle formalism, these two physical constraintts require a similar thing. The restriction of the cross-section $\sigma(x,t)|_{t=t_0}$ at a constant $t = t_0$ (for every moment of time t_0) needs to be a diffeomorphism of the manifold $F = M$. Of course, this is also possible because the bundle is trivial, and there is a canonical diffeomorphism between any two fibers at any two points.

Let us ignore for a second the deep geometrical implications of the existence of the group of diffeomorphisms, and let us just look at these conditions locally, in terms of coordinates. For some more insight into this topic, we recommend for example [11, 212, 215]. This condition is equivalent to the vector field to be divergence free. This means that the infinitesimal generator of this diffeomorphisms is a divergence-free vector field, or in other words that the flow is incompressible.

In addition, the specific cross-section form should result from a solution of the dynamic equations of motion, for example Euler (10.15) or Navier–Stokes (10.13) equations, under some additional boundary, initial or regularity conditions which may be required, too. This constraint will be addressed in the next chapters. For an explicit discussion of this topics, see for example [215, Theorem 2.1] and reference herein.

In the local coordinates of a given fiber, $y(x,t) \in F_{(x,t)}$ represents the spatial position of the particle x at moment t, $(x,t,y)\sigma(x,t)$. The path lines are the restrictions of the cross-section $\boldsymbol{r}_L(x_0,t) = \sigma|_{x=(x_0,t)}$ for fixed point in the space part of the base space. The tangent vectors to these curves can be expressed in two ways. If we write $\boldsymbol{v}_L(x,t) = \partial\sigma^v(x,t)/\partial t$ we have the Lagrangian (material) velocity field. The superscript v (as in vertical) represents the components of the cross-section along the fiber. The Lagrangian velocity field is actually represented by the last three components of the cross-section in the first jet bundle $d\sigma$. Namely $j^1\sigma = (\sigma, \partial_i\sigma, \boldsymbol{v}_L)$.

Conversely, if we invert the equation $y(x,t)$ with respect to y, we can express the velocity field in coordinates $\boldsymbol{v}_L(x(y),t) = \boldsymbol{v}_E(y,t)$, which is nothing but the Eulerian velocity field. So, even if locally the Eulerian and Lagrangian velocities coincide at the same point of the fiber bundle $\mathcal{F}$, they are vector fields in different spaces. The Eulerian velocity is a vector space defined on the standard fiber manifold F. Indeed, because the fiber at any point $F_{(x,t)}$ is diffeomorphic with the standard fiber F, according to the *minimal constraintts*, we can map vectors tangent to any fiber into vectors tangent to the standard fiber $F = M$. So, a cross-section σ in $\mathcal{F}$ generates a vector field on F at any moment of time, the Eulerian flow. The integral curves of this field are, at every moment of time, the collections of time-dependent stream lines, they lie in the standard fiber, and they have no special assigned parameter (the stream lines collection is also called flow net [111]). Contrary to the stream lines, the path lines are time parametrized, hence constant, and they lie in the fiber bundle. Again, the collection of path lines do not coincide with the flow net in general (they coincide if the flow is stationary). It is also true that the path lines never cross the flow net lines.

If we come back to Fig. 9.4, we understand now the trihedron presented there as the base space, and the horizontal planes as fibers at different points, with their associated Eulerian fields of velocities. The reunion of all path lines forms the cross-section σ.

Since $\sigma(M, t_0) \simeq M$ is a diffeomorphisms because of the *minimal constraintts*, the image of any compact set in M is a compact set in $F_{(x,t)}$. Such sets are the particle structures that remain "stable" to this extent. If such a set is a submanifold of dimension 1, we call it particle line or material line or circuit line, or filament. Once identified in the reference fluid container, this line conserves its topological proprieties in time. If the submanifold is two dimensional, it is a particle surface, or free fluid surface, etc., and so on. We noticed above that the particle paths are restrictions of the cross-sections describing the dynamics for constant x. Similarly, particle lines are restrictions of the cross-section for constant time, and on subsets of the M manifold: $\sigma(x,t)|_{(x \in D, t=t_0)} = \hat{\sigma}(x)|_{x \in D}$.

There is another interesting approach about the path lines as orbits of a group of diffeomorphisms of the spatial part of the base space. Actually, any such diffeomorphism (any flow) can be understood as a relabeling operation of the fluid particles. Such a relabeling operation is connected with a continuous symmetry of the system. If we consider the fluid a Lagrangian system and the flow is incompressible, the Noether current associated to this symmetry is the fluid momentum conservation, see Fig. 9.8.

In the following, we give an interpretation of the transformation between variation of Eulerian and Lagrangian quantities (9.1), (9.6), or (9.7) in terms of a connection.

Let us consider again the fiber bundle $\mathcal{F}$ representing a fluid confined in a fixed space domain identified by the manifold $M \ni (x^i)$, where $i, j = 1, \ldots, 3$ and $\mu = 0, \ldots, 3$. The base space is the direct product $X = M \times \mathbb{R} \ni (x^\mu) = (x^i, x^0 = t)$. We choose the fiber $F = M$, a trivial identity structure group $G = \{e\}$, the projection Π, $F_x = \Pi^{-1}(x)$ and a cross-section $\sigma : X \to \mathcal{F}$. The cross-section

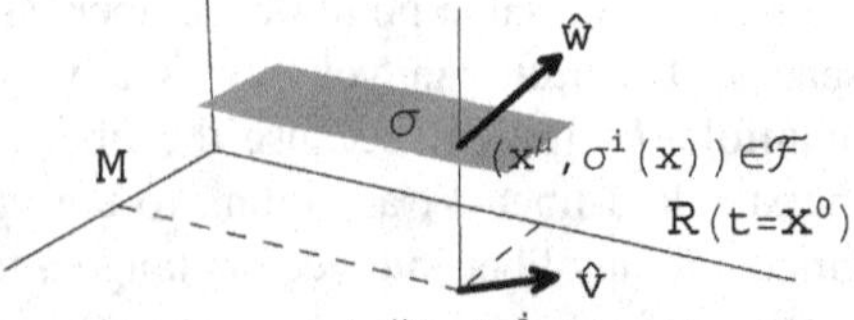

Fig. 9.8 Structure of the fiber bundle associated with a fluid. The axes here are the base space (M), the fiber (F), and the time ($t = x^0$)

maps $x = (x^\mu) \to \sigma^a = (x, \sigma^j(x))$, and its differential $d\sigma : TX \to T\mathcal{F}$ maps $T_xX \ni \hat{v}(x) = (\boldsymbol{v}, v^0) = (v^i, v^0) = (v^\mu) \to \hat{w} = (\boldsymbol{w}, w^0, \bar{w}) = (w^i, w^0, \bar{w}^j) \in T_{\sigma(x)}\mathcal{F}$, with $a = (\mu, j)$. In components, the action of the differential, which is a section in the first jet fiber bundle over $\mathcal{F}$, reads

$$d\sigma(\hat{v}) = \left(\frac{\partial \sigma^a}{\partial x^\mu} v^\mu\right) = \left(\frac{\partial \sigma^\nu}{\partial x^\mu} v^\mu, \frac{\partial \sigma^j}{\partial x^\mu} v^\mu\right) = \left(\frac{\partial x^\nu}{\partial x^\mu} v^\mu, \frac{\partial \sigma^j}{\partial x^i} v^i + \frac{\partial \sigma^j}{\partial t} v^0\right)$$

$$= \left(v^\nu, \left(v^i \frac{\partial}{\partial_i}\right)\sigma^j + v^0 \frac{\partial \sigma^j}{\partial t}\right) = \left(\boldsymbol{v}, 1, (\boldsymbol{v} \cdot \nabla)\boldsymbol{\sigma} + v^0 \frac{\partial}{\partial t}\boldsymbol{\sigma}\right), \tag{9.8}$$

according to (4.4). If we restrict ourselves on curves being path lines in the time parametrization, the tangent vectors are $\hat{v} = (\boldsymbol{v}, 1)$, i.e., $v^0 = 1$. The interpretation of (9.8) is as follows. Spatial part $\boldsymbol{\sigma}$ of vectors in the tangent space to the base is in one-to-one correspondence with vectors in the tangent space to the fiber, by the triviality of $\mathcal{F}$. So $\boldsymbol{\sigma}$ is actually a fiber vector, i.e., an "Eulerian" vector in a local space frame. This Eulerian vector is mapped to a vector in the tangent space to the bundle, which is a "Lagrangian" vector

$$TM \ni \boldsymbol{v} \to \left[(\boldsymbol{v} \cdot \nabla) + \frac{\partial}{\partial t}\right]\boldsymbol{\sigma}, \text{ with } \hat{\sigma} = (x, \boldsymbol{\sigma}) \in T\mathcal{F}. \tag{9.9}$$

If we put $\boldsymbol{v}_E = \boldsymbol{\sigma}$, (9.9) reads $d\sigma(\boldsymbol{v}_E) = \boldsymbol{v}_L$, i.e., the well-known transformation between the partial time derivative and the material (total) derivative. In this sense, (9.9) describes a connection in $\mathcal{F}$ in the first jet bundle J^1 (for example, see Olver's book [242]). Coming down to the $\mathcal{F}$ bundle, we note that the only possible connection is a trivial one, with zero coefficients. This is because the bundle is trivial, so the only admissible infinitesimal transformations are translations. The situation is different if the shape of the fluid container is allowed to change in time.

Even if we used such a complicated fiber bundle construction for the transformation of the time derivatives, the Eulerian–Lagrangian transformation formula (9.9) is useful so far only for the tangent vectors (i.e., tangent to the path lines), and it cannot be applied to more general vector fields, not mentioning higher rank mixed tensorial fields.

9.2.5 Free Surface Fiber Bundle

If the shape of the reference fluid container changes with time (boundaries not fixed anymore), the fiber F_x depends on the point $(x^i, t) \in X$ through the time dependence and the bundle is not anymore a global cartesian product. Consequently, it has a nontrivial structure group G. If the fluid has only one compact free surface, the fiber bundle $\mathcal{F}$ has a different structure than the one described in Sect. 9.2.4.

We consider the fluid "drop" as a connected, simple-connected domain $\mathcal{D}_\Sigma \simeq D_3 \subset \mathbb{R}^3$ with smooth boundary (shape) $\partial\mathcal{D} = \Sigma$, and under no external forces or torques. By $\simeq D_3$ we mean a diffeomorphisms with the three-dimensional disc $x^2 + y^2 + z^2 <= 1$. The drop has a set of possible shapes. If we can parameterize the set of all possible shapes with coordinates, we could set the structure of a manifold M. The shape coordinates can be determined by the expansion in spherical harmonics, for example, and we can associate to M the $l_2(\mathbb{C})$ space structure with the topology induced by the norm. We call M the *shape space* of the drop. The base space will be, like in the previous case, $X = M \times \mathbb{R} \ni (\Sigma, t)$.

For any shape we choose a trihedron fixed in this shape, for example the origin in the center of mass, and the axes directed toward the positions of some chosen zeros of the spherical harmonics. The configuration of the fluid within the given shape Σ will be referred to this trihedron. For a given shape Σ, all possible configurations of the fluid particles $\{\boldsymbol{r} | \boldsymbol{r} \in \mathcal{D}_\Sigma\}$ can be described by the set of diffeomorphic (shape invariant) transformation of $\mathcal{D}_\Sigma$ onto itself. These transformations form a Lie group of diffeomorphisms Diff_Σ. Any element g_Σ of this group maps some distribution of particles inside this shape into another distribution of particles within the same shape. So, by the *minimal constraintts*, the fiber over $x = (\Sigma, t) \in M$ is represented by the group of diffeomorphisms of the shape $\Pi^{-1}(\Sigma, t) = \mathit{Diff}_\Sigma$. The structure group is the group of diffeomorphisms of the three-dimensional disc, Diff_{D_3}, which is the group model for all the other diffeomorphisms groups. Consequently, $\mathcal{F}$ is a principal bundle, and the coordinate on the fiber over (Σ, t) is a certain group transformation $\mathit{Diff}_\Sigma \ni g_\Sigma : \mathcal{D}_\Sigma \to \mathcal{D}_\Sigma$.

This construction must be carried out for all possible shapes. Thus, the total configuration space of the fluid $\mathcal{F}$ is a fiber bundle over the base X, of fiber Diff_Σ. A shape evolution will be identified by a (time-like) curve $\gamma \in X$, i.e., a regular curve of shapes $\Sigma(t)$ parametrized by time. For any particular shape, we have to integrate a set of dynamical equations $\Delta(\Sigma, \boldsymbol{r}, t)$ to find the positions of the particles associated to that shape. The shape at any moment of time determines the position of particles within the fiber. Hence, a cross-section $\sigma : X \to \mathcal{F}$ represents the evolution of the drop, namely in components $t \to \Sigma(t) \to \boldsymbol{r}_L(\boldsymbol{r}_0, t) = g_{\Sigma(t)}(\boldsymbol{r}_0)$. From the geometrical point of view, the dynamical equations of the free surface fluid are equations for this section. These are basically the equation of continuity, equations for momentum conservation (Euler or Navier–Stokes equations), and energy transfer equation.

For any shape in M, we need to specify its fixed reference trihedron and its reference (we may call it initial) distribution of particles $\boldsymbol{r}_0$. This choice is not unique, and the freedom involved is a typical gauge freedom. A similar gauge freedom is encountered in electromagnetism when we study magnetic monopoles, in the dynamics of elastic bodies or in the study of the geometric phase change of the wave function for time variable Hamiltonian (Berry's phase). Making a choice for the trihedron orientation and the reference particle distribution with respect to any shape is nothing but a cross-section in $\mathcal{F}$. However, the physical results should be independent of this choice, i.e., gauge invariant.

Translation of the drop center of mass could be eliminated from the beginning, but the shapes should also conserve total angular momentum. Angular momentum can be changed by deformations (motion in the base space) and also by particle rotations (motions in the fiber). We need to "synchronize" the succession of deformations with a unique succession of rotations, such that total angular momentum to be constant. In that, we can introduce a new type of connection, different from that one introduced above between Eulerian and Lagrangian approach on tangent vectors (9.9).

For any given smooth curve γ in the base space M, we need to lift it to a curve γ' in the total space $\mathcal{F}$ in a unique way. Remember that a lift is a map $\gamma \in M \rightarrow \gamma' \in \mathcal{F}$ such that $\Pi(\gamma') = \gamma$. However, the lift of a path is not unique by definition. The mathematical tool needed to make it unique is the connection [74, 101]. A connection, or better said its differential expression, would assign to any tangent vector $v(x) \in T_x M$, an element in TF_x, which is the Lie algebra of the group Diff_Σ. Globally, when we move along a closed path in M the corresponding lifted path in $\mathcal{F}$ may not be closed. That is for $\gamma(x_0) = \gamma x_1$ we may have $\gamma'(x_0) = \gamma' x_1$ in F. Two different points on the same fiber mean a relabeling of the particles, or a motion inside the drop. Such a relabeling could be associated with a finite nonzero rotation of the drop. The drop begins to move by changing its shape and ends up to the same initial shape after a finite amount of time. But during this motion, it actually undergoes a net rotation.

A similar situation happens when we build the configuration space of a deformable body. Again, we choose for any shape a trihedron fixed in this shape. The orientation of the body, ignoring free translations of the center of mass, could be described by a proper rotation matrix $\hat{R} \in SO(3)$ which maps the body-fixed trihedron to a space frame contained in the ambient space in which the drop is constrainted to move, i.e., $\mathbb{R}^3$. Thus, the total configuration space $\mathcal{F}$ is a fiber bundle over the base $M \times \mathbb{R}$, of fiber $SO(3)$.

Like in the case of the drop, the angular momentum of the body can be changed by deformations (motion in the base space) and also by rotations (motions in the fiber). In this example, the connection assigns to any tangent vector $v(x) \in TM$, an element in $TSO(3)$, which is nothing but the Lie algebra $so(3)$. When we move along a closed path in M the corresponding lifted path is not closed in general. Two different points on the same fiber mean a change in the orientation, a rotation. The body moves and changes its shape, but during this motion, it undergoes a rotation. However, because the $SO(3)$ Lie group is not commutative, there are problems in integrating this lifted path in the fiber. The problem is solved, for example in gauge field theory, by the so-called *Wilson integral*. In [101] there is an eloquent example, namely the falling cat problem. The cat is dropped from an upside down position, but it lands on its feet, even if it is isolated. The cat manages to deform its body during the flight, such that all in all involves a net rotation of the body, to conserve its angular momentum, see also [278]. Similar examples of free deformable compact shapes occur in the theory of swimming of microorganisms in zero Reynolds number [231]. In that case the systems are investigated by using the theory of a gauge field over the space of shapes. The topics of fiber bundles in hydrodynamics have plenty of online and printed resources out of which we mention for example [11, 141, 212–215].

9.2.6 *How Does the Time Derivative of Tensors Transform from Euler to Lagrange Frame?*

In Sects. 9.2.4 and 9.2.5, we have seen that changing the frame from the Eulerian to Lagrangian is actually mapping vectors from the tangent space of the base space to the tangent space of the fiber. To transform higher-order tensors we need to introduce a new time derivative through a covariant formalism. Equations (9.1) and (9.5)–(9.7) are not covariant because the time is not explicitly included in the metric, yet the Lagrangian → Eulerian transformation $\omega(x,t) \to \Omega(\sigma,t)$ is a time-dependent coordinate change. Consequently, the partial time derivative does not transform like a tensor because of the time-dependent basis vectors, the same reason that ordinary derivatives are not covariant (see for example in Sect. 4.10 the comment right after (4.45)).

The traditional material derivative is covariant just for the coordinates, the velocity vector, and (obviously) for scalars, as we know from (9.1) and (9.5)–(9.7), and it was proved geometrically in (9.9), because the velocity belongs to the tangent space. Let us have an (r,s) Lagrangian tensor $\omega(x,t)$ depending on the Lagrangian coordinates (x,t). Its time derivative, i.e., the rate of change $d\omega/dt$ of the tensor while keeping the Lagrangian coordinates constant, does not transform into the time derivative of the corresponding Eulerian tensor, $\omega(x,t) \to \Omega(\sigma,t)$).

$$\frac{\partial \omega}{\partial t}(x,t) \nrightarrow \frac{\partial \Omega}{\partial t}(\sigma,t).$$

To provide a covariant time derivative for arbitrary vector fields and higher-order tensors, we need to calculate the pull-back transformation of (9.9), and make sure that the result is a tensor of the same type. That is, to introduce a *covariant time derivative* operator (e.g., [314] where it is called convected or convective) which describes the change in time for a certain geometrical quantity ω along (or with respect to) the flow lines of the fluid, in the Eulerian frame (σ^1,t). The covariant variation of this quantity is the sum of its internal time variation described by the partial derivative, and the Lie derivative of ω with respect to the flow described by the vector field $\boldsymbol{v}_E = (v^i)$

$$\frac{d_c \Omega(\sigma,t)}{dt} = \frac{\partial \Omega}{\partial t} + \boldsymbol{v}_E(\Omega). \tag{9.10}$$

For scalars, (9.10) reduces to the well-known formula (9.1) or (9.8). We will refer in the following to (4.19) and (4.20), describing the action of the Lie derivative on various geometrical objects.

For example, the time covariant derivative acts on a contravariant vector field $\boldsymbol{A}(\sigma,t) = (A^i)$ defined in the Eulerian frame, according to the form (4.19)

$$\frac{d_c \boldsymbol{A}}{dt} = \frac{\partial \boldsymbol{A}}{\partial t} + [\boldsymbol{v}_E, \boldsymbol{A}]. \tag{9.11}$$

The covariant time derivative action on a covariant vector $\omega = (A_i)$ is given by the sum between the partial derivative with respect to time and the Lie derivative with respect to $\boldsymbol{v}_E$ acting on the 1-form (4.20)

$$\frac{d_c \Omega_i}{dt} = \frac{\partial \Omega_i}{\partial t} + v^k \frac{\partial \Omega_i}{\partial \sigma^k} + \Omega_k \frac{\partial v^k}{\partial \sigma^i}, \tag{9.12}$$

The action on an Eulerian tensor of rank $(0,2)$ is

$$\frac{d_c \Omega_{ij}}{dt} = \frac{\partial \Omega_{ij}}{\partial t} + v^k \frac{\partial \Omega_{ij}}{\partial \sigma^k} + \omega_{kj} \frac{\partial v^k}{\partial \sigma^i} + \Omega_{ik} \frac{\partial v^k}{\partial \sigma^j}, \tag{9.13}$$

and so on. The physical signification of the covariant derivative on the LHS of all (9.11)–(9.13) is the following. First, we calculate the partial time derivative of a Lagrangian tensor, then we transform this quantity into the Eulerian frame. This transformed Eulerian object is not anymore the simple partial derivative of the Eulerian tensor, but the covariant time derivative of the Eulerian tensor.

To exemplify (9.10) in a direct and even more intuitive way, we obtain the transformation of the time derivative for a tensors of rank $(1,1)$ for example by a simple matrix transformation formalism based on formula (4.46). Similar calculations in components are done in [10, Chap. 8]. We write the tensor transformation of components of ω when changing frame from Lagrangian to Eulerian

$$\Omega = J\omega J^{-1}, \text{ that is } \Omega^p_q = \frac{\partial \sigma^p}{\partial x^i} \frac{\partial x^j}{\partial \sigma^q} \omega^i_j. \tag{9.14}$$

By time differentiation of (9.14) with respect to time, we have

$$\frac{d\Omega}{dt} J + \Omega \frac{dJ}{dt} = \frac{dJ}{dt} \omega + J \frac{d\omega}{dt}.$$

Since Ω is Eulerian we have $\Omega(\boldsymbol{\sigma}, t)$ and further $\Omega(\boldsymbol{\sigma}(\boldsymbol{x}, t), t)$, so

$$\frac{d\Omega^p_q}{dt} = \frac{\partial \Omega^p_q}{\partial t} + v^j \frac{\partial \Omega^p_q}{\partial \sigma^j}.$$

Moreover, we can write

$$\frac{dJ^j_i}{dt} = \frac{\partial v^j}{\partial x^i} = \frac{\partial v^j}{\partial \sigma^k} \frac{\partial \sigma^k}{\partial x^i},$$

and define the matrix of gradients of velocity

$$\gamma^j_i = \frac{\partial v^j}{\partial \sigma^i}.$$

With these notations we have

$$\Omega\gamma J + \frac{d\Omega}{dt}J - \gamma J\omega = J\frac{d\omega}{dt},$$

and by using $dJ/dt = \gamma J$ and by multiplication with J^{-1} to the right, we obtain

$$J\frac{d\omega}{dt}J^{-1} = \frac{d\Omega}{dt} + [\Omega, \gamma] \equiv \frac{d_c\Omega}{dt}, \tag{9.15}$$

where the commutator on the RHS arises from $\Omega\gamma - \gamma(J\omega J^{-1})$. Equation (9.15) represents the transformation of the time derivative $d\omega/dt$, and since the RHS is an operator applied to the Eulerian tensor Ω, we define the LHS as the *covariant* (or convected) time derivative. In components it reads

$$\left(J\frac{d\omega}{dt}J^{-1}\right)_i^j \equiv \frac{d_c\Omega_i^j}{dt} = \frac{\partial\Omega_i^j}{\partial t} + v^k\frac{\partial\Omega_i^j}{\partial\sigma^k} + \Omega_k^j\frac{\partial v^k}{\partial\sigma^i} - \Omega_i^k\frac{\partial v^j}{\partial\sigma^k}, \tag{9.16}$$

where we used the notation d_c/dt for this covariant derivative. It is easy to check that (9.16) is in agreement with the general formulation from (9.12) and (9.13). For the action of the covariant time derivative on other types of tensors, see Exercises 4 and 5 at the end of the chapter. Also the action of d_c/dt can be expressed entirely in terms of covariant derivatives [10]. For example for a (0, 2)-tensor, we have

$$\frac{d_c\Omega_{ij}}{dt} = \frac{\partial\Omega_{ij}}{\partial t} + (v^k\nabla_k)\Omega_{ij} + (\nabla_j v^k)\Omega_{ik} + (\nabla_i v^k)\Omega_{kj}. \tag{9.17}$$

Let us choose a simple example to understand how (9.16) works. We consider a stationary viscous flow next to a rigid wall at $\sigma^3 = 0$ (or simply $z = 0$) with velocity $\mathbf{v}_E = (0, v, 0)$. The velocity is subjected to a boundary layer effect and it depends on the distance to the wall, $v = v(\sigma^3)$. In the Lagrangian (convected) frame the pressure is constant in time and so is its gradient, having nonzero component in the σ^3-direction, $\nabla P = (0, 0, \partial P/\partial\sigma^3) = (\alpha_1, \alpha_2, \alpha_3)$. The time derivative of this gradient, which is a (0, 1) covariant vector, is zero. However, in the Eulerian frame by using (9.16), we have a nonzero material time derivative

$$\frac{d_c\nabla P}{dt} = \left(0, \frac{\partial v_E}{\partial\sigma^3}(\nabla P)_3, 0\right).$$

There is a change in time for the gradient in the Eulerian frame even if the same gradient is constant in the Lagrangian frame, and this contribution comes from the last term in the RHS of (9.16), and not from the first two traditional terms on the same RHS. Physically, it means that the gradient is initially vertical, but because of the horizontal shearing of the layers of fluid, this gradient is "tilted" more and more horizontally.

This treatment presented above is not the only way to introduce a covariant time derivative. For example in [145] the authors introduce a *corotational* derivative where the local vorticity of the flow is incorporated into the derivative. However, the covariant time derivative defined by (9.15) and (9.16) is the most familiar one, and it was initially introduced in [241] in formulating rheological equations of state. This derivative was used in [292] to develop a theory of fluid motion on an interface, and later was geometrically extended in [10, 314]. In this last citation there are enumerated some disadvantages of the covariant time derivative. For example, it is not compatible with the metric tensor, and it involves gradients of the velocity so it is not directional. On the other hand, the importance of the covariant time derivative (9.15) and (9.16) is not only mathematical. Many nonlinear transport and mixing processes are described by advection–diffusion equations [314], consisting in a material time derivative for the concentration of the quantity advected, and a divergence of the diffusivity tensor. In the Lagrangian frame (along the direction of compression of fluid elements) the advected terms drop out, and the governing equation reduces to a simple diffusion equation, much more tractable. Moreover, because of the formalism presented in this section, this simplified diffusion equation is still covariant. This allows the introduction of a Riemannian metric on the tangent space to the coordinate space, and allows in principle the use of spectral approximation procedures.

9.3 Path Lines, Stream Lines, and Particle Contours

In this section, we present a parallel between the Eulerian and Lagrangian approaches from the point of view of the flow box theorem (see Sect. 4.4). We discus here only finite time flows with $t \in [t_1, t_2], -\infty < t_1 < t_2 < \infty$. We begin our construction with the fluid initial reference container, i.e., a domain $D_0 \subset \mathbb{R}^3$. We construct the base space $X = \mathbb{R}^3 \times [t_1, t_2]$, and we assign a local coordinate system in $\boldsymbol{r}_0 \in D_0$. We assume that we are given the fluid flow as smooth homeomorphisms $\boldsymbol{r}_L : D_0 \times [t_1, t_2] \to \mathbb{R}^3$ such that the restriction $\boldsymbol{r}_L|_{D_0 \times \{t\}}$ is injective for any fixed $t \in [t_1, t_2]$. In coordinates this reads $(\boldsymbol{r}_0, t) \to \boldsymbol{r}_L(\boldsymbol{r}_0, t)$. The family of curves $L = \{\gamma_L \Rightarrow \boldsymbol{r}_L(\boldsymbol{r}_0, t) | \boldsymbol{r}_0 \in D_0\}$ is the particle paths, with tangents $\dot{\boldsymbol{r}}_L = \boldsymbol{v}_L$ and metric $g_L = v_L^2$. These curves can be lifted in the base space and mapped into a family $\tilde{L} = \{\gamma_L' \Rightarrow (\boldsymbol{r}_L(\boldsymbol{r}_0, t), t) \in \mathbb{R}^3 \times [t_1, t_2] | \boldsymbol{r}_0 \in D_0\}$. The metric of γ_L' is $\tilde{g}_L = v_L^2 + 1$. Both γ_L and γ_L' are Lagrangian path lines viewed in different spaces.

For any $t \in [t_1, t_2]$ we can construct $D_t = \boldsymbol{r}_L(D_0, t) \subset \mathbb{R}^3$. A particle contour is a parametrized curve $\Gamma_0 = \{\gamma_0(s) \subset D_0, s \in I\} \subset D_0$. The question is what happens to such a particle contour in time. Is $\Gamma_t = \{\gamma(s, t) = \boldsymbol{r}_L(\gamma_0(s), t)\}$ a regular curve with the same topology as Γ_0? We have the following result.

Lemma 7. *The set Γ_t defined by $\gamma(s, t)$ as above is a* regular parametrized curve *if*

$$\hat{J}(\boldsymbol{r}_L(\boldsymbol{r}_0, t))|_{t=const.} \cdot \boldsymbol{t}_{\Gamma_0} \neq 0,$$

for $\forall s \in I, t \in [t_1, t_2]$. *Here* $\boldsymbol{t}$ *is the tangent vector to a curve.*

Proof. We have

$$\frac{\partial \boldsymbol{r}}{\partial s}(s,t) = \frac{\partial x_L^i}{\partial \gamma^j}\frac{d\gamma^j}{ds} = \frac{\partial x_L^i}{\partial x_0^j} \cdot \boldsymbol{t}_{\Gamma_0}^j(s), \tag{9.18}$$

which represents the requested inequality. □

In other words, a particle contour at the initial moment of the flow remains a regular curve while transported by the flow in time if the unit tangent of this initial curve is not in the kernel of the Jacobian matrix of the Lagrangian path function of the initial coordinates (the flow). If conditions in Lemma 7 are fulfilled, the particle contour Γ_0 remains a regular curve during the flow, so one can apply circulation or other types of theorems on it. The Jacobian matrix plays a basic role in hydrodynamics [331]. It allows the determination of the main flow parameters and the geometrical characteristics, in particular the metric properties.

As an application, we can use Lemma 7 criterium in Example 2. The initial vertical particle contour (for example $x_0 = 0, y_0 \in [-a, a]$) will breakup at a certain moment of time t if, according to (9.18),

$$\frac{\partial \boldsymbol{r}_L}{\partial y_0}(t) = 0,$$

where we consider y to be the vertical axis in Fig. 9.3. Obviously, from the continuity of the cylinder contour, the coordinates of all path lines depend on y, so (even it looks hard to believe) the above derivative is nonzero everywhere and consequently the path lines will not disrupt.

The question is whether the set $\cup_{t \in [t_1, t_2]} D_t$ is a submanifold of $\mathbb{R}^3$. If it is, we can assign local coordinates for its points in the form $p = (\boldsymbol{r}_0, t)$. In other words, if the reunion of all path lines over a certain finite interval of time is dense enough to form a topological space. The answer can be given at least locally, by using the flow box theorem (Theorem 6). Obviously, the Lagrangian velocity field of any particle $\boldsymbol{v}_L$ fulfills the conditions for the existence of flow boxes on X. Indeed, for any $t \in [t_1, t_2]]$ and any point $p = (\boldsymbol{r}, t) \subset D_t$, we can find a neighborhood $V(\boldsymbol{r})$ and $t \pm \delta t$ such that it exists $a > 0$ and the triple

$$((V(\boldsymbol{r}), (t - \delta t, t + \delta t))a, \gamma_L(\boldsymbol{r}_L(\boldsymbol{r}_0, t), t + \lambda)),$$

is a flow box.

Moreover, we assume that the fluid flows in such a way that X is a topological space with the product topology of $\mathbb{R}^3 \times \mathbb{R}$. We also assume that the fluid flows in a bounded region (bounded fixed region or free compact surface), so the Lagrangian velocity field has compact support in X. Consequently $\gamma_L(\boldsymbol{r}_0, t)$ are maximal integral curves and form a foliation of X (see Sect. 4.4). Since the field of velocities of particles has compact support, according to Lemma 2, it is complete, and any of its integral curves can be extended so that its domain of parameter becomes $\mathbb{R}$.

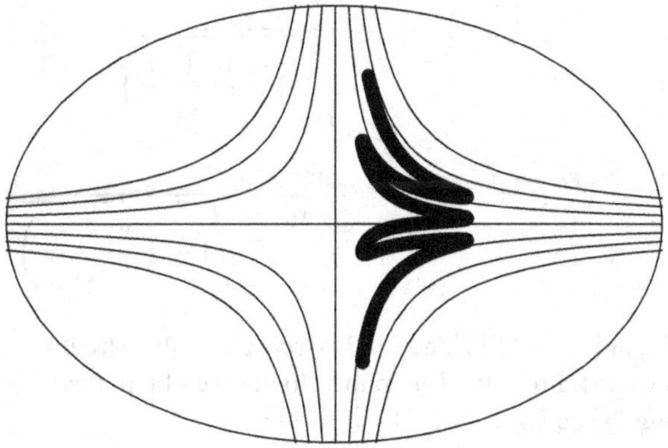

Fig. 9.9 Cross-section into a spherical drop of incompressible inviscid fluid in oscillation in an $l = 2$ mode. The *thin curves* are the stream lines, while the *thick curve* is an example of a path line

So the Lagrangian paths $\gamma_L(\boldsymbol{r}_0)$ form a foliation of the manifold D_t which is homeomorphic with D_0. We mention again that inside each D_t, we have $\boldsymbol{v}_E(\boldsymbol{r}_L(\boldsymbol{r}_0, t), t) \equiv \dot{\boldsymbol{r}}_L(\boldsymbol{r}_0, t)$, but inside the same D_t the integral curves of $\dot{\boldsymbol{r}}_L$ are not the γ_L curves.

There are of course differences and similarities between the stream and path lines.

Example 3. In Fig. 9.9 we present a cross-section into a spherical drop of incompressible inviscid fluid in oscillation with an $l = 2$ mode. The thin lines are the stream lines and the thick line is a path line.

Example 4. To illustrate better these differences, we present a simple example of a two-dimensional flow. We assume that we know the flow of this two-dimensional fluid in the Eulerian frame, and hence we know the Eulerian velocities $\boldsymbol{v}_E(\boldsymbol{r}, t)$ at every point and every moment of time. For example let us choose

$$\boldsymbol{v}_E(x, y, t) = (x, y + \epsilon t), \tag{9.19}$$

where ϵ is an arbitrary parameter. The stream lines, lying in the instantaneous plane $\mathbb{R}^2$, are obtained by integrating

$$\frac{dx}{x} = \frac{dy}{y + \epsilon t}, \tag{9.20}$$

resulting in the implicit equation

$$y_E = \frac{y_0 + \epsilon t}{x_0} x_E - \epsilon t, \tag{9.21}$$

or in the parametric form $\boldsymbol{r}_E(s; x_0, y_0; t)$

$$x = \frac{s}{\sqrt{1 + \left(\frac{y_0 + \epsilon t}{x_0}\right)^2}}$$

$$y = \frac{y_0 + \epsilon t}{x_0} \frac{s}{\sqrt{1 + \left(\frac{y_0 + \epsilon t}{x_0}\right)^2}} - \epsilon t. \tag{9.22}$$

Equations (9.21) and (9.22) represent the stream line passing through a point (x_0, y_0). From the Eulerian velocity we obtain the Lagrangian velocity by integrating the equations

$$\frac{dx_L}{dt} = x_L(x_0, y_0, t)$$

$$\frac{dy_L}{dt} = y_L(x_0, y_0, t) + \epsilon t.$$

The lifted path lines in parametric form have the expression $\gamma_L(x_L(x_0, y_0, t), y_L(x_0, y_0, t), t)$ with

$$x_L(x_0, t) = x_0 e^t$$

$$y_L(x_0, y_0, t) = (y_0 + \epsilon t)e^t - \epsilon(t + 1), \tag{9.23}$$

and in implicit form read

$$y_L(x_0, y_0, t) = (y_0 + \epsilon)\frac{x_L}{x_0} - \epsilon\left(\ln \frac{x_L}{x_0} + 1\right). \tag{9.24}$$

Of course the path lines and the stream lines have different expressions, not forgetting the fact that they belong to different spaces. For a check, we notice that if we eliminate the time dependence by setting $\epsilon = 0$, these lines (9.21)–(9.24) have the same expression. In stationary flow the stream lines and the path lines coincide in the horizontal space. We can also check the definition condition $\boldsymbol{v}_L(t) = \boldsymbol{v}_E(\boldsymbol{r}_L(t), t)$. Indeed, we can write

$$v_{Ex} = x_E|_{\boldsymbol{r}_L(t)} = x_L(t) = x_0 e^t = v_{xL}(t),$$

and from (9.23)

$$v_{Ly}(t) = (y_0 + \epsilon)e^t - \epsilon = y_E|_{\boldsymbol{r}=\boldsymbol{r}_L(t)} + \epsilon t = v_{Ey}.$$

Another check is to verify the relation between the Eulerian and Lagrangian

$$\frac{dv_{Ly}}{dt} = (y_0 + \epsilon)e^t = y + \epsilon t + \epsilon$$

$$\frac{\partial v_{Ey}}{\partial t} + (\boldsymbol{v}_E \cdot \nabla) v_{Ey} = \epsilon + x \frac{\partial (y + \epsilon t)}{\partial x} + (y + \epsilon t) \frac{\partial (y + \epsilon t)}{\partial t} = y + \epsilon t + \epsilon, \quad (9.25)$$

and a similar equation for v_x.

For any t, the stream lines (9.22) form a family of curves $\gamma_E(s; \boldsymbol{r}_0; t)$ labeled by the points $\boldsymbol{r}_0 \in \gamma_E$, parameterized by the arc-length s. These curves provide foliations of each horizontal space $\mathbb{R}^2$, for each moment of time. The vector field $\boldsymbol{v}_E(\boldsymbol{r}, t)$ generates also a family of integral curves in the base space $\mathbb{R}^3 = \mathbb{R}^2 \times \mathbb{R}^{time}$ determined by the equations

$$\frac{dx}{x} = \frac{dy}{y + \epsilon t} = \frac{dt}{1}. \quad (9.26)$$

At $t = 0$ we have

$$\gamma_E(s; \boldsymbol{r}_0; 0) = \frac{s}{\sqrt{x_0^2 + y_0^2}}(x_0, y_0) \quad (9.27)$$

and the solutions of (9.26) and (9.27) coincide modulo a reparameterization. This means that the Eulerian stream lines are the projections of the lifted Lagrangian path lines in the horizontal planes *only* at $t = 0$. The above example is also shown in Fig. 9.10.

In Fig. 9.11, we present the same flow described by (9.21) and (9.23) in the base space (a three-dimensional representation, where time is the vertical axis).

9.4 Eulerian–Lagrangian Description for Moving Curves

This section is very short, and its purpose is to recall that the idea of establishing a Lagrangian–Eulerian change of frames in lower-dimensional flows is not quite trivial. We elaborated a little about Eulerian–Lagrangian coordinates and velocities in Sects. 8.2 and 8.3 together with the introduction of the convective velocity. Here we just mention one possibility to introduce Eulerian coordinates on a moving curve, like for example a thin vortex filament in motion. We can consider that the Lagrangian coordinates along a curve of length L are given by the arc-length parameterized form of the curve $\boldsymbol{r}(s, t)$. The curve is in motion, and the velocity can be expressed in its Serret–Frenet local frame $\{\boldsymbol{t}, \boldsymbol{n}\}$ in the form $\boldsymbol{V}(s, t) = U(s, t)\boldsymbol{n} + W(s, t)\boldsymbol{t}$. We introduce the mapping $e : [0, L] \to \mathbb{C}$

$$e(s, t) = \int^s e^{i\theta(s', t)} ds',$$

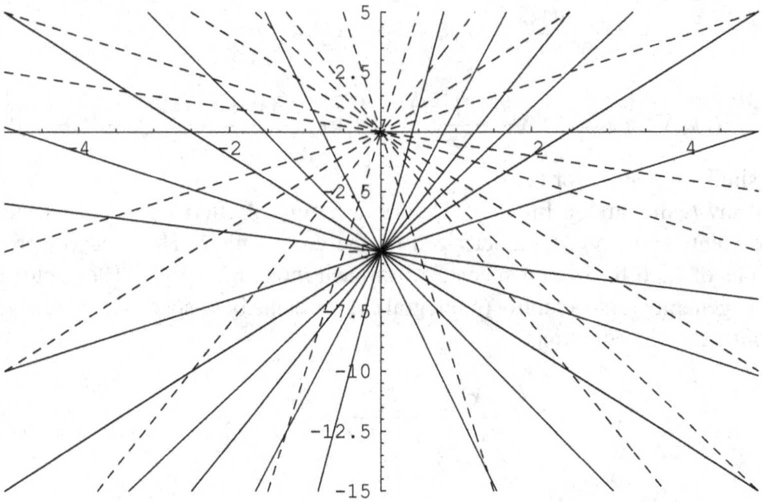

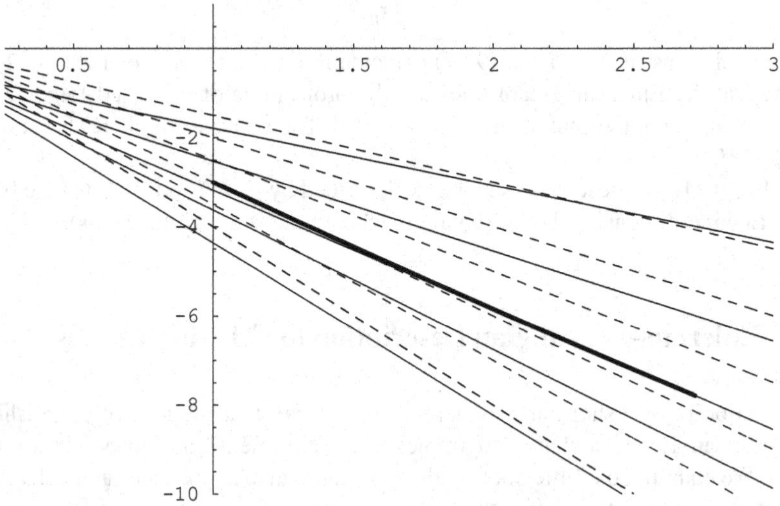

Fig. 9.10 Two dimensional plot $(x - y)$ of flow lines. *Upper graphic*: stream lines $\gamma_E(t)$ in the horizontal plane generated by (9.21) at $t = 0$ (*dashed lines*) and $t = 1$ (*continuous lines*). *Lower graphic*: a region of the same flow, with stream lines at $t = 0$ (*dashed*) and $t = 1$ (*smooth*), and a path line (*thick line*) of a particle moving from $t = 0$ to $t = 1$. The path line is tangent to $\boldsymbol{v}_E(t = 0)$ (*dashed line*) at its upper left end, and tangent to $\boldsymbol{v}_E(t = 1)$ (*smooth line*) at its lower right end, respectively

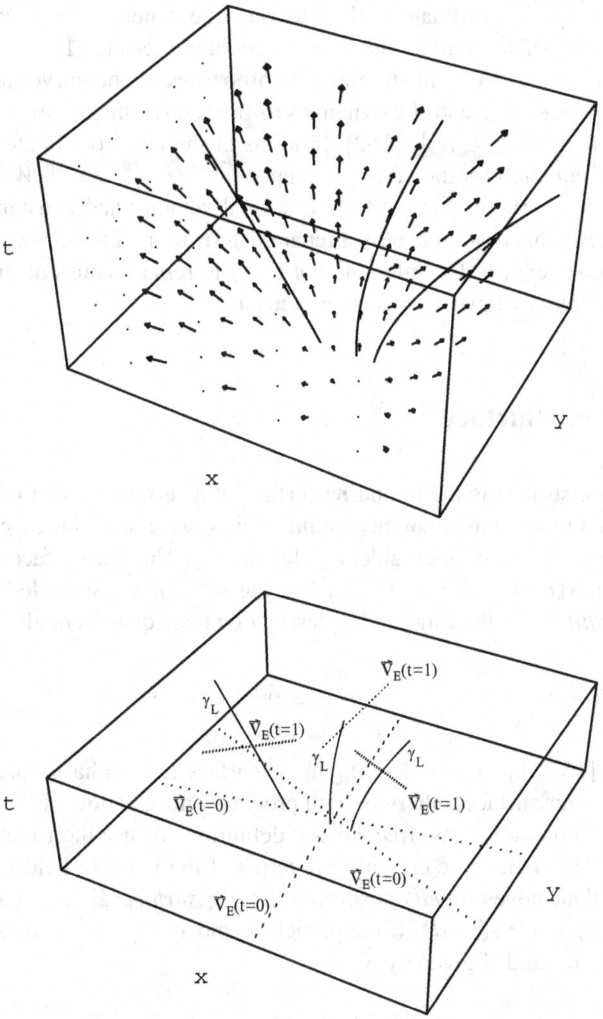

Fig. 9.11 *Upper box*: Lagrangian velocity field represented in the base space with *arrows*. Three Lagrangian paths as particular integral curves of this field are shown. *Lower box*: same Lagrangian paths γ_L (*continuous line*). If we project the unit tangent of each such Lagrangian path onto the horizontal plane, we obtain the Eulerian velocity field $\boldsymbol{v}_E$. The *dotted lines* are integral curves of this Eulerian field. The three longer dotted lines on the base of the box are three such stream lines, intersecting the three Lagrangian path lines at $t = 0$, respectively. The other three dotted (shorter) lines in the upper plane are other three stream lines, occurring at $t = 1$, and intersecting the same three Lagrangian path lines at $t = 1$, respectively

from the Lagrangian coordinate to the Eulerian one, where $\theta = \int^s \kappa(s', t) ds'$ is the tangent angle of the curve, and κ is its curvature (Sect. 5.1). In the Eulerian coordinate, we can express all the intrinsic properties of the curve, namely $\theta = -i \ln(e_s), \kappa = -i e_{ss}/e_s$, and the dynamics of the transformation of coordinates is given by $e_{st} = [(W - iU)e_s]_s$ [152]. In terms of the new coordinate e and time, the dynamical equation for the velocity components is $\theta_t e^{i\theta} = e^{2i\theta}(W - iU)_e$. Let us choose now a curve motion with zero normal velocity and constant tangential velocity. Since such a motion is only a reparameterization of the curve, i.e., it is not a real motion, we expect the Eulerian coordinate to remain constant. Indeed, from the above relations we have $e_{st} = 0$ so $e =$const.

9.5 The Free Surface

Physically, free surface is the bounding surface of a certain amount of fluid under consideration. From the mathematical point of view, we consider the free surface Σ to be a piecewise smooth, orientable, regular surface. The free surface is described by the relation $S(\boldsymbol{r}, t) = 0$. This free surface has to fulfill the so-called *free surface kinematic condition*. In the Lagrangian description this equation reads

$$\frac{dS}{dt} = 0, \tag{9.28}$$

which means [167] that a particle lying in the surface can not have normal velocity with respect to this surface, otherwise will produce a normal flow of fluid across the surface, which contradicts the free surface definition. To use the Eulerian picture, and to express the kinematic condition in terms of the velocity field $\boldsymbol{v}$, we choose a particle P that moves *together with the moving surface* Σ. The particle has a velocity $\boldsymbol{v}_{P\Sigma}(t) = d\boldsymbol{r}_P(t)/dt$. If the particle P moves *together* with Σ, there is a relation between $\boldsymbol{v}$ and S given by

$$\boldsymbol{v}_{P\Sigma} \cdot \nabla S + \frac{\partial S}{\partial t} = 0. \tag{9.29}$$

It is easy to prove this equation if we assume that the particle is contained in the surface at an arbitrary moment t and also at $t + \delta t$. That is: if $S(\boldsymbol{r}_{P\Sigma}(t), t) = 0$, then $S(\boldsymbol{r}_{P\Sigma}(t + \delta t), t + \delta t) = 0$. Equation (9.28) can also be written as

$$\left(\boldsymbol{v} \cdot \nabla S + \frac{\partial S}{\partial t} \right)_{\Sigma} = 0,$$

and this is a possible form for the free surface kinematics condition. The Σ subscript means that this equation is taken only on Σ, or in other words that, in this equation $(\boldsymbol{r}, t)$ have to fulfill $S(\boldsymbol{r}, t) = 0$. This form is more useful if the surface equation S

is provided explicitly. For example if $S = 0 \rightarrow z = \eta(x, y, t)$, we have

$$\frac{d\eta}{dt} = v_z = \frac{\partial \eta}{\partial t} + v_x \frac{\partial \eta}{\partial x} + v_y \frac{\partial \eta}{\partial y}. \tag{9.30}$$

We would like to comment that, in some literature, this free surface kinematics condition is explained as "a fluid particle originally on the boundary surface will remain on it." This is not, in general, true. The P particle may sink inside the fluid (like in the case of dragging of the capillary surface by adherence forces) or evaporate. A more general physical statement would be that, for any particle lying at moment t in the surface, its velocity is tangent to the surface at that moment. From the mathematical point of view, this problem is equivalent to the fact that $d\boldsymbol{r}/dt$ is not well defined at the surface, because the set of points forming a geometrical surface Σ admits many mappings into itself. To eliminate this ambiguity, one can use just the normal velocity, as it is suggested by Meyer [220]. We can define the unit normal to the regular surface $S(\boldsymbol{r}, t) = 0$ by $\boldsymbol{n} = \nabla S / |\nabla S|$. The normal component of the velocity of Σ is

$$v_n = \left(\boldsymbol{n} \frac{d\boldsymbol{r}}{dt} \right) \bigg|_{\Sigma} \cdot \boldsymbol{n} = -\boldsymbol{n} \frac{\partial S}{\partial t} \frac{1}{|\nabla S|}.$$

By using (9.28) for S, we have

$$v_n = -\frac{\frac{\partial S}{\partial t}}{|\nabla S|} = -\frac{\frac{dS}{dt} - (\boldsymbol{V} \cdot \nabla) S}{|\nabla S|} = \frac{(\boldsymbol{V} \cdot \nabla) S}{|\nabla S|},$$

where the last RHS is nothing but the velocity field along the normal to the surface $\boldsymbol{V} \cdot \boldsymbol{n}$. So we have obtained

$$v_n = V_n, \tag{9.31}$$

which is the most compact (and precise) form of the free surface kinematic condition: the normal component of the Lagrangian fluid particle velocity is equal, in any point of the surface, with the normal component of the Eulerian velocity.

9.6 Equation of Continuity

In Sects. 9.6.1 and 9.6.2, we analyze the equation of continuity. There are two reasons for choosing this topic. The first reason is that this equation provides a simple working application of the basic theorems of existence and uniqueness of the solutions of (linear or nonlinear) PDE. The second reason is that the equation of continuity has variable coefficients and it represents also a good toy model for such type of equations. However, it is still linear PDE, yet interesting in some of its particular solutions so it makes a "smooth" pedagogical transition from linear to nonlinear.

9.6.1 Introduction

In the nonrelativistic approximation mass is neither created nor destroyed, so we have the law of conservation of mass, i.e., a positive invariant

$$m = \int_D \rho d\mathcal{V} > 0,$$

integrated on the closure of the domain D filled with fluid. From its invariance we find the so-called equation of continuity integral or differential form

$$\int_D \left(\frac{\partial \rho}{\partial t} + div(\rho \boldsymbol{V}) \right) d\mathcal{V} = 0, \quad \frac{\partial \rho}{\partial t} + div(\rho \boldsymbol{V}) = 0, \tag{9.32}$$

in either integral or differential form. $\boldsymbol{V}(\boldsymbol{r}, t)$ is the velocity field and $\mathcal{V}$ is the volume. In fluid mechanics, the equation of continuity is coupled with other equations for conservation of momentum (Euler or Navier–Stokes) and for energy or entropy transfer, such that in total we have five scalar PDEs for the five scalar fields for the problem: ρ, $\boldsymbol{V}$, and p the pressure (by scalar we mean here also a component of a vector field). The continuity equation alone is not useful for physics, and some of its solutions do not have physical signification, unless coupled with the other dynamical equations. However, we present in the followings a theorem of existence and uniqueness, and some applications for (9.32). Such examples are not usually analyzed in books of fluid dynamics, but they can work as a good exercise of mathematical physics.

We study the equation of continuity when the velocity field is given, and we integrate it to find the density distribution. The continuity equation (9.32) is a homogenous linear PDE of order 1, with variable coefficients, defined in a certain domain $\mathcal{D} \subset \mathbb{R}^4$ of space–time. The main tool we need is the Cauchy–Kovalevskaya theorem for existence and uniqueness of the solutions of a general (not necessarily linear) PDE [63]. According to this theorem, the continuity equation has one unique real analytic solution $\rho(\boldsymbol{r}, t)$ for a given analytic velocity field $\boldsymbol{V}(\boldsymbol{r}, t)$ and given Cauchy condition provided by $\rho(\boldsymbol{r}, t)|_\Sigma = g(\xi_1, \xi_2, \xi_3)$, where g is an analytic function defined on a regular hypersurface $\Sigma \subset \mathbb{R}^4$. The Cauchy–Kovalevskaya theorem can be applied to any nonlinear PDE, for arbitrary Cauchy conditions expressed in terms of analytic functions, if one of the highest order derivative of the PDE can be explicitly written as an analytic function depending on the other terms and variables in the PDE. For example in (9.32), PDE of order 1, we can write the time derivative of the unknown function ρ on the LHS, and express it as an analytic function of the variable coefficients V_i and partial derivatives of ρ with respect to the other coordinates x_i, on the RHS (named generically $f(\boldsymbol{r}, t, \rho, \partial\rho/\partial x_i, \ldots)$)

$$\frac{\partial \rho}{\partial t} = f - \sum_{i=1}^{3} \frac{\partial(\rho V_i)}{\partial x_i}.$$

The function f is analytical because the finite sum and multiplication preserve analyticity, so we are in the frame of the Cauchy–Kovalevskaya theorem. In general, if the PDE is of order m we need m Cauchy conditions, one for each derivative of order 0 to $m-1$ of the unknown function, with respect to a nontangent direction on the Cauchy hypersurface.

Theorem 23 (Theorem of Existence and Uniqueness Cauchy–Kovalevskaya). *If a PDE of order m in the unknown function $u(x_1,\dots,x_n)$ can be written in the form*

$$\frac{\partial^m u}{\partial x_1^m} = f\left(x_1,\dots,x_n,u,\frac{\partial u}{\partial x_1},\dots,\frac{\partial^m u}{\partial x_1^{m_1}\dots\partial x_n^{m_n}}\right), \tag{9.33}$$

where $m = m_1 + \dots + m_n$ and where the term $\frac{\partial^m u}{\partial x_1^m}$ does not appear on the RHS, then the Cauchy problem attached to this PDE:

$$\left.\frac{\partial^j u}{\partial l^j}\right|_{\Sigma} = g_j, \quad j = 0,1,\dots,m-1 \tag{9.34}$$

with functions g_j defined on the $(n-1)$-dimensional regular hypersurface $\Sigma \subset \mathbb{R}^n$, where l is an arbitrary not tangent direction on Σ, admits a unique analytical solution u, if the functions f, g_j are analytical on their domains of definition.

For a proof see [63,64,274,317]. This theorem states the existence and uniqueness of an analytic solution, but this does not exclude the existence of other, nonanalytical solutions of the same Cauchy problem. However, if the PDE is linear (Holmgren uniqueness theorem) there are no solutions except the analytical ones. This last result shows that possible compact supported solutions or very localized solutions (like solitons, compactons, peakons, etc.), which of course are not analytical functions, could not arise from a linear PDE. High localization is strictly related, or generated, by the nonlinearity in the PDE. We remind here that there is one special case in which linear equations provide compact supported solutions, i.e., the discrete wavelets 2-scale equation [336]. For example, the Haar scaling function (the step function), defined as 1 on $[0,1]$ and zero in the rest of real axis, is a solution of the finite difference equation $\Phi(x/2) = \Phi(x) + \Phi(x-1)$. This result reveals a possible deeper connection between linear finite difference equations (or infinite-order linear PDE equations) and nonlinear PDE.

Returning to the continuity equation we prove the existence and uniqueness theorem for its Cauchy problem. In the course of this proof we use special Cauchy condition defined on the hyperplane $t = t_0$. However, it is easy to generalize the following proof for general Cauchy conditions on an arbitrary hypersurface. This is because any arbitrary Cauchy hypersurface is regular, and hence we can find a local change of coordinates $(\boldsymbol{x},t) \to (\boldsymbol{x}',t')$, such that the hypersurface in the new coordinates is determined by the equation $t' = t'_0$, without any loss of generality or analyticity. Choosing the Cauchy condition on the hyperplane $t = t_0$ means knowing the density at the initial moment in the whole space, or in the domain of definition of the position vector. In the general Cauchy hypersurface case, the condition can be both initial condition and boundary condition, for example if Σ is defined by $\Sigma = \{(\boldsymbol{x}, t) | t = t_0 \text{ and } \boldsymbol{x} \in D\} \bigcup \{(\boldsymbol{x}, t) | t \le t_0 \text{ and } \boldsymbol{x} \in \partial D\}$, etc. Moreover, we can always reduce any Cauchy condition to a null Cauchy condition. If the function $\tilde{\rho}$ is a solution of the equation

$$\frac{\partial \tilde{\rho}}{\partial t} = -div(\tilde{\rho} V) - div(g V) \tag{9.35}$$

under the null Cauchy condition $\tilde{\rho}(\boldsymbol{r}, t_0) = 0$, then $\rho = \tilde{\rho} + g(\boldsymbol{r})$ is a solution of the continuity equation (9.32) for the same V, and the general Cauchy condition $\rho(\boldsymbol{r}, t_0) = g(\boldsymbol{r})$. The analyticity of the functions involved is not changed by this functional substitution. In the following, we use a generic function f instead of the RHS of the PDE under consideration, no matter if it is (9.32), (9.33), or (9.35).

The sketch of the proof of existence and uniqueness of the solution of the continuity equation can be presented briefly as follows. We construct the Taylor series of a hypothetic analytic solution ρ of (9.32), by using the initial condition and the equation itself. If such a solution exists, then by construction it is unique. To prove its existence, we construct an upper bound function f^{ub} for the RHS of (9.32). Such a construction is always possible, and the good news is that its associate solution, i.e., the solution of $\partial\rho/\partial t = f^{ub}$, is an upper bound function for ρ. By using the comparison criterium, $\rho \ll \rho^{ub}$, it results that ρ is uniformly convergent, hence analytical. This concludes the proof. Now we proceed with the detailed discussion.

To construct the Taylor series we use the following.

Lemma 8. *If the velocity field $V(\boldsymbol{r}, t)$ and the Cauchy condition $\rho(\boldsymbol{r}, t_0) = g(\boldsymbol{r})$ are analytic in a neighborhood $\mathcal{V}(\boldsymbol{r}_0, t_0)$, then the Cauchy problem for* (9.32) *admits one unique analytic solution in $\mathcal{V}$.*

Proof. Since this hypothetic solution is analytic, we can construct it as a Taylor series in the form

$$\rho(\boldsymbol{r}, t) = \rho(\boldsymbol{r}_0, t_0) + (t - t_0) \left.\frac{\partial \rho}{\partial t}\right|_0 + \sum_{i=1}^{3} (x_i - x_{i0}) \left.\frac{\partial \rho}{\partial x_i}\right|_0$$

$$+ \frac{1}{2!} \left[(x_i - x_{i0})(x_j - x_{j0}) \sum_{i,j=0}^{3} \left.\frac{\partial^2 \rho}{\partial x_i \partial x_j}\right|_0 \right.$$

$$+\sum_{i}^{3}(x_i - x_{i0})(t - t_0)\frac{\partial^2 \rho}{\partial x_i \partial t}\bigg|_0 + (t - t_0)^2 \frac{\partial^2 \rho}{\partial t^2}\bigg|_0 \Bigg]$$

$$+\frac{1}{3!}\Bigg[\sum_{i,j,k=0}^{3}(x_i - x_{i0})(x_j - x_{j0})(x_k - x_{k0})\frac{\partial^3 \rho}{\partial x_i \partial x_j \partial x_k}\bigg|_0 + \cdots\Bigg] + \cdots, \tag{9.36}$$

where by subscript 0 we understand that the value is taken in the point $(\boldsymbol{r}_0, t_0)$. Substitute in this series the initial Cauchy and the equation itself

$$\rho(\boldsymbol{r}_0, t_0) = g(\boldsymbol{r}_0)$$

$$\frac{\partial \rho}{\partial t}\bigg|_0 = -div(\rho \boldsymbol{V})|_0 = -div(g\boldsymbol{V})$$

$$\frac{\partial \rho}{\partial x_i}\bigg|_0 = \left(\frac{\partial}{\partial x_i}\rho(\boldsymbol{r}, t_0)\right)_{\boldsymbol{r}_0} = \frac{\partial g}{\partial x_i}(\boldsymbol{r}_0)$$

$$\frac{\partial^I \rho}{\partial x_{i_1} \partial x_{i_2} \ldots \partial x_{i_n}}\bigg|_0 = \frac{\partial^I g}{\partial x_{i_1} \partial x_{i_2} \ldots \partial x_{i_n}}(\boldsymbol{r}_0)$$

$$\frac{\partial^2 \rho}{\partial x_i \partial t}\bigg|_0 = -div \frac{\partial}{\partial x_i}(g\boldsymbol{V}(\boldsymbol{r}, t_0))_{\boldsymbol{r}_0}, \text{etc.}, \tag{9.37}$$

and so on, for all terms. The hypothetic analytic solution is now fully determined, which proves its uniqueness. To prove its existence, we need to introduce the concept of *upper bound function* in general in $\mathbb{R}^n$. □

Definition 61. Let $x_0 \in \mathbb{R}^n$ and f is an analytic function defined on a neighborhood $\mathcal{V}(x_0)$, such that

$$f(x) = \sum_{i_1, i_2, \ldots i_n} F_{i_1, i_2, \ldots, i_n}(x_1 - x_{01})^{i_1} \cdots (x_n - x_{0n})^{i_n},$$

for $x \in \mathcal{V}(x_0)$. We define an analytic function on $\mathcal{V}(x_0)$

$$f^{ub}(x) = \sum_{i_1, i_2, \ldots i_n} G_{i_1, i_2, \ldots, i_n}(x_1 - x_{01})^{i_1} \cdots (x_n - x_{0n})^{i_n},$$

called *upper bound* of f, if $\forall i_1, \ldots i_n$ we have:

1. $|F_{i_1,\ldots,i_n}| < G_{i_1,\ldots,i_n}$.
2. $0 \leq G_{i_1,\ldots,i_n}$.

The notation is $f \ll f^{ub}$. The next step is to find an upper bound function for the RHS term of the continuity equation.

Theorem 24. *For any function*

$$f = \sum_{i_1,i_2,\ldots i_n} F_{i_1,i_2,\ldots i_n}(x_1 - x_{01})^{i_1} \cdots (x_n - x_{0n})^{i_n},$$

analytic on a neighborhood $\mathcal{V}(x_0)$, there is a neighborhood $\mathcal{W}(x_0) \subset \mathcal{V}(x_0)$ where f has an analytic upper bound function of the form

$$f^{ub}(x) = \frac{M}{1 - \frac{\sum_{i=1}^{n}(x_i - x_{0i})}{\alpha}} + C, \tag{9.38}$$

where $M > 0, \alpha \in \mathbb{R}$, and C is a constant.

Proof. Obviously, $\exists \xi \in \mathcal{W}$ such that the numeric series

$$\sum_{i_1,i_2,\ldots i_n} F_{i_1,i_2,\ldots i_n}(\xi_1 - x_{01})^{i_1} \cdots (\xi_n - x_{0n})^{i_n},$$

is uniformly convergent, which implies that the sequence $F_{i_1,i_2,\ldots i_n}(\xi_1 - x_{01})^{i_1} \cdots (\xi_n - x_{0n})^{i_n} \to 0$, so it is bounded, i.e., $\exists M > 0$ such that

$$|F_{i_1,i_2,\ldots i_n}(\xi_1 - x_{01})^{i_1} \cdots (\xi_n - x_{0n})^{i_n}| < M.$$

Then

$$M \sum_{i_1,\ldots i_n} \frac{(x_1 - x_{01})^{i_1} \cdots (x_n - x_{0n})^{i_n}}{|(\xi_1 - x_{01})^{i_1} \cdots (\xi_n - x_{0n})^{i_n}|},$$

is an upper bound for f on $\mathcal{W}$, according to Definition 7. Since the above series is also a geometric progression, we can calculate its sum. Then we can find an upper bound function f^{ub} for this progression in the form

$$\frac{M}{\left(1 - \frac{x_1 - x_{01}}{|\xi_1 - x_{01}|}\right) \cdots \left(1 - \frac{x_n - x_{0n}}{|\xi_n - x_{0n}|}\right)} < \frac{M}{1 - \frac{\sum_{i=1}^{n}(x_i - x_{0i})}{\alpha}} + \text{cst.} = f^{ub}(x), \tag{9.39}$$

with $\alpha = \min\{|\xi_1 - x_{01}|, \ldots |\xi_n - x_{0n}|\}$. The next step is to take this type of upper bound function in $n = 4$ and use it in the RHS of the continuity equation, instead of its original RHS, with an appropriate choice of the arbitrary constant cst.

$$\frac{\partial \rho^{ub}}{\partial t} = \frac{M}{1 - \frac{t + x + y + z + \rho + \sum_{i=1}^{3} \frac{\partial \rho}{\partial x_i}}{\alpha}} - M. \quad \square \tag{9.40}$$

Lemma 9. *The null Cauchy problem for* (5.14) *has a unique analytic solution* ρ^{ub} *in a neighborhood of* 0*, whose Taylor series has all coefficients nonnegative.*

Proof. We introduce the variable $\chi = t + x + y + z$ and we look for solutions of (5.14) of the form $\rho(t, x, y, z) = u(\chi)$ under the initial condition $u(0) = 0$. The PDE (5.14) reduces to an ODE

$$u'(\alpha - \chi - 3M) - uu' - 3(u')^2 - Mu - M\chi = 0,$$

and according to the Peano theorem (remember, it is based on the fixed point theorem [160]) this equation has a unique analytical solution in the initial condition $u(0) = 0$. When $\chi = 0$ we have a possible solution $u'(0) = 0$. By differentiating the ODE one more time, and by calculating it again in $\chi = 0$, we have $u''(0) = M/(\alpha - 3M)$. If we choose $\alpha \geq 3M$ it results $u^{(k)}(0) \geq 0$ for $k = 0, 1, 2$. In general, after n successive differentiations, we have

$$u^{(n)}(0) = \frac{1}{\alpha - 3M}\left(\sum_{k,j=0}^{n} |C_{kj}| u^{(k)}(0) u^{(j)}(0) + (\alpha M + n) u^{(n)}(0)\right).$$

It results, by induction, that $\forall k, u^{(k)}(0) \geq 0$ if $\alpha > 3M$. This result proves that the null Cauchy problem for (9.40) has always an unique analytic solution, whose Taylor series coefficients are nonnegative:

$$\rho^{ub}(\boldsymbol{r}, t) = \sum |C_{i_0,i_1,i_2,i_3}| t^{i_0} x^{i_1} y^{i_2} z^{i_3}. \tag{9.41}$$

There is no loss of generality by choosing null Cauchy conditions in Lemma 4. We proved in (9.35) that any null Cauchy conditions can be changed into arbitrary Cauchy conditions, so Lemma 4 is general. Now we attack the final step of our proof.

The uniqueness of the Cauchy problem for (9.32) was proved in Lemma 3, so we just need to prove the existence of analytic solution ρ. Since the actual RHS term of the continuity equation is analytic in all its variables, we can find an upper bound function for the PDE in the form of (9.38). We solved this auxiliary PDE (Lemma 9) and its solution ρ^{ub} has the property: $\rho \ll \rho^{ub}$. This is true because we build the solutions term by term, by using the functions f, f^{ub}, and the Cauchy data g (like we did in (9.36) and (9.37)). The upper bound property transfers from the fs to the ρs. Consequently, all the coefficients (partial derivatives in 0) of the Taylor series for ρ are upper bounded by the corresponding coefficients (corresponding partial derivatives in 0) of ρ^{ub}. Since the series in (9.41) is analytic, by the comparison criterium, it results the analyticity of the series ρ (see (9.36) and (9.37)). But this is the actual solution of (9.32), which proves the whole theorem.

We briefly present the above proof in the equation (9.42)

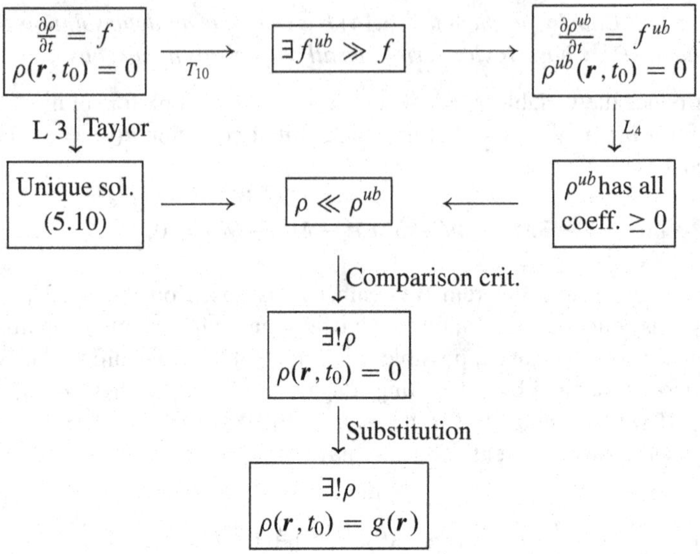

$$\boxed{\begin{array}{l}\frac{\partial \rho}{\partial t} = f \\ \rho(\boldsymbol{r}, t_0) = 0\end{array}} \xrightarrow{\;T_{10}\;} \boxed{\exists f^{ub} \gg f} \longrightarrow \boxed{\begin{array}{l}\frac{\partial \rho^{ub}}{\partial t} = f^{ub} \\ \rho^{ub}(\boldsymbol{r}, t_0) = 0\end{array}}$$

$$\text{L 3}\downarrow\text{Taylor} \qquad\qquad\qquad\qquad\qquad\qquad\qquad \downarrow L_4$$

$$\boxed{\begin{array}{c}\text{Unique sol.} \\ (5.10)\end{array}} \longrightarrow \boxed{\rho \ll \rho^{ub}} \longleftarrow \boxed{\begin{array}{c}\rho^{ub}\text{ has all} \\ \text{coeff.} \geq 0\end{array}}$$

$$\downarrow \text{Comparison crit.} \qquad\qquad\qquad \Box \quad (9.42)$$

$$\boxed{\begin{array}{l}\exists! \rho \\ \rho(\boldsymbol{r}, t_0) = 0\end{array}}$$

$$\downarrow \text{Substitution}$$

$$\boxed{\begin{array}{l}\exists! \rho \\ \rho(\boldsymbol{r}, t_0) = g(\boldsymbol{r})\end{array}}$$

9.6.2   Solutions of the Continuity Equation on Compact Intervals

In Sect. 9.6.1 we discussed the general conditions under which the continuity equation has a unique analytical solution. In this section we investigate some special one-dimensional situations having exact solutions. That is a Cauchy one-dimensional problem for $\rho(x,t)$ for given $V(x,t)$. We focus especially on the behavior of the solutions at the boundaries of a compact interval of length $2L$. The one-dimensional version of the continuity equation reads

$$\frac{\partial \rho}{\partial t} + \rho \frac{\partial V}{\partial x} + V \frac{\partial \rho}{\partial x} = 0, \tag{9.43}$$

for $x \in [-L, L]$, $t \geq 0$. At the boundaries of the interval, we should have no flow of matter so we impose the BC $v(\pm L, t) = 0$, in addition to the Cauchy condition. It is easy to build the general solution from the Fourier expansions

$$\rho(x,t) = \sum_{n \geq 0} \rho_n(t) e^{\frac{i\pi n x}{L}}, \quad V(x,t) = \sum_{n \geq 0} V_n(t) e^{\frac{i\pi n x}{L}}, \tag{9.44}$$

and from the BC we have

$$\sum_{n \geq 0} (-1)^n V_n(t) = 0. \tag{9.45}$$

If we plug the formulas from (9.44) in the continuity equation (9.43), we obtain a recursion relation

$$\rho_k'(t) = -\frac{i\pi k}{L}\sum_{n=0}^{k} \rho_n V_{k-n}. \tag{9.46}$$

With the notation

$$\mathbb{V}^k(t) \equiv e^{-\frac{i\pi k}{L}\int_0^t V_0(t')dt'},$$

we have (9.46), the new recursion relation

$$\rho_k(t) = \mathbb{V}^k(t)\left(\rho_k(0) - \frac{i\pi k}{L}\int_0^t \mathbb{V}^{-k}(t')\sum_{n=0}^{k-1}\rho_n(t')V_{k-n}(t')dt'\right), \tag{9.47}$$

where $\rho_k(0)$ are determined by the initial condition through the inverse Fourier transform

$$\rho_n(0) = \frac{1}{2\pi}\int_{-L}^{L} \rho_{initial}(x)e^{-\frac{in\pi x}{L}}dx. \tag{9.48}$$

We choose a simple physical example, where the initial density is the same everywhere within the compact $[-L, L]$, and zero outside. That is $\rho(x,0) = m/(2L)$, where m is the total mass of the fluid inside the bounded segment. It results $\rho_0(0) = m/(2L)$ and $\rho_n(0) = 0$ for $n > 0$. We also choose a simple configuration for the velocity, namely $V(x,t) = a\sin(\omega t)\left(e^{\frac{i\pi x}{L}} + e^{\frac{2i\pi x}{L}}\right)$. That is $V_1(t) = V_2(t)$. This is a stationary (longitudinal) oscillation in velocity along the segment, with zero velocity in the ends. We have $V_n(t) = 0$ for $n = 0, 3, \ldots$. By substituting these expressions for the velocity components in (5.23), we obtain $\mathbb{V}^{\pm k} = 1$ and

$$\rho_k(t) = -\frac{i\pi k a}{L}\int_0^t \sin(\omega t')(\rho_{k-1} + \rho_{k-2})dt', \quad k = 1, 2, \ldots \tag{9.49}$$

This recursion provides the unique solution for $k \geq 1$.

Apparently, finding general solutions for the continuity equation in one-dimensional, $\rho_t + \rho V_x + \rho_x V = 0$, is a simple procedure (subscripts represent, again, differentiation). However, there is a hidden problem at the boundaries, produced by the zeros of the coefficients in the PDE. At the ends of the interval, we have to assume no flow of fluid, so $V(\pm L, t) = 0$. In a neighborhood $(L - \epsilon, L)$ of the right boundary for example, we can test the behavior of a Fourier component of the solution $\rho_\omega(x,t) = r(x)e^{i\omega t}$, and we obtain

$$\frac{d(ln\ r_\omega)}{dx}V_\omega = -\left(\frac{dV_\omega}{dx} + i\omega\right), \tag{9.50}$$

which means that in this neighborhood, even if $V_x = 0$, we still have the RHS nonzero. But, when $V \to 0$, it seems that $d(ln\ r_\omega)/dx \to \infty$. So, the zeros of velocity at boundaries may introduce singularities in density (by reciprocity, in the inverse problem, isolated zeros of density can also introduce singularities in velocity). Let us suppose that the velocity approaches the zero as a power law $V(L-\epsilon,t) \simeq \epsilon^a$, $a > 0$. If $a < 1$ we have $lim_{x\to L}(\rho) < +\infty$. But if $a > 1$ we expect $lim_{x\to L}(\rho) = +\infty$. If V is a rapidly decreasing function in that neighborhood, we can neglect the third term in (9.43) and use the approximation

$$\frac{\partial\rho}{\partial t} \simeq -\rho\frac{\partial V}{\partial x},$$

to investigate the behavior of ρ. By direct integration we obtain

$$\rho(L-\epsilon,t) \simeq \rho_L e^{-\int_0^t V_x(L-\epsilon,t')dt'},$$

where ρ_L is a constant. This asymptotic solution is a very rapidly increasing function toward L, but it is not anymore a singularity.

Let us illustrate with examples. We take a simple form for velocity in a compact interval $x \in [-L, L]$

$$v(x,t) = V_0 \sin\omega t \cos kx,$$

as stationary oscillations, where $k = (2n+1)\pi/(2L)$, n arbitrary integer and V_0, ω are constants. The solution can be easily obtained by the procedure indicated above or by simple separation of variables. The general solution is a real integral over the label λ of the following components

$$\rho(x,t,\lambda) = \rho_0 e^{-\frac{\lambda}{\omega}\cos\omega t}\frac{\left(\cos\frac{kx}{2}+\sin\frac{kx}{2}\right)^{a-1}}{\left(\cos\frac{kx}{2}-\sin\frac{kx}{2}\right)^{a+1}},$$

where $a = -\lambda/(kV_0)$, and ρ_0 are constants. Obviously this solution has singularities within $[-L, L]$, provided by the trigonometric zeros of the denominator. The reason is the cancellation of velocity in different points (function of how large is n) including the boundaries. Velocity approaches zero by following a quadratic law: $V(L-\epsilon,t) \simeq k^2\epsilon^2/2$.

What can be done to eliminate these singularities? Of course, by coupling the continuity equation with Euler and energy conservation equations, the nonphysical solutions will be eliminated. However, one simple possibility to eliminate the singularity in density is to introduce an artificial constant term in velocity

$$V = V_0(\sin\omega t \cos kx + V_1).$$

From the physical point of view, it means that we have a little ($V_1 \ll 1$) constant "leakage" of fluid at the boundaries. With this new expression for velocity we have

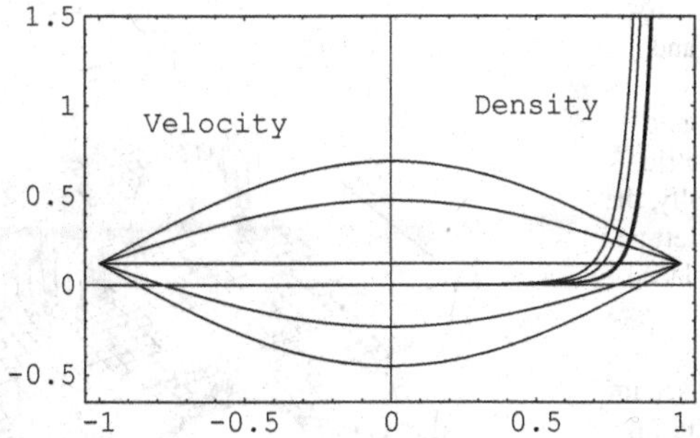

Fig. 9.12 Plot of velocity and density from one-dimensional continuity equation on an interval $[-1, 1]$. Velocity has stationary oscillations – up and down in this figure means motion of the fluid to right and left – and the fluid is accumulating in the right end. The density has itself push–pull oscillations

$$\rho(x,t,\lambda) = \rho_0 e^{-\frac{\lambda}{\omega}\cos\omega t} \left(\frac{1 + \dfrac{V_1 - 1}{\sqrt{1 - V_1^2}} \tan \dfrac{(2n+1)\pi x}{4L}}{1 - \dfrac{V_1 - 1}{\sqrt{1 - V_1^2}} \tan \dfrac{(2n+1)\pi x}{4L}} \right)^{\frac{2L\lambda}{(2n+1)\pi V_0 \sqrt{1-V_1^2}}}$$

$$\cdot \frac{1}{V_1 + \cos \dfrac{(2n+1)\pi x}{2L}}.$$

The solution is not anymore singular in $\pm L$ and it is illustrated in Fig. 9.12.

Global longitudinal oscillations of the fluid induce oscillations in the amount of fluid accumulated to the right end of the domain.

It is interesting to check the reverse phenomenon, namely if zeros in density provide singularities in velocity. For the stationary oscillating density inside $[-L, L]$

$$\rho(x,t) = \rho_1 \sin kx \sin \omega t,$$

with ω, ρ_1 constants and k defined as above, we compute the velocity in the form

$$V(x,t) = V_0 \cot \omega t \frac{C_1 + \omega \rho_1 \cos kx}{k(\rho_0 + \rho_1 \sin kx)},$$

where V_0, C_1, and ρ_0 are constants. In Fig. 9.13 we plot both the velocity and the density for this example for $L = 1$. Indeed, the density-isolated zeros provided by $\sin kx$ result in singularity in velocity given by the cot function.

Another example is presented for a semi-infinite domain $x \in (-\infty, 0]$. We choose the velocity of the form

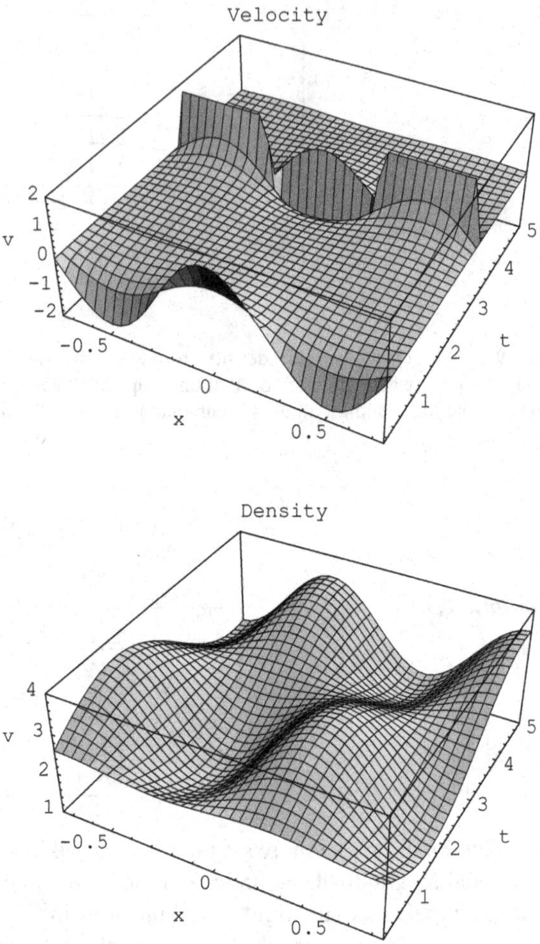

Fig. 9.13 At $t = 0$ density is uniformly distributed, and velocity has a positive maximum centered around $x = 0$, and two symmetric negative minima. Initially, the matter is pushed from left and right into two points, placed with approximation at $x = 0.25$ and $x = -1$. Around $t = 2$ one can see in the density plot the resulting accumulation of fluid in these two points. At this moment the velocity is almost zero and we have quasiequilibrium. Next, the velocity changes the sign, and the fluid is pushed toward two other centers, namely $x = 0$ and $x = 1$. As a result, at $t = 5$ we have more accumulation of fluid in these points. About $t = 4$ velocity has its singularity

$$V(x,t) = -\frac{ax}{at + \rho_0 \cosh \frac{tx}{b}},$$

where a, b, and ρ_0 are arbitrary constants. Around zero the velocity behaves like $V(0) \simeq x$ which provides a "milder" type of singularity for ρ. The corresponding solution for density is

$$\rho(x,t) = \rho_0 + at\,\mathrm{sech}\frac{tx}{b}.$$

The results are presented in Fig. 9.14.

In the last example, we present some localized traveling wave solutions along the axis. We assume the propagation of a KdV solitary wave on the free surface of a one-dimensional channel

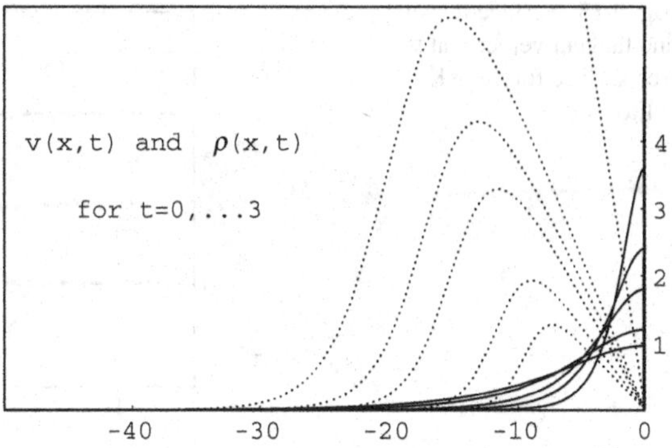

Fig. 9.14 Velocity (*dotted lines*) and density (*continuous lines*) for a one-dimensional semi-infinite axis. The velocity has a localized bump which pushes the fluid against the right wall, creating a fluid accumulation

$$\eta(x,t) = A\text{sech}_2 \frac{x - vt}{L},$$

where A is the wave amplitude, L the half-width, and v the group velocity. The tangent velocity of the fluid at the free surface is given by

$$V(x,t) = -\frac{2A}{L}\text{sech}^2 \frac{x - vt}{L} \tanh \frac{x - vt}{L}.$$

We neglect that the KdV equation for shallow water was deduced in the incompressibility approximation, at least for a very thin layer on the surface [169]. Let us presume that this layer is compressible (like a surfactant layer on the surface of the incompressible fluid) and the density in it is the solution of the continuity equation for the velocity given above. The density reads

$$\rho(x,t) = \rho_0 \frac{1}{v - V(x,t)},$$

where ρ_0 is the equilibrium density in the absence of the wave. Density has no singularities in this example. We present the results in Fig. 9.15. We can obtain a similar result for an MKdV soliton. We choose the velocity profile as a modulated breather [169]

$$V(x,t) = V_0\text{sech} \frac{x - vt}{L} \sin \omega(x - vt).$$

The density profile is given by a similar equation as in the KdV case

$$\rho(x,t) = \frac{\rho_0}{V(x) - v},$$

see Fig. 9.15.

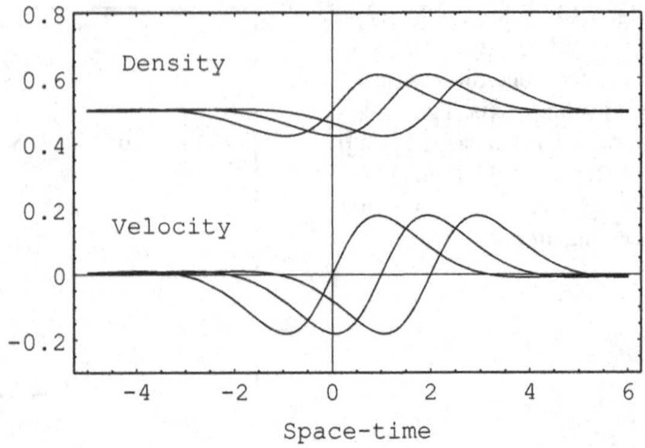

Fig. 9.15 Surface density and tangent velocity at the free surface for an MKdV soliton

9.7 Problems

1. Show that the free surface condition, i.e., the path of a fluid particle $\boldsymbol{r}_L$ does not leave a surface Σ (see (9.5), (9.28), and (9.29)), is the equivalent of requesting the Lagrangian path of the particle to belong to the time variable surface, both described in extended space $\mathbb{R} \times \mathbb{R}^3$ for time and positions.
2. Consider a sphere of radius R at rest surrounded by inviscid, incompressible, and irrotational fluid of density ρ. The fluid moves past the sphere such that the velocity at infinite distance from the sphere is a constant and uniform field $\boldsymbol{v}_\infty = (0, 0, -u)$. Find the Eulerian velocity, the pressure field and the stream lines. Find the Lagrangian paths and compare them with the stream lines.
3. Let us have the following field of Eulerian velocity

$$\boldsymbol{v}_E(\boldsymbol{r}, t) = (a_1(t)x^{\alpha_1}, a_2(t)y^{\alpha_2}, a_3(t)z^{\alpha_3}),$$

   where $a_i(t)$ are arbitrary smooth functions and $\alpha_i \in \mathbb{R}$. Find the equations of the stream lines and the path lines. Show that if $a_i(t)$ are constant, the stream and path lines coincide for an appropriate choice of integration constants.
4. Consider the Lagrangian paths of some fluid particles $\boldsymbol{r}_L(\boldsymbol{r}_0, t)$ as a one-parameter t group of diffeomorphisms mapping the initial positions of the particles into the current ones $\boldsymbol{r}_0 \to \boldsymbol{r}_L$, acting in $\mathbb{R}^3$. Consider a time-dependent physical quantity Ω described by a differentiable 1-form ω defined on $T^*_{\boldsymbol{r}_L}\mathbb{R}^3$. Prove that the Lie derivative of this 1-form with respect to the tangent directions to the diffeomorphism transformations

$$L_{\boldsymbol{r}_L(\boldsymbol{r}_0,t)}(\omega) = \lim_{dt \to 0} \frac{d\boldsymbol{r}^*_L(\omega) - \omega}{dt} = \frac{d}{dt}\left(\omega_j \frac{\partial x^j_L}{\partial x^i_0} - \omega_i\right) dx^i$$

   provides the Eulerian–Lagrangian law of transformation for Ω.

5. Equations (9.12) and (9.13) were obtained by using the Lie derivative with respect to the fluid flow. Try to find the same equations from a different approach, namely a new law of covariant differentiation on a four-dimensional manifold (σ^0, σ^i) with a linear connection. The last two and three terms, respectively, in the RHS of (9.12) and (9.13) could be understood as connection coefficients with the Christoffel symbols of the second kind fulfilling

$$\Gamma^i_{k0} = -\frac{\partial v^i}{\partial \sigma^k}.$$

Hint: we need to introduce a metric on this manifold, $g_{\mu\nu}$, with $\mu, \nu = 0, 1, \ldots, 3$. The Christoffel symbols of first and second kind are related by $\Gamma^\alpha_{\beta\gamma} = g^{\delta\alpha}\Gamma_{\beta\delta\gamma}$, and the last one is defined by the metric

$$\Gamma_{\alpha\beta\gamma} = \frac{1}{2}\left(\frac{\partial g_{\gamma\beta}}{\partial \sigma^\alpha} + \frac{\partial g_{\beta\alpha}}{\partial \sigma^\gamma} - \frac{\partial g_{\alpha\gamma}}{\partial \sigma^\beta}\right),$$

see for example [10, 19, 158, 181, 299]. A possible hypothesis could be $g_{i0} = 0$, g_{00} =const. The remaining PDE equations for g_{ij} may result in an exponential matrix solution. It is interesting to relate the skew-symmetry property of this PDE in the metric coefficients with the fact that the integral curves of a rotational flow are singular.

6. Prove that the covariant time derivative (9.12) and (9.13) has the following actions

$$\frac{d_c A}{dt} = \frac{dA}{dt} + \gamma^t A, \text{ on covariant vectors,}$$

$$\frac{d_c A}{dt} = \frac{dA}{dt} - \gamma A, \text{ on contravariant vectors,}$$

$$\frac{d_c \Omega}{dt} = \frac{d\Omega}{dt} - \gamma\Omega - \Omega\gamma^t, \text{ on (2.0) tensors,}$$

$$\frac{d_c \Omega}{dt} = \frac{d\Omega}{dt} + \gamma^t\Omega + \Omega\gamma, \text{ on (0.2) tensors.}$$

Chapter 10
Dynamics of Hydrodynamics

The mathematical description of the states of a fluid is based on the study of three fields defined on the domain occupied by the fluid: the velocity field V, the density ρ, and the pressure field P. These three "unknowns" are determined by integrating other five scalar equations, namely the mass conservation (continuity equation), the three components of the equation of momentum balance (Euler or Navier–Stokes), and the energy balance. This last equation needs in addition information about the thermodynamics of the fluid, so it may need to be supplied with some equation of state. In addition to these five equations, we request regularity, asymptotic and, if it is the case, boundary conditions, to provide a unique solution. When we study the dynamics of the fluid confined in a compact domain with free boundaries, the system is slightly more complicated, and we have to add the kinematical equation of the free surface, as well as equations of momentum balance at the surface. If we take into account the nonlinear terms in the dynamical equations, and in the associated curved geometry, some interesting solutions occur. Special nonlinear effects related to fluids on compact domains with free surface could be Gibbs–Marangoni effect, dividing the flow in cells (Bènard effect), couplings between different modes, collective effects, separation of flow in layer (boundary layer, turbulence), standing traveling surface waves, etc. In this chapter, we introduce some elements of general hydrodynamics which we will use later on in the book, boundary conditions especially at free surfaces, surface pressure theory, and representation theorems.

10.1 Momentum Conservation: Euler and Navier–Stokes Equations

The continuity equation for fluid dynamics (9.32) was derived in Sect. 9.6 and it has the form

$$\frac{\partial \rho}{\partial t} + \nabla \cdot (\rho V) = 0, \tag{10.1}$$

A. Ludu, *Nonlinear Waves and Solitons on Contours and Closed Surfaces*,
Springer Series in Synergetics, DOI 10.1007/978-3-642-22895-7_10,

where $V = (V_i)$ is the Lagrangian or material velocity of the fluid particle, and ρ is the fluid density. Because we study the fluid in the three-dimensional Euclidean space of flat metric, there is no difference between covariant and contravariant character of the Euclidean vectors, so we will place the label as subscripts as a rule in this section. The momentum of the unit of fluid volume is given by

$$p_i \equiv \frac{\partial}{\partial t}(\rho V_i) = f_i = \frac{F_i}{V}, \tag{10.2}$$

where $\boldsymbol{f} = (f_i)$ is the volume force density, derived for the total force field in the fluid $\boldsymbol{F}$. From (10.1) and (10.2), we have

$$\frac{\partial V_i}{\partial t} + V_k\left(\frac{\partial}{\partial x_k}V_i\right) = -\frac{\partial}{\partial x_j}(P\delta_{ij} + \rho V_i V_j) \equiv -\frac{\partial}{\partial x_j}\Theta_{ij}, \tag{10.3}$$

where P is the pressure, and we define the fluid symmetric *momentum flux tensor* as $\hat{\Theta}$. In the inviscid case, where we have no loss of momentum in viscosity and internal frictions, this tensor has the property

$$f_i = \frac{\partial p_i}{\partial t} = \frac{\partial}{\partial t}(\rho V_i) = -\frac{\partial}{\partial x_i}\Theta_{ij}^{inviscid}. \tag{10.4}$$

If we draw an imaginary smooth surface with unit normal $\boldsymbol{N}$, (10.4) can be written in the form

$$\hat{\Pi}^{inviscid} \cdot \boldsymbol{N} = P\boldsymbol{N} + \rho \boldsymbol{V}(\boldsymbol{V} \cdot \boldsymbol{N}), \tag{10.5}$$

which represents the balance of *reversible* momentum. The LHS term represents how much momentum is transferred per unit of time and cross-section area in the direction $\boldsymbol{N}$, the first term on the RHS is the change of momentum by molecular motion and interaction, and the last term is the change of momentum by bulk flow only.

If we consider the viscosity, η, we have to extend the momentum flux tensor with an extra term, namely

$$\Theta_{ij}^{inviscid} \to \Theta_{ij} = P\delta_{ij} + \rho V_i V_j - \sigma'_{ij}. \tag{10.6}$$

In literature [50, 111, 167, 171, 220, 224, 305], authors use another tensor, namely the fluid *stress tensor* $\hat{\sigma}$, inspired from the study of elasticity, representing the total momentum transferred by molecular motion both reversible and irreversible, and defined by

$$\sigma_{ij} = -P\delta_{ij} + \sigma'_{ij}, \tag{10.7}$$

so that

$$\Theta_{ij} = -\sigma_{ij} + \rho V_i V_j. \tag{10.8}$$

So far we took for granted that these stress tensors are symmetric. The proof is based on the judgment that the total torque, $dM_i = \epsilon_{ijk} x_j \partial\Theta_{kl}/\partial x_l dV$, produced by fluid forces in an infinitesimal domain depends only on the surface of the domain, because inside forces between different elements cancel each other in action–reaction pairs. From the Green theorem applied on this domain, we obtain that $\epsilon_{ijk}\Theta_{jk} = 0$, where from $\Theta_{ij} = \Theta_{ji}, \sigma_{ij} = \sigma_{ji}$.

To have an expression for the stress tensor, we need to use the *Newtonian fluid* hypothesis, namely the part of the momentum flux tensor which results from frictional interaction of the fluid in relative motion (represented by the viscous stress tensor σ') depends only on the instantaneous gradient of fluid velocity. In addition, this dependence is approximated to be linear. If we keep the general dependence on the gradient, the fluid is called *Stokesian fluid*, but the hypothesis need to be supplemented by requiring smoothness, isotropy, and homogeneity [10, 294]. So, we can write

$$\sigma'_{ij} = C_{ijkl}\frac{\partial V_k}{\partial V_l}. \tag{10.9}$$

To determine the tensor C, we note that a global rotation of the fluid should not introduce any stress, so we have $C_{ijkl} = C_{ijlk}$. In addition we require C to be an *isotropic tensor*, namely invariant to any rotation. We know that the only rotational invariant tensors of rank 0 is a scalar, of rank 1 there is none, of rank 2 is the Kronecker symbol δ_{ij}, and of rank 3 is the Levi–Civita tensor ϵ_{ijk}. The number of linear independent isotropic tensors of rank k is given by the Motzkin recursion formula

$$N_k = \frac{k-1}{k+1}(2N_{k-1} + 3N_{k-2}), \quad k = 1, 2, \ldots,$$

from where it results $N_4 = 3$ [304]. To obtain the general formula for the C tensor, we can use a theorem from elasticity [143, 256]. This theorem states that a rank 2 symmetric tensor (i.e., $\hat{\sigma}'$) generated by all possible linear combinations between another rank 2 tensor ∇V and a rank 4 isotropic tensor $\hat{C}$ with the above listed properties is a linear combination of the symmetric part of ∇V and the Kronecker tensor times the trace of ∇V. That is

$$\hat{\sigma} = -P\hat{I} + \eta(\nabla V + (\nabla V)^t) + \lambda \text{Tr}(\nabla V)\hat{I}, \tag{10.10}$$

where $(\hat{I}) = \delta_{ij}$, and where the second term on the RHS is the symmetric part of ∇V (containing the transpose), also called the *rate of deformation* (or *rate of strain*), and $\text{Tr}(\nabla V) = \nabla \cdot V$ is called *rate of expansion* [50, 167]. The last assumption on the stress tensor (Stokes' assumption) namely $\hat{\sigma}'$ makes no contributions to the mean normal stress, so we have $\lambda = -2/3$ from here. It results

$$\begin{aligned} \hat{\sigma} &= -P\hat{I} + \eta\left[(\hat{\nabla V} + (\nabla V)^t) - \frac{2}{3}\text{Tr}(\nabla V)\right]\hat{I} \\ &= -P\delta_{ij} + \eta\left[\frac{\partial V_i}{\partial x_j} + \frac{\partial V_j}{\partial x_i} - \frac{2}{3}\frac{\partial V_k}{\partial x_k}\delta_{ij}\right]. \end{aligned} \tag{10.11}$$

If we neglect the Stokesian assumption, and we also consider the contribution of a *dilatational viscosity*, we correct (10.11) into

$$\sigma_{ij} = -P\delta_{ij} + \eta\left[\frac{\partial V_i}{\partial x_j} + \frac{\partial V_j}{\partial x_i} - \frac{2}{3}\frac{\partial V_k}{\partial x_k}\delta_{ij}\right] + \zeta\frac{\partial V_k}{\partial x_k}\delta_{ij}, \tag{10.12}$$

where ζ is the *coefficient of dilatational viscosity*. In the non-Newtonian fluid, we have $\eta, \zeta = f(\partial v_i/\partial x_k)$.

We can rewrite (10.12) in a vectorial form, such that the dynamical equation for a viscous fluid reads

$$\rho\left[\frac{\partial \boldsymbol{V}}{\partial t} + (\boldsymbol{V}\nabla)\boldsymbol{V}\right] = -\nabla P + \rho \boldsymbol{f} + \eta\Delta\boldsymbol{V} + \left(\zeta + \frac{\eta}{3}\right)\nabla(\nabla\cdot\boldsymbol{V}), \tag{10.13}$$

which is the famous Navier–Stokes equation of a fluid in the presence of a volume density force $\boldsymbol{f}$. In the case of incompressible fluid, (10.14) becomes

$$\frac{\partial \boldsymbol{V}}{\partial t} + (\boldsymbol{V}\nabla)\boldsymbol{V} = -\frac{1}{\rho}\nabla P + \boldsymbol{f} + \frac{\eta}{\rho}\Delta\boldsymbol{V}, \tag{10.14}$$

which reduces to the Euler equation in absence of viscosity

$$\frac{\partial \boldsymbol{V}}{\partial t} + (\boldsymbol{V}\nabla)\boldsymbol{V} = -\frac{1}{\rho}\nabla P + \boldsymbol{f}. \tag{10.15}$$

10.2 Boundary Conditions

Boundary conditions at the surface of a fluid Σ can be of three types: separation between two fluids (fluid interface), free surface of a fluid in a rarefacted gaseous atmosphere (or vacuum), and contact with rigid surfaces. The expressions of the conditions of continuity in each case depend if the fluid (fluids) is viscous or inviscid. Basically, we can write a general continuity condition for the separation of two fluids (say fluids 1 and 2), and this condition can be modified for the other two cases.

The continuity of the velocity at the interface is a relation strongly dependent on the model (viscous or not, slipping interface or not, etc.), so we will use it for every situation in particular. Nevertheless, we can write a provisional continuity condition in the form $\boldsymbol{V}_1|_\Sigma = \boldsymbol{V}_2|_\Sigma$ or

$$V_{n,1}|_\Sigma = V_{n,2}|_\Sigma, \quad V_{\parallel,1}|_\Sigma = V_{\parallel,2}|_\Sigma, \tag{10.16}$$

where the two components are the normal and the parallel one to the surface. In many models, it is more practical to rewrite the continuity conditions (10.16) in another form,

$$V_{n,1}|_\Sigma = V_{n,2}|_\Sigma,$$
$$N \cdot (\nabla_\Sigma \cdot V_1|_\Sigma) = N \cdot (\nabla_\Sigma \cdot V_2|_\Sigma),$$
$$N \cdot (\nabla_\Sigma \times V_1|_\Sigma) = N \cdot (\nabla_\Sigma \times V_2|_\Sigma), \tag{10.17}$$

namely the continuity of the normal components of the velocity, of the divergence and the curl of the velocity. The last one is nothing but the continuity of the normal component of the vorticity $\boldsymbol{\omega} = \nabla \times \boldsymbol{V}$. The operator ∇_Σ is the surface gradient. Basically, it represents the gradient expressed in surface curvilinear coordinates, acting on vectors in the tangent plane to Σ. Its rigorous definition and properties are described in Sect. 6.5. Equations (10.17) represent mixed Dirichlet and von Neumann boundary conditions, and guarantee the uniqueness of the solution of the (elliptic type partial differential equations) Euler or Navier–Stokes equations (see (10.13) and (10.15)).

In the case of rigid surface in contact with the fluid, because of the cohesive forces, we ask $V|_\Sigma = 0$. Such a relation cannot be fulfilled by the Euler equation (it would generate zero solutions all over the space), but it can be fulfilled at least for the normal components in the case of inviscid fluids (or actually the normal component of fluid velocity should be equal to the local velocity of the rigid surface), while $V_\parallel \neq 0$ for ideal fluids. Consequently, the separation between the fluid and the rigid boundary is a special zone, so-called "vortex-sheet" or "boundary layer" where we model the discontinuity for the tangent velocity. In the boundary layer the vorticity is nonzero, but because the equation for vorticity in the viscous case is a diffusion type of equation

$$\frac{\partial \boldsymbol{\omega}}{\partial t} = \nu \Delta \boldsymbol{\omega},$$

where we eliminate the volume forces for simplification, we expect the vorticity to decay toward the bulk of the fluid, away from the boundary layer. This also implies that out of the boundary layer the velocity is almost potential.

The balance of the momentum across the surface is

$$F_1|_\Sigma = F_2|_\Sigma \rightarrow N_i \sigma^1_{ik}|_\Sigma = N_i \sigma^2_{ik}|_\Sigma \tag{10.18}$$

or in tensor form

$$(\hat{\sigma}^1 - \hat{\sigma}^2) \cdot N = 0, \quad \text{on } \Sigma. \tag{10.19}$$

For a free surface, (10.18) reduces to

$$N_i \sigma'^1_{ik}|_\Sigma = P|_\Sigma N_k. \tag{10.20}$$

In tensor form the continuity condition across a free surface reads

$$(\hat{\sigma}' \cdot N)_\Sigma = P|_\Sigma \cdot N = 2\sigma H,$$

$$(\hat{\sigma}' \cdot \boldsymbol{t}_{a,b})_\Sigma = 0, \tag{10.21}$$

where $\boldsymbol{t}_{a,b}$ form a basis in the tangent space of the surface, σ is the *coefficient of surface tension*, and H is the mean curvature of the surface. These equations will be elaborated in detail in Sect. 10.4. In this case of an isolated droplet, the driving force (the surface tension) acts always perpendicularly to the free surface. Therefore, the tangential stress on the surface vanishes, and the normal stress is the driving force. In Chap. 8, we have noticed that there are a lot of other interactions at the interface between two fluids, especially if the surface is material and it is moving.

If the surface of separation carries some material properties, for example it has mass distribution, internal viscoelastic forces, etc. (in this case the separation is called an interface), the continuity equations for the stress (10.19) and (10.21) change correspondingly

$$(\hat{\sigma}^1 - \hat{\sigma}^2) \cdot \boldsymbol{N}|_\Sigma = \boldsymbol{F}_{\text{net},\Sigma}, \tag{10.22}$$

where the RHS is the net force per unit of surface area acting upon the physical surface, sometimes denoted $\boldsymbol{\sigma}_\Sigma$. This surface density force, $\boldsymbol{F}_{\text{net}} = F_n\boldsymbol{N} + \boldsymbol{F}_\parallel$, contains the surface tension and many other terms related to the existence of surface elasticity, viscosity, shear, surfactants, mass transfer, etc. Its expression is obtained on differential geometry grounds in Sect. 8.4 (see (8.41) and (8.56)).

10.3 Circulation Theorem

This subject was initially investigated by Thomson [315] and Helmholtz [125]. Some different proofs of the theorems on vortex motion were given later by Lord Kelvin [153]. The circulation theorem states that:

Theorem 25 (Kelvin Circulation Theorem). *The line integral of the fluid velocity* $\boldsymbol{v}$ *along a closed circuit* Γ *(the circulation of the velocity) which moves together with the fluid is constant in time if the fluid is perfect*

$$C_{v,\Gamma} = \oint_\Gamma \boldsymbol{v} \cdot \boldsymbol{t} ds = const. \tag{10.23}$$

Here $\boldsymbol{v}$ is calculated in the Lagrangian frame and $\boldsymbol{t}$ is unit tangent to Γ.

By perfect fluid we understand here inviscid isentropic flow, governed by Euler (10.15) in the presence of only potential external forces

$$\boldsymbol{a} = \frac{d\boldsymbol{v}}{dt} = \frac{\partial \boldsymbol{v}}{\partial t} + (\boldsymbol{v} \cdot \nabla)\boldsymbol{v} = -\frac{1}{\rho}\nabla P - \nabla U, \tag{10.24}$$

where $\boldsymbol{a}$ is the Lagrangian acceleration and U is the potential of external forces acting on the fluid. This result is important both for vortex motion and potential motion. However, in spite of the fact that the concept of closed circuit *moving with the fluid* is intuitive, and it is based on the Lagrangian point of view, this concept is not quite rigorously defined geometrically. In the following, we give two proofs for the circulation theorem differing in the degree of rigorousness and geometry involved [167, 171, 224].

Proof 1. Equation of State Approach. The rate of change of the circulation is

$$\frac{dC_{v,\Gamma}}{dt} = \oint_{\Gamma} \boldsymbol{a} \cdot \boldsymbol{t} ds + \oint_{\Gamma} \boldsymbol{v} \cdot d\left(\frac{d\boldsymbol{r}}{dt}\right). \tag{10.25}$$

The second integral on the RHS is a total differential ($\boldsymbol{v} d\boldsymbol{v}$) and it provides zero contribution on the closed circuit. According to the hypotheses, the acceleration is given by the Euler (10.15). If the flow is isentropic, the Lagrangian variation of the entropy of the unit of mass of the fluid is zero, $d\left(\frac{S}{m}\right) = ds = 0$. Consequently, we can write the variation of the enthalpy of the unit of mass

$$dh = \frac{VdP + TdS}{m} = \frac{1}{\rho} dP, \tag{10.26}$$

where P is the pressure. In this way the acceleration becomes a gradient $\boldsymbol{a} = -\nabla(h + U)$, and the first integral in (10.25) is also zero. The circulation of velocity on any closed circuit moving with the fluid is indeed constant. □

In other approaches (for example [224]) Theorem 25 is formulated with a different hypothesis. It is stated that in the inviscid fluid the density is either constant or function of pressure only (barotropic flow). The equivalence of the two formulations is obvious: if the fluid is isentropic, then the constancy of entropy provides an equation of state in terms of density and pressure only, $s = s(p, \rho)$, from where the requested dependence [171].

It is interesting to observe that, for inviscid fluids which are not isentropic (not barotropic fluids) and for which the circulation is not conserved, the acceleration has the property

$$\nabla \times \boldsymbol{a} = \nabla P \times \nabla \frac{1}{\rho}. \tag{10.27}$$

This means that the rate of change of circulation can be expressed through the Stokes theorem in the form

$$\frac{dC_{v,\Gamma}}{dt} = \int_{\Sigma} \left(\nabla P \times \nabla \frac{1}{\rho}\right) \cdot \boldsymbol{N} dA, \tag{10.28}$$

where Σ is a surface bounded by the circuit Γ. That means that the average (over a small surface) rate of change of the circulation is directed along the intersection between isobaric surfaces and surfaces of constant density. A lot of convection effects, including for example the surface vs. bottom salted water current between the Black Sea and the Mediterranean Sea, are generated by this mechanism [224].

On the other hand, the circulation Theorem 25 helps to understand the permanent character of the potential flow: once the curl of velocity is zero in some region and at some initial moment of time, the velocity will be irrotational in any region of the space and at any later moment, by circulation (zero in this case) conservation. The irrotational character of the flow is transported by physical fluid particles in all the flow region.

Proof 2. Free Surface Approach. The physical hypotheses are the same: ideal inviscid isentropic fluid with potential external forces. We need to work with the concept of moving particle circuit, i.e., the closed curve of particles moving with the fluid. In other words a closed contour always consists of the same fluid particles. For a rigorous geometric definition of particle lines and circuits in terms of fiber bundles, the reader can return to the Sects. 9.2, 9.2.2, 9.2.4, and 9.2.5.

We prepare the proof of the Kelvin theorem by using traditional definitions of path lines and particle contours, like those introduced in Sects. 9.1.2, 9.2.3, and 9.3. Later on we reformulate the theorem in terms of differential geometry. Let us choose at $t = 0$ a compact, connected, and simply connected surface Σ made by fluid particles, and consider its boundary the closed curve $\Gamma = \partial\Sigma$. We call Γ a particle circuit. The existence and stability in time of such a curve are discussed in the above-mentioned sections. We parametrize this curve with the equation $\boldsymbol{r}_0(s)$, where s labels the fluid particles in the circuit. At a later moment of time, within some finite time interval $t \in [0, T]$, we construct a diffeomorphic deformation of Σ into Σ', i.e., the fluid flow. This mapping induces a diffeomorphic deformation of Γ into Γ', described by $\boldsymbol{r}_0(s) \to \boldsymbol{r}(t,s)$. The $\boldsymbol{r}(t,s)$ function represents the position of the s fluid particle at moment t. When time runs, the diffeomorphism generates a family of curves (particle circuits moving with the fluid) each one parameterized by the same label s. The set of these closed curves is called a *tube of flow* based on the particle sheets Σ and Σ'. The question is if this tube of flow described by the curves $\boldsymbol{r}(t,s)$ is a *regular surface*. The answer is given by Theorem 26.

Theorem 26. *Let $\boldsymbol{a}(\boldsymbol{r})$ be a differential vector field on an open domain $\mathcal{D} \subset \mathbb{R}^3$ and $\Gamma \subset \mathcal{D}$ be an arc-length parameterized regular simple closed curve of equation $\boldsymbol{r}_\Gamma(s)$ with $s \in [0, L_\Gamma]$ and $\boldsymbol{r}_\Gamma(0) = \boldsymbol{r}_\Gamma(L_\Gamma)$. For every $s \in [0, L_\Gamma]$ we build a regular simple parameterized curve Γ_s of equation $\boldsymbol{r}(\sigma, s)$ with $\sigma \in [0, \sigma_{max}]$ as follows:*

1. *The equation $\boldsymbol{r}(\sigma, s) = \boldsymbol{r}_\Gamma(s)$ has one and only one solution $\sigma = 0$.*
2. *If $\boldsymbol{t}_{\Gamma_s}(\sigma)$ is the unit tangent for each Γ_s curve, then $\forall \sigma \in [0, \sigma_{max}]$*

$$\frac{\partial \boldsymbol{r}}{\partial \sigma}(\sigma, s) \equiv \boldsymbol{t}_{\Gamma_s}(\sigma) = \boldsymbol{a}(\boldsymbol{r}(\sigma, s)),$$

$$\boldsymbol{a}(\boldsymbol{r}(\sigma, s)) \times \frac{d\boldsymbol{r}_\Gamma}{ds}(s) \neq 0.$$

$\boldsymbol{r}(\sigma, s)$ is a regular parameterized surface $\Sigma^\Gamma_{[0,\sigma_{max}]}$ for $\sigma \in [0, \sigma_{max}]$, $s \in [0, L_\Gamma]$.

Proof. See Fig. 10.1. Since the field $\boldsymbol{a}$ is differentiable, the curves Γ_s are its integral curves and depend smoothly on their natural arc-length parameter σ. Also, from the Frobenius existence and uniqueness theorem (Theorem 5), all these curves depend smoothly on their initial data, i.e., the s parameter (see also [46, Theorem 1, p. 176]). Consequently $\boldsymbol{r}(\sigma, s)$ is a differentiable function. From the hypotheses each integral curve intersects the contour only one time. The Jacobian matrix

$$\hat{J}\boldsymbol{r}(\sigma, s) = \left(\frac{\partial x^i}{\partial \sigma}, \frac{\partial x^i}{\partial x^j_\Gamma}\frac{dx^j_\Gamma}{ds}\right) = (a^i(\boldsymbol{r}(\sigma, s)), \delta_{ij} t^j_\Gamma(s)) \neq 0$$

is nonzero by hypothesis. The Jacobian has rank 2 and hence the tangent map $d\boldsymbol{r}$ is one-to-one. Consequently $\boldsymbol{r}(\sigma, s)$ is a regular parametrized surface. □

From Theorem 26 we know that moving particles arranged in a closed contour Γ generate a tube of flow $\boldsymbol{r}(t, s)$ based on Γ and Γ'. Now we can come back to the second proof of the Kelvin circulation theorem. We write (10.23) in the form

$$\oint_{\Gamma} \boldsymbol{v} \cdot \boldsymbol{t} ds = \oint_{\Gamma'} \boldsymbol{v} \cdot \boldsymbol{t} ds,$$

where Γ, Γ' represent the particle contour at two different moments of time.

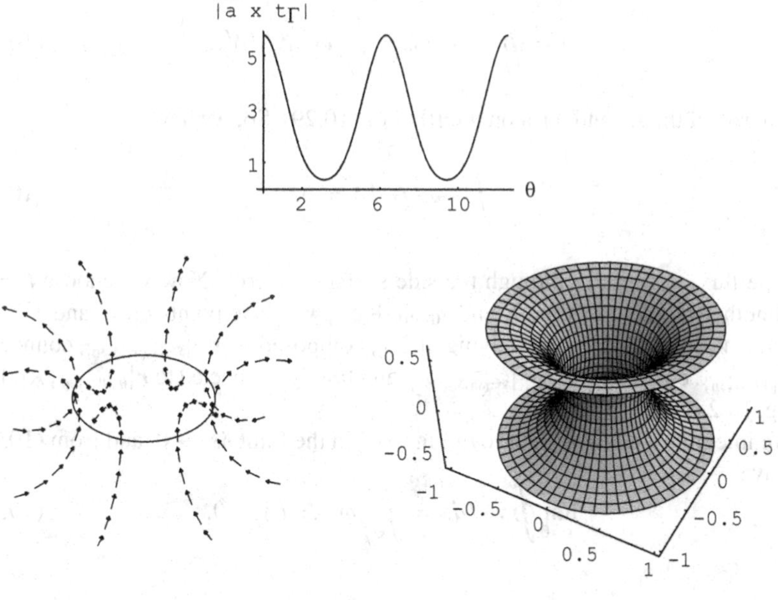

Fig. 10.1 *Left*: particle circuit Γ (*horizontal circle*) and corresponding particle paths (Γ_s, *arrows*). *Right*: resulting tube of flow $\Sigma^{\Gamma}_{[0,\sigma_{max}]}$. *Top*: the regularity condition in Theorem 26 is fulfilled, i.e., $\boldsymbol{a}(\boldsymbol{r}(\sigma, s)) \times \boldsymbol{r}'_\Gamma(s) \neq 0$

The vorticity $\boldsymbol{\omega} = \nabla \times \boldsymbol{v}$ has the property $\nabla \cdot \boldsymbol{\omega} = 0$ which means that, for any domain $\mathcal{D}$, we have

$$\int_{\mathcal{D}} \nabla \cdot \boldsymbol{\omega} dV = \oint_{\partial \mathcal{D}} \boldsymbol{\omega} \cdot \boldsymbol{N} dA = 0,$$

where dV, dA are the volume and area elements and $\boldsymbol{N}$ is the unit normal to Σ. We choose $\mathcal{D}$ to be the inside of a tube of flow bounded by Σ, Σ' and a side area described by the flows $\boldsymbol{r}(t, s)$, denoted in the following Σ_f. We have

$$0 = \oint_{\Sigma \cup \Sigma' \cup \Sigma_f} \boldsymbol{\omega} \cdot \boldsymbol{N} dA = \int_{\Sigma_f} \boldsymbol{\omega} \cdot \boldsymbol{N} dA + \int_{\Sigma \cup \Sigma'} \boldsymbol{\omega} \cdot \boldsymbol{N} dA. \tag{10.29}$$

Because Σ, Σ' are particle surfaces, we have

$$\boldsymbol{v}|_{\Sigma} \times \boldsymbol{N}_{\Sigma} = 0, \boldsymbol{v}|_{\Sigma'} \times \boldsymbol{N}_{\Sigma'} = 0, \tag{10.30}$$

and hence we have $\boldsymbol{v}|_{\Gamma} \cdot \boldsymbol{t}_{\Gamma} = \boldsymbol{v}|_{\Gamma'} \cdot \boldsymbol{t}_{\Gamma'} = 0$.[1] Consequently

$$0 = \oint_{\Gamma} \boldsymbol{v} \cdot \boldsymbol{t} ds = \int_{\Sigma} \boldsymbol{\omega} \cdot \boldsymbol{N} dA$$

$$0 = \oint_{\Gamma'} \boldsymbol{v} \cdot \boldsymbol{t}' ds = \int_{\Sigma'} \boldsymbol{\omega} \cdot \boldsymbol{N}' dA', \tag{10.31}$$

which cancel the second term on the RHS of (10.29). So, we have

$$\int_{\Sigma_f} \boldsymbol{\omega} \cdot \boldsymbol{N} dA = 0, \tag{10.32}$$

i.e., the flux of vorticity through the side surface is zero.[2] Now we choose $t = 0$ and another moment of time t, and $s_0, s_0 + \delta s$ two close points on Γ and Γ'. We integrate $\boldsymbol{v}$ along a closed curve lying in Σ_f, composed by $\boldsymbol{r}_{\Gamma}|_{s \in [s_0 + \delta s, s_0]}$, connected to $\boldsymbol{r}|_{[0,t] \times \{s_0\}}$, connected to $\boldsymbol{r}_{\Gamma'}|_{s \in [s_0, s_0 + \delta s]}$, and finally connected to $\boldsymbol{r}|_{[t,0] \times \{s_0 + \delta s\}}$, like in Fig. 10.2.

We integrate $\boldsymbol{v}$ along the curve in Fig. 10.2 in the limit $\delta s \to 0$, and from (10.32) we have

$$\lim_{\delta s = 0} \oint \boldsymbol{v} \cdot \boldsymbol{t} ds = \int_{\Sigma_f} \boldsymbol{\omega} \cdot \boldsymbol{N} dA = 0. \tag{10.33}$$

[1]For the proof of these relations, see Problem 5 at the end of this chapter.

[2]The fact that the flux of vorticity is zero on a tube of flow surface is an interesting result by itself. For more discussions, also see Problem 5 at the end of this chapter.

But

$$\oint \boldsymbol{v}\cdot\boldsymbol{t}ds = \int_{\Gamma} \boldsymbol{v}\cdot\boldsymbol{t}ds + \int_{\boldsymbol{r}(s_0,0)}^{\boldsymbol{r}(s_0,t)} \boldsymbol{v}\cdot\boldsymbol{t}ds - \int_{\Gamma'} \boldsymbol{v}\cdot\boldsymbol{t}ds + \int_{\boldsymbol{r}(s_0+\delta s,t)}^{\boldsymbol{r}(s_0+\delta s,0)} \boldsymbol{v}\cdot\boldsymbol{t}ds. \quad (10.34)$$

In the limit $\lim_{\delta s=0}$, the second and the fourth terms in the RHS of (10.34) cancel each other, and by using (10.33) we prove the Kelvin circulation theorem.

Traditional proofs of the same theorem can be found, for example, in Article 146 from [167], in Sect.. 3.51 from [224], or in Sect. 8 from [171].

Comment. There is a geometrical way to prove (10.32). Since we work only on the fluid particle surface, it is natural to use the surface differential operators instead of the full three-dimensional ones. We apply the surface divergence theorem (6.61), where we substitute $\boldsymbol{A} = \boldsymbol{v}\times\boldsymbol{N}$. From the formula (6.69) in the problems at the end of Chap. 6, we have $\nabla_{\Sigma_f}\cdot(\boldsymbol{v}\times\boldsymbol{N}) = \boldsymbol{N}\cdot(\nabla_{\Sigma_f}\times\boldsymbol{v}) - \boldsymbol{v}\cdot(\nabla_{\Sigma_f}\times\boldsymbol{N})$ and this reduces to $\boldsymbol{N}\cdot(\nabla_{\Sigma_f}\times\boldsymbol{v})$ because of the property of the normal from in (6.54). It results

$$\iint_{\Sigma_f} \nabla_{\Sigma_f}\cdot(\boldsymbol{v}\times\boldsymbol{N})dA = \oint \boldsymbol{\omega}\cdot\boldsymbol{N}dA,$$

where the contour integral is taken along the curve in Fig. 10.2. Both RHS terms in the surface divergence theorem formula cancel. On one hand we have

$$\oint (\boldsymbol{v}\times\boldsymbol{N})\cdot\boldsymbol{t}^{\perp}ds = \oint (\boldsymbol{t}^{\perp}\times\boldsymbol{v})\cdot\boldsymbol{N}ds = 0,$$

because $\boldsymbol{v} \parallel \boldsymbol{t}^{\perp}$ by the definition of Σ_f. The second term on the RHS of the divergence theorem formula cancels by construction

$$-2\iint H(\boldsymbol{v}\times\boldsymbol{N})\cdot\boldsymbol{N}dA = 0,$$

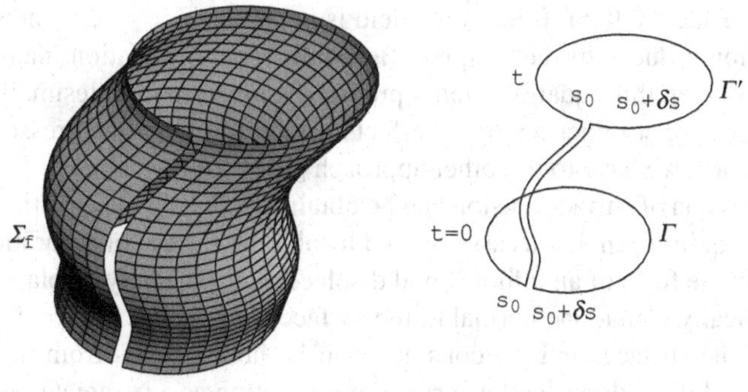

Fig. 10.2 Closed contour of integration on a tube of flow

so it results (10.32). The reason we wanted to mention this geometric amendment is related to (10.30). In Proof 2, these equations are somehow postulated on physical grounds (i.e., particles contained in the surface move together with the surface), however in this comment they result automatically as a rigorous consequence.

10.4 Surface Tension

10.4.1 *Physical Problem*

In this section, we study certain phenomena that occur in the neighborhood of a closed surface of separation between two continuous media that do not mix. In reality, the two systems in contact are separated by a thin boundary layer having special properties. However, in the following, we neglect the internal structure of this transition layer, and we assimilate it with an infinite thin geometric surface. In the neighborhood of a curved surface of separation, the pressure in the two media is different, and we call this pressure difference *surface tension*. In Sect. 8.4 (see (8.32)), we introduce the same surface tension in another manner, starting from dynamical considerations. Here, we assume that the free energy of this state of tension (the stress between two adjacent elements of surface) depends only on the area of the common boundary, on the nature of the two media, and on temperature. The special case of additional electric, acoustic, etc., fields, or presence of surfactants will be discussed later in another chapter. For a more detailed discussion on the topic, see Article 265 in [167]. Although, the original first treatment of the problem belongs to Lagrange who first determined a *minimal* surface in 1760. A review on the topics of capillarity is presented in [259] and references herein.

In the stationary case $\boldsymbol{v} = 0$ for a fluid with free boundary S, the Euler equation reads

$$-\frac{1}{\rho}\nabla P + \boldsymbol{f} = 0, \tag{10.35}$$

where ρ is the fluid density, P is the pressure, and $\boldsymbol{f}$ is the mass density of the force field acting inside the fluid. If the force field is potential, $\boldsymbol{f} = -\Delta u$, the stationary Euler equation reduces to the simplest Bernoulli type of equation, namely $P = P_0 - \rho u$. However, this equation cannot predict the pressure infinitesimally close to the surface, where stronger nonlinear effects occur. To obtain the pressure next to the fluid surface, we have to use other approach [171].

The expression of surface tension can be obtained by using the equations of thermodynamic equilibrium. Let us assume that locally the surface of separation suffers a variation in the form of an infinitesimal displacement. The only displacement that counts physically is that one normal to the surface, because we neglect the internal structure of the surface, and we consider it to be homogenous from the physical point of view. Let us describe the surface of separation as a parameterized regular geometrical surface $\boldsymbol{r}(u, v) : U \to S$ (see Chap. 18) with unit normal $\boldsymbol{N}(u, v)$.

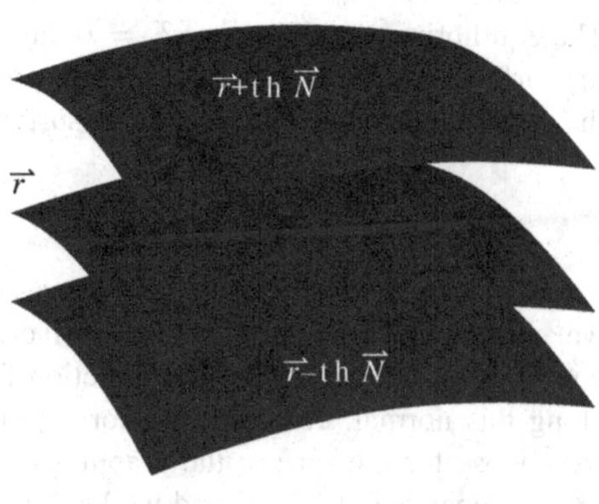

Fig. 10.3 A normal variation of $\boldsymbol{r}(U)$

We define the *normal variation* of the surface S as the function

$$\boldsymbol{r}^t(u,v,t) = \boldsymbol{r}(u,v) + t\ h(u,v)\boldsymbol{N}(u,v), \tag{10.36}$$

where $(u,v) \in U$, $t \in (-\varepsilon, \varepsilon)$ is a parameter, and $h(u,v)$ is a differential real function defined on U. For each t, the map $\boldsymbol{r}^t : U \times (-\varepsilon, \varepsilon) \to \mathbb{R}^3$ is a regular parameetrized surface (see Fig. 10.3). For $t = 0$, the normal variation reduces to the original surface.

We assume that the original surface suffered a normal variation determined by the $h(u,v)$ function, and it is not anymore in thermodynamic equilibrium. The elementary volume of an infinitesimal element of space bounded by the original surface and by the graphs of the function $\boldsymbol{r}^t$ is $t\ h(u,v)dA(u,v)$, where dA is the elementary area of the original surface, $dA = \sqrt{EG - F^2}dudv$ (from Definition 52). We denote by P_1 and P_2 the pressures in the medium 1 and medium 2, respectively, separated by S, in the neighborhood of the surface, and we choose the direction from 1 to 2 in the direction of the unit normal $\boldsymbol{N}$. The work produced by a compression upon this elementary volume, which is also the change in its free energy $\mathcal{F}$, is

$$W_{vol} = \delta\mathcal{F}_{vol} = t \iint_{\bar{U}} (P_2 - P_1)h\sqrt{EG - F^2}dudv. \tag{10.37}$$

The total change in the free energy of the system is given by δW_{vol} plus the work associated with the variation of the area of the separation surface, i.e., the superficial (or surface) energy. In a simple model, this second part of the free energy is given by the product between a constant σ and the variation of the area δA. The constant σ is called *surface tension coefficient* and depends on the nature of the two media, and on temperature. The total variation in the free energy becomes

$$\delta\mathcal{F} = t \iint_{\bar{U}} (P_2 - P_1)h\sqrt{EG - F^2}dudv + \sigma\delta A. \tag{10.38}$$

The equilibrium condition is $\delta\mathcal{F} = 0$, and from here we obtain the expression of the surface tension, $P|_S = P_2 - P_1$. We prove in Sect. 10.4.2 that the expression of the surface tension at a point $\boldsymbol{r}$ on the surface is

$$P_2 - P_1 = P_{\boldsymbol{r}\in S} = \sigma(\kappa_1 + \kappa_2),$$

where $\kappa_{1,2}$ are the two principal curvatures of the surface at p. In all our examples, we choose the orientation of the surfaces such that the normal is toward the convexity of the curve, and the direction from medium 1 to medium 2 is chosen along this normal. To check the correct sign of the surface pressure expression, we choose for the surface the graphics of a differential function $z = \eta(x)$. The profile depends only on x, and we have full symmetry along the other coordinate y. In this one-dimensional case, we have just one principal curvature nonzero, this $\kappa_1 = \kappa$ ($\kappa_2 = 0$) is called the curvature of the function η, and it has the expression $\kappa = \frac{\eta''}{(1+\eta'^2)^{\frac{3}{2}}}$. If we choose a convex function with $\eta'' < 0$, we have $\kappa < 0$ and consequently $P_1 > P_2$. That pressure P_1 inside the concavity is larger, as it should be. A more geometrical definition of the surface tension can be found in Sect. 8.4 or in [10, 292].

10.4.2 *Minimal Surfaces*

To find the explicit expression for the surface tension in the most general situation, we need to calculate the RHS term in (10.38). The coefficients of the first fundamental form of the modified surface $\boldsymbol{r}^t$ are

$$\begin{aligned} E^t &= E + 2th\boldsymbol{r}_u \cdot \boldsymbol{N}_u + t^2h^2\boldsymbol{N}_u \cdot \boldsymbol{N}_u + t^2(h_u)^2, \\ F^t &= F + th(\boldsymbol{r}_u \cdot \boldsymbol{N}_v + \boldsymbol{r}_v \cdot \boldsymbol{N}_u) + t^2h^2\boldsymbol{N}_u \cdot \boldsymbol{N}_v + t^2h_uh_v, \\ G^t &= G + 2th\boldsymbol{r}_v \cdot \boldsymbol{N}_v + t^2h^2\boldsymbol{N}_v \cdot \boldsymbol{N}_v + t^2(h_v)^2. \end{aligned} \tag{10.39}$$

By using the definition relations for the second fundamental form of the surface (see Chap. 18)

$$e = -\boldsymbol{r}_u \cdot \boldsymbol{N}_u, \;\; f = -(\boldsymbol{r}_u \cdot \boldsymbol{N}_v + \boldsymbol{r}_v \cdot \boldsymbol{N}_u)/2, \;\; g = -\boldsymbol{r}_v \cdot \boldsymbol{N}_v$$

and the definition of the mean curvature of a surface (6.10)

$$H = \frac{Eg - 2fF + Ge}{2(EG - F^2)}, \tag{10.40}$$

we obtain

$$\begin{aligned} E^tG^t - (F^t)^2 &= EG - F^2 - 2th(Eg - 2fF + Ge) + \mathcal{O}(t) \\ &= (EG - F^2)(1 - 4thH) + \mathcal{O}(t), \end{aligned} \tag{10.41}$$

where $\mathcal{O}(t)$ is a term that approaches zero more rapidly than t when $t \to 0$. From (10.41), it results that, if ε is small enough, the surface $\boldsymbol{r}^t$ is a regular parameterized surface. Just now we can use $\boldsymbol{r}^t$ as the equation of a surface in the calculation of the free energy and surface tension. The area $A(t)$ of $\boldsymbol{r}^t(\bar{U})$ is given by

$$A(t) = \iint_{\bar{U}} \sqrt{E^t G^t - (F^t)^2} du dv$$

$$= \iint_{\bar{U}} \sqrt{1 - 4thH + \frac{\mathcal{O}(t)}{EG - F^2}} \sqrt{EG - F^2} du dv. \qquad (10.42)$$

It follows that, in the limit of small ε, $A(t)$ is differentiable with respect to t, and its derivative at $t = 0$ is

$$\frac{dA}{dt}(0) = -2 \iint_{\bar{U}} hH\sqrt{EG - F^2} du dv = -\int hH dA. \qquad (10.43)$$

So, the variation of the area during this deformation parameterized by the parameter t is $\delta A = (dA/dt)dt$. At $t = 0$ we have

$$\delta A = -2 \iint_{\bar{U}} h\, H\sqrt{EG - F^2} du\, dv\, dt = -\int (\varkappa_1 + \varkappa_2) h dA dt, \qquad (10.44)$$

where $\varkappa_{1,2}$ are the principal curvatures of the surface at the point of coordinates (u, v) (see Chap. 18). Equation (10.44) can provide an interesting interpretation of the mean curvature, in terms of the minimal surfaces. We can define the *mean curvature vector* by $\boldsymbol{H} = H\boldsymbol{N}$, and by choosing $h = H$ in (10.44) we can write

$$\delta A = -2 \iint_{\bar{U}} \boldsymbol{H} \cdot \boldsymbol{H} \sqrt{EG - F^2} du\, dv\, dt. \qquad (10.45)$$

Equation (10.45) means that the area of the deformed surface $\boldsymbol{r}^t(U)$ always decreases if we deform it in every point toward the direction of the mean curvature vector. For a given surface, the mean curvature vector points toward the direction where this surface tends to become a minimal surface. For example, in the case of an infinitesimal normal variation of a spherical surface, the mean curvature is still negative (the corrections in the first order in ε are smaller than 1) and since the normal is directed outside the sphere and $H < 0$, the vector $\boldsymbol{H}$ points toward the center. This is indeed the direction along which the area of an elementary spherical surface would become smaller, by flattening toward a plane.

The unit normal field for S is a divergence-free vector field. This comes from the fact that the mean curvature is related to the normal direction of the surface by the equation

$$H = -\frac{1}{2} \nabla_S \cdot \boldsymbol{N},$$

from Proposition 5 (Sect. 6.5.2), where $\nabla_S \cdot$ is the surface divergence operator. From here it results

Proposition 9. *For a minimal surface the normal vector field is surface divergence free.*

Coming back to the dynamics of the surface, if we consider the variation of the original area from $t = 0$ to a certain small value of t, we have $dt = t$, and introducing (10.44) in (10.38), we have the condition of equilibrium in the form

$$\iint_{\bar{U}} (P_2 - P_1 - \sigma(\varkappa_1 + \varkappa_2))t\, h\sqrt{EG - F^2}du dv = 0.$$

Since the function h is arbitrary, we have to fulfill

$$P_2 - P_1 = \sigma(\varkappa_1 + \varkappa_2) = 2\sigma H \tag{10.46}$$

which determines the expression of the surface pressure (Laplace formula for capillarity). H is the mean curvature. For a more physical proof the reader can check (8.55). If, for example, the principal curvatures are positive, it results that $P_1 > P_2$, i.e., the pressure is larger in the medium located inside the concavity of the surface.

We end this section with a property of minimal surfaces which results as a consequence of the divergence integral theorem (6.61). From the relation

$$\nabla_S \times \boldsymbol{r} = 0,$$

where $\nabla_S \times$ is the surface curl and $\boldsymbol{r}$ is the position vector, we can write two integral conditions valid for any closed curve Γ on any minimal surface S

$$\oint_{\Gamma} \boldsymbol{t}^{\perp} ds = 0 \tag{10.47}$$

$$\oint_{\Gamma} \boldsymbol{r} \times \boldsymbol{t}^{\perp} ds = 0, \tag{10.48}$$

where $\boldsymbol{t}^{\perp} = \boldsymbol{N} \times \boldsymbol{t}$ with $\boldsymbol{t}, \boldsymbol{r}$ having their regular interpretation and s being the arc-length along Γ. These two equations can be regarded as the dynamical equilibrium conditions for the minimal surface. The first one represents force balance, and the second one represents the momentum balance of a domain of S surrounded by Γ.

10.4.3 Application

To have a better intuition of the direction of the surface tension gradient, we present in the following a simpler example. Let us choose a parameterized surface S as the graph of a differential function $z = h(x, y)$ and U is an open set of the xOy $\mathbb{R}^2$ plane. The parameterizations of the surface are $\boldsymbol{r} = (u, v, h(u, v))$ with $u = x$ and $v = y$. We have

$$\boldsymbol{N}(x, y) = \frac{(-h_x, -h_y, 1)}{(1 + h_x^2 + h_y^2)^{1/2}} \tag{10.49}$$

and

$$H = \frac{(1 + h_x^2)h_{yy} - 2h_x h_y h_{xy} + (1 + h_y^2)h_{xx}}{(1 + h_x^2 + h_y^2)^{1/2}}. \tag{10.50}$$

For a more concrete example, we consider the surface of a semicylinder having the axis along Ox and its points at $z = f(x, y) > 0$. If it rains from above, this cylinder will not keep the water. Close to the top of the cylinder, we have $\boldsymbol{N} \simeq (0, 0, 1)$, and the normal is oriented upward, toward positive z. It means medium 1 (we choose medium 1 to be liquid) is under the cylinder, inside its concavity, and medium 2 (we choose medium 2 to be air) is above the cylinder. We also assume that the cylinder radius R is large enough so we can neglect nonlinear terms in the expression of the mean curvature. At points close to the top of this cylinder ($x \simeq 0, z \simeq R$), we have, according to (10.46) and (10.50)

$$P_2 - P_1 = \sigma(\varkappa_1 + \varkappa_2) \simeq h_{yy}, \tag{10.51}$$

and because at this points $h_{yy} < 0$ it results $P_2 < P_1$, so the liquid is under more pressure than the ambient atmosphere, which is in agreement with the Laplace law of capillarity.

We can use the condition (10.46) to find the equilibrium free surface S for $P_1 = P_2 =$ constant. This is a system subjected to the same internal and external pressure in all its points, i.e., a system consisting only in free surfaces, like soap films in microgravity. The total free energy of this system is proportional to the area of the surface, and attains its minimum when the area is minimal. The surface equation $\boldsymbol{r}$ is a *minimal* surface (i.e., $H = 0$) if and only if $\delta A = 0$, i.e., when $A'(t = 0) = 0$, for all normal variations of the surface S. Indeed, if the surface is minimal, $H = 0$ and according to (10.43), $A' = 0$. Conversely, let us assume that $A' = 0$ but let us make the hypothesis that $H \neq 0$, at least in a certain open subset of U. Then, we can always choose $h = H$ in that open set, and zero elsewhere, and it results that $A' < 0$ which contradicts the hypothesis.

To understand the role of surface tension in the geometry of the free surface, we analyze a region of fluid, in the stationary case, and in absence of any external (bulk) forces. The Euler equation reduces to $\nabla P = 0$, so the pressure is the same everywhere inside the fluid (Pascal principle). Because the pressure outside of the liquid P_0 is also considered to be the same, we find the equilibrium condition

$$P - P_0 = (P - P_0)_S = -2\sigma H = -2\sigma(\kappa_1 + \kappa_2) = \text{const.} \tag{10.52}$$

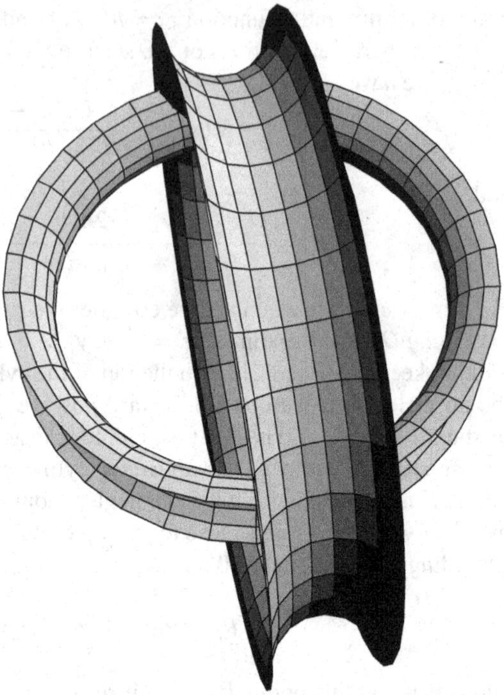

Fig. 10.4 Simulation of an experimental minimal surface produced by dipping a 4-circles wire frame into a soap solution

Consequently, the free boundary of a stationary, isolated (no external forces) drop of liquid should have the mean curvature constant all over it. If the mean curvature is constant and there are no other superficial constraints, the surface is spherical. The $H =$ const. condition is not dependent on the compressibility of the fluid, as far as the forces are absent. However, if the free surface is supported by a fixed curve, the shape is much more complicated (see for example Fig. 10.4). In the case of rigid boundaries for the free surface, the parameterized surface is not anymore regular. In the general case there will be singularities along the rigid boundaries. This problem was first formulated in the following form: for any given closed curve $\alpha \in \mathbb{R}^3$, there is a surface S of minimum area with α as boundary. There is a special case when this problem becomes simpler, namely when the liquid forms itself one or more very thin layers, like the above-mentioned soap films, suspended by some closed rigid curves, and exposed to the same external pressure P_0 in every point. Actually, no matter how thin the films are, there are always three-dimensional regions of liquid bounded by these surfaces. Because the liquid region is very thin compared to its overall dimensions, we can describe the liquid film as being bounded by two identical surfaces, separated by a very small distance along the common unit normal. We consider locally these two surfaces as two identical copies of the same surface, separated by a very small normal displacement. By local we mean here any open domain of the surfaces which do not intersect the boundary curves.

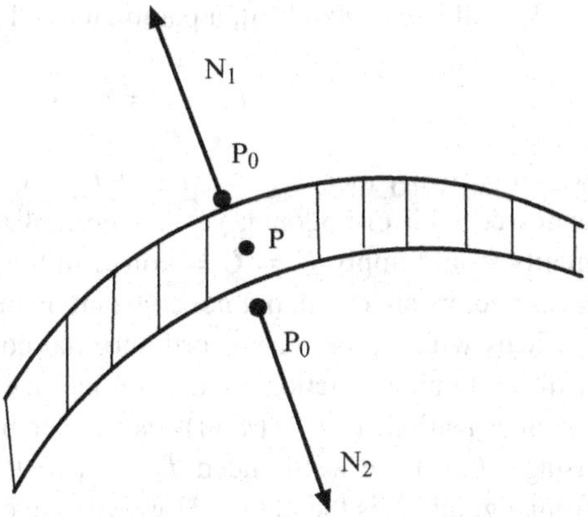

Fig. 10.5 The pressure inside a thin liquid film

On every such open domain, the unit normals $H_{1,2}$ of these two surfaces have the same support, except they point in opposite directions (Fig. 10.5).

Any point inside the fluid is infinitesimally close to any of these two identical surfaces, so we can write the surface tension condition as

$$P - P_0 = -2\sigma H_1 = -2\sigma H_2 = 2\sigma H_1. \quad (10.53)$$

It results that the only possibility is to have zero mean curvature in all points. In conclusion, in the absence of forces and in the stationary case, the surface tension and the mean curvature of the free surface are either constant for a free regular surface surrounding the liquid or zero for a thin liquid film. When $H = 0$ we call these surface *minimal*, because they have indeed the minimum area under given constraints. Some of the properties of the minimal surfaces also apply to surfaces of constant mean curvature [246].

10.4.4 Isothermal Parametrization

According to (10.40) and (10.44), the local criterium for the existence of minimal surfaces is played by the PDE $H = Eg - 2fF + Ge = 0$. The structure of this equation simplifies considerably if the coordinate system on the surface S is orthogonal, namely $F = \boldsymbol{r}_u \cdot \boldsymbol{r}_v = 0$. It is always possible to choose such an *orthogonal parametrization* (also called orthogonal curvilinear system of coordinates) for a regular surface. Indeed, for any point $p \in S$ there is a parametrization $\boldsymbol{r}(u, v)$ in a neighborhood of p, $\mathcal{V}(p)$, with the property that the curves $u =$ const. and $v =$ const. are perpendicular. For example, if we choose two differentiable vector fields on S defined by $\boldsymbol{w}_1 = \boldsymbol{r}_u$ and $\boldsymbol{w}_2 = -\frac{F}{E}\boldsymbol{r}_u + \boldsymbol{r}_v$. Moreover, if the vectors of the local basis have equal norms, $E = G$, then the minimal surface local condition reduces to a Laplace equation.

We call *isothermal* [46], a parameterized surface $\boldsymbol{r}(u, v)$ fulfilling the conditions

$$\boldsymbol{r}_u \cdot \boldsymbol{r}_u = \boldsymbol{r}_v \cdot \boldsymbol{r}_v, \quad \boldsymbol{r}_u \cdot \boldsymbol{r}_v = 0, \tag{10.54}$$

which basically means $E = G$ and $F = 0$. Isothermal parameterized surfaces are endowed with orthogonal, yet not normalized, curvilinear coordinates. Orthonormality would imply $E = G =$ const. In the isothermal case the norms of the local basis vectors are equal, but not constant on the surface. It is not easy to parameterize surfaces with isothermal or orthonormal coordinates. For example, the graphs of a differentiable function as a parameterized surface in the independent variable parametrization, $(u, v, f(u, v))$, can never be an isothermal surfaces, because, by using (10.50), we would need $f_u = f_v = 0$ (the only isothermal surface emerging from a graphics is the plane). However, we can provide the following result.

Theorem 27. *Given a parameterized surface $\boldsymbol{r}(u, v)$, we can change the parametrization $(u, v) \to (\alpha, \beta)$ by the map $(u, v) = \Phi(\alpha, \beta) : W \subset \mathbb{R}^2 \to U \subset \mathbb{R}^2$ such that $(\tilde{\boldsymbol{r}} \circ \Phi)(\alpha, \beta)$ is isothermal.*

Proof. We have $\Phi(u(\alpha, \beta), v(\alpha, \beta))$ and

$$\tilde{\boldsymbol{r}}_\alpha = \tilde{\boldsymbol{r}}_u u_\alpha + \tilde{\boldsymbol{r}}_v v_\alpha, \quad \tilde{\boldsymbol{r}}_\beta = \tilde{\boldsymbol{r}}_u u_\beta + \tilde{\boldsymbol{r}}_v v_\beta,$$

and we request $\tilde{\boldsymbol{r}}_\alpha \cdot \tilde{\boldsymbol{r}}_\beta = 0$ and $\tilde{\boldsymbol{r}}_\alpha \cdot \tilde{\boldsymbol{r}}_\alpha = \tilde{\boldsymbol{r}}_\beta \cdot \tilde{\boldsymbol{r}}_\beta$. These conditions are equivalent with the following system of two nonlinear PDE

$$\begin{cases} E u_\alpha u_\beta + F(u_\alpha v_\beta + u_\beta v_\alpha) + G v_\alpha v_\beta = 0 \\ E u_\alpha^2 + 2F u_\alpha u_\beta + G u_\beta^2 = E v_\alpha^2 + 2F v_\alpha v_\beta + G v_\beta^2 \end{cases}. \tag{10.55}$$

The two solutions of this PD system of equations $u(\alpha, \beta), v(\alpha, \beta)$ should also fulfill the compatibility conditions $u_{\alpha,\beta} = u_{\beta,\alpha}, v_{\alpha,\beta} = v_{\beta,\alpha}$. By using the theorem of existence and uniqueness from Sect. 4.3, we can always find solutions for (10.55) defined in a neighborhood, under Cauchy arbitrary conditions. Consequently, we can always provide the given parameterized surface with new isothermal curvilinear coordinates. □

For example, if $S = \{(x, y, z) \in S_2 \subset \mathbb{R}^3 | z > 0\}$, $x = u, y = v$, originally parameterized as the graphics of the function $z = f(u, v) = \sqrt{1 - u^2 - v^2}$, we have $\boldsymbol{r} = (u, v, f(u, v))$

$$\boldsymbol{r}_u = (1, 0, f_u), \quad \boldsymbol{r}_v = (0, 1, f_v), \quad N = \frac{(-f_u, -f_v, 1)}{\sqrt{1 + f_u^2 + f_v^2}},$$

and $E = 1 + f_u^2$, $G = 1 + f_v^2$, and $F = f_u f_v$. Obviously this surface is not isothermal, but if we map u, v into spherical coordinates θ, φ we have $\tilde{\boldsymbol{r}} = (\sin(\theta)\cos(\varphi), \sin(\theta)\sin\varphi, \cos(\theta))$. The new first fundamental form reads $\tilde{E} = 1$,

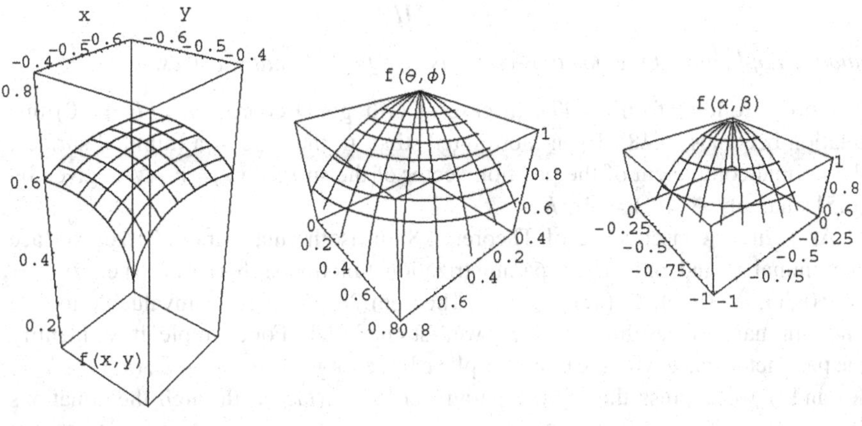

Fig. 10.6 From left to right: a domain of a sphere represented in cartesian coordinates, in spherical coordinates, and in the α, β coordinates

$\tilde{F} = 0$, and $\tilde{G} = \sin^2\theta$. We need to map these new coordinates into a new set of curvilinear coordinates, α, β, which have to fulfill again the isothermal conditions (10.54), i.e.,

$$\begin{cases} \theta_\alpha\theta_\beta + \sin^2\theta\varphi_\alpha\varphi_\beta = 0 \\ \theta_\alpha^2 + \sin^2\theta\theta_\beta^2 = \varphi_\alpha^2 + \sin^2\theta\varphi_\beta^2 \end{cases}.$$

A possible solution of the above system is provided by $\varphi = \beta$ and $\theta(\alpha) = 2\arctan C_0 e^{\pm\alpha}$, with arbitrary constant C_0. In Fig. 10.6 we present a subset of the surface S in all these three parameterizations.

The main result of this section can be expressed by the following affirmation regarding minimal isothermal surfaces.

Theorem 28. *If the parameterized surface* $\mathbf{r}(u, v)$ *is isothermal, we can write*

$$\boldsymbol{H} = H\boldsymbol{N} = \frac{1}{2E}\Delta\boldsymbol{r}, \qquad (10.56)$$

where $\Delta = \partial_{uu} + \partial_{vv}$ *is the Laplace operator in the surface curvilinear coordinates, and we introduce the mean curvature vector* $\boldsymbol{H}$.

Proof. By differentiating $\boldsymbol{r}_u \cdot \boldsymbol{r}_v = 0$ and $\boldsymbol{r}_u \cdot \boldsymbol{r}_u = \boldsymbol{r}_v \cdot \boldsymbol{r}_v$ with respect to u and v, we obtain $\boldsymbol{r}_v \cdot \Delta\boldsymbol{r} = \boldsymbol{r}_u \cdot \Delta\boldsymbol{r}$, so $\Delta\boldsymbol{r}$ is parallel to $\boldsymbol{N}$. On the other side, we have $H = (e+g)/(2E) = \boldsymbol{N}\cdot\Delta\boldsymbol{r}/(2E)$ so $\boldsymbol{H} = \boldsymbol{N}(\boldsymbol{N}\cdot\Delta\boldsymbol{r})/(2E)$. □

Theorem 28 has a different expression if instead of the full three-dimensional Laplace operator we use the surface Laplace operator Δ_S defined in Sect. 6.5.3. In the surface differential operator case, we have

Proposition 10. *On a surface* Σ *parameterized with orthogonal coordinates, we have*

$$\Delta_S \boldsymbol{r} = 2H\boldsymbol{N},$$

and the Laplacian of the position vector is zero for minimal surfaces.

The proof follows from (6.47). In case of orthogonal coordinates ($F = 0$) this relation becomes (6.48). Even more interesting, in the case of a minimal surface, the normal component of the position vector of the surface $r_n = \boldsymbol{r} \cdot \boldsymbol{N}$ is given by (6.51), namely $\Delta_S(r_n) = 2r_n K$.

As a direct consequence of Theorem 28, an isothermal parameterized surface is minimal if and only if its parametrization function is harmonic (i.e., $\Delta \boldsymbol{r} = (\Delta x(u,v), \Delta y(u,v), \Delta z(u,v)) = 0$). Theorem 28 provides an invaluable tool to find minimal surfaces through a very well-studied PDE. For example, if we identify the parameter space with the complex plane by setting $z = u+iv \in \mathbb{C}$, $(u,v) \in U \subset \mathbb{R}^2$ and if we express the regular parameterized surface $\boldsymbol{r}$ through the equations $\varphi_j = \frac{\partial x_j}{\partial u} - i\frac{\partial x_j}{\partial v}$, $j = 1,2,3$, then, the parameterized surface $\boldsymbol{r}$ is isothermal if and only if $\varphi_1 + \varphi_2 + \varphi_3 = 0$ and this surface is minimal if and only if the three complex functions φ_j are analytic. Indeed, analyticity implies harmonicity of the coordinate functions by the Cauchy–Riemann conditions. In Fig. 10.7 we present some traditional examples of minimal surfaces. The Scherk's surface [46] is such an example of complex surface.

In addition to their simplification over the minimal surfaces equation, the isothermal surfaces ($E = G, F = 0$) have another interesting property related to the Laplace operator. The Gaussian curvature is $K = \frac{1}{2E}\Delta \log E$ [299].

10.4.5 Topological Properties of Minimal Surfaces

Minimal surfaces have a lot of interesting topological properties. The zeros of the Gaussian of a minimal surface are isolated, meaning that if a minimal surface has planar or parabolic points, they are isolated. In other words, there is no straight escaping line along a minimal surfaces, they are really "very twisted." Also, there are no compact minimal surfaces. This is easy to prove, because all the points of a regular minimal surface are hyperbolic. If a minimal surface S is compact (bounded and closed), we can find an S_2 sphere of radius R containing S. We can choose R such that $S_2 \cap S = \varnothing$. Then, we decrease R continuously until the intersection between S and the sphere becomes nonempty. If the intersection is an open set for the first time, this set should be homeomorphic to an open part of S_2, having all its points elliptic points, which is forbidden by $H = 0$. If the intersection consists in only isolated points $q \in S \cap S_2$, we can find neighborhoods of these points $\mathcal{V}(q) \subset S$ lying both inside and outside S_2, contradicting hence the hypothesis. So, all (regular) minimal surfaces are unbounded, hence noncompact. We remember here that compact regular surfaces have at least one elliptic ($K > 0$) point.

If S is a regular closed minimal surface which is not a plane, the image of the Gauss map is dense in the sphere S_2. When a point moves along the surface, the

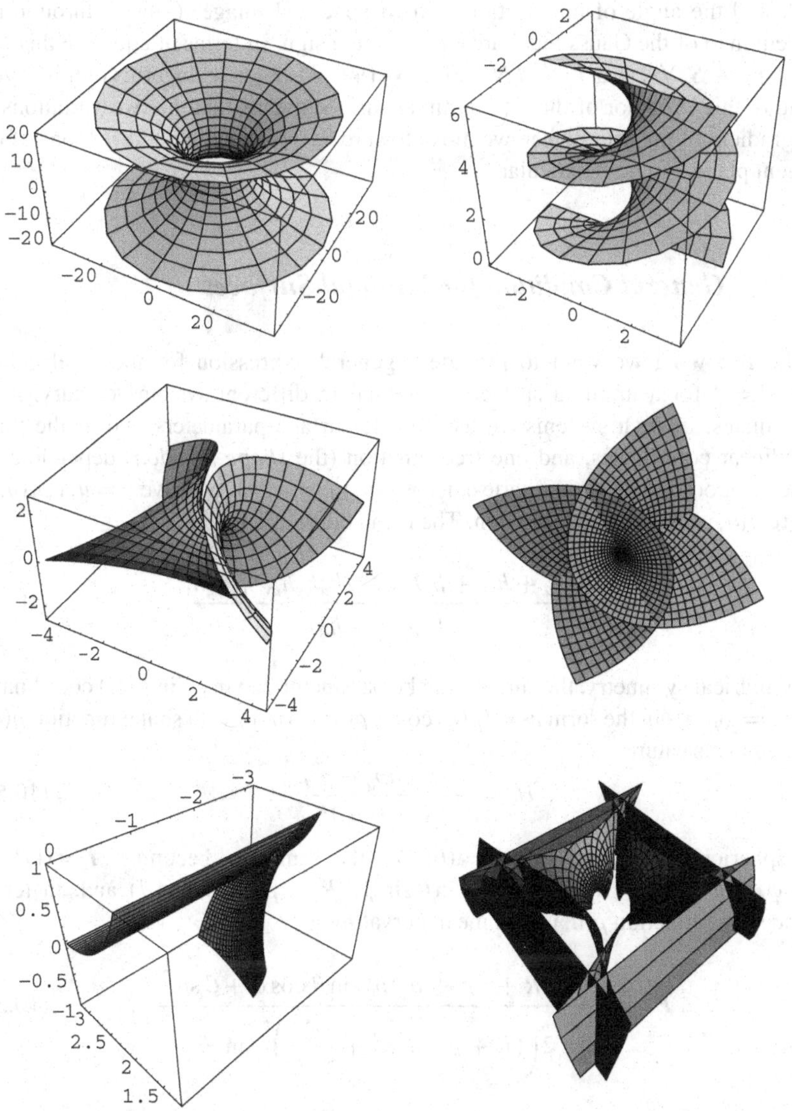

Fig. 10.7 Examples of minimal surfaces. *Upper line*: catenoid and helicoid. *Middle line*: Enneper's polynomial surface. *Lower line*: Scherk's periodical surface from complex analysis

normal $\boldsymbol{N}$ takes "almost" all possible orientations in $\mathbb{R}^3$. That is, for every arbitrary direction $\boldsymbol{N}_0$, there are open sets of points on S, such that the corresponding normal of these points approaches the given direction as close as we want.

We also mention another property of the minimal surfaces. If S is minimal and has no planar points ($K \neq 0$ on S), then the angle of intersection of any two curves on S and the angle of intersection of their spherical images (images through the tangent map of the Gauss map) are equal up to a sign. In terms of equation this fact reads $\forall p \in S, \forall \boldsymbol{v}, \boldsymbol{w} \in T_p S, dN_p(\boldsymbol{v}) \cdot dN_p(\boldsymbol{w}) = -K_p \boldsymbol{v} \cdot \boldsymbol{w}$. In terms of thin layers of fluid, this behavior of the free minimal surface means that the two variations of the gradient of pressure, when we move toward two perpendicular directions of the tangent plane, are perpendicular.

10.4.6 General Condition for Minimal Surfaces

In the following we want to provide a general expression for the local condition $H = 0$ for a minimal surface, expressed in different systems of curvilinear coordinates. In such systems we use for the surface parameters two of the three curvilinear coordinates, and one free function (the *shape function*) depending on these two coordinates. In the cartesian case $(u, v) = (x, y)$, we have $\boldsymbol{r} = u, v, h(u, v)$ where $h(u, v)$ is the shape function. The mean curvature is

$$H = \frac{h_{uu} + h_{vv} + h_v^2 h_{uu} - 2h_u h_v h_{uv} + h_u^2 h_{vv}}{(1 + h_u^2 + h_v^2)^{\frac{3}{2}}}. \tag{10.57}$$

In cylindrical symmetry, the surface can be parameterized in cylindrical coordinates $((u, v) = (\varphi, z))$ in the form $\boldsymbol{r} = (\rho(u) \cos u, \rho(u) \sin u, v)$ with shape function $\rho(u)$. The mean curvature is

$$H = \frac{\rho^3 + 2\rho\rho_u^2 - \rho^2 \rho_{uu}}{(\rho^2 + \rho_u^2)^2}. \tag{10.58}$$

In spherical symmetry $(u, v) = (\theta, \varphi)$, the surface becomes $\boldsymbol{r} = ((R + \rho(\theta, \varphi)) \sin\theta \cos\varphi, (R + \rho(\theta, \varphi)) \sin\theta \sin\varphi, (R + \rho(\theta, \varphi)) \cos\theta)$ and, in terms of the shape function $\rho(u, v)$, the mean curvature is

$$H = \frac{\mathcal{B} - ((R + \rho)^2 + \rho_\theta^2)\rho_\theta \sin\theta \cos\theta + \mathcal{C} \sin^2\theta}{2\left((R + \rho)^2 + \rho_\theta^2 + \frac{\rho_\varphi^2}{\sin^2\theta}\right)^2 \sin^2\theta}, \tag{10.59}$$

where

$$\begin{aligned}
\mathcal{B} &= 3R\rho_\varphi^2 + 3\rho\rho_\varphi^2 - R^2\rho_{\varphi\varphi} - 2R\rho\rho_{\varphi\varphi} - \rho^2\rho_{\varphi\varphi} - \rho_\theta^2\rho_{\varphi\varphi} \\
&\quad + 2\rho_\theta\rho_\varphi\rho_{\theta\varphi} - \rho_{\theta\theta}\rho_\varphi^2 - 2\rho_\theta\rho_\varphi^2 \cot\theta, \\
\mathcal{C} &= (R + \rho)(2(R + \rho)^2 + 3\rho_\theta^2 - R\rho_{\theta\theta} - \rho\rho_{\theta\theta}).
\end{aligned}$$

If the shape function is small compared to the radius, $\rho \ll R$, we have the following hierarchy of orders of smallness in ρ/R for H

$$\mathcal{O}(0) = -\frac{1}{R},$$

$$\mathcal{O}(1) = \frac{\rho}{R^2} + \frac{1}{2R^2}\Delta_\Omega \rho,$$

$$\mathcal{O}(2) = -\frac{\rho^2}{R^3} + \frac{\rho_\theta^2}{2R^3} - \frac{\rho\rho_\theta}{R^3} + \frac{\rho_\varphi^2}{2R^3 \sin^2\theta} - \frac{\rho\rho_{\varphi\varphi}}{R^3 \sin^2\theta} - \frac{\rho\rho_\theta \cot\theta}{R^3}, \quad (10.60)$$

where

$$\Delta_\Omega = \rho_{\theta\theta} + \cot\theta \rho_\theta + \frac{\rho_{\varphi\varphi}}{\sin^2\theta}$$

is the angular part of the Laplace operator in spherical coordinates.

In all these examples, the expression of H is very close to the Laplacian of the free function describing the surface in the corresponding curvilinear coordinates. If the curvilinear coordinates are isothermal, the mean curvature equation is precisely the Laplace equation, and this behavior is natural in view of (10.56). It is interesting to check how does the Laplacian of $\Delta \boldsymbol{r}$ reduce to the Laplacian of the shape scalar function, Δh or $\Delta \rho$, like in the examples above. In general, orthogonal curvilinear coordinates are not isothermal, so we expect H to contain in addition to the Laplacian of the free function, also some other terms. The question is: to what extent, in some given curvilinear coordinates, we can approximate the minimal surface equation $H = 0$ and the surface pressure expression, with the Laplace equation of the curvilinear coordinates? It would be of practical application to find the approximate expression of the surface tension for surfaces that are small deviation from an isothermal, or at least orthogonally parameterized surface.

10.4.7 Surface Tension for Almost Isothermal Parametrization

We consider a thin liquid surface S, initially in "equilibrium," parameterized by isothermal coordinates, $\boldsymbol{r}_0(u, v)$ defined in an open set $(u, v) \in U$, with $E = G, F = 0$. Next to this surface, the pressure is the surface tension and it has the expression provided by (10.52) and (10.56)

$$P = \frac{2\sigma}{2E}|\Delta \boldsymbol{r}|.$$

We consider that some external interaction occurs (like the presence of a force field or a nonuniform change in temperature) and produces a deformation of this surface. This deformation, or variation, is defined as a new parameterized surface $\boldsymbol{r}(u, v) = \boldsymbol{r}_0(u, v) + \epsilon \boldsymbol{\rho}(u, v)$. We consider this new surface to be a small variation of the original isothermal one if $\epsilon \max_{(u,v)\in U}\{|\boldsymbol{\rho}|\} \ll |\boldsymbol{r}_0|$. In the following we denote any quantity that refers to the original isothermal surface with a zero label, like for example $\boldsymbol{r}_{0u} \cdot \boldsymbol{r}_{0u} = \boldsymbol{r}_{0v} \cdot \boldsymbol{r}_{0v} = E_0 = G_0$ and $\boldsymbol{r}_{0u} \cdot \boldsymbol{r}_{0v} = F_0 = 0$. The surface tension expression

$$P(u, v, \epsilon, \rho(u, v)) = \sigma \frac{Eg - 2fF + Ge}{(EG - F^2)} \tag{10.61}$$

reduces in the limit $\lim_{\epsilon=0} P = P_0 = 2\sigma H_0 = \sigma(g_0 + e_0)/E_0$. For small variations we work in the first linear approximation of ϵ and we neglect $\mathcal{O}(\epsilon^2)$.

In the following we choose a *normal variation* $\boldsymbol{\rho} = \rho(u, v)\boldsymbol{N}_0(u, v)$. There is no loss of generality in this choice, because any arbitrary deformation can be reduced to a normal one by a reparameterization. Besides, in the case of orthogonal curvilinear coordinates, the deformed surface is always normal, since the deformation occurs along the orthogonal parameter. For example in the spherical case, $\boldsymbol{r}_0 = (R \sin u \cos v, R \sin u \sin v, R \cos u)$ with $R = \text{const.}$, the usual variation of the coordinate surface has the form $\boldsymbol{\rho} = \epsilon\rho(u, v)(\sin u \cos v, \sin u \sin v, \cos u)$, which means $\boldsymbol{r}_0 \perp \boldsymbol{\rho}$, and consequently the variation is normal.

Since we are interested in surfaces close to the isothermal one, we follow the calculations just in the first order in ϵ. From the definition of the normal variation, and from $E_0 = G_0$, $F_0 = 0$, we obtain

$$\boldsymbol{r}_u = \boldsymbol{r}_{0u} + \epsilon\rho_u \boldsymbol{N}_0 + \epsilon\rho \boldsymbol{N}_{0u},$$
$$\boldsymbol{r}_v = \boldsymbol{r}_{0v} + \epsilon\rho_v \boldsymbol{N}_0 + \epsilon\rho \boldsymbol{N}_{0v},$$

and consequently we have the coefficients of the first fundamental form of the deformed surface in the first order in ϵ

$$E = E_0 - 2\epsilon\rho e_0, \quad G = E_0 - 2\epsilon\rho g_0, \quad F = -2\epsilon\rho f_0. \tag{10.62}$$

We notice that it is impossible to have, in general, a surface and its infinitesimal normal variation, simultaneously isothermal, $F_0 = F = 0$. This is possible in the linear approximation only if $f_0 = 0$. The unit normal has the form

$$\boldsymbol{N} = \boldsymbol{N}_0 - \frac{\epsilon}{E_0}(\rho_u \boldsymbol{r}_{0u} + \rho_v \boldsymbol{r}_{0v}) + \mathcal{O}(\epsilon^2).$$

The second fundamental form has the coefficients

$$e = e_0 + \epsilon\left(\rho_{uu} - \frac{1}{2E_0}(\rho_u E_{0u} - \rho_v E_{0v}) - \frac{\rho}{E_0}(e_0^2 + f_0^2)\right),$$
$$g = g_0 + \epsilon\left(\rho_{vv} - \frac{1}{2E_0}(\rho_v E_{0v} - \rho_u E_{0u}) - \frac{\rho}{E_0}(g_0^2 + f_0^2)\right),$$
$$f = f_0 + \epsilon\left(\rho_{uv} - \frac{1}{2E_0}(\rho_u E_{0v} + \rho_v E_{0u}) - \frac{\rho f_0}{E_0}(e_0 + g_0)\right).$$

By introducing all these coefficients in (10.40), we obtain

$$H = \frac{e_0 + g_0}{2E_0} + \epsilon\frac{\rho(e_0^2 + g_0^2)}{2E_0^2} + \epsilon\frac{\Delta\rho}{2E_0} + \mathcal{O}(\epsilon^2), \tag{10.63}$$

which describes the mean curvature of the infinitesimal normal variation of an isothermal surface in the linear approximation. This form is a linear operator in ρ with variable coefficients, and the surface tension may be written as

$$P_S = -2\sigma(\mathcal{A} + \epsilon\mathcal{B}\rho + \epsilon\mathcal{C}\Delta\rho) + \mathcal{O}(\epsilon^2), \tag{10.64}$$

where the three variable coefficients $\mathcal{A}$, $\mathcal{B}$, and $\mathcal{C}$ can be identified from (10.63).

Such a simple form as (10.63) for the surface pressure is not always available. In practical situations one uses orthogonal curvilinear coordinates which are not necessarily isothermal, mainly because $E_0 \neq G_0$. In the following we obtain a similar first-order approximation of the mean curvature for a normal deviation starting from an orthogonal parameterized surface.

Definition 62. Three families of smooth (of rank 3) surfaces are a *triply orthogonal* system in an open $U \subset \mathbb{R}^3$ if one unique surface of each family passes through any point $P \in U$, and if the three surfaces that pass through each point $p \in U$ are pairwise orthogonal.

The second constraint means that $\boldsymbol{r}_u, \boldsymbol{r}_v$, and $\boldsymbol{r}_w$ are always orthogonal. The curves of intersection of any pair of surfaces from different system are lines of curvature in each of the respective surfaces, i.e., the intersection lines are principal directions. The traditional 12 systems of curvilinear coordinates are the examples (cartesian, cylindric, spherical, elliptic, parabolic, bowls, etc.). In the case of orthogonal parametrization, the coefficients of the first fundamental form are similar to (10.62). The normal is different

$$\boldsymbol{N} = \boldsymbol{N}_0 - \epsilon \begin{pmatrix} \rho_u \boldsymbol{r}_{0u} \\ \rho_v \boldsymbol{r}_{0v} \end{pmatrix}.$$

The coefficients of the second fundamental form are different

$$e = e_0 + \epsilon\left(\rho_{uu} - \frac{1}{2E_0G_0}(\rho_u E_{0u} G_0 - \rho_v E_{0v} E_0) - \rho\frac{e_0^2 G_0 + f_0^2 E_0}{E_0 G_0}\right) + \mathcal{O}(\epsilon^2),$$

$$g = g_0 + \epsilon\left(\rho_{vv} - \frac{1}{2E_0G_0}(\rho_v G_{0v} E_0 - \rho_u G_{0u} G_0) - \rho\frac{f_0^2 G_0 + g_0^2 E_0}{E_0 G_0}\right) + \mathcal{O}(\epsilon^2),$$

$$f = f_0 + \epsilon\left(\rho_{uv} - \frac{1}{2E_0G_0}(\rho_v G_{0u} E_0 + \rho_u E_{0v} G_0) - \rho f_0\frac{e_0 G_0 + g_0 E_0}{E_0 G_0}\right) + \mathcal{O}(\epsilon^2).$$

In the end, the form for the mean curvature of the deformed surface in the first order of approximation is

$$H = \frac{e_0 G - 0 + g_0 E_0}{2E_0 G_0} + \epsilon\left(\frac{G_0\rho_{uu} + E_0\rho_{vv}}{2E_0G_0} - \frac{\rho_u E_{0u}}{4E_0^2} - \frac{\rho_v G_{0v}}{4G_0^2}\right.$$
$$\left. + \frac{\rho_u G_{0u} + \rho_v E_{0v}}{4E_0 G_0} + \frac{\rho g_0^2}{2G_0^2} + \frac{\rho e_0^2}{2E_0^2} + \frac{3\rho f_0^2}{2E_0 G_0}\right) + \mathcal{O}(\epsilon^2).$$

It is easy to check that (10.65) reduces to the particular cases discussed above for spherical, cartesian, etc., coordinates. Still this expression is a linear second-order differential operator acting on ρ with variable coefficients.

10.5 Special Fluids

There are important differences between Newtonian (traditional or small molecule) fluids obeying Newtonian fluid dynamics and "polymeric" (macromolecular) fluids. The features of the macromolecular architecture influence the flow behavior. Polymeric fluids have molecular weights several orders of magnitude higher than normal fluids, and besides, this molecular weight is not uniformly distributed in the mass of the fluid. In addition, the polymers have a huge number of metastable configurations at equilibrium, and consequently the flow is altered in time and space by the local stretching and alignment of macromolecules. In high concentration polymers (melts), the macromolecules can form entanglement networks, and the number of entanglement junctions can change with the flow conditions. In [22] there is a detailed discussion of such types of flow. The most important property of macromolecular fluids is the non-Newtonian viscosity, i.e., the fact that the viscosity of the fluid changes with the shear rate. In *viscoplastic* (or *dilatant*) fluids, there is present the phenomenon of shear thickening, namely the viscosity of the fluid increases with the shear rate. Such fluids will not flow at all unless acted on by at least some critical shear stress, called *yield stress*. In some other polymeric fluids, we have the phenomenon of elasticity and memory of the flow, called the *viscoelastic* property. After the external pressure is removed, the fluid begins retreating in the direction from which it came. The fluid, however, does not return all the way to its original position (like an ideal rubber band for example), since its temporary entanglement junctions have a finite lifetime, and they are continuously being created and destroyed by the flow. Such a viscoelastic fluid behaves like having a *fading* memory.

10.6 Representation Theorems in Fluid Dynamics

10.6.1 Helmholtz Decomposition Theorem in $\mathbb{R}^3$

Theorem 29 (Helmholtz Theorem for the Whole Space). *Any single-valued continuous vector field $\mathbf{v}(\mathbf{r}) : \mathbb{R}^3 \to \mathbb{R}^3$ satisfying*

$$\nabla \cdot \mathbf{v} \to 0, \quad \nabla \times \mathbf{v} \to 0, \quad \textit{when } r \to \infty,$$

$$\exists \epsilon > 0, |\mathbf{v}| < \frac{1}{r^{1+\epsilon}}, \quad \textit{when } r \to \infty,$$

may be written as the sum of an irrotational (or conservative or lamellar) part and a solenoidal part

$$\boldsymbol{v} = \nabla \Phi + \nabla \times \boldsymbol{A},$$

such that

$$\Phi(\boldsymbol{r}) = \frac{1}{4\pi} \iiint_{\mathbb{R}^3} \frac{\nabla' \cdot \boldsymbol{v}(\boldsymbol{r}')}{|\boldsymbol{r} - \boldsymbol{r}'|} d^3\boldsymbol{r}'$$

$$\boldsymbol{A}(\boldsymbol{r}) = \frac{1}{4\pi} \iiint_{\mathbb{R}^3} \frac{\nabla' \times \boldsymbol{v}(\boldsymbol{r}')}{|\boldsymbol{r} - \boldsymbol{r}'|} d^3\boldsymbol{r}' \text{ and } \nabla \cdot \boldsymbol{A} = 0.$$

For a proof of the theorem see [10, 34, 50].

Usually, the Helmholtz theorem is formulated as "source plus condition at infinity" problem. Given the source fields $\rho(\boldsymbol{r}), \boldsymbol{j}(\boldsymbol{r})$ defined on $\mathbb{R}^3$ with the regularity propriety at $|\boldsymbol{r}| \to \infty$, $\rho, \boldsymbol{j} \to 0$, and the vector field equation

$$\nabla \cdot \boldsymbol{v} = \rho, \quad \nabla \times \boldsymbol{v} = \boldsymbol{j},$$

there is a unique solution for the unknown vector field $\boldsymbol{v} = \nabla \Phi + \nabla \times \boldsymbol{A}$, with the *potentials* $\rho, \boldsymbol{A}$ solutions of the equations

$$\Delta \Phi = \rho, \quad \Delta \boldsymbol{A} = \boldsymbol{j}, \quad \nabla \cdot \boldsymbol{A} = 0.$$

Also, the potentials are not uniquely determined up to their gauge transformations. Namely, Φ is defined modulo addition of an arbitrary harmonic function $\Phi \to \Phi + f(\boldsymbol{r})$, $\Delta f = 0$, and $\boldsymbol{A}$ is defined modulo addition of the gradient of an arbitrary function $\boldsymbol{A} \to \boldsymbol{A} + \nabla g(\boldsymbol{r})$.

The Helmholtz theorem (Theorem 29) can be extended by using a Neumann–Debye decomposition [34]. Instead of using one scalar Φ and one vector function $\boldsymbol{A}$ plus the divergence constraint (i.e., $1 + 3 - 1 = 3$ degrees of freedom), we can use three scalar functions. If the field $\boldsymbol{v}$ is continuous and single-valued, and it fulfills the same regularity conditions at ∞ as in the Helmholtz theorem, we have the following decomposition

$$\boldsymbol{v} = \nabla \Phi + \nabla \times (\boldsymbol{r} \Psi) + \nabla \times (\nabla \times \boldsymbol{r} \chi) = \nabla \Phi + \boldsymbol{L} \Psi + \boldsymbol{Q} \chi, \tag{10.65}$$

where the operators are $\boldsymbol{L} = -\boldsymbol{r} \times \nabla$ (angular momentum) and $\boldsymbol{Q} = \nabla \times \boldsymbol{L}$. The functions Ψ, χ are the so-called *Debye potentials* and are related to the operators by the equations

$$\Phi(\boldsymbol{r}) = \frac{1}{4\pi} \iiint_{\mathbb{R}^3} \frac{\nabla' \cdot \boldsymbol{v} d^3\boldsymbol{r}'}{|\boldsymbol{r} - \boldsymbol{r}'|}$$

$$\Psi(\boldsymbol{r}) = \frac{1}{4\pi} \iiint_{\mathbb{R}^3} \boldsymbol{r}' \cdot (\nabla' \times \boldsymbol{v}) \ln(1 - \boldsymbol{r}' \cdot \boldsymbol{r}) d^3\boldsymbol{r}'$$

$$\chi(\boldsymbol{r}) = \frac{1}{16\pi^2}\iiint_{\mathbb{R}^3} d^3\boldsymbol{r}'\ln(1-\boldsymbol{r}\cdot\boldsymbol{r}')(\boldsymbol{r}'\cdot\nabla')\iiint_{\mathbb{R}^3}\frac{\nabla''\cdot\boldsymbol{v}(\boldsymbol{r}'')}{|\boldsymbol{r}'-\boldsymbol{r}''|}d^3\boldsymbol{r}''$$
$$-\frac{1}{4\pi}\iiint_{\mathbb{R}^3}\ln(1-\boldsymbol{r}\cdot\boldsymbol{r}')\boldsymbol{r}'\cdot\boldsymbol{v}d^3\boldsymbol{r}'.$$

The operators involved in this generalized Helmholtz theorem fulfill interesting algebraic relations. The angular momentum operator is closed under commutation relation and spans the $su(1,1)$ Lie algebra by $[L_i, L_j] = \mathcal{E}_{ijk}L_k$. The operator $\boldsymbol{Q}$ is a left ideal of this algebra $[L_i, Q_J] = \mathcal{E}_{ijk}Q_k$, and the Laplace operator is the Casimir element of this algebra $[\boldsymbol{L}, \Delta] = [\boldsymbol{Q}, \Delta] = 0$.

A very useful version of the Neumann–Debye (10.65) is related to the linear Navier–Stokes fluid dynamics equation in absence of external forces

$$\frac{\partial \boldsymbol{V}}{\partial t} = -\frac{1}{\rho}\nabla P - \nu\nabla\times(\nabla\times\boldsymbol{V}), \tag{10.66}$$

where the fluid velocity field $\boldsymbol{V}(\boldsymbol{r},t)$ is a smooth nonsingular time-dependent (Euclidean) vector field defined on a domain $D \subset \mathbb{R}^3$ with values in $T\mathbb{R}^3$; ρ and ν are positive constants, density and viscosity, respectively, and $P(\boldsymbol{r},t)$ is the pressure scalar field, also defined on $D \subset \mathbb{R}^3$. If we ask for the velocity field to be divergence free on D, i.e., to have no net sources of fluid,

$$\nabla\cdot\boldsymbol{V} = 0, \tag{10.67}$$

it is possible to apply the representation theorem (10.65) for solutions of (10.67) in D. We have

Theorem 30. *Let us define a vector field*

$$\boldsymbol{V} = \nabla\times(\boldsymbol{Q}\beta) + \nabla\times\nabla\times(\boldsymbol{Q}b) + \nabla c$$

and the scalar field

$$P = -\rho\frac{\partial c}{\partial t},$$

where $\boldsymbol{Q}(\boldsymbol{r},t)$ is an arbitrary smooth vector field on $D\times\mathbb{R}$, and β, b, c are arbitrary smooth scalar fields depending on $(\boldsymbol{r},t) \in D\times\mathbb{R}$. Then $\boldsymbol{V}, P$ defined above are solutions for the Navier–Stokes equations (10.66) *in the divergence-free condition* (10.67) *if the following conditions are fulfilled on $D\times\mathbb{R}$*

$$\nu\Delta\beta = \frac{\partial\beta}{\partial t},\quad \nu\Delta b = \frac{\partial b}{\partial t},\quad \Delta c = 0,\quad \boldsymbol{Q} = C_0\boldsymbol{r},$$

with C_0 an arbitrary constant.

The proof of the theorem is by direct calculation. Details and applications can be found in [16, 31].

An interesting version of the Helmholtz theorem in a domain $\mathcal{D}$ with boundary $\partial\mathcal{D} \neq \emptyset$ is presented in Chorin and Marsden's book [54], under the name of Helmholtz–Hodge theorem. In this formalism, a vector field $\boldsymbol{v}$ is decomposed into a potential field $\nabla\Phi$ and an incompressible vector field $\boldsymbol{u}, div\boldsymbol{u} = 0$ which is parallel to the boundary of $\mathcal{D}$, $(\boldsymbol{u} \cdot \boldsymbol{N})_{\partial\mathcal{D}} = 0$. The existence of the Helmholtz–Hodge decomposition is guaranteed by the existence of a solution to the Neumann-associated problem for Φ. Uniqueness is guarantied by the fact that the two terms of the decomposition are orthogonal in an average taken through an integration over $\mathcal{D}$. Indeed, $\int_{\mathcal{D}} \boldsymbol{u}\cdot grad\Phi = 0$ through Gauss formula and because of the properties of $\boldsymbol{u}$. Consequently, any two distinct Helmholtz–Hodge decompositions must have same $\boldsymbol{u}$ and same Φ, up to an additive constant. In this form the theorem is more adapted to hydrodynamics problems where one has incompressible fluid in a bounded region. Because the velocity is divergence free and vanishes on the boundary, the Navier–Stokes equation can be projected into a divergence-free component which does not contain the pressure, i.e., the gradient term.

Hydrodynamics is perhaps one of the best-studied fields of application of nonlinear equations, waves, and their solutions, and we have barely touched the subject. A very comprehensive and extended treatment of hydrodynamics in general, toward the nonlinear problems open at the time when the book was written, is [167]. The book is dense in solved examples and problems in almost any field of basic hydrodynamics. The book goes hand in hand with mathematical physics text books like [64, 317] or in the same style. The calculations are detailed and comprehensive, very much relying on expansions in series of functions and independent mode analysis. A book which complements Lamb's book on hydrodynamics and is written in the same *grand* style is [50], especially for magnetohydrodynamics and fluid and plasma stability problems. Another comprehensive book on hydrodynamics, where very special problems are solved in very original ways, is [171]. If the reader is more concerned about mathematical rigorousness, toward functional analysis and operator approach in hydrodynamics, a good lecture would be [305]. More restrictive topics, yet presented on a fundamental basis and mathematical rigorous, are approached in [10, 111, 220]. In this last mentioned spectrum, more oriented toward mathematics is the attractive and clear book of Chorin and Marsden [54], or more toward applied mathematics [315]. For specific topics on waves in general and nonlinear waves in fluids, the reader may consider to consult [169, 174].

10.6.2 Decomposition Formula for Transversal Isotropic Vector Fields

This special decomposition works for axially and/or translational symmetric vector fields. It is particularly useful in convective hydrodynamics stability calculations, and in general in physical systems exhibiting transport and transformation processes. It is also useful in the dynamics of viscous drops submerged in viscous fluids [223]. This decomposition formula was introduced for a particular axisymmetric field in [50, Sect. 61], and later, for spherical surfaces and even for more general situations in [286]. The big advantage of this decomposition consists in the fact that the vector field $\boldsymbol{v}$ can be expressed as function of the radial component v_r, the divergence $div\boldsymbol{v}$ and the radial component of the vorticity, ω_r, where $\boldsymbol{\omega} = \nabla \times \boldsymbol{v}$. When the flow is incompressible, and the velocity field has spherical symmetry, this decomposition becomes very useful because of its simplicity. Moreover, for solenoidal fields, like vorticity, this divergence term is also canceled and the vector field can be constructed from the radial components only.

In general, the formula works for any curvilinear orthogonal system of coordinates of the form (r, q_1, q_2) with a local basis $\{\boldsymbol{e}_r, \boldsymbol{q}_1, \boldsymbol{q}_2\}$, where $r =$ const. describes closed coordinate surface homotopic to the sphere S_2. At the same time, we can expand any vector field $\boldsymbol{v}(r, q_1, q_2)$ in an orthogonal basis of functions defined on the compact surface $r =$ const. This surface S, being homotopic to S_2, allows the existence of an $L_2(S)$ Hilbert space with countable basis of harmonic polynomials defined on S_2. In the case of spherical coordinates, these are the spherical harmonics $Y_{l,m}$. In the following we introduce this vector decomposition in spherical coordinates (r, θ, φ). For the calculation of components and operator action, we refer to Sect. 18.3.

Any vector field, like for example the velocity field $\boldsymbol{v}$, can be decomposed in its normal (radial for spherical) and parallel components

$$\boldsymbol{v} = v_r \boldsymbol{e}_r + \boldsymbol{v}_{\parallel}, \tag{10.68}$$

and also the gradient and Laplace operators can be decomposed in a similar way

$$\nabla_{\parallel} = \nabla - \boldsymbol{e}_r(\boldsymbol{e}_r \cdot \nabla) = \nabla - \boldsymbol{e}_r \frac{\partial}{\partial r}, \quad \Delta = \Delta_r(r, \partial/\partial r) + \Delta_{\parallel}(\theta, \partial/\partial\theta, \varphi, \partial/\partial\varphi,). \tag{10.69}$$

From vector analysis we have the formula

$$\begin{aligned} \Delta \boldsymbol{v}_{\parallel} &= \nabla(\nabla \cdot \boldsymbol{v}_{\parallel}) - \nabla \times (\nabla \times \boldsymbol{v}_{\parallel}) \\ &= \nabla_{\parallel}(\nabla \cdot \boldsymbol{v}_{\parallel}) - [\nabla \times (\nabla \times \boldsymbol{v}_{\parallel})]_{\parallel}, \end{aligned} \tag{10.70}$$

where we retain on the RHS only the parallel terms (the normal terms cancel each other), because the LHS in (10.70) contains by definition only parallel terms. We have

$$\Delta \boldsymbol{v}_{\parallel} = \nabla_{\parallel}(\nabla \cdot \boldsymbol{v}) - \nabla_{\parallel} \mathcal{D} v_r - [\nabla \times (\nabla \times \boldsymbol{v}_{\parallel})]_{\parallel}, \tag{10.71}$$

where $\mathcal{D} = \frac{1}{r^2}\frac{\partial}{\partial r}(r^2)$, i.e., the radial part of the div operator in the curvilinear coordinates.

We can expand the vector field $\boldsymbol{v}(r,\theta,\varphi)$ in spherical harmonics. We have

$$\boldsymbol{v} = v_r \boldsymbol{e}_r + \boldsymbol{v}_\parallel = \sum_{l,m} \boldsymbol{v}_{l,m}(r,t) Y_{l,m}(\theta,\varphi), \tag{10.72}$$

where $\boldsymbol{v}_{l,m} = \boldsymbol{e}_r v_{r,lm} + \boldsymbol{v}_{\parallel,lm}$. With these notations we obtain

$$\Delta \boldsymbol{v}_\parallel = \Delta_r \boldsymbol{v}_\parallel + \Delta_\parallel \boldsymbol{v}_\parallel = \frac{1}{r^2}\frac{\partial}{\partial r}\left(r^2 \frac{\partial}{\partial r}\boldsymbol{v}_\parallel\right) + \Delta_\Omega \boldsymbol{v}_\parallel, \tag{10.73}$$

where Ω is the angular (parallel) part of the Laplace operator (see Sect. 18.3). For any l, m component we can write

$$\Delta \boldsymbol{v}_{\parallel,lm} = \frac{1}{r^2}\frac{\partial}{\partial r}\left(r^2 \frac{\partial}{\partial r}\boldsymbol{v}_{\parallel,lm}\right) - \frac{l(l+1)}{r^2}\boldsymbol{v}_{\parallel,lm}, \tag{10.74}$$

accordingly to the action of the angular Laplacian operator on spherical harmonics. It results

$$\boldsymbol{v}_{\parallel,lm} = \frac{r^2}{l(l+1)}(\Delta_r \boldsymbol{v}_\parallel - \Delta \boldsymbol{v}_\parallel). \tag{10.75}$$

In the following equations, we skip the labels l, m, but we refer to the l, m component, unless otherwise stated. From (10.68), (10.71), and (10.75), we have the following preliminary form for the decomposition

$$\boldsymbol{v} = v_r \boldsymbol{e}_r + \frac{r^2}{l(l+1)}\Big(\Delta_r \boldsymbol{v}_\parallel + \nabla_\parallel \mathcal{D} v_r - \nabla_\parallel(\nabla \cdot \boldsymbol{v}) + [\nabla \times (\nabla \times \boldsymbol{v}_\parallel)]_\parallel\Big). \tag{10.76}$$

In the following, we focus on the first and fourth term in the RHS parenthesis in (10.76). We have

$$\Delta_r \boldsymbol{v}_\parallel + [\nabla \times (\nabla \times \boldsymbol{v}_\parallel)]_\parallel = \Delta_r \boldsymbol{v}_\parallel + (\nabla \times \boldsymbol{\omega})_\parallel - [\nabla \times (\nabla \times u_r \boldsymbol{e}_r)]_\parallel, \tag{10.77}$$

where $\boldsymbol{\omega} = \nabla \times \boldsymbol{v}$ is the vorticity field. We also notice that $\omega_r = \boldsymbol{e}_r \cdot (\nabla_\parallel \times \boldsymbol{v}_\parallel)$. This is possible because of the relation

$$\nabla \times \boldsymbol{v} = \nabla_\parallel \times \boldsymbol{v}_\parallel + \boldsymbol{e}_r(\boldsymbol{e}_r \cdot \nabla) \times \boldsymbol{v}_\parallel - \nabla_\parallel \times \boldsymbol{e}_r v_r - \boldsymbol{e}_r(\boldsymbol{e}_r \cdot \nabla) \times \boldsymbol{e}_r v_r,$$

where all the last three terms are perpendicular on $\boldsymbol{e}_r$, hence they have only parallel components. The only normal component in the RHS of the equation above is contained the first term. We also notice the identity [286, Equation (H1.12)]

$$\nabla \times \boldsymbol{v} = \boldsymbol{e}_r \omega_r + \boldsymbol{e}_r \times \left(\frac{1}{r}\frac{\partial}{\partial r}(r\boldsymbol{v}_\parallel) - \nabla_\parallel u_r\right). \tag{10.78}$$

From (10.77) and (10.78), we have

$$\Delta_r \boldsymbol{v}_\parallel + (\boldsymbol{e}_r(\boldsymbol{e}_r \cdot (\nabla_\parallel \times \boldsymbol{\omega}_\parallel))) + \boldsymbol{e}_r \left[\frac{1}{r}\frac{\partial}{\partial r}(r\boldsymbol{\omega}_\parallel) - \nabla_\parallel \omega_r \right]_\parallel - [\nabla \times (\nabla \times v_r \boldsymbol{e}_r)]_\parallel$$
$$= \Delta_r \boldsymbol{v}_\parallel + \boldsymbol{e}_r \times \left[\frac{1}{r}\frac{\partial}{\partial r}(r\boldsymbol{\omega}_\parallel) \right] - [\nabla \times (\nabla \times v_r \boldsymbol{e}_r)]_\parallel - \boldsymbol{e}_r \times \nabla_\parallel \omega_r$$
$$= -\boldsymbol{e}_r \times \nabla_\parallel \omega_r. \tag{10.79}$$

The last equality holds because the first three terms in the second line of (10.79) cancel each other, as one can check by direct calculations in spherical coordinates components. Consequently we have

$$\Delta_r \boldsymbol{v}_\parallel + [\nabla \times (\nabla \times \boldsymbol{v}_\parallel)]_\parallel = -\boldsymbol{e}_r \times \nabla_\parallel \omega_r. \tag{10.80}$$

From (10.76) to (10.80), we can write the final decomposition formula

$$\boldsymbol{v} = v_r \boldsymbol{e}_r + \frac{r^2}{l(l+1)} \left[\nabla_\parallel \mathcal{D} v_r - \nabla_\parallel (\nabla \cdot \boldsymbol{v}) - \boldsymbol{e}_r \times \nabla_\parallel \omega_r \right]. \tag{10.81}$$

That is, we can express the velocity field function of its radial component, and function of the radial component of the vorticity and the divergence of velocity.

10.6.3 Solenoidal–Toroidal Decomposition Formulas

Another version of the above decomposition formula can be obtained for an axisymmetric solenoidal vector field. We use a cylindrical system of coordinates (r_c, φ, z), and the axis of symmetry is taken in the z-direction. In this case the field can be expressed as a superposition of a poloidal and toroidal field in terms of two azimuth-independent scalar functions $U(r_c, z)$ and $V(r_c, z)$ [50]

$$\boldsymbol{v} = -r_c \frac{\partial U}{\partial z} \boldsymbol{e}_{r_c} + r_c V \boldsymbol{e}_\varphi + \frac{1}{r_c}\frac{\partial}{\partial r_c}(r_c^2 U)\boldsymbol{e}_z. \tag{10.82}$$

An equivalent and unified way of writing (10.82) and the curl of velocity is

$$\begin{aligned} \boldsymbol{u} &= \boldsymbol{e}_z \times \boldsymbol{r} V + \nabla \times (\boldsymbol{e}_z \times \boldsymbol{r} U), \text{ and} \\ \nabla \times \boldsymbol{u} &= -\boldsymbol{e}_z \times \boldsymbol{r} \Delta_5 U + \nabla \times (\boldsymbol{e}_z \times \boldsymbol{r} V), \end{aligned} \tag{10.83}$$

where Δ_5 is the Laplacian operator in a five-dimensional Euclidean space in cylindrical coordinates. According to Chandrasekhar [50, Sect. 61], there is a particular advantage of this representation in that no matter of how many times one applies curl operator to the velocity and vorticity fields, the representations in (10.83) have the same type of expression.

In spherical coordinates, the Chandrasekhar poloidal–toroidal decomposition of an axisymmetric solenoidal field has the form

$$\boldsymbol{u} = -\frac{1}{\sin\theta}\frac{\partial}{\partial\theta}(\sin^2\theta U)\boldsymbol{e}_r - \frac{\sin\theta}{r}\frac{\partial}{\partial r}(r^2 U)\boldsymbol{e}_\theta + r\sin\theta V\boldsymbol{e}_\varphi. \tag{10.84}$$

The interpretation of the scalars U, V is straightforward. Since fields derived only from the scalar U have components only in the meridional planes, it results that the U field is nothing but the Stokes' stream function for motions in these planes (meridional motions). The field V defines motions which are entirely rotational. Another advantage is this types of representations reciprocity: a poloidal field has toroidal vorticity and, conversely, a toroidal field has poloidal vorticity.

10.7 Problems

1. In Sect. 10.3 we conjecture (10.30) and (10.32) by using the physical intuition that particles contained in particle surface move together with the surface, and never tangent to it. Prove this affirmation on a more geometrical background. Hint: use the integral formulas in Sect. 6.5.
2. *Monge's potential representation*: show that an arbitrary differentiable vector field $\boldsymbol{v}$ can be always represented as

$$\boldsymbol{v} = \nabla\varphi + \psi\nabla\chi,$$

where the first term on the RHS is irrotational field, and the second term has the property of being perpendicular to its curl, $(\psi\nabla\chi)\cdot(\nabla\times\psi\nabla\chi)$. Such fields are called *complex lamellar* fields [10].

Chapter 11
Nonlinear Surface Waves in One Dimension

In this chapter, we present some examples of nonlinear evolution equations in one space dimension. We re-discuss the traditional Korteweg–de Vries (KdV) equation for the shallow water long channel case, and its cnoidal waves and soliton solutions. Then we briefly present the MKdV equation and some nonlinear dispersion extension of it. In the last sections, we discuss some possible dynamical generalizations of the shallow water models on compact intervals, for any depth of the fluid. The resulting equation is an infinite-order differential one, and it reduces to a finite difference differential equation. We show that this generalized KdV equation approaches the KdV, MKdV, and Camassa–Holm limiting equations, both at the equation and at the solution level, in the appropriate physical conditions. In the last part we discuss the Boussinesq equations on a circle.

11.1 KdV Equation Deduction for Shallow Waters

The one-dimensional KdV equation for shallow water and infinite long channels, and its cnoidal waves and soliton solution, represent a well-established model for water waves [7, 135]. The KdV equation has the dimensionless form

$$u_t + 6uu_x + u_{xxx} = 0,$$

and has an infinite set of conservation laws, out of which the first two are

$$u_t + (3u^2 + u_{xx})_x = 0, \quad (u^2)_t + (4u^3 + 2uu_{xx} - u_x^2)_x = 0.$$

The basic configuration is presented in Fig. 11.1 The model consists of an infinite long channel along Ox axis, filled with stationary liquid, in normal gaseous atmosphere, up to a height h measured along Oy axis. The fluid velocity is $V = (u, v)$ and the free surface Σ is described by the equation $y|_{\Sigma} = h + \eta(x, t)$.

A. Ludu, *Nonlinear Waves and Solitons on Contours and Closed Surfaces*,
Springer Series in Synergetics, DOI 10.1007/978-3-642-22895-7_11,

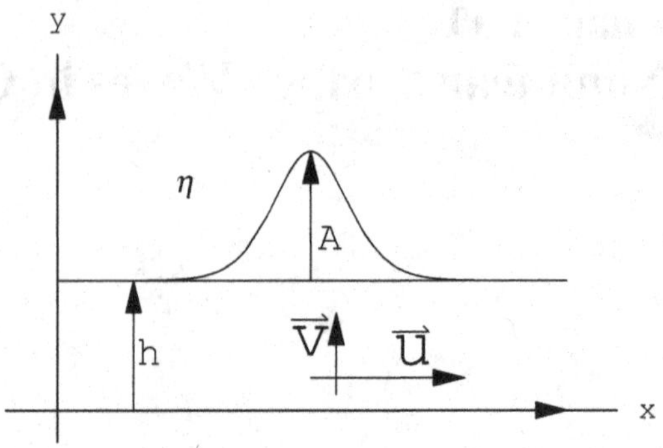

Fig. 11.1 Shallow water model and traveling localized disturbance on the free surface

We denote $a = \max|\eta|$. In this section, the subscript Σ means that the quantity is evaluated at the free surface. The KdV one-dimensional infinite long model is obtained [2, 169], under the following hypotheses:

1. Incompressible fluid $\rho(x, y, t) = \text{const}$.
2. Irrotational flow $\nabla \times \boldsymbol{V} = 0$.
3. Inviscid fluid. Dynamics is governed by Euler equation.
4. At a certain point in the demonstration we need to make some approximations based on the size of the disturbance. It is the so-called "shallow water, long waves" approximation, and basically consists in the introduction of two smallness dimensionless parameters, ϵ and δ, and an expansion of the equations in terms of order of magnitude of these two parameters.
5. At a certain point we will introduce an approximation based on time scales, also in terms of ϵ, δ.
6. There are two interactions taken into account. One is an external vertical uniform field of force (e.g., gravitation $\boldsymbol{g} = (0, -g)$), and the other is the surface pressure at the free surface in contact with the atmosphere.

From hypotheses (1) and the equation of continuity we have div $\boldsymbol{V} = 0$. From this relation, hypothesis (2) and the Helmholtz representation theorem (Theorem 29) in Sect. 10.6.1, we have a velocity field potential Φ, $\boldsymbol{V} = \nabla\Phi$ which fulfills the Laplace equation

$$\Delta\Phi = 0. \tag{11.1}$$

In principle one should be careful while using this Helmholtz representation, since it is sensitive to the topology of the domain of definition. Because the roots of the Helmholtz theorem are in the Poincaré Lemma (Sect. 4.8), it inherits the restrictions of this lemma, namely about the domain on which it applies: it should be star shaped, or contractible to a point. In other words, if the flow space has holes or compact obstacles, one has the rethink the representation of the potential as a multiform function.

The kinematic condition at the free surface (see Sect. 9.5) reads

$$\frac{d(y|_\Sigma)}{dt} = v|_\Sigma = \frac{d\eta}{dt} = \frac{\partial \eta}{\partial t} + \frac{\partial \eta}{\partial x}\frac{dx}{dt}. \tag{11.2}$$

The surface pressure is given by [171, 174, 271]

$$P|_\Sigma = -\frac{\sigma}{\mathcal{R}} = -\sigma \frac{\eta_{xx}}{[1+(\eta_x)^2]^{3/2}}, \tag{11.3}$$

where σ is the coefficient of surface pressure of the fluid (material constant), $\mathcal{R}$ is the local radius of curvature of the curve describing the surface Σ, and the x labels represent differentiation (from now on in this section, it is easier to use such labeling for derivatives). The Euler equation (hypothesis (3)) together with hypothesis (6) provide the equation

$$\frac{d\boldsymbol{V}}{dt} = \frac{\partial \boldsymbol{V}}{\partial t} + (\boldsymbol{V}\cdot\nabla)\boldsymbol{V} = -\frac{1}{\rho}\nabla P + \boldsymbol{g}, \tag{11.4}$$

where P is the pressure. The Euler equation can be written in terms of the potential of flow, and then integrated once with respect to the gradient. It results

$$\Phi_t + \frac{1}{2}(\nabla\Phi)^2 = -\frac{P}{\rho} - gy, \tag{11.5}$$

where an arbitrary additive function of time resulting from the space integration can be neglected because it represents just a gauge transformation for the velocity field. Next step we evaluate (11.4) and (11.5) on the surface Σ and then, we differentiate it with respect to x. Then we express the potential flow derivatives in terms of the components of the velocity field, and use (11.3) for pressure at the free surface. All in all we obtain

$$\left(u_t + uu_x + vv_x - \left[\frac{\sigma\eta_{xx}}{\rho[1+(\eta_x)^2]^{3/2}}\right]_x + g\eta_x \right)_\Sigma = 0, \tag{11.6}$$

which is the Euler equation of momentum conservation at the free surface Σ. The sign in front of the pressure surface term is minus because the surface pressure acts toward inside the fluid if the curvature is negative. Same sign analysis is discussed in [271] ((5B.24) and Chap. 5.4 and Appendices 5B and 5C), [174] ((48), p. 223), and [171] (p. 298). Practically, the dynamics of the surface is obtained by solving (11.2) and (11.6).

The domain of the Laplace equation is bounded by the rigid bed $y = 0$ and by the free surface $y = h + \eta$. If the functions involved are analytical, we can solve the Dirichlet problem for Φ and obtain a unique analytical solution, for example in the form of a Taylor series

$$\Phi(x, y, t) = \sum_{n\geq 0} y^n \Phi_n(x, t). \tag{11.7}$$

By substituting (11.7) in (11.1) we obtain the recursion relations

$$\Phi_{k,xx} + (k+1)(k+2)\Phi_{k+2} = 0. \tag{11.8}$$

The rigid bed condition $v(x,0,t) = \Phi_y(x,0,t) = 0$ results in annihilation of the odd coefficients $\Phi_{2n+1}(x,t) = 0$, and expresses all the even coefficients functions of $\Phi_{0,x} = f$ which we denote by a new symbol, $f(x,t)$. For example, we have the velocities

$$u = f - \frac{y^2}{2} f_{xx} + \frac{y^4}{24} f_{xxxx} - \cdots,$$
$$v = -y f_x + \frac{y^3}{6} f_{xxx} - \frac{y^5}{120} f_{xxxxx} + \cdots. \tag{11.9}$$

In the following, to fulfill hypotheses (4) and (5), we introduce substitutions which provide dimensionless quantities:

$$x = lx' \quad y = hy' \quad \partial_x = \frac{1}{l}\partial_{x'}$$
$$t = \frac{l}{c_0}t' \quad \partial_t = \frac{c_0}{l}\partial_{t'}$$
$$\eta = a\eta'$$
$$u = \epsilon c_0 u'$$
$$v = \epsilon \delta c_0 v'$$
$$f = \epsilon c_0 f'. \tag{11.10}$$

Here, l is an arbitrary length which should be of the same order of magnitude as the half-width (wavelength) of the localized solutions. The speed of sound in fluid is $c_0 = \sqrt{gh}$ [167, 171, 174]. We can also introduce the Bond number

$$B_o = \frac{\sigma}{\rho g h^2} = \left(\frac{l_c}{h}\right)^2, \quad \text{where } l_c = \sqrt{\frac{\sigma}{\rho g}}, \quad \text{(capillary length)}. \tag{11.11}$$

The two dimensionless parameters are $\epsilon = \frac{a}{h}$ and $\delta = \frac{h}{l}$. So far the model has two free parameters (a, l) and two physical parameters (h, g).

Hypothesis (4) requests that $\epsilon \ll 1$ and $\delta \ll 1$. As a consequence we will approximate all equations to the first orders in ϵ, δ. With this notations, and with velocities expressed in (11.9), we can rewrite the dynamical equations (11.2) and (11.6) in the form

$$\eta'_{t'} + f'_{x'} + \epsilon \eta' f'_{x'} + \epsilon \eta'_{x'} f' - \frac{\delta^2}{6} f'_{x'x'x'} = \mathcal{O}(\epsilon\delta^2, \ldots), \tag{11.12}$$

$$f'_{t'} + \eta'_{x'} + \epsilon f' f'_{x'} - \frac{\delta^2}{2} f'_{x'x't'} - \delta^2 B_o \eta'_{x'x'x'} = \mathcal{O}(\epsilon\delta^2, \ldots). \tag{11.13}$$

The next step is the logical consequence of the linearization of equations. If we neglect all nonlinear terms, we obtain $f' = \eta'$, so it is natural to expand f' in a series of orders in the two parameters. We have

$$f'(x', t') = \eta' + \epsilon f^{(1)} + \delta^2 f^{(2)} + \mathcal{O}(\epsilon\delta^2). \tag{11.14}$$

By introducing the approximation (11.14) in (11.12) and (11.13), we obtain

$$\eta'_{t'} + \eta'_{x'} + \epsilon(f^{(1)}_{x'} + 2\eta'\eta'_{x'}) + \delta^2 \left(f^{(2)}_{x'} - \frac{1}{6}\eta'_{x'x'x'} \right) = \mathcal{O}(\epsilon\delta^2, \ldots) \tag{11.15}$$

$$\eta'_{t'} + \eta'_{x'} + \epsilon(f^{(1)}_{t'} + \eta'\eta'_{x'}) + \delta^2 \left(f^{(2)}_{t'} - \frac{1}{2}\eta'_{x'x't'} - B_o \eta'_{x'x'x'} \right) = \mathcal{O}(\epsilon\delta^2, \ldots). \tag{11.16}$$

By subtracting (11.16) from (11.15), and by identifying the coefficients of the same orders of magnitude (we assume here that different powers of the smallness parameters, corresponding to different scale phenomena, are independent), we obtain the series for the f' function

$$f'(x', t') = \eta' - \frac{1}{4}(\eta')^2 + \frac{1}{12}\eta'_{x'x'} - \frac{B_o}{2}\eta'_{x'x'} - \frac{1}{4}\eta'_{x't'} + C + \mathcal{O}(\epsilon\delta^2, \ldots), \tag{11.17}$$

where C is an arbitrary constant of integration. The last step is to plug back this final expression for f' into (11.15), and to come back to original physical quantities. The result, approximated up to the orders ϵ and δ^2, is one of the forms of the well-known KdV equation

$$\eta_t + c_0 \eta_x + \frac{3}{2}\frac{c_0}{h}\eta\eta_x + \frac{c_0 h^2}{2}\left[-\left(\frac{1}{6} + B_o \right) \eta_{xxx} - \frac{1}{2c_0}\eta_{xxt} \right] = 0. \tag{11.18}$$

If we apply the fifth hypothesis, we have to make an order of smallness evaluation of space and time derivatives. By using the traveling wave reduction

$$(x, t) \to \xi = x - Vt,$$

and by keeping V as a free parameter, and changing $\partial/\partial t \to -Vd/d\xi$ we have $\eta'_{x'x't'} = -\eta'_{x'x'x'} + \mathcal{O}_4(\eta')$ so we obtain the second (and most used) form for the KdV equation

$$\eta_t + c_0 \eta_x + \frac{3}{2}\frac{c_0}{h}\eta\eta_x + \frac{c_0 h^2}{2}\left(\frac{1}{3} - B_o \right) \eta_{xxx} = 0. \tag{11.19}$$

In the end of this section we present a flow chart of the full deduction of the KdV equation for shallow channels. This chart is useful for the reader who wants to understand the logical steps, and the play of the approximations. It is also useful in the following sections, where we will use the same chart to find a generalized KdV equation, valid for any height, and more importantly, valid on compact intervals (finite rectangular tanks of water). In the blocks of the chart we wrote briefly the operation done and the number of the resulting equation from the text. Next to the arrows we indicated the operation performed from one block to another.

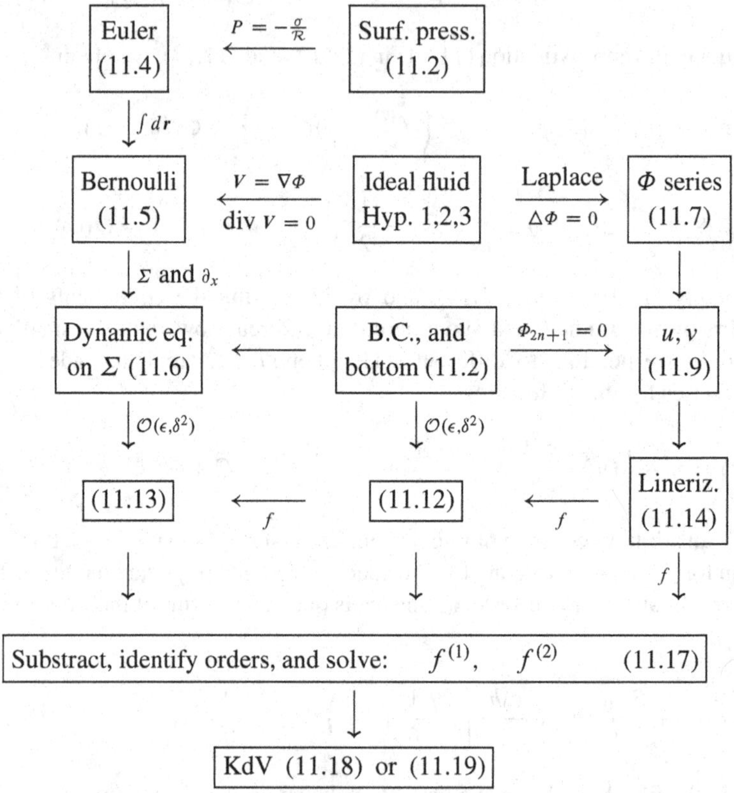

(11.20)

11.2 Smooth Transitions Between Periodic and Aperiodic Solutions

To find the one-soliton solution of the KdV equation (also called the steady-state solution) we convert the KdV PDE into an ODE by the traveling wave substitution

$$(x, t) \to \xi = x - Vt,$$

keep V as a free parameter, and use the substitutions $\partial/\partial t \to -V d/d\xi$. We transform (11.19) into

$$A\eta_\xi + B\eta\eta_\xi + C\eta_{\xi\xi\xi} = 0, \tag{11.21}$$

where $A = c_0 - V$, $B = 3c_0/2h$, and $C = c_0h^2(1-3B_0)/6$. After integrated once, multiplied with η_ξ, and integrated again, (11.21) becomes

$$\frac{C(\eta_\xi)^2}{2} = -\frac{B\eta^3}{6} - \frac{A\eta^2}{2} + C_1\eta + C_2 = 0, \tag{11.22}$$

where $C_{1,2}$ are constants of integration. We can always factorize (11.22) in the form

$$(\eta_\xi)^2 = -4(\eta - a_1)(\eta - a_2)(\eta - a_3), \tag{11.23}$$

where all roots are real because the RHS is a positive function. Without any loss of generality we assume $a_1 < a_2 < a_3$. We substitute $\eta - a_3 = -(a_3 - a_2)f(\xi)$ and we have

$$(f_\xi)^2 = (a_3 - a_1)(1 - f^2)(1 - k^2 f^2), \tag{11.24}$$

with

$$k = \frac{a_3 - a_2}{a_3 - a_1}.$$

Equation (11.24) is nothing but the ODE for the cnoidal sine function (18.7). So, one steady traveling solution for the KdV equation is

$$\eta(x - Vt) = a_3 - (a_3 - a_2)\mathrm{sn}^2[\sqrt{a_3 - a_1}(x - Vt)|k]. \tag{11.25}$$

This solution represents the *cnoidal wave* KdV solution, which is a periodic function of period $T = 2K(k)(a_3 - a_1)^{-1/2}$, where $K(k)$ is the complete elliptic integral of the first kind (Sect. 18.3). When $a_3 = a_2$, $k = 0$ and in principle the solution should reduce to a linear wave. However, because the amplitude of the cnoidal wave is equal to the numerator of k, the solution reduces to a trivial constant in this case. This is a consequence of the fact that the KdV equation is nonperturbational. Of course, for $a_3 \sim a_2$ the solution behaves very close to a small amplitude linear oscillation. If $a_2 = a_1$, we have $k = 1$ and the cnoidal wave reduces to one-soliton solution

$$\eta_{sol}(x - Vt) = a_2 + (a_3 - a_2)\mathrm{sech}^2[\sqrt{a_3 - a_2}(x - Vt)]. \tag{11.26}$$

This smooth transition effect from a periodic function to a nonperiodic one is a very peculiar property of the nonlinear equations. This limiting process is actually responsible for a transition from discrete to continuous, and from compact to noncompact. For example, if such a traveling cnoidal wave is obtained on a circle ($x \to \varphi$), the one-soliton solution cannot exist because of nonperiodicity condition, but the cnoidal wave, even if close enough to a soliton, could fulfill the periodicity constrain if it exists an integer n such that

$$nK\left(\frac{a_3 - a_2}{a_3 - a_1}\right) = \pi\sqrt{a_3 - a_1}.$$

The form of the KdV equation in (11.22) is called the *potential picture* associated to the KdV equation. We can interpret the LHS term as a kinetic energy of an abstract point in one-dimensional motion, whose law of motion is $\eta = \eta(\xi)$. The RHS has the interpretation of minus the potential energy associated to this point. An analysis of the consequences of this interpretation on the KdV equation is given in [86].

The general cnoidal solution in (11.25) and the soliton in (11.26) are not written in a practical form in terms of the roots a_i. By substituting these equations back into a general form of a KdV equation like

$$\eta_t + a\eta_x + b\eta\eta_x + c\eta_{xxx} = 0, \tag{11.27}$$

we can write the cnoidal solution of (11.27) in the form

$$\eta(x,t) = A\mathrm{sn}^2\left(\frac{x - Vt}{L}\bigg| k\right) + B, \tag{11.28}$$

with

$$L = 2\sqrt{-\frac{3ck}{Ab}}, \quad V = \frac{Ab}{3}\left(1 + \frac{1}{k}\right) + a + bB. \tag{11.29}$$

Also, a soliton solution of the same (11.27) reads

$$\eta_{sol}(x,t) = A\mathrm{sech}^2\left(\frac{x - Vt}{L}\right), \tag{11.30}$$

with

$$L = 2\sqrt{\frac{3c}{Ab}}, \quad V = \frac{Ab}{3} + a. \tag{11.31}$$

The classical one-soliton solutions for the KdV (11.18) and (11.19) have the form

$$\eta(x,t) = A\mathrm{sech}\frac{x - Vt}{L}, \quad \text{with } V = \frac{Ac_0}{2h} + c_0, \tag{11.32}$$

This soliton profile is valid for both versions of the KdV equations, while the difference is made by the half-width. For (11.19) we have

$$L = 2\sqrt{\frac{h^3(1 - 3B - o)}{3A}}, \tag{11.33}$$

and for (11.18) we have

$$L = \sqrt{\frac{3Ah^2 + 4h^3 - 12B_o h^3}{3A}}. \tag{11.34}$$

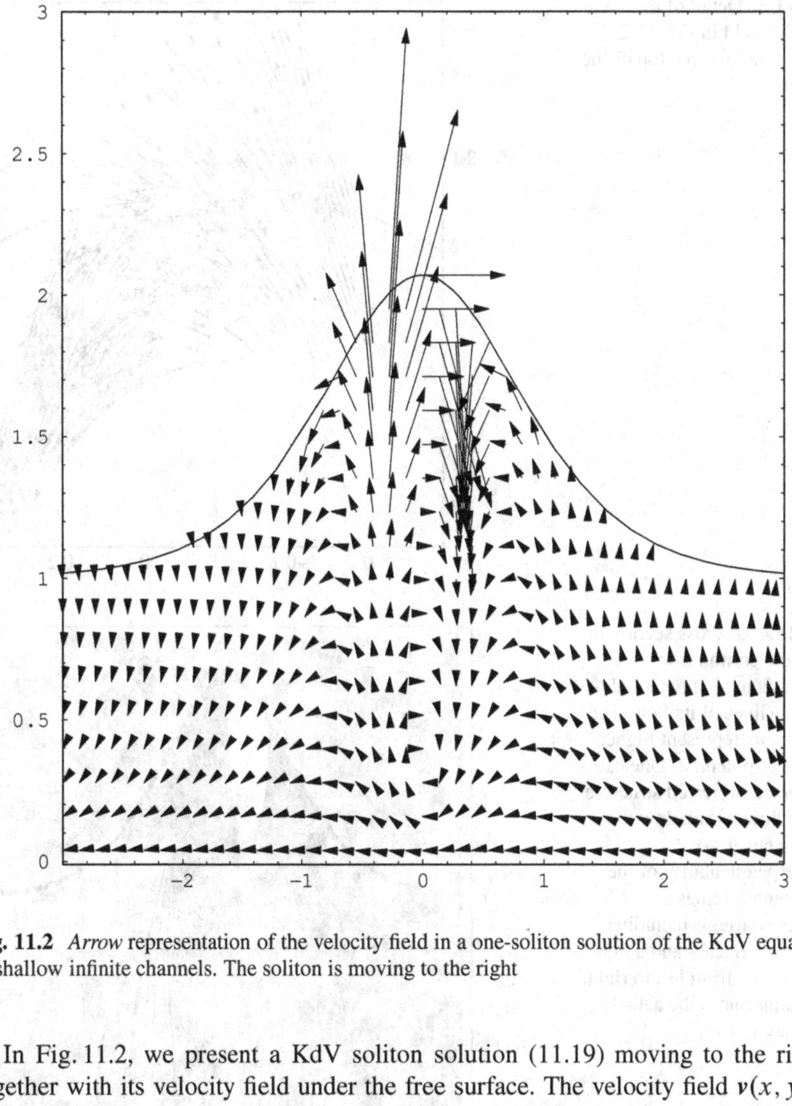

Fig. 11.2 *Arrow* representation of the velocity field in a one-soliton solution of the KdV equation in shallow infinite channels. The soliton is moving to the right

In Fig. 11.2, we present a KdV soliton solution (11.19) moving to the right, together with its velocity field under the free surface. The velocity field $v(x, y, t)$ is calculated on the base of (11.9), (11.17), and (11.19). We notice the horizontal velocity of the envelope on the top of the soliton. Basically, the fluid is rising in front of the soliton wave (right of $x = 1$), then it performs almost a closed loop in the inverse trigonometric sense (clockwise), for $0 < x < 1$, and has a symmetric behavior to the left of the maximum. Around $x = 0$, in a vertical section there is a strong sheer in the flows. Although, apparently there are vortices in this flow, actually the contours are open. One can check directly to calculate $\nabla \times V$ from (11.9), (11.17), and (11.19), and note that the flow is indeed irrotational up to the order $\epsilon\delta^2$. A more detailed illustration of the flow next to the top of the soliton is presented in Fig. 11.3. In Fig. 11.4, we present a contour plot of the potential $\Phi(x, y, t)$ lines for

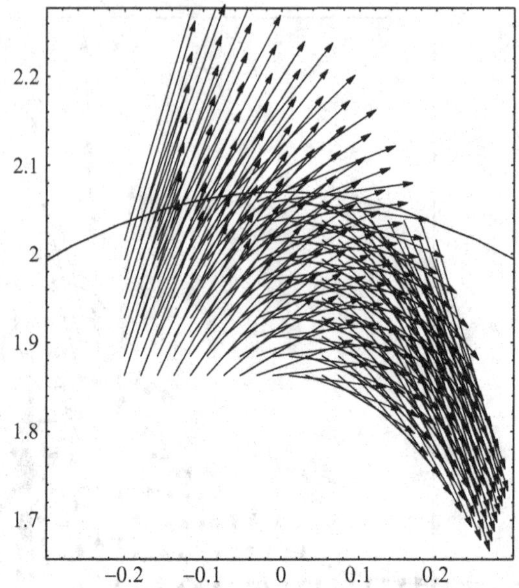

Fig. 11.3 Detail of the velocity field in Fig. 11.2, zoomed around the top of the soliton

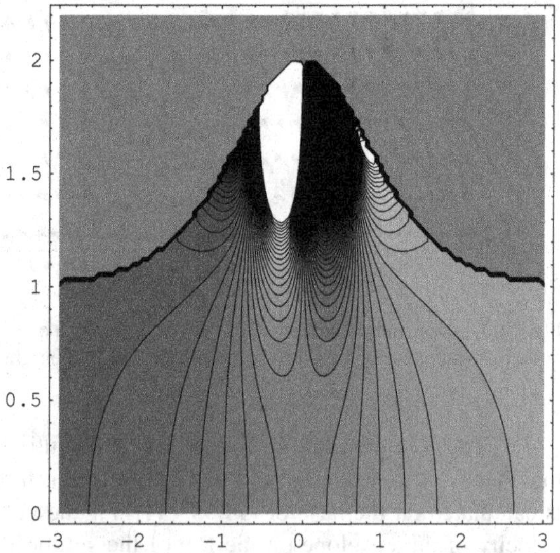

Fig. 11.4 A cross section in the KdV soliton from Fig. 11.2 and the potential Φ contour lines of its flow. The *light areas* represent higher values of potential. One can see the complicated structure of the flow right below the soliton envelope, the perpendicularity of the equipotential lines at the boundaries – including the free surface – and a bias of potential from left to right, which produces the actual translation

the flow for the same soliton solution. In Fig. 11.5, we present the hydrodynamic pressure inside the fluid in the case of a soliton wave, calculated with (11.5), where we plugged (11.7), (11.8), (11.17), (11.21), and (11.22). Multisoliton solutions of the KdV equations are obtained by using the inverse scattering theory (IST) [2, 78, 79, 169, 271]. In this book we do not intend to elaborate on the IST method, since we rather focus on identifying nonlinear integrable models for compact systems.

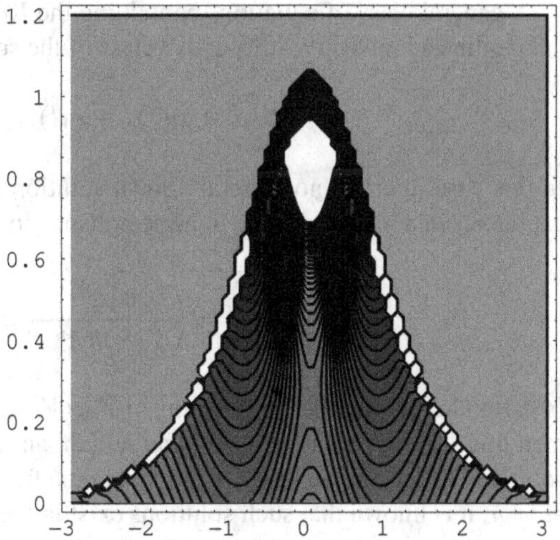

Fig. 11.5 Isobaric contours of the pressure distribution in a KdV shallow water soliton, including the surface pressure effect

11.3 Modified KdV Equation and Generalizations

The modified KdV equation (MKdV) is of the form:

$$u_t + 6u^2 u_x + u_{xxx} = 0 \tag{11.35}$$

and is a model that appears in the context of ion acoustic solitons, van Alfvén waves in collisionless plasma, Schottky barrier transmission lines, models of traffic congestion as well as phonons in anharmonic lattices among others (see, e.g., [155] and references therein). By the help of *Miura transformation* we know to map any solution $v(x,t)$ of the MKdV equation, into a solution $u(x,t) = -v^2 - v_x$ of the KdV equation. The MKdV equation has also an infinite number of laws of conservation. The conserved densities of the first three of them have the form

$$C_1(x,t) = -v^2 - v_x,\ C_2(x,t) = -(v^2+v_x)_x,\ C_3(x,t) = -(v^2+v_x)_{xx} + (v^2+v_x)^2.$$

The solitary wave solution of (11.35) is given by $u = A\operatorname{sech}[(x-Vt)/L]$, with $L = 1/A$ and $V = A^2$. There are also different classes of solutions of the equation, like solutions with compact support (the so-called MKdV *compactons*) of the form [165]

$$u(x,t) = \frac{\sqrt{32}k\cos[k(x-4k^2t)]^2}{3\left(1 - \dfrac{2}{3}\cos[k(x-4k^2t)]^2\right)}. \tag{11.36}$$

A general class of equations containing the KdV and MKdV as special cases is the nonlinear convective–dispersive class of the so-called K(m,n) [276]:

$$u_t + (u^m)_x + (u^n)_{xxx} = 0. \tag{11.37}$$

In this case, there is no general known solution for arbitrary combinations of the exponents m and n. From the nonlinear dispersion relations results

$$L = \left(\frac{n^3 A^{n-1}}{V - mA^{m-1}} \right)^{1/2}. \tag{11.38}$$

In particular, this predicts the scaling of V as $V \sim A^{m-1}$ and indicates that if $m = n$, then this scaling results in a *constant* length, an indication of compactly supported solutions which are well known to exist in the frame of the K(m,n) [276]. In fact for $m = n$, it is known that such solutions exist in the form:

$$u = A \left[\cos \left(\frac{x - Vt}{L} \right) \right]^{\frac{2}{n-1}} \tag{11.39}$$

for $|x - vt| \leq 2n\pi/(n-1)$ (and $u = 0$ otherwise). For these solutions $L = 4n/(n-1) =$ *constant* and $A = (2Vn/(n+1))^{n-1}$, in agreement with the predictions of the nonlinear dispersion relation. Compactons of a parabolic profile such as, e.g., $u = [37.5V - (x - Vt)^2]/30$ for $m = 3$ and $n = 2$ may also exist [276].

11.4 Hydrodynamic Equations Involving Higher-Order Nonlinearities

In the majority of liquid models for solitons one uses two main approximations. The first one is the about small perturbation waves, i.e., the amplitude of the wave is small compared to other geometrical parameters of the perturbation or environment, like the wavelength or depth of the liquid layer, for example $\eta_0 \ll L, h$. The second approximation requires that the depth of the fluid is either very small compared to the wavelength (shallow water, or long waves approximations), or the depth of the fluid is much larger than the wavelength (deep water approximations), i.e., $h \ll L$ [2, 169]. In the shallow water case, beyond the traditional KdV model, there are several other models that try to extend the two limits mentioned above. For example the Boussinesq, Gear–Grimshaw, Benjamin–Onno, Bona–Smith–Chen, Whitham-2, Camassa–Holm equations, and their generalizations [3, 17, 25, 26, 42, 73, 80, 90, 100, 105, 121, 163, 313, 340, 341]. In the limit of deep waters the traditional equation is the nonlinear Schrödinger equation (NLS) and various versions or extensions of it [3, 17, 26, 121, 163]. Some models try to unify the two limits for the depth, but the integrability of such equations is questionable both from the point of view of the physical values of the coefficients of the equation, and from the point of view of "near"-integrability in numerical procedures [245]. Basically, all these models start from the Euler equations and perform different types of truncations or approximations based on scaling criteria.

In this section we present a somehow different model for a generalization of the KdV equation for liquids of arbitrary depth, and more importantly, for flow in a one-dimensional compact domain [187]. In this model it is used only one of the two smallness conditions for the KdV equation obtained from a one-dimensional liquid free surface problem, i.e., only the smallness of the amplitude of the soliton η_0 with respect to the depth of the channel, h. This condition, $\epsilon = \eta_0/h \ll 1$ is the only one used. Here h is taken to be arbitrary parameter, especially when compared to wavelength. Moreover, the model does not limit to infinite long channels, and we study the evolution of the surface in a finite dimensional "tank." Therefore, we study the nonlinear dynamics of a fluid of arbitrary depth in a bounded domain. These different constraints lead to a new type of equation which generalizes in some sense the KdV equation and the other above mentioned models. Such type of one-dimensional nonlinear models find their applications not only in fluid dynamics, but also in biological systems dynamics (swimming of motile cells, [193], nerve pulse propagation, [200, 226]), and mesoscopic superconductivity [197, 199].

11.4.1 A Compact Version for KdV

Let us have a one-dimensional inviscid incompressible irrotational fluid layer of depth h and density ρ under uniform gravity g. The Laplace equation for the potential Φ of the velocity field $V = (u, v)$ is solved for appropriate boundary conditions, i.e., within a two-dimensional domain $x \in [x_0 - L, x_0 + L]$ (as the "horizontal" coordinate) and $y \in [0, \xi(x,t)]$ (as the "vertical" coordinate), where x_0 is arbitrary so far, L is the arbitrary length of the tank, and $\xi(x,t)$ is the shape of the free surface of the fluid. We have rigid boundary conditions for the lateral walls $u|_{x=x_0\pm L, y\in[0,\xi(x,t)]} = 0$ and the bottom $v|_{y=0, x\in[x_0-L, x_0+L]} = 0$, and the kinematic condition for the free surface

$$v|_\Sigma = (\xi_t + \xi_x u)|_\Sigma, \tag{11.40}$$

where Σ the free surface of equation $y = \xi(x,t)$, and the subscript indicates the derivative. By taking into account the boundary conditions we can write a general solution for the Laplace equation in the form

$$\Phi(x,y,t) = \sum_{k\geq 0} \alpha_k(t) \frac{\cosh \dfrac{k\pi y}{L}}{\cos \dfrac{k\pi x_0}{L}} \cos \frac{k\pi}{L}(x - x_0), \tag{11.41}$$

where α_k are time dependent coefficients, so far arbitrary. Apparently, this solution should be an even function in x, because it is symmetric with respect to x_0. However, the physical part of the domain is just $[x_0, x_0 + L]$ so the symmetry is just an "artificial" mirroring. Consequently, the BC at vertical wells reduces to aperiodicity condition

$$u\Big|_{x=x_0} = 0, \quad u\big|_{x=x_0+jL} = 0, \quad j \in Z. \tag{11.42}$$

Actually instead of even solutions on $x \in [x_0 - L, x_0 + L]$ we look for periodic solutions on $x \in [x_0 + jL, x_0 + (j+1)L]$ with j integer. In the infinite channel limit, $L \to \infty$ we can choose $x_0 = 0$ without loss of generality. With this potential the velocities read

$$u(x, y, t) = \Phi_x = \sum_{k\geq 0} \alpha_k(t) \frac{k\pi}{L} \frac{\cosh \dfrac{k\pi y}{L} \sin \dfrac{k\pi(x_0 - x)}{L}}{\cos \dfrac{k\pi x_0}{L}} \tag{11.43}$$

$$v(x, y, t) = \Phi_y = \sum_{k\geq 0} \alpha_k(t) \frac{k\pi}{L} \frac{\sinh \dfrac{k\pi y}{L} \cos \dfrac{k\pi(x_0 - x)}{L}}{\cos \dfrac{k\pi x_0}{L}}, \tag{11.44}$$

and indeed we have $u(x = x_0) = u(x = x_0 + L) = 0$ and $v(y = 0) = 0$ We introduce the test function

$$f(x, t) = \sum_{k\geq 0} \alpha_k(t) \frac{k\pi}{L} \frac{\sin \dfrac{k\pi(x_0 - x)}{L}}{\cos \dfrac{k\pi x_0}{L}}, \tag{11.45}$$

such that the velocity field can be formally written

$$\begin{aligned} u &= \Phi_x = \cos(y\partial) f(x, t) \\ v &= \Phi_y = -\sin(y\partial) f(x, t). \end{aligned} \tag{11.46}$$

where, for simplicity, the operator ∂ represents the partial derivative with respect to the x coordinate. It easy to check that these expressions for velocity fulfill the irrotational condition $\nabla \times (u, v) = 0$, and again the same requested BC. Equations (11.46) do not depend on L, therefore any approach toward the long channel limit must include the $L \to \infty$ (unbounded) limit. We also notice the relation

$$f(x, t) = \Phi_x|_{y=0} = u|_{y=0}, \tag{11.47}$$

which provides the physical interpretation of the test function f: it is the horizontal velocity of fluid along the rigid bed (bottom) level. In other words, if we introduce a complex potential flow $\omega(x, y, t) = u + iv$, we have $\omega_{bottom} = u(x, 0, t)$, so at the bottom the potential becomes pure real, and we have

$$e^{iy\partial_x}\omega_{bottom} = \omega, \tag{11.48}$$

which means that the velocity field at any height is given by a local translation through a one-parameter $\epsilon = iy$ complex Lie group of transformations. Possibly, this happens because the general Lie group of diffeomorphisms that conserves density and irrotational flow for the general Euler equation reduces for this rectangular one-dimensional geometry to this one-parameter subgroup.

Rigorous treatment of the functional operators in (11.46) is found in the exponentiation theory, or in the theory of formal Taylor series of operators, or finally in the theory of pseudodifferential operators [136, 146, 164, 251]. In all these formalisms, the exponential of a functional coefficient differential operator is approached in the sense of a continuous representation of a Lie group in a complex Banach space. Basically, one uses the associate formal Taylor series of the trigonometric functions and act with it term by term on the functions. The domain of definition of such trigonometric functions of operators is provided by $\mathbf{L}_2$ differentiable functions. Real problems occur when one needs to handle algebraically such operator equations, because ∂_x and $\eta(x,t)\partial_x$ do not commute and do not close any finite dimensional Lie algebra. The algebraic relations have to be substituted by the Baker–Campbell–Hausdorff (BCH) type of formulas. Even the classical BCH formula in terms of noncommuting linear operators need to be replaced in this case by a generalized commuting relation where instead of the infinite series of commutators one has the exponential of the adjoint operator. However, in the following, we work only on the set of differentiable functions $\eta(x,t)$ defined on compact interval $[x_0, x_0 + L] \times [0, T]$, $T < \infty$, and as a consequence we can use the regular BCH formula. Also, we use a simplified functional formalism by treating the exponential operator as a formal Taylor series, and its action upon functions is taken term by term. The region of convergence of the series, as well as the domain of definition of the operators should be studied for any specific choice of solutions.

11.4.2 Small Amplitude Approximation

In the following we describe the resulting equations for small height compared to the depth, but not necessarily for large wavelengths. Consequently, we write all equations for $y = \xi(x,t) = h + \eta(x,t)$, and expand all equations in formal Taylor series in the first order in η. This approximation is valid in the limit $\max_{x\in[x_0,L]} |\eta| \ll h$. In this situation (11.46) need to be approximated accordingly to the formula

$$\cos(y\partial) = \cos{(h+\eta)\partial} = \cos{(h\partial + \eta\partial)}$$

$$\cos{(h\partial)}\cos{(\eta\partial)} - \sin{(h\partial)}\sin{(\eta\partial)} \simeq \cos{(h\partial)} - \sin{(h\partial)}\,\eta\partial,$$

but exact calculation needs more functional analysis. The algebraic part of the problem can be solved by using the BCH formula [43, 44, 124] which expresses the product of two noncommuting operators as an infinite sum of repeated commutators. We actually use the Zassenhaus formula (derived by Magnus in 1954 [204] citing unpublished work by Zassenhaus) which is the dual of the BCH formula, and expresses the product of two noncommuting exponential operators as an infinite product of their repeated commutators. The classical BCH formula for two noncommuting operators A, B can be expressed in the form given by Wilcox [343]

$$e^A e^B = e^{(\sum_{i\geq 1} D_i)}, \tag{11.49}$$

where D_i are polynomials in A, B of degree i. We have

$$\begin{aligned}
D_1 &= A + B \\
D_2 &= \frac{1}{2}[A, B] \\
D_3 &= \frac{1}{12}\Big([A, [A, B]] - [B, [A, B]]\Big) \\
D_4 &= -\frac{1}{24}[A, [B, [A, B]]],
\end{aligned} \tag{11.50}$$

and more terms can be found in the recent analysis [230]. The Zassenhaus formula can be given in the form

$$e^{A+B} = e^A e^B \prod_{i\geq 2} e^{C_i}, \tag{11.51}$$

where again C_i are polynomials of degree i in A, B. Such formulas are treated by Feynnman's method for disentangling noncommutative operators [98]. The first four coefficients are calculated [64, 230]

$$\begin{aligned}
C_2 &= \frac{1}{2}[B, A] \\
C_3 &= \frac{1}{3}[[B, A], B] + \frac{1}{6}[[B, A], A] \\
C_4 &= \frac{1}{8}\Big([[[B, A], B], B] + [[[B, A], A], B]\Big) + \frac{1}{24}[[[B, A], A], A].
\end{aligned} \tag{11.52}$$

We mention that there are numerical procedures attempting to get past the obstacle of noncommuting operators. One possible approximation is provided by the split-step Fourier numerical method [95, 144]. However, in the following we follow an analytical approach. On one hand we have the exact expression of the operators in (11.46) defined on the free surface

$$\cos(h\partial + \eta\partial) = \frac{e^{ih\partial + i\eta\partial} + e^{-ih\partial - i\eta\partial}}{2}, \tag{11.53}$$

and a similar expression for sin. To apply a smallness criterium we need to bring this equation as close as possible to the form

$$\cos(h\partial)\cos(\eta\partial) - \sin(h\partial)\sin(\eta\partial) = \frac{e^{ih\partial}e^{i\eta\partial} + e^{-ih\partial}e^{-i\eta\partial}}{2}. \tag{11.54}$$

We have the commuting relations $[ih\partial, i\eta\partial] = -\eta_x\partial$, $[ih\partial, [ih\partial, i\eta\partial]] = -ih\eta_{xx}$ and $[i\eta\partial, [ih\partial, i\eta\partial]] = i(\eta_x^2\partial - \eta\eta_{xx})$, which do not entitle us to use the simplified version of the BCH formula, i.e., the finite one. By using (11.51) and (11.52) for $A = ih\partial, B = i\eta(x)\partial$ we obtain

$$\begin{aligned} C_2 &= -\frac{h}{2}\eta_x\partial \\ C_3 &= -\frac{ih}{3}\left(\eta_x^2 - \eta\eta_{xx} - \frac{h}{2}\eta_{xx}\right)\partial, \end{aligned} \tag{11.55}$$

and so on. Finally, we obtain

$$\begin{aligned} \cos(h+\eta)\partial = &\Big(\cos(h\partial) - \sin(h\partial)\eta\partial\Big)\left(1 - \frac{h}{2}\eta_x\partial + \frac{h^2}{4}(\eta_x\eta_{xx}\partial + \eta_x^2\partial^2) + \dots\right) \\ &- \frac{ih}{3}\Big(\sin(h\partial) - \cos(h\partial)\eta\partial\Big)\left((\eta_x^2 - \eta\eta_{xx} - \frac{h}{2}\eta_{xx})\partial + \dots\right). \end{aligned} \tag{11.56}$$

By plugging (11.55) and (11.56) in (11.46), we have a first-order approximation of the operator series, with η/h being the smallness parameter

$$\begin{aligned} u(x,\xi(x,t),t) &= [\cos(h\partial) - \eta(x,t)\partial\sin(h\partial)]\, f(x,t) \\ v(x,\xi(x,t),t) &= -[\sin(h\partial) + \eta(x,t)\partial\cos(h\partial)]\, f(x,t). \end{aligned} \tag{11.57}$$

The advantage of these approximations is that instead of having a complicated differential operator function with variable coefficients, we reduced it, in the limit of small waves, to series of differential operators with constant coefficients.

The dynamics of the fluid is described by the Euler equation at the free surface. The equation that results is written on the surface Σ in terms of the potential and differentiated with respect to x. By imposing the condition $y = \xi(x,t)$, and by using a constant force field we obtain the form

$$u_t + uu_x + vv_x + g\eta_x + \frac{1}{\rho}P_x = 0, \tag{11.58}$$

where g represents the force field constant and P is the surface pressure. Following the same approach as used in the calculation of surface capillary waves [43], we have for our one-dimensional case

$$P|_{\Sigma} = -\frac{\sigma}{\mathcal{R}} = \frac{\sigma \eta_{xx}}{(1+\eta_x^2)^{3/2}} \simeq -\sigma \eta_{xx}, \quad \text{for small } \eta, \tag{11.59}$$

where $\mathcal{R}$ is the local radius of curvature of the surface (in this case, the curvature radius of the curve $y = \xi(x,t)$) and σ is the surface pressure coefficient. Inside the fluid the pressure is given by the Euler equation. Consequently, we have a system of two differential equations (11.40) and (11.58) for two unknown functions: $f(x,t)$ and $\eta(x,t)$ since u and v depend only on η and f through (11.46) or (11.57). With f and η determined we can come back and find the expressions for the velocities u and v, which completely solves the problem.

11.4.3 Dispersion Relations

From the linearization of the free surface kinematic condition (11.40) $v = \xi_t = \eta_t$ we obtain

$$-\sin(h\partial) f = \eta_t, \tag{11.60}$$

and from the linearization of (11.58)–(11.60), we have

$$\cos(h\partial) f_t + g\eta_x - \frac{\sigma}{\rho}\eta_{xxx} = 0. \tag{11.61}$$

By defining the speed of sound $c_0 = \sqrt{gh}$, applying the operator $\sin(h\partial)$ to (11.62), and using (11.61) we obtain the general dispersion relation for our system

$$\cos(h\partial)\eta_{tt} = \sin(h\partial)\left[\frac{c_0^2}{h}\eta_x - \frac{\sigma}{\rho}\eta_{xxx}\right]. \tag{11.62}$$

In the zeroth approximation $\cos(h\partial) \simeq 1, \sin(h\partial) \simeq h\partial$ we have

$$\eta_{tt} = c_0^2 \eta_{xx} - \frac{\sigma h}{\rho}\eta_{xxxx}. \tag{11.63}$$

By using the notations $l_c = \frac{\sigma}{\rho g}$ for the "capillary length," and $B_o = \frac{\sigma}{\rho g h^2} = \left(\frac{l_c}{h}\right)^2$ for the Bond number, we reobtain the linear surface water wave for our system

$$\frac{1}{c_0^2}\eta_{tt} = \eta_{xx} - B_o h^2 \eta_{xxxx}. \tag{11.64}$$

In the following, we discuss the dispersion relation for some limiting situations. For example, if the surface pressure is negligible ($B_o \simeq 0$) we reobtain the simplest dispersion relation of a linear wave

$$\omega = c_0 k. \tag{11.65}$$

In the presence of capillary effects we obtain from the same equations

$$\omega = \sqrt{gh + \frac{\sigma}{\rho} hk^2}, \tag{11.66}$$

which is precisely the dispersion relation for shallow water equations. In the general case, by choosing $\eta = \eta_0 e^{i(\omega t - kx)}$, and by noting that

$$\cos(h\partial)\eta = \cosh(hk)\eta, \quad \sin(h\partial)\eta = -i \sinh(hk)\eta,$$

we obtain the dispersion relation

$$\omega = \sqrt{\left(gk + \frac{\sigma k^3}{\rho}\right) \tanh(hk)}, \tag{11.67}$$

which is the same dispersion relation obtained directly from the Euler equations in more particular cases [174, 271]. Of course, from (11.68) we can recover both the shallow water dispersion relation, in the limit $h \to 0$ (11.67), and, in the limit $h \to \infty$, the dispersion relation for the deep water waves

$$\omega = \sqrt{\frac{g}{k} + \frac{k\sigma}{\rho}}. \tag{11.68}$$

11.4.4 The Full Equation

To obtain our final system of equations, we introduce the velocity operators (11.57) in the first approximation for small amplitude waves in the physical equations. Namely from (11.40) for the free surface BC we obtain

$$[-\sin(h\partial) - \eta\partial\cos(h\partial)]f - \eta_t - \eta_x[\cos(h\partial) - \eta\partial\sin(h\partial)] = 0 \tag{11.69}$$

and from the Euler (11.58) and (11.60) in presence of gravity (uniform vertical field) and surface pressure (11.59) we obtain

$$\begin{aligned}
&[\cos(h\partial) - \eta\partial\sin(h\partial)]f_t - \eta_t\partial\sin(h\partial)f + g\eta_x - \frac{\sigma}{\rho}\eta_{xxx} \times [\cos(h\partial) \\
&\quad - \eta\partial\sin(h\partial)]f \cdot [(\cos(h\partial) - \eta\partial\sin(h\partial))f_x - \eta_x\partial\sin(h\partial)]f - [\sin(h\partial) \\
&\quad + \eta\partial\cos(h\partial)]f \cdot \{[-\sin(h\partial) - \eta\partial\cos(h\partial)]f_x - \eta_x\partial\cos(h\partial)\}f = 0.
\end{aligned} \tag{11.70}$$

These two equations are a set of infinite-order nonlinear PDE in η, f. In the following we use a system of dimensionless quantities defined by the rules $\partial = \partial'/l, \partial_t = (c_0/l)\partial'_{t'}, \eta = a\eta', h\partial = \delta\partial', f = \epsilon c_0 f'$ where $\epsilon = a/h, \delta = h/l$ and a, l are free parameters. In principle we have the restriction $max|\eta| < a$. We also introduce two more notations of operators

$$A = \cos(\delta\partial') - \epsilon\delta\eta'\partial'\sin(\delta\partial')$$
$$D = -\sin(\delta\partial') - \epsilon\delta\eta'\partial'\cos(\delta\partial').$$

With this notations the dynamics (11.69) and (11.70) in the first order in ϵ read

$$Af'_{t'} - \epsilon\delta\eta'_{t'}f'_{x'} + \epsilon(Af')\cdot(Af'_{x'}) + \epsilon(Df')\cdot(Df'_{x'}) + \eta'_{x'} - \delta^2 B_o\eta'_{x'x'x'} = \mathcal{O}(\epsilon^2), \tag{11.71}$$

$$\eta'_{t'} + \frac{1}{\delta}\sin(\delta\partial')f' + \epsilon\eta'\cos(\delta\partial')f'_{x'} + \epsilon\eta'_{x'}\cos(\delta\partial')f' = \mathcal{O}(\epsilon^2). \tag{11.72}$$

Next we introduce the same type of hypothesis as in the classical theory of the KdV equation, namely

$$f' = \eta' + \epsilon f^{(1)} + \delta^2 f^{(2)}, \tag{11.73}$$

where the functions $f^{(1,2)}$ are to be determined. The last steps in obtaining the main dynamic equation consist in: introducing f given by (11.73) in (11.71) and (11.72) and subtract them one from another. Identify coefficients of the terms having the same orders, expand operators A, D in Taylor series, and find the functions $f^{(1,2)}$. After subtracting (11.71) and (11.72) we obtain

$$\begin{aligned}(1-A)\eta'_{t'} &+ \left(\frac{1}{\delta}\sin(\delta\partial') - \partial'\right)\eta' + \epsilon\Big[\frac{1}{\delta}\sin(\delta\partial')f^{(1)} + (\eta'\cos(\delta\partial')\eta')_x \\ &- Af^{(1)}_{t'} - \frac{1}{2}(A\eta')^2_{x'} - \frac{1}{2}(D\eta')^2_{x'}\Big] + \delta^2\Big[\frac{1}{\delta}\sin(\delta\partial')f^{(2)} - f^{(2)}_{t'} + B_o\eta'_{x'x'x'}\Big] \\ &- \epsilon\delta\eta'_{t'}\eta'_{x'} + \epsilon\delta^2\Big[(\eta'\cos(\delta\partial')f^{(2)})_{x'} - [(A\eta')\cdot(Af^{(2)}) \\ &+ (D\eta')\cdot(Df^{(2)})]_{x'}\Big] - \epsilon\delta^3\eta'_{t'}f^{(2)}_{x'} - \frac{\epsilon\delta^4}{2}[(Af^{(2)})^2 + (Df^{(2)})^2]_{x'} = \mathcal{O}(\epsilon^2).\end{aligned} \tag{11.74}$$

After the identification of coefficients of the same order and solving f we can write the final dynamical equation up to the eighth order of smallness

$$\begin{aligned}\eta'_{t'} &+ \eta'_{x'} + \epsilon\left(\frac{3}{2}\eta'\eta'_{x'}\right) + \delta^2\left(-\frac{1}{4}\eta'_{x'x't'} - \frac{1}{12}\eta'_{x'x'x'} - \frac{1}{2}B_o\eta'_{x'x'x'}\right) \\ &+ \epsilon^2\left(C_1\eta'_{x'} - \frac{3}{4}\eta'^2\eta'_{x'}\right) + \epsilon\delta^2\Big(C_2\eta'_{x'} + \frac{1}{8}\eta'_{t'}\eta'_{x'x'} + \frac{7}{12}\eta'_{x'}\eta'_{x'x'} - \frac{5}{12}\eta'\eta'_{x'x'x'}\end{aligned}$$

$$
\begin{aligned}
&-\frac{1}{2}B_o\eta'\eta'_{x'x'x'}-\frac{1}{2}B_o\eta'_{x'}\eta'_{x'x'}-\frac{1}{2}\eta'_{x'}\eta'_{x't'}-\frac{5}{8}\eta'\eta'_{x'x't'}\biggr)+\epsilon^2\delta^2\biggl(C_4\eta'_{x'}+\frac{1}{8}\eta'_{t'}(\eta'_{x'})^2\\
&+\frac{2}{3}(\eta'_{x'})^3-\frac{5}{8}\eta'\eta'_{x'}\eta'_{x't'}+\frac{1}{8}\eta'\eta'_{t'}\eta'_{x'x'}+\frac{5}{3}\eta'\eta'_{x'}\eta'_{x'x'}-\frac{3}{8}\eta'^2\eta'_{x'x't'}+\frac{1}{6}\eta'^2\eta'_{x'x'x'}\biggr)\\
&+\epsilon\delta^4\biggl(\frac{1}{8}\eta'_{x'x'}\eta'_{x'x't'}-\frac{1}{4}\eta'_{x'}\eta'_{x'x't't'}-\frac{13}{36}\eta'_{x'x'}\eta'_{x'x'x'}+\frac{13}{48}\eta'_{x'}\eta'_{x'x'x't'}-\frac{1}{48}\eta'\eta'^{(5)}\\
&+\frac{5}{24}B_o\eta'\eta'^{(5)}\frac{1}{8}B_o\eta'\eta'^{(4,1)}+\frac{1}{8}B_o\eta'_{x'}\eta'_{x'x'x't'}+\frac{1}{6}\eta'\eta'^{(4,1)}-\frac{1}{4}\eta'\eta'^{(3,2)}-\frac{1}{48}\eta'_{t'}\eta'^{(4)}\\
&-\frac{19}{144}\eta'_{x'}\eta'^{(4)}+\frac{5}{24}B_o\eta'\eta'^{(4)}\biggr)+\epsilon\delta^6\biggl(\frac{1}{8}\eta'_{x'}\eta'^{(4,2)}+\frac{1}{96}\eta'_{x'}\eta'^{(5,1)}-\frac{1}{16}B_o\eta'_{x'}\eta'^{(5,1)}\\
&+\frac{1}{8}\eta'\eta'^{(5,2)}-\frac{1}{288}\eta'_{x'}\eta'^{(6)}+\frac{1}{48}B_o\eta'_{x'}\eta'^{(6)}+\frac{1}{96}\eta'\eta'^{(6,1)}-\frac{1}{16}B_o\eta'\eta'^{(6,1)}\\
&-\frac{1}{288}\eta'\eta'^{(7)}+\frac{1}{48}B_o\eta'\eta'^{(7)}\biggr)+\delta^4\biggl(\frac{1}{144}\eta'^{(5)}+\frac{1}{24}B_o\eta'^{(5)}-\frac{1}{4}\eta'^{(3,2)}-\frac{1}{48}\eta'^{(4,1)}\\
&+\frac{1}{8}B_o\eta'^{(4,1)}\biggr)+\delta^6\biggl(\frac{1}{24}\eta'^{(5,2)}+\frac{1}{288}\eta'^{(6,1)}-\frac{1}{48}B_o\eta'^{(6,1)}-\frac{1}{864}\eta'^{(7)}\\
&+\frac{1}{144}B_o\eta'^{(7)}\biggr)=0, \qquad (11.75)
\end{aligned}
$$

where C_i are arbitrary integration constants and the superscripts represent differentiation, like for example $\eta'^{(5,2)} = \eta'_{x'x'x'x'x't't'}$, etc. Equation (11.75) is an approximation of the general system described by the pair (11.70). This long equation is still very general. It can be reduced, for different ranges of ϵ and δ, to several well-known equations in nonlinear fluid dynamics.

11.4.5 Reduction of GKdV to Other Equations and Solutions

Equation (11.75) is unreasonably complicated, and to understand it, we plot in Fig. 11.6 a diagram of the relative contribution of the terms in this equation, function of the ranges of smallness of the two free parameters

$$
\epsilon = \frac{\max|\eta|}{h},\ \delta = \frac{\max L}{h},
$$

describing the amplitude and the wavelength of the solution with respect to the depth of the finite liquid layer. Along the vertical direction we provide different ranges for ϵ, δ, represented in the first two columns. Each row represents the coefficients of the terms in (11.75), written in decreasing order from left to right, for that specific values of the parameters ϵ, δ written in the first two left boxes. In the figure we denoted $\epsilon \to e$ and $\delta \to d$ for graphical purpose. The arrows show how different

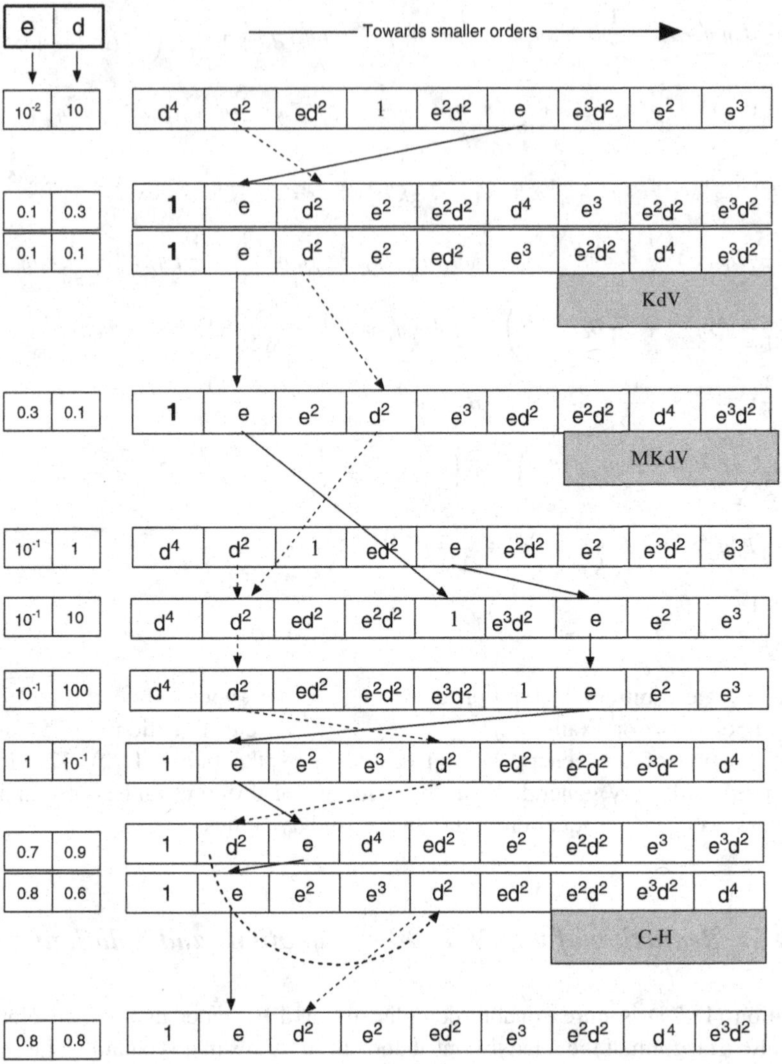

Fig. 11.6 The relative contribution of the terms in the general (11.75), as function of the relative smallness of ϵ, δ. The *arrows* show how the traditional KdV main terms ϵ (*full arrow*) and δ^2 (*dotted arrow*) change their importance in the smallness hierarchy in (11.75) function of their values. We denoted $\epsilon \rightarrow e$ and $\delta \rightarrow d$ for graphical purpose

terms change their importance for different ranges of the parameters. We emphasize in the gray boxes the reduction of (11.75) to KdV and MKdV, respectively.

From Fig. 11.6, it is easier to note how (11.75) can be reduced toward different other models, function of the choice of the parameters ϵ, δ in different ranges, and by neglecting the rest of smaller terms. For example, if $\epsilon \lesssim \delta$, by taking into account only the first four orders of magnitude, we reobtain the KdV equation, and its generalization with the corresponding higher-order dispersion terms. In the normal dimensional form it reads

$$\eta_t + c_0\eta_x + \frac{3c_0}{2h}\eta\eta_x + \frac{c_0h^2}{2}\left(\frac{1}{3} - B_o\right)\eta_{xxx} + \frac{C_2ahc_0}{l^2}\eta_x$$

$$+\frac{c_0h}{2}\left(\frac{23}{12} - B_o\right)\eta_x\eta_{xx} + \frac{c_0h}{2}\left(\frac{5}{12} - B_o\right)\eta\eta_{xxx} - \frac{c_0h^4}{6}\left(\frac{1}{3} + \frac{B_o}{2}\right)\eta_{xxxxx} = 0, \tag{11.76}$$

where C_2 (and consequently the scaling factor l) are arbitrary integration constants. We notice the exact version of the KdV equation. The extra higher-order dispersion terms are useful in the research of the transition between soliton and antisoliton solutions, when the fourth term in (11.76) vanishes. From the physical point of view, this range for ϵ, δ provides the shallow water regime.

In the range $\delta \lessapprox \epsilon$ we obtain a combination between the KdV and the MKdV equation. This range of the parameters can model the deep water situation. Moreover, if we take into account the first five orders of magnitude we obtain the Camassa–Holm equation [42, 80] for fluids.

It is interesting to investigate the behavior of soliton solutions of the GKdV equation in the $B_o = 1/3$ limit, when the coefficient of the main dispersion term η_{xxx} vanishes. Experimental investigation have been performed for this range with Mercury [90]. The range of the parameters is $\epsilon \in [0.1, 0.03], \delta \in [0.57, 0.26]$. In this range, we can approximate the GKdV equation (11.75) by a KdV equation plus a fifth-order dispersion term (η_{xxxxx}), and two nonlinear dispersion terms of the form $\eta_x\eta_{xx}$ and $\eta\eta_{xxx}$. However, in this parameter range the term in δ^2 is small, so we can neglect it, too. Also, for both soliton range (positive amplitude soliton) for $\epsilon \simeq 0.1, \delta \simeq 0.57$ and antisoliton range (negative amplitude) $\epsilon \simeq 0.03, \delta \simeq 0.26$ we notice that we can also neglect the $\eta\eta_{xxx}$ term. Consequently, for this transition regime, the equation becomes

$$\eta'_{t'} + (1 + C_2\epsilon\delta^2)\eta'_{x'} + \frac{3\epsilon}{2}\eta'\eta'_{x'} - \delta^4\frac{11}{36}\eta'_{x'x'x'x'x'} + \epsilon\delta^2\frac{19}{24}\eta'_{x'}\eta'_{x'x'} = 0,$$

and we found two types of solutions. One is a traveling linear wave in the form

$$\eta'(x,t) = a'\cos\frac{x' - V't'}{L'}, \quad L' = \frac{\sqrt{19}}{6}\delta, \quad V' = \frac{361C_1 - 396}{361},$$

in the dimensionless notation, and with *no restriction on the amplitude a'*. Another solution is a compacton-like

$$\eta' = a'\cos^2\left(\frac{x' - V't'}{L'}\right), \quad L' = \frac{\sqrt{19}}{\epsilon}\delta, \quad V' = \frac{1083\epsilon a' + 1444C_1 - 1584}{1444}.$$

In the following, we discuss the existence of different exact solutions for different ranges of parameters. For example, if we neglect the order ϵ^2, we have linear traveling solutions in the form

$$\eta = a\cos\frac{x - Vt}{L}, \quad L = \frac{1}{3}\sqrt{7 - 6B_o\delta} \tag{11.77}$$

$$V = \frac{136 - 201B_o + 36B_o^2 + 196C_2\delta^2\epsilon - 336B_oC_2\delta^2\epsilon + 144B_o^2C_2\delta^2\epsilon}{4(6B_o - 7)^2}.$$

as well as compacton type of solutions

$$\eta = a\cos^2\left(\frac{x - Vt}{L}\right), \quad L = \frac{2}{3}\sqrt{7 - 6B_o\delta}, \tag{11.78}$$

$$V = [544 - 804B_o + 144B_o^2 + 483a\epsilon - 666a\epsilon B_o + 216a\epsilon B_o^2 + 784C_2\epsilon\delta^2 - 1344B_oC_2\epsilon\delta^2 + 576B_o^2C_2\epsilon\delta^2] \times [16(6B_o - 7)^2]^{-1},$$

where we note that these last two solutions exist also in the limit $B_o = 1/3$. In the same limit we have a cnoidal wave solution in the form

$$\eta = Acn^2\left(\frac{x - Vt}{L}, L\right) \tag{11.79}$$

for

$$A = -\frac{12(31m - 15B_om + 36B_0^2m)}{\epsilon(2m - 1)(55 - 192B_o + 144B_o^2)},$$

and

$$L = \sqrt{\frac{10(-10\delta^2 + 9B_o\delta^2 + 36B_0^2\delta^2 + 20m\delta^2 - 18B_om\delta^2 - 72B_0^2m\delta^2)}{9(31 - 15B_o + 36B_0^2)}}.$$

In the following, we present exact solutions for (11.62), which represent another approximation range of the generalized (11.75). We can rewrite (11.62) in the form

$$\cos(h\partial) = \frac{g}{h}\sin(h\partial)(1 - B_o(h\partial)^2)(h\partial)\eta. \tag{11.80}$$

The solution describes a combination of incident and reflected waves in the form

$$\eta(x,t) = \eta_0 e^{i\omega t}(e^{-ikx} + e^{-2ikx_0}e^{ikx}), \tag{11.81}$$

where ω and k fulfill the dispersion relation

$$\omega^2 = gk(1 + B_oh^2k^2)\tanh(kh),$$

and the velocity potential has the form

$$\Phi(x, y, t) = 2i\omega\eta_0 \frac{\cosh(ky)}{\sinh(kh)} \cos(k(x - x_0))e^{i(\omega t - kx_0)}.$$

This solution can be generalized to

$$\eta(x, t) = \sum_{k\in\mathbb{Z}} \left(a_k e^{\frac{(3+4k)\pi x}{2h} + \frac{c_0 t}{h}} + b_k e^{-\frac{(3+4k)\pi x}{2h} - \frac{c_0 t}{h}} \right), \tag{11.82}$$

with the coefficients a_K and b_k determined by initial conditions.

The generalized equation (11.75) can be also related to some classes of evolution equations of the general form

$$u_t + u_x + (f(u))_x - Lu_x = 0, \qquad x \in (-\infty, \infty), \quad t \geq 0,$$

$$u_t + u_x + (f(u))_x + Lu_t = 0, \qquad x \in (-\infty, \infty), \quad t \geq 0,$$

where f is typically a polynomial and L is a Fourier multiplier operator with symbol α, and L and α are related by $\widehat{Lv}(\xi) = \alpha(\xi)\widehat{v}(\xi)$ for all wave numbers ξ. Such equations arise in the description of waves in quite a number of physical situations [18, 255]. The circumflex connotes the Fourier transform (with respect to the spatial variable x) of the function In [27, 33, 118, 313] Bona and Chen discuss the existence of periodic traveling wave solutions of the above equations which is the analog of the cnoidal wave solutions of the KdV equation. They show that the solutions in question have the form

$$u(x, t) = u(x - ct) = \sum_{n=-\infty}^{\infty} u_n e^{i\frac{n\pi}{l}(x-ct)}$$

where $l > 0$ and $c > 0$ are constants. The theory depends on topological methods coupled with degree theory of positive operators. The coefficients u_n are shown to fulfill a set of nonlinear recursion relation very similar to those we have obtained for some solutions of (11.75) (see Sect. 18.4). The generalized action of the trigonometric operators given in (11.46) on the free surface of the fluid, namely $y \to h + \eta(x, t)$, is developed in Appendix 18.4, too. One merit of such types of generalized equations consists in providing a larger number of higher-order dispersion terms that survive in the $B_o = 1/3$ limit. This behavior can be used to investigate the soliton antisoliton transition in variable depth waters.

11.4.6 The Finite Difference Form

Confining a nonlinear differential equation in a compact domain, like in the model we discussed in the last three sections (or like the oscillating drops) results in a behavior similar with the behavior of solutions of finite difference equations. This is similar to the quantization effects (i.e., spectrum becomes discrete) happening with linear differential systems when they are compactified by boundary conditions (Sturm–Liouville, or bilocal, or simply compact eigenvalue problems). We discussed in Sect. 1.1 an example.

We present a property of the free surface nonlinear condition (11.69) with f given by (11.73), valid for any depth h of the liquid tank. We work this in the first order in ϵ, which is one of the two KdV limits. If we consider only traveling solutions of the form $\eta(x,t) = \eta(x + Ac_0 t) = \eta(X)$ where $A \in \mathcal{R}$ and $X = x + Ac_0 t$. Equation (11.69) can be written in the form

$$A h \eta_X(X) + \frac{\eta(X+ih) - \eta(X-ih)}{2i} + \eta_X(X)\frac{\eta(X+ih) + \eta(X-ih)}{2}$$
$$+ \eta(X)\frac{\eta(X+ih) + \eta(X-ih)}{2} = 0, \tag{11.83}$$

if we suppose that η is an analytic function. We study rapidly decreasing solutions at infinity, and we make the substitution $v = e^{Bx}$ for $x \in (-\infty, 0)$ and $v = e^{-Bx}$ for $x \in (0, \infty)$, where B is a positive constant. By introducing $\eta(X) = -hA + f(v)$ we obtain a differential-finite difference equation for the function $f(v)$

$$f(v)\frac{\delta f_v^2(v)}{\delta f_v(v)} + f(v)\frac{\delta f^2(v)}{\delta f(v)} + 2\frac{\sin(Bh)}{B}\delta f(v) = 0, \tag{11.84}$$

where we define the finite difference operator as

$$\delta f(v) = \frac{f(e^{iBh}v) - f(e^{-iBh}v)}{e^{iBh}v - e^{-iBh}v}. \tag{11.85}$$

We can write the solution of (11.83) (or (11.84)) as a power series in v

$$f(v) = \sum_{n=0}^{\infty} a_n v^n, \tag{11.86}$$

and we choose $a_0 = hA$ to have $\lim_{x\to\pm\infty} \eta(x) = 0$. Equation (11.83) results in a nonlinear recursion relation for the coefficients a_n, i.e.,

$$\left[Ahk + \frac{1}{B}\sin(Bhk) \right] a_k$$
$$= -\sum_{n=1}^{k-1} n \left(\cos\left(Bh\left(k-n\right)\right) + \cos\left(Bh\left(k-1\right)\right)\right) a_n a_{k-n}. \tag{11.87}$$

By taking $k = 1$ in the above relation, we obtain $a_1 \left[Ah + \frac{1}{B} \sin(Bh) \right] = 0$. Without loss of generality and because of the arbitrariness of B we can write

$$A = -\frac{\sin(Bh)}{Bh}. \tag{11.88}$$

This relation fixes the velocity of the envelope of the perturbation if its asymptotic behavior is fulfilled. To have $A \neq 0$, we need $Bh \neq k\pi$ for k integer. Under this condition a_1 is still arbitrary and by writing $a_k = \alpha_k a_1^k$ we have $\alpha_1 = 1$ and the recursion relation

$$\alpha_k = \frac{2B \cos \frac{Bh(k-1)}{2}}{k \sin(Bh) - \sin(kBh)} \sum_{n=1}^{k-1} n \cos \frac{Bh\,(2k-n-1)}{2} \alpha_n \alpha_{k-n}, \tag{11.89}$$

for $k \geq 2$. This recursion relation gives the coefficient for k in terms of those for $k-1$ and lesser values. For a smooth behavior of the solution $\eta(X)$ at $X = 0$, i.e., continuity of its derivative, we must introduce the condition

$$f_v(1) = \sum_{n=1}^{\infty} n \alpha_n a_1^{n-1} = 0, \tag{11.90}$$

or require that the derivative of the power series $f(v)$ with coefficients given in (11.83) to be zero in $z \in R$, $z = a_1$. This sets the value for a_1. In the following we study a limiting case of the relation (11.87), by replacing the sin and cos expressions with their lowest nonvanishing terms in their power expansions

$$\alpha_k = \frac{6}{B^2 h^3 k\,(k^2 - 1)} \sum_{n=1}^{k-1} n \alpha_n \alpha_{k-n}. \tag{11.91}$$

It is straightforward exercise to prove that

$$\alpha_k = \left(\frac{1}{2B^2 h^3} \right)^{k-1} k, \tag{11.92}$$

is a solution of the recursion equation. This can be done using mathematical induction and by taking into account the relations

$$\sum_{n=1}^{k-1} n^2 = \frac{k(k-1)(2k-1)}{6},$$

$$\sum_{n=1}^{k-1} n^3 = \left(\frac{k(k-1)}{2} \right)^2. \tag{11.93}$$

We can write the power expansion

$$g(z) = \sum_{k=1}^{\infty} k \left(\frac{1}{2B^2h^3} \right)^{k-1} z^k, \tag{11.94}$$

which has the radius of convergence $\mathcal{R} = 2B^2h^3$ (due to the Cauchy–Hadamard criteria). The function $g(z)$ can be written in the form

$$g(z) = z \left(\frac{1}{1 - \frac{z}{2B^2h^3}} \right)_z 2B^2h^3 = -\frac{z}{\left(1 - \frac{z}{2B^2h^3}\right)^2}. \tag{11.95}$$

Conditions (11.91) and (11.92) result in $a_1 = -2B^2h^3$ and

$$\alpha_k = k \left(\frac{1}{2B^2h^3} \right)^{k-1} \left(-2B^2h^3\right)^k = 2B^2h^3 \, (-1)^k, \tag{11.96}$$

which provides

$$\eta(x) = 2B^2h^3 \sum_{k=1}^{\infty} k \left(-e^{-B|X|} \right)^k$$

$$2B^2h^3 \frac{e^{-B|X|}}{\left(1 + e^{-B|X|}\right)^2} = \frac{B^2h^3}{2} \frac{1}{\left(\cosh\left(\frac{BX}{2}\right)\right)^2}. \tag{11.97}$$

As expected, this solution is exactly the single-soliton solution of the KdV equation and it was indeed obtained by assuming h small in the recursion relation (11.87).

11.5 Boussinesq Equations on a Circle

Another type of integrable nonlinear PDE defined on a compact interval is the Boussinesq equation on a circle for $u(t, \varphi)$

$$u_{tt} - u_{\varphi\varphi} + u_{\varphi\varphi\varphi\varphi} + (f(u))_{\varphi\varphi} = 0, \tag{11.98}$$

where $\varphi \in [0, 2\pi)$ is the angular coordinate in the unit circle, $t \in \mathbb{R}$, and $f(u)$ is a polynomial function depending on u, $|u|$. The solution is supposed to satisfy the initial conditions

$$u(0, \varphi) = u_0(\varphi), \;\; u_t(0, \varphi) = u_1(\varphi). \tag{11.99}$$

Such equations can model water waves (the original Boussinesq model), nonlinear strings, or shape-memory alloys, see [92]. The linearized version of (11.98)

$$u_{tt} - u_{\varphi\varphi} + u_{\varphi\varphi\varphi\varphi} = 0,$$

has solutions that are periodic in space but aperiodic in time, namely the solutions are linear combinations of functions with different noninteger periods. In contrast with the Boussinesq equations on the real axis, (11.98) has no dispersion and no decay in the time variable. For comparison, see experiments described in Sect. 17.2 about similar long-life nonlinear dispersionless excitations on glass spheres. The Boussinesq equation on the circle can be written as a Hamiltonian system in the form

$$u_t = v_\varphi, \quad v_t = u_\varphi - u_{\varphi\varphi\varphi} - (f(u))_\varphi. \tag{11.100}$$

The system in (11.100) has at least two conserved quantities, namely the energy

$$E = \frac{1}{2}\int_0^{2\pi} [v^2 + u^2 + (u_\varphi)^2 - 2F(u)]d\varphi,$$

with $F_\varphi = f$ and $F(0) = 0$, and the momentum

$$P = \int_0^{2\pi} uv\, d\varphi.$$

If the energy is positive defined, its conservation can lead to global existence of stable solutions on the circle. If the energy is not positive defined the solutions may blowup in finite time.

Traveling solutions along the circle for (11.98) read

$$u(t,\varphi) = \eta(\varphi - ct) = \eta(\xi),$$

and fulfill the equation

$$\eta_{\xi\xi} + (c^2 - 1)\eta - f(\eta) = 0.$$

This equation has also a quadrature in the form

$$(\eta_\xi)^2 + (c^2 - 1)\eta^2 + 2F(\eta) = 0,$$

which enables us to determine the conditions on F such that (11.98) possess solitary wave solutions. According to [92] the allowable functions are

$$f(u) = \pm u^2, \quad f(u) = |u|^{p-1}u, \text{ with } |c| < 1, p > 1,$$

or more general

$$f(u) = \lambda |u|^{q-1}u - |u|^{p-1}u,$$

for $\lambda > 0, 1 < q < p$. Solitary wave solutions on the circle are described by the following existence theorems. If the initial data of the problem (11.99) satisfy $u_0 \in H^1, u_t \in H^{-1}$, and if we choose $f(u) = \lambda|u|^{q-1}u - |u|^{p-1}u$ with $1 < q < p$ and λ real, the solution $u(t, \varphi)$ of (11.98) is unique, and it exists for all time. Here H^s are Sobolev spaces, i.e., normed spaces of functions obtained by imposing on a function u and its weak derivatives up to some order the condition of finite L_s norm.

Chapter 12
Nonlinear Surface Waves in Two Dimensions

Two-dimensional flow is a very useful model for practising applications of differential geometry in fluid dynamics. This flow still contains all the special features of the compact three-dimensional flow but is simpler in calculations. In addition, it is not just an idealization, because there are systems that can be modeled with two-dimensional drop systems. Examples of such systems are highly flattened droplets in gravity moving frictionless on rigid surfaces, cell motility and division, electron drops in high magnetics field, long wavelength jets emitted from orifices, evolution of oil spots surrounded by water in oil extraction or ecologic accidents, or closed polymer chains surrounding water bodies. In the following, we discuss some general geometrical properties of two-dimensional flow, and then we study a model of a two-dimensional drop in oscillation, both theoretical and experimental (see Fig. 12.1).

12.1 Geometry of Two-Dimensional Flow

In this section, we introduce few elements of two-dimensional ideal flow, and we discuss their differential geometry interpretation. If we consider a flow in the $\mathbb{R}_2$ plane $\boldsymbol{v} = (u, v)$, we can use the Helmholtz theorem of representation (Theorem 29). That is, if the velocity field is single-valued, continuous, and if $div\, \boldsymbol{v}, curl\, \boldsymbol{v} \to 0$ when $r = \sqrt{x^2 + y^2} \to \infty$, we can always represent the flow in the form

$$\boldsymbol{v} = \nabla \Phi + curl\, \boldsymbol{\Psi}, \tag{12.1}$$

with

$$div\, \boldsymbol{v} = \Delta \Phi, \quad \text{and} \quad curl\, \boldsymbol{v} = grad\, div\, \boldsymbol{\Psi} - \Delta \boldsymbol{\Psi}. \tag{12.2}$$

The two functions are called: Φ, the velocity field potential, and $\boldsymbol{\Psi}$, the *stream function*. Of course, in the two-dimensional case $\boldsymbol{\Psi} = (0, 0, \Psi)$. Also $\boldsymbol{\xi} = curl\boldsymbol{v}$ is the *vorticity*, and it has the property that its Lagrangian time derivative is zero. These functions are not unique, modulo a *gauge transformation*: $\Phi \to \Phi +$ const.,

A. Ludu, *Nonlinear Waves and Solitons on Contours and Closed Surfaces*, Springer Series in Synergetics, DOI 10.1007/978-3-642-22895-7_12,

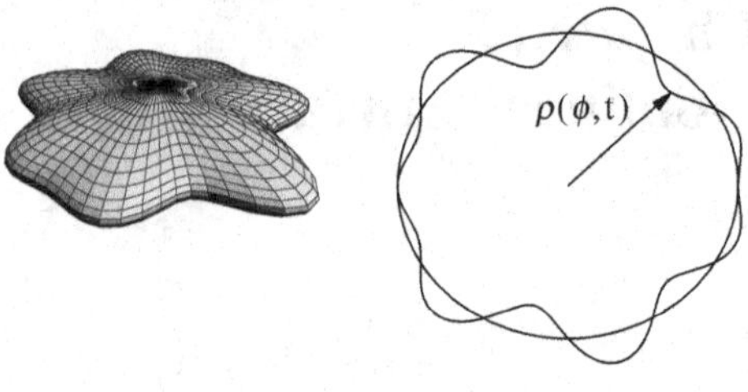

Fig. 12.1 Two-dimensional drop with free surface

$\boldsymbol{\Psi} \to \boldsymbol{\Psi} + \nabla f$. The arbitrary gauge function f can be always chosen such that *div* $\boldsymbol{\Psi} = 0$, and hence we have *curl* $\boldsymbol{v} = -\Delta\boldsymbol{\Psi}$. Among different types of flows in two dimensions, there are two ideal cases which allow a special treatment. All in all, in the two-dimensional case we have

$$u = \Phi_x + \Psi_y, \quad v = \Phi_y - \Psi_x,$$

with

$$\Phi(x, y) = \iint_{\mathbb{R}_2} \frac{div'\boldsymbol{v}}{4\pi|\boldsymbol{r} - \boldsymbol{r}'|} d^2\boldsymbol{r}', \quad \Psi(x, y) = \iint_{\mathbb{R}_2} \frac{curl'\boldsymbol{v}}{4\pi|\boldsymbol{r} - \boldsymbol{r}'|} d^2\boldsymbol{r}'.$$

We define an *irrotational* flow if $curl\,\boldsymbol{v} = \nabla \times \boldsymbol{v} = 0$, and we defined an *incompressible* flow if $div\,\boldsymbol{v} = \nabla \cdot \boldsymbol{v} = 0$. A flow is *potential* if $\exists\Phi, \boldsymbol{v} = \nabla\Phi$, and a flow is *rotational* (or solenoidal) if $\exists\boldsymbol{\Psi}, \boldsymbol{v} = curl\,\boldsymbol{\Psi}$. According to these definitions, for a flow under the Helmholtz theorem hypotheses, the velocity can be always written as the sum of a potential flow and a rotational flow. Obviously, a potential flow is also irrotational, and a rotational flow is also incompressible. The converse affirmations are controlled by the Poincaré Lemma (Sect. 4.8), and consequently by the topology of the domain of definition of the velocity field. If this domain is the whole space, or just simply connected domain, the converse theorems are true: for a flow in a simply connected domain, the properties of potential and irrotational are equivalent. Since we are interested in this book in the dynamics of fluid with free boundary, we will assume in this section that the flow exists in a simply connected domain. If a flow is both irrotational and incompressible, it is called *Laplacean* and we have $\Delta\boldsymbol{v} = 0$. Such a flow is a very special situation, but is an important tool for understanding more complicated flows. Usually, irrotational and incompressible flows can coexist in the same domain, especially in the presence of boundaries. For example, there are flows when vorticity is concentrated in some thin layers, or even in some points of the fluid, while the flow is irrotational outside these layers or points. This is the so-called *almost potential flow* and a good example is worked out in [54, Sect. 2]. For such types of flow, the mechanism for producing vorticity is the interaction of fluid with rigid boundaries. For example, in the case of a flow past a rigid obstacle, the flow is irrotational everywhere, except two streamlines emanating from the body, in which vorticity is not zero. To provide a geometrical interpretation for the potential and the stream function, we consider the graphics of these two functions, namely the parametrized surfaces S_Φ and S_Ψ defined by $\boldsymbol{r}_\Phi(x, y) = (x, y, \Phi) : \mathbb{R}_2 \to \mathbb{R}_3$ and $\boldsymbol{r}_\Psi(x, y) = (x, y, \Psi) : \mathbb{R}_2 \to \mathbb{R}_3$, respectively. The curvatures of these surfaces are in interesting relations with the corresponding types of flow.

In the irrotational case the flow is vortex free ($\boldsymbol{\xi} = 0$), and the circulation of the velocity on any closed curve is zero. The forces acting upon the fluid are only conservative forces. In the irrotational flow the velocity field behaves exactly like an electric field, and $\boldsymbol{v} = \nabla\Phi$ or $u = \Phi_x, v = \Phi_y$ (some books prefer the $\boldsymbol{v} = -\nabla\Phi$ notation). From $curl\,\boldsymbol{v} = 0$ we have $u_y = v_x$ and both the velocity potential and the stream function are Laplacean fields, $\Delta\boldsymbol{\Psi} = \Delta\Phi = 0$. This type of motion is of particular interest for this book since the flow pattern depends solely on the boundary conditions. If the fluid has no free surface, the flow pattern depends only on the motion of the boundaries, and it is independent of the external fields of force. In this case the fields of force affect only the pressure field. The proof is based on a uniqueness theorem via the Gauss formula for kinetic energy of the fluid. Moreover, if all the boundaries are at rest, or if the fluid has zero velocity at infinity, then in the irrotational flow the fluid must be in equilibrium at rest. Sometimes the single-valued irrotational flow is also called *acyclic* [315] to make the distinction with the multivalued irrotational flow which may hold in multiply connected regions (i.e., *cyclic*). In irrotational flow the maximum values of the speed occur on the boundary. If in addition the flow is stationary, from Bernoulli's theorem, the pressure has its minimum values on the boundaries. The proof is immediate because the velocity potential is a harmonic function, and its maxima must occur on its boundaries.

The incompressible flow $div\,\boldsymbol{v} = 0$ occurs when density is constant (from continuity equation we have the divergence zero condition) and the velocity field behaves like a magnetic field. We have $\boldsymbol{v} = curl\,\boldsymbol{\Psi}$, $\Phi = \text{const.}$, i.e., $u = -\Psi_y$, $v = \Psi_x$. The stream function fulfills the equation

$$\frac{\partial}{\partial t}\Delta\boldsymbol{\Psi} + (\nabla\Psi \times \nabla\Delta\boldsymbol{\Psi})_z = 0, \tag{12.3}$$

where $\boldsymbol{\Psi} = (0, 0, \Psi)$, and the subscript z shows that we take only the third component of the resulting vector. In geometric notation this equation means that the directional derivative of the stream function along the velocity field is zero, $D_{\boldsymbol{v}}\boldsymbol{\Psi} = 0$. That is, the stream function is constant along the streamlines. In the stationary case, $d\Psi/dt = 0$, from (12.3) it results that the fields $\nabla\Psi$ and $\nabla(\Delta\Psi)$ are parallel on the surface S_Ψ. Since in the linear approximation the mean curvature is almost the Laplacean of the stream function, $H(S_\Psi) \simeq \Delta\Psi$, it results that the streamlines lie in the level lines of the stream function parametrized surface $\boldsymbol{r} = (x, y, \Psi(x, y))$. That is, in stationary rotational two-dimensional flow, in the linear approximation, the fluid flows along the lines of constant mean curvature of the surface S_Ψ. An incompressible flow cannot start from a fluid at rest (only irrotational flow can start from a stationary state), and the rotational flow is permanent (in absence of dissipation). This is because $d\boldsymbol{\xi}/dt = 0$.

If we consider the parametrized surface associated with the velocity potential, S_Φ, defined by $\boldsymbol{r}_\Phi(x, y) = (x, y, \Phi) : \mathbb{R}_2 \to \mathbb{R}_3$, we notice that the velocity "source" field $\rho(\boldsymbol{r}) = div \boldsymbol{v} = u_x + v_y = \Delta\Phi$ is equal to the mean curvature of S_Φ, in the linear approximation $H_\Phi(x, y) \simeq \Delta\Phi$. Same thing happens for parametrized surfaces of vorticity S_Ψ, defined by $\boldsymbol{r}_\Psi(x, y) = (x, y, \Psi) : \mathbb{R}_2 \to \mathbb{R}_3$, since $H_\Psi(x, y) \simeq \Delta\Psi$. This fact has the following interpretation. We know that a minimal surface ($H = 0$) cannot be compact (see Sect. 10.4.2). So, in the case of irrotational flow, because S_Φ is nearly minimal, it either must extend to infinity or must have singularities. The same thing happens in the case of incompressible flow: because S_Ψ is almost minimal, the flow either should extend to infinity or has singularities.

In a two-dimensional irrotational flow, we call *stagnation point* a point where $\boldsymbol{v} = 0$. At a stagnation point the velocity potential surface S_Φ is quadratic, and it has its first fundamental form coefficients $(1, 1, 0)$

$$E = \Phi_x^2 + 1 \to 1, G = \Phi_y^2 + 1 \to 1, F = \Phi_x\Phi_y \to 0.$$

Let us assume that we choose the origin of the plane in the stagnation point. It results that there is a neighborhood of the origin $\mathcal{V}(O)$, where the surface S_Φ has $E \simeq G, F \simeq 0$, so it is nearly isothermal (see Sect. 10.4.2). In other words, $\forall\epsilon > 0$, $\exists\delta(\epsilon)$ such that

$$|E - G| < \epsilon, \ |F| < \epsilon \text{ if } x^2 + y^2 < \delta(\epsilon)^2.$$

In this neighborhood we can expand the velocity potential in Taylor series

$$\Phi|_{\mathcal{V}(O)} \simeq \frac{1}{2}(x^2\Phi_{xx} + 2xy\Phi_{xy} + y^2\Phi_{yy}) + \mathcal{O}(3),$$

and its expression is just the Hessian of Φ. On the other hand, the Hessian of a function defining a surface is just the second fundamental form on the surface (see Definition 54 in Chap. 6), so $\Phi \simeq \Pi_O$. If we map the velocity field lines from $\mathcal{V}(O)$ on the surface S_Φ, and the resulting curves have unit tangent $\boldsymbol{t}$, we can write $\Pi_O(\boldsymbol{t}, \boldsymbol{t}) = \kappa_n(\boldsymbol{t})$, where κ_n is the normal curvature (Chap. 6) of S_Φ. That provides a nice geometrical interpretation: in the vicinity of a stagnation point, the velocity potential (of a two-dimensional irrotational flow) is equal to the normal curvature of the potential surface S_Φ, up to third-order terms in the Taylor expansion. Similar configurations are presented in Figs. 12.3 and 12.4, except we change the values of the curvatures. The flow in these figures is just nearly incompressible. There is another connection between irrotational incompressible two-dimensional flow and the curvature of the potential surface. Being divergence free, the velocity potential is harmonic, $\Delta\Phi = 0$, and in the linear approximation the mean curvature is zero, $H_\Phi \simeq \Delta\Phi = 0$. If the mean curvature of the potential surface S_Φ is zero at a nonplanar point ($K \neq 0$), then this point has two orthogonal *asymptotic directions*. An asymptotic direction is a direction in the tangent plane to the surface such that the normal curvature is zero along it. From the flow point of

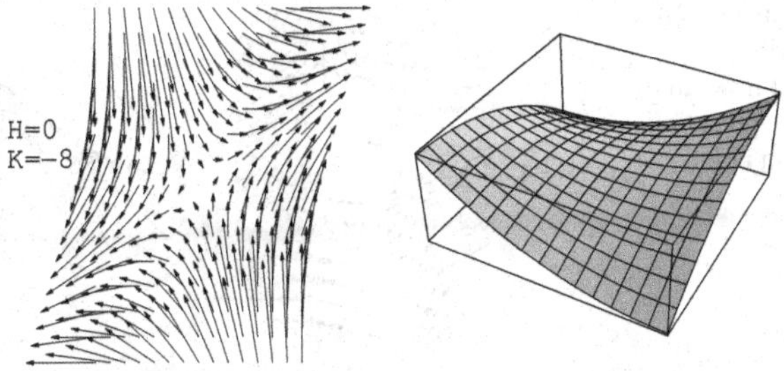

Fig. 12.2 A stagnation point for a potential and incompressible two-dimensional flow. *Left*: The two-dimensional velocity flow $\mathbf{v}(x, y)$. *Right*: plot of the velocity potential $\Phi(x, y)$. The mean curvature of the surface $\Phi(x, y)$ is zero at the stagnation point, and it is approximatively equal to $H \sim \Delta\Phi$. The Gaussian curvature at the stagnation point is $K = -8$

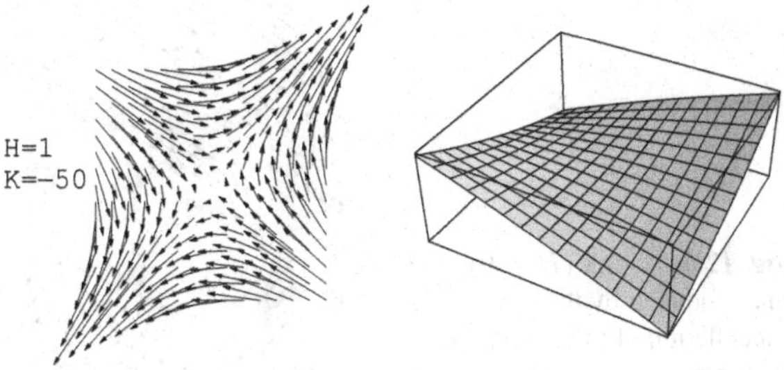

Fig. 12.3 Two-dimensional nearly incompressible $H = 1$ potential flow, and plot of the velocity potential around a stagnation point

view, such a direction in the velocity potential surface is equivalent to a reflection at a "rigid" wall. If the fluid is nearly incompressible ($|div\, \mathbf{v}| \leq \epsilon$), the Laplacean of the potential is not zero, but still of the same order of smallness as ϵ. Because the mean curvature is not zero anymore, the principal curvatures are not equal and opposite, and hence the Gaussian curvature is not necessarily negative. The sign of the Gaussian curvature K_{S_Φ} can be anything and describes somehow the degree of compressibility. We have $\Delta\Phi = \Phi_{xx} + \Phi_{yy} = \epsilon \ll 1$, and consequently

$$K \simeq \Phi_{xx}\Phi_{yy} - \Phi_{xy}^2 = \Phi_{xx}(\epsilon - \Phi_{xx}) - \Phi_{xy}^2 = \epsilon\Phi_{xx} - (\Phi_{xx}^2 + \Phi_{xy}^2) < 0,$$

the real stagnation points are always hyperbolic. Since $K < 0$ we have a hyperbolic point, and there is a real flow where the fluid passes by the stagnation points. If K would be positive, we should have an elliptic point, and the flow must have a sink point, or a source as stagnation point.

In Figs. 12.2–12.8, we present the two-dimensional potential flow of velocity and the graphic of the S_Φ surface in different such situations. In Fig. 12.2 we have potential incompressible flow with $\Delta\Phi \simeq H = 0$ and negative Gaussian curvature. The fluid passes by a stagnation point placed in the origin, where the velocity field has a singular point of index $I = -1$.

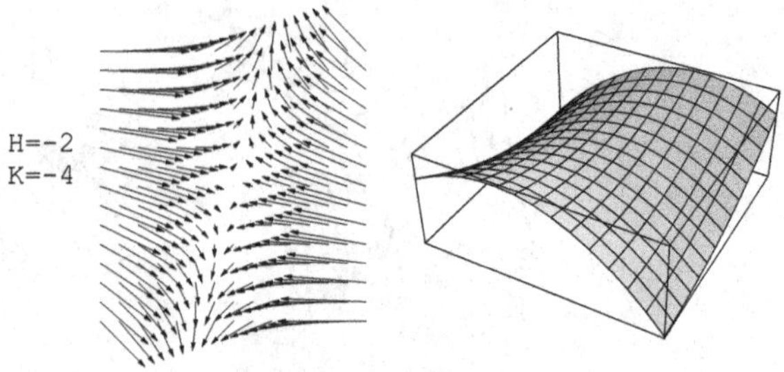

Fig. 12.4 Two-dimensional nearly incompressible $H = -2$ potential flow, and plot of the velocity potential around a stagnation point

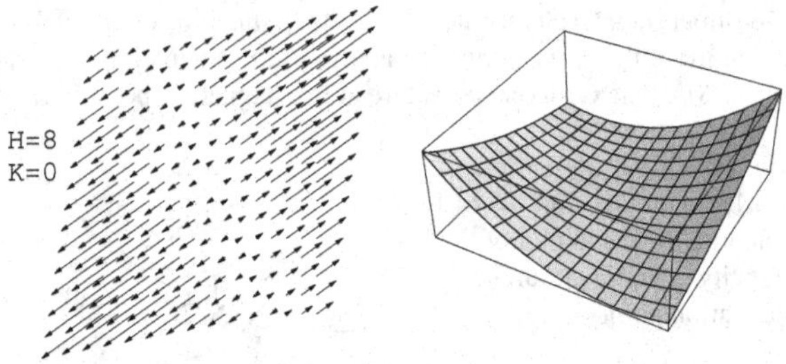

Fig. 12.5 A source ($H > 0$) line produced by the cancellation of the Gaussian curvature

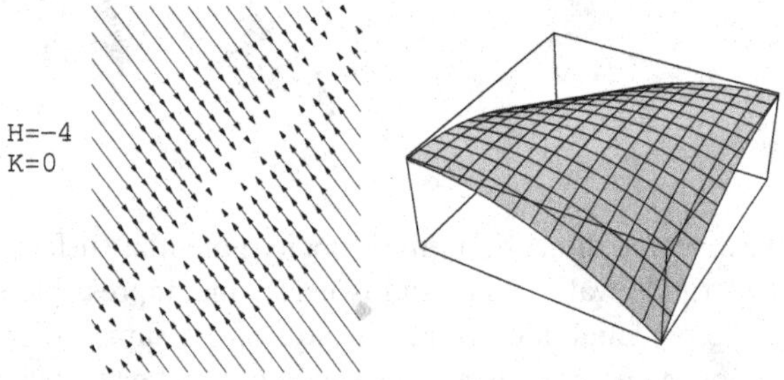

Fig. 12.6 A sink ($H < 0$) line produced by the cancellation of the Gaussian curvature

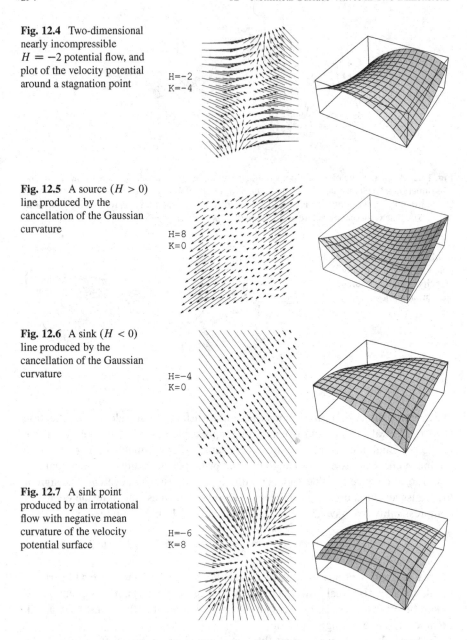

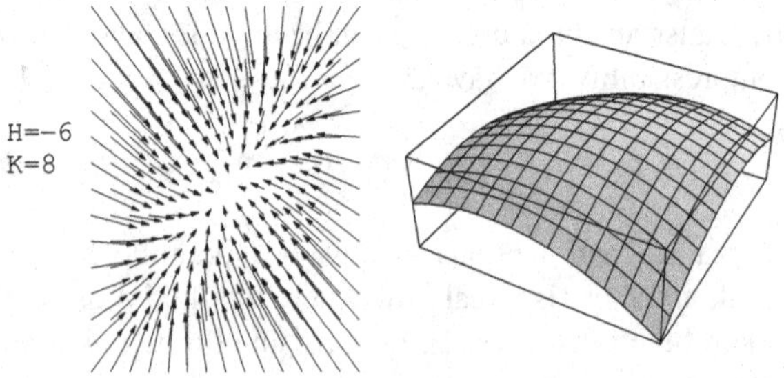

Fig. 12.7 A sink point produced by an irrotational flow with negative mean curvature of the velocity potential surface

The index $I(\boldsymbol{v}, P)$ of an isolated singular point P of a vector field $\boldsymbol{v}$ defined on a surface is, in general, the number of full 2π rotations performed by $\boldsymbol{v}$ when it runs along an infinitesimal simple closed regular curve around P. The stagnation point is hyperbolic, and it creates two asymptotic directions along which the fluid runs away. If the Gaussian curvature of the velocity potential surface is zero (parabolic

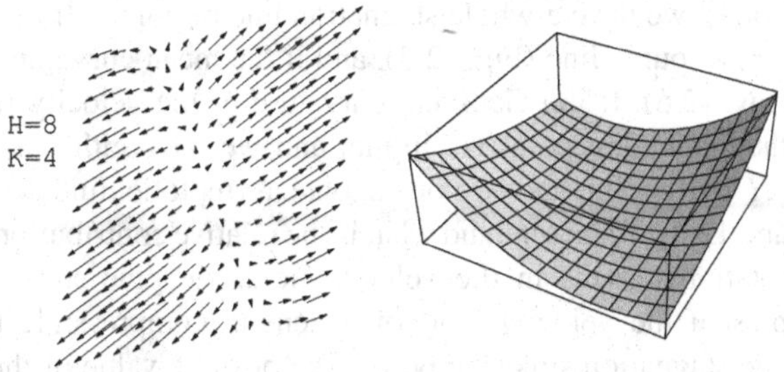

Fig. 12.8 A source point produced by an irrotational flow with positive mean curvature of the velocity potential surface

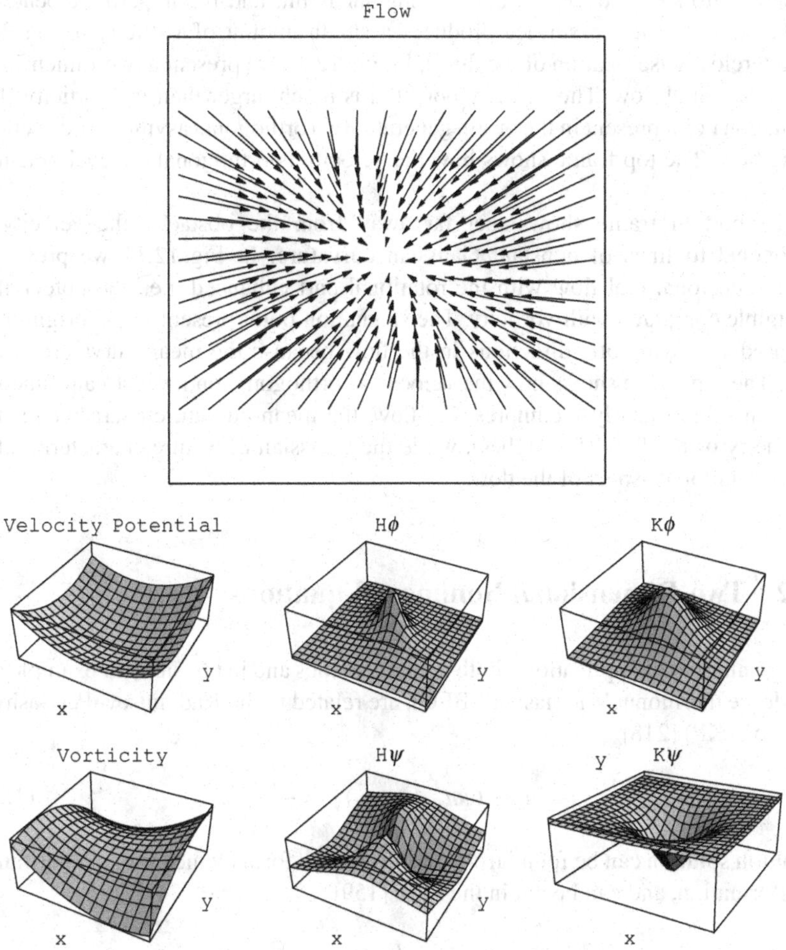

Fig. 12.9 Velocity field for a general two-dimensional flow. The potential and stream function are plotted, together with the graphics of their mean and Gaussian curvatures

point), we have a whole stagnation line of points. If the mean curvature is positive, it is a source line (Fig. 12.5), and if the mean curvature is negative, it is a sink line (Fig. 12.6). If the Gaussian curvature of the velocity potential surface is positive, the stagnation point is elliptic, and we have either a sink ($H < 0$) or a source ($H > 0$) point, and two orthogonal asymptotic fluid "escape" directions (Figs. 12.7 and 12.8). To understand what is the relative contribution of each of the potential and rotational terms in the velocity field, we give some examples. In Fig. 12.9 we present the velocity field of a general compressible two-dimensional flow with one stagnation sink-like point. The positive value of the Gaussian curvature of the potential surface produces an elliptic singular point, and the hyperbolic behavior of the stream function surface produces a small amount of vorticity to the flow (counterclockwise rotation of the fluid). In Fig. 12.10 we present a two-dimensional nearly potential flow. The velocity potential is much larger than the vorticity. The stagnation point present in the origin generates two orthogonal asymptotic directions in the flow. The top frame shows how the velocity is orthogonal on level potential lines.

The bottom frame shows that far away from the obstacle, the velocity is orthogonal to lines of constant Gaussian curvature. In Fig. 12.11 we present a two-dimensional real flow with the rotational part enhanced, i.e., the potential is negligible compared with vorticity. The stagnation point present in the origin does not produce asymptotic directions in the flow because the mean curvature is not zero. The top frame shows how the velocity is orthogonal on level stream function lines. In such a nearly incompressible flow, the mean curvature characterizes the symmetry of the flow (four lobes), while the Gaussian curvature characterizes the global rotational aspect of the flow.

12.2 Two-Dimensional Nonlinear Equations

A large number of applications, both in mathematics and in physics (for example hot and dense thermonuclear plasmas, BEC), are related to the Kadomtsev–Petviashvili equation (KP) [218]

$$(-4u_t + 6uu_x + u_{xxx})_x = -3u_{yy}. \tag{12.4}$$

A soliton solution can be found from the Wronskian form by means of a logarithmic transformation, and can be put in the form [159]

$$u(x, y, t) = \frac{(k_1 - k_2)^2}{2} \operatorname{sech}^2 \frac{\theta_1 - \theta_2}{2}, \tag{12.5}$$

where the phase functions are given by

$$\theta_j = -k_j x + k_j^2 y - k_j^3 t + \theta_j^0.$$

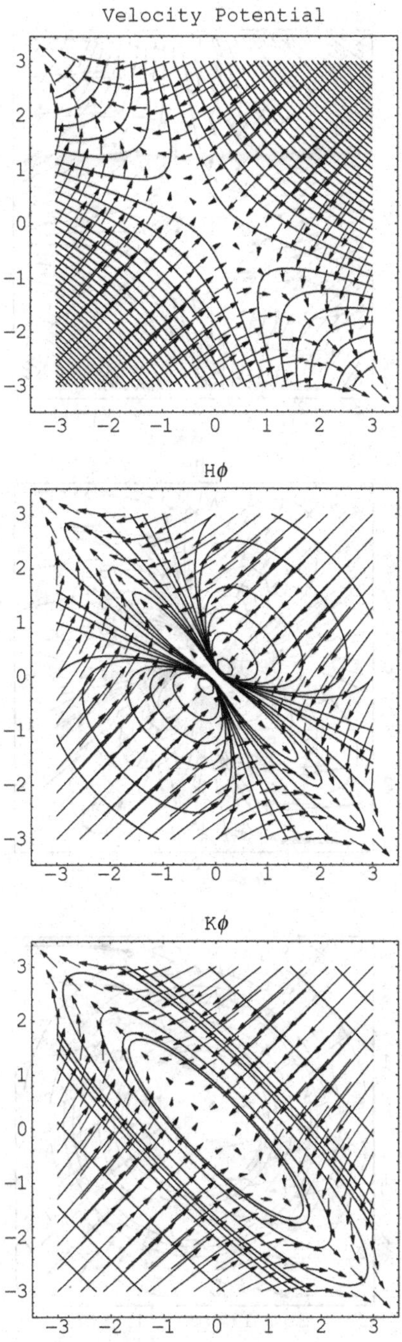

Fig. 12.10 Real two-dimensional flow with negligible vorticity, around a stagnation point. Level lines from top to bottom: velocity potential Φ, mean curvature, and Gaussian curvature of the velocity potential graphics S_Φ. All superimposed on the velocity field

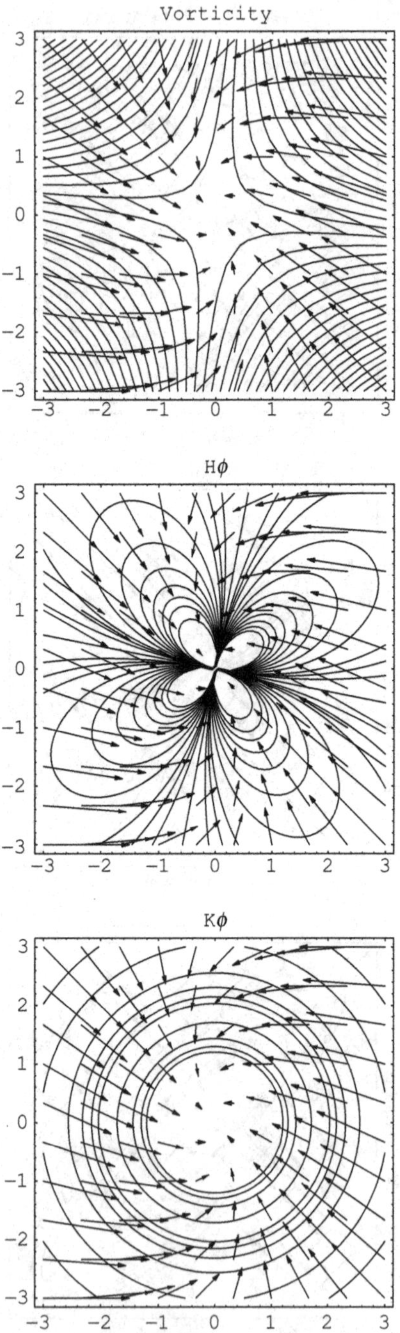

Fig. 12.11 Real two-dimensional flow with enhanced vorticity. Level lines from top to bottom: stream function Ψ, mean curvature, and Gaussian curvature of the stream function graphics S_{Ψ}. All superimposed on the velocity field

By denoting

$$A = \frac{(k_1 - k_2)^2}{2}, \quad L = \frac{2}{k_2 - k_1}, \quad V = -(k_1^2 + k_1 k_2 + k_2^2),$$

we can write the solution in a soliton form

$$u = A \operatorname{sech}^2 \frac{x - (k_1 + k_2)y - Vt}{L}.$$

The nonlinear dispersion relation analysis requests to choose two possible space–time scales for the two different directions. To remain as general as possible, we make the following hypothesis

$$u_t = -V_1 u_\xi - V_2 u_\eta,$$

where $\xi = x - V_1 t, \eta = y - V_2 t$ is the transformation of coordinates into a arbitrarily diagonally moving frame. It results

$$L^2 = \frac{1}{3A + 4V_1 + 3V_2^2},$$

which describes pretty much the real behavior of the dispersion relation for the exact soliton solution in (12.5). However, the solutions have long tails, and they are not of interest for the following topics of this chapter.

12.3 Two-Dimensional Fluid Systems with Boundary

We consider a bounded two-dimensional variable domain $\mathcal{D}(t)$ in $\mathbb{R}^2$ with moving frontier described by a smooth closed curve $\Gamma(t) = \partial\mathcal{D}(t)$ of equation $\boldsymbol{r} = \boldsymbol{r}(\alpha, t)$, where α is a time-invariant parameter along the curve $\alpha \in [0, \alpha_{max}]$. We define for this curve the metrics $g(\alpha, t)$, its Serret–Frenet local frame $\boldsymbol{t}, \boldsymbol{n}$, the curvature $k(\alpha, t)$, and the local velocity of the curve $V(\alpha, t) = U\boldsymbol{n} + W\boldsymbol{t}$, namely (5.3). The frontier curve (also called contour or free boundary) has a length and encloses an area, provided by (7.36) and (7.40), with the flows given by

$$\frac{\partial L}{\partial t} = -kUds, \quad \frac{\partial A}{\partial t} = -Uds. \tag{12.6}$$

In the following we want to relate the normal and tangent velocities (which are defined in terms of the $\boldsymbol{r}(\alpha, t)$ equation for the contour) to the free surface kinematic condition (9.5), (9.29), and (9.30), which is expressed in terms of the equation $S(\boldsymbol{r}, t) = 0$. The first formalism represents the Lagrangian point of view, where we describe the motion of a certain entity (the arc-length of the curve), and we can establish a correspondence between a certain value of α and a fluid particle lying on the surface. The second approach in terms of the function S tells us [167] that the normal velocity of a particle inside the surface is equal to the normal velocity of the surface itself, $V_{n,particle} = v_{n,S} = -U$, because by definition we have no flux of particles across the surface Γ. Like we proved in Sect. 9.5, from $S(\boldsymbol{r},t) = 0$ we infer $S(\boldsymbol{r} + \boldsymbol{n}\delta u, t + \delta t) = 0$, where δu is the displacement of the surface toward its normal direction. From here $\boldsymbol{n} = \frac{\nabla S}{|\nabla S|}$, and we can write

$$v_{n,surface} = -\frac{1}{|\nabla S|}\frac{\partial S}{\partial t} = -U, \quad \text{or} \quad \frac{1}{|\nabla S|}\frac{\partial S}{\partial t} = \frac{\partial \boldsymbol{r}}{\partial t} \cdot \boldsymbol{r}. \tag{12.7}$$

When the contour is parameterized by $\boldsymbol{r} = \boldsymbol{r}(\alpha, t)$, we have

$$|\nabla S| = \frac{\sqrt{\left(\frac{\partial x}{\partial \alpha}\right)^2 + \left(\frac{\partial y}{\partial \alpha}\right)^2}}{\frac{\partial x}{\partial \alpha}}$$

and

$$\frac{\partial S}{\partial t} = \frac{|\frac{\partial \boldsymbol{r}}{\partial t} \times \boldsymbol{t}|}{g^{1/2}\frac{\partial x}{\partial \alpha}}.$$

These two equations check $\frac{\partial \boldsymbol{r}}{\partial t} = U\boldsymbol{n} + W\boldsymbol{t}$.

In the following we use polar coordinates for the expression of the contour function, in the form

$$x = (R + \xi(\phi, t))\cos\phi, \quad y = (R + \xi(\phi, t))\sin\phi,$$

where Φ is a time-invariant parameter, $\alpha = \phi \in [0, 2\pi)$, R is a fixed radius and ξ describes the perturbation of a circle into the actual contour. The metric is

$$g(\phi, t) = (R + \xi)^2 + \left(\frac{\partial \xi}{\partial \phi}\right)^2, \quad ds = \sqrt{(R + \xi)^2 + \left(\frac{\partial \xi}{\partial \phi}\right)^2}.$$

We have the Lagrangian velocity of the contour

$$\boldsymbol{v}(\phi, t) = \frac{\partial \boldsymbol{r}}{\partial t} = (\xi_t \cos\phi, \xi_t \sin\phi) = U\boldsymbol{n} + W\boldsymbol{t} = v_r \boldsymbol{e}_r + v_\phi \boldsymbol{e}_\phi, \tag{12.8}$$

where subscripts denote differentiation, and $\boldsymbol{e}_{r,\phi}$ are the polar unit vectors and velocity components in the radial and angular directions. In polar coordinates, $\boldsymbol{v} = (v_r, v_\phi)$ and $\nabla = (\partial_r, \partial_\phi / r)$. The free surface kinematic condition reads in polar coordinates

$$v_r\Big|_\Gamma = \left(\frac{\partial \xi}{\partial t} + \frac{\partial \xi}{\partial \phi}\frac{v_\phi}{R + \xi}\right)_\Gamma. \tag{12.9}$$

The tangent to the contour has the expression

$$t = g^{-\frac{1}{2}} \frac{\partial \boldsymbol{r}}{\partial \phi} = \frac{(\xi_\phi \cos\phi - (R+\xi)\sin\phi, \xi_\phi \sin\phi + (R+\xi)\cos\phi)}{\sqrt{(R+\xi)^2 + \xi_\phi^2}}, \quad (12.10)$$

and the curvature reads

$$k = \frac{(R+\xi)^2 + 2\xi_\phi^2 - (R+\xi)\xi_{\phi\phi}}{((R+\xi)^2 + \xi_\phi^2)^{3/2}}. \quad (12.11)$$

From (12.8), (12.10), and (12.11), we obtain the relations between the local normal and tangent components of the velocity of the curve, and its polar components

$$v_r = -\xi_t \frac{(R+\xi)^2 - \xi_\phi^2}{g}, \quad v_\phi = \xi_\phi \xi_t \frac{2(R+\xi)}{g}, \quad (12.12)$$

and

$$U = -\xi_t \frac{R+\xi}{g^{1/2}}, \quad W = \frac{\xi_t \xi_\phi}{g^{1/2}}. \quad (12.13)$$

Finally, we write the length of the contour, and the area inside it

$$L(t) = \int_0^{2\pi} \sqrt{(R+\xi)^2 + \xi_\phi^2 d\phi}, \quad A(t) = \frac{1}{2}\int_0^{2\pi} (R+\xi)^2 d\phi. \quad (12.14)$$

To find the linear oscillations limit, we assume the variable contour to be very close to a circle of radius R, i.e., $r(\phi,t) = R + \xi(\phi,t)$ with $\max|\xi| \ll R$. The calculation of the pressure surface needs the expression of the infinite small variation of the arc-length. We use $\delta dL = -kU\delta t ds$, where k is given in (12.11). However, to understand how the polar coordinates work in this case, we double-check the arc-length variation formula, by obtaining it again, through variational calculations directly in polar coordinates. We introduce an arbitrary infinitesimal variation of the contour shape $\delta\xi$, and we have

$$\delta L = L(t, \xi + \delta xi) - L(t,\xi) = \int_0^{2\pi} \left[\frac{R+\xi}{\sqrt{(R+\xi)^2 + \xi_\phi^2}} \delta\xi + \frac{\xi_\phi}{\sqrt{(R+\xi)^2 + \xi_\phi^2}} \delta\xi_\phi \right] d\phi.$$

After an integration by parts we have

$$\delta L = \int_0^{2\pi} (R+\xi) k \delta\xi d\phi. \quad (12.15)$$

This result is in perfect agreement with previous expressions for δL as it can be checked by substituting $\delta\xi$ into $k(R+\xi)\delta\xi d\phi = -kU\delta t ds$.

In the second order of approximation with respect to $\delta\xi$, we have

$$\delta L = \int_0^{2\pi} \left(1 - \frac{\xi_{\phi\phi}}{R} + \frac{2\xi\xi_{\phi\phi} + \xi_{\phi^2}}{2R^2} + \mathcal{O}(3) \right) \delta\xi d\phi. \tag{12.16}$$

Using the same variational approach, we obtain the infinitesimal variation of the area

$$\delta A = \int_0^{2\pi} (R + \xi)\delta\xi d\phi. \tag{12.17}$$

12.4 Oscillations in Two-Dimensional Liquid Drops

We consider a very flatted drop of equilibrium radius R_0 on a horizontal surface, described in spherical coordinates by the radial coordinate

$$r(\theta, \varphi, t) = R_0 \sin\theta \sqrt{(1 + \epsilon f(\varphi, t))^2 + \delta^2 \cot^2\theta}, \tag{12.18}$$

where ϵ is the ratio between the maximum planar perturbation of the drop from a circular shape, and δ is the ratio between the vertical height of the drop and R_0. That is $\delta = 0$ will describe a totally flat drop, and $\delta = 1$ will describe an axisymmetric three-dimensional shape. In the following, ϵ and δ are free small (much less than 1) parameters in this formulation. The dynamics of the drop is described by oscillations and waves along the contour Γ of the drop, i.e., $r(\pi/2, \varphi, t) = R_0(1 + \epsilon f(\varphi, t))$, so the problem is solved if we find the $f(\varphi, t)$ shape function. To account for the surface tension effects, we need to estimate the mean curvature of this drop. From (10.60) we can write in the first order of smallness in ϵ, δ

$$H(\theta, \varphi, t) = -\frac{1}{R_0} + \frac{r}{R_0^2} - \frac{r^2}{R_0^3} + \frac{r_\varphi^2}{2R_0^3} - \frac{r r_{\varphi\varphi}}{R_0^3} - \frac{r r_\theta}{R_0^3} + \frac{r_\theta^2}{2R_0^3}$$
$$+ \frac{r_{\varphi\varphi} + r_{\theta\theta}}{2R_0^2} + O_3(r/R_0), \tag{12.19}$$

with r is the general shape of the droplet, and subscripts denote differentiation. If we substitute r from (12.18) we obtain in the first order in ϵ

$$H \simeq \frac{\delta^2 - 3}{2R_0} - \epsilon \frac{(3 + \delta^2) f + f_{\varphi\varphi}}{R_0}, \tag{12.20}$$

and consequently the surface tension at the boundary of the drop is given by (10.53)

$$P = \sigma \left(\frac{\delta^2 - 3}{2R_0} - \epsilon \frac{(3 + \delta^2) f + f_{\varphi\varphi}}{R_0} \right) + P_0 + O_2(\epsilon, \delta). \tag{12.21}$$

In the following we assume, for simplicity, that the drop is incompressible and inviscid, and the flow is irrotational, so the velocity is obtained from the velocity potential $\Phi(r,\theta,\varphi,t)$. We assume that the horizontal surface of the drop is flat, so there will be no contribution to the potential energy from this part. The only important region is Γ, the closed contour of the drop parameterized by φ, where $\theta \sim \pi/2$. The dynamics is hence controlled by the Laplace equation for potential in the bulk, the Euler equation for the contour, and boundary conditions: free liquid surface on one side and rigid core (if it is the case) on the other side. The general three-dimensional treatment of the associated linear problem for the dynamics of the free surface will be given in Chap. 13, Sect. 13.1, and here we will follow the same procedure. So, in this section we just mention the guiding lines for this simpler two-dimensional system. We assume the two-dimensional approximation, so the potential is chosen

$$\Phi(r,\varphi,t) = \sum_{l=0}^{\infty} f_l(r)\cos(l\varphi)e^{-i\omega_l t}. \tag{12.22}$$

From $\Delta_{r,\varphi}\Phi = 0$, we have

$$f_l(r) = A_l r^l + \frac{B_l}{r^l}. \tag{12.23}$$

The dynamic equation (Euler equation on the contour) and the two boundary conditions (the first one from (9.30) and second on the rigid core) are, respectively,

$$\left.\frac{\partial\Phi}{\partial t}\right|_{\Gamma} = -\left.\frac{P}{\rho}\right|_{\Gamma}, \quad \left.\frac{\partial\Phi}{\partial r}\right|_{\Gamma} = -\left.\frac{\partial f}{\partial t}\right|_{\Gamma}, \quad \left.\frac{\partial\Phi}{\partial r}\right|_{r=a} = 0, \tag{12.24}$$

where the last condition requests zero normal velocity on a rigid core of radius a. From these equations we obtain $B_l = A_l a^{2l}$ and

$$\omega_l^2 = \frac{\sigma l(l^2-3)\left[1-\left(\frac{a}{R_0}\right)^{2l}\right]}{\rho R_0^3\left[1+\left(\frac{a}{R_0}\right)^{2l}\right]}. \tag{12.25}$$

Equation (12.25) describes the linear modes of oscillations of this two-dimensional ideal drop, for $l = 1,\ldots$. We can make a few remarks. First, the modes $l = 1,2$ are not forbidden like in the three-dimensional case. For no core, or for inner core and external free surface, this equation gives good results. However, for *inner modes*, i.e., when the rigid surface is exterior and the drop becomes a two-dimensional shell with inner free surface, (12.25) does not work so well because the frequencies become imaginary. This means that all such internal oscillation modes should be damped. This equation cannot predict traveling waves along the inner free surface, which is actually the experimental situation (see Sect. 12.6). To explain the existence of traveling modes along the inner contour, one should introduce both viscosity and vorticity. Indeed, if we keep the irrotational hypothesis, change the potential structure into

$$\Phi(r, \varphi, t) = \sum_{l=0}^{\infty} f_l(r) g_l(\varphi, t),$$

and try to bring more nonlinear terms into the mean curvature (10.60), still the Laplace equation will force the angular dependence to be linear, i.e., $g_l(\varphi, t) \to \cos(l\varphi + \beta_l(t))$, and to have nonzero vorticity, we need the viscosity.

The viscous, yet irrotational, two-dimensional case was recently modeled, for example in [323], by a numerical boundary integral method. The dynamical equation used by these authors was an unsteady Bernoulli equation for the potential flow in the form

$$\left.\frac{d\Phi}{dt}\right|_{\Gamma} = \frac{1}{2}\left[-\left(\frac{\partial \Phi}{\partial s}\right)^2 = V_n^2\right] - \kappa(s), \tag{12.26}$$

where the LHS is the material derivative of the contour, the contour is parameterized by the arc-length s, V_n is the normal velocity at the contour, $\boldsymbol{n}$ is the normal unit vector to the contour, and κ is the curvature of Γ. The equation is coupled with area and energy conservation

$$A = \frac{1}{2}\oint_{\Gamma} \nabla \cdot \boldsymbol{r} dA = \frac{1}{2}\int_0^L \boldsymbol{n} \cdot \boldsymbol{r}(s) ds, \;\; K = \frac{1}{2}\oint_{\Gamma} |\nabla \Phi|^2 dA = \int_0^L \Phi(s) V_n(s) ds. \tag{12.27}$$

However, even with this improvements, and even by taking into account the shear viscosity and the surface dilatational viscosity (see Sect. 8.4), the solution does not provide stable localized traveling waves like those obtained experimentally and presented in Sect. 12.5. Only by introducing the vorticity, one can explain such nonlinear effects. We describe such a nonlinear model in Sect. 16.6 [340, 341] when we refer to application of contour nonlinear waves in microscopic systems, so we do not repeat the calculations here. We just mention that the two-dimensional liquid drop nonlinear approach can predict solitons on the surface of the droplets, and can even work in the presence of rigid cores, inside or outside the two-dimensional drop or shell, respectively. The additional condition is given in (16.86).

12.5 Contours Described by Quartic Closed Curves

An interesting application of the contour dynamics is the planar flow of a drop of incompressible homogenous viscous fluid through a porous medium [323]. In this situation the Bernoulli equation is replaced by the Darcy's law

$$V = -c\nabla P, \tag{12.28}$$

where $c > 0$ is a constant, inversely proportional to the dynamic viscosity μ. Since the potential of flow is harmonic, we can represent it as being generated by a finite sum of sources and sinks of coordinates $\boldsymbol{r}_j$ and intensities

$$q_j = \int_{\Gamma_j} V \cdot \boldsymbol{n} ds,$$
$$\Phi(\boldsymbol{r}, t) = \sum_j \frac{q_j}{2\pi} \ln |\boldsymbol{r} - \boldsymbol{r}_j| + \Phi_0(\boldsymbol{r}, t), \tag{12.29}$$

where Φ_0 is a smooth function defined in the domain, Γ_j are contours surrounding the sinks and sources, and $\boldsymbol{n}_j$ are the principal normals of these contours. The problem is to find the motion of the boundary of the fluid saturating this porous surface. The problem is nonlinear, and the solutions, that are the evolution of the boundary of the planar drops, have an interesting soliton property. Namely, there is an infinite series of conservation laws associated with this flow. The proof of this property can be obtained through the Richardson's integrability theorem [273]. Namely, for any time variable domain $D(t)$ of viscous fluid under the hypotheses enounced above, and for any arbitrary harmonic function in the plane $u(x, y)$, there is the relation

$$\frac{d}{dt} \int_{D(t)} u dx dy = \sum_{j=1}^{n} q_j u(\boldsymbol{r}_j). \tag{12.30}$$

As a simple example, for $u = 1$, the above relation assures the conservation of the area. The Richardson's problem is useful mainly since one can reconstruct the shape of the domain by using these first integrals.

12.6 Surface Nonlinear Waves in Two-Dimensional Liquid Nitrogen Drops

Some experiments with fluid in rotating vessels, [142], prove the existence of very interesting nonlinear patterns in almost two-dimensions. Liquid nitrogen ($\rho = 808\,\text{kg}\,\text{m}^{-3}$, $\sigma = 31\,\text{dyn}\,\text{cm}^{-1}$ compared to water $\sigma = 72\,\text{dyn}\,\text{cm}^{-1}$, $T = -195.8°\text{C}$ at $P = 1$ atm, $\mu = 0.15$ cP compared to water 1 cP) is an ideal system for testing the theory of nonlinear two-dimensional oscillations of liquid drops. Having very low viscosity but large enough surface tension, and being always "coated" by a layer of vapors, it can be considered pretty isolated for the container walls. On a horizontal flat surface, a droplet will take a radius of about 3–15 cm and a height of about 2 mm which qualify it for a two-dimensional model.

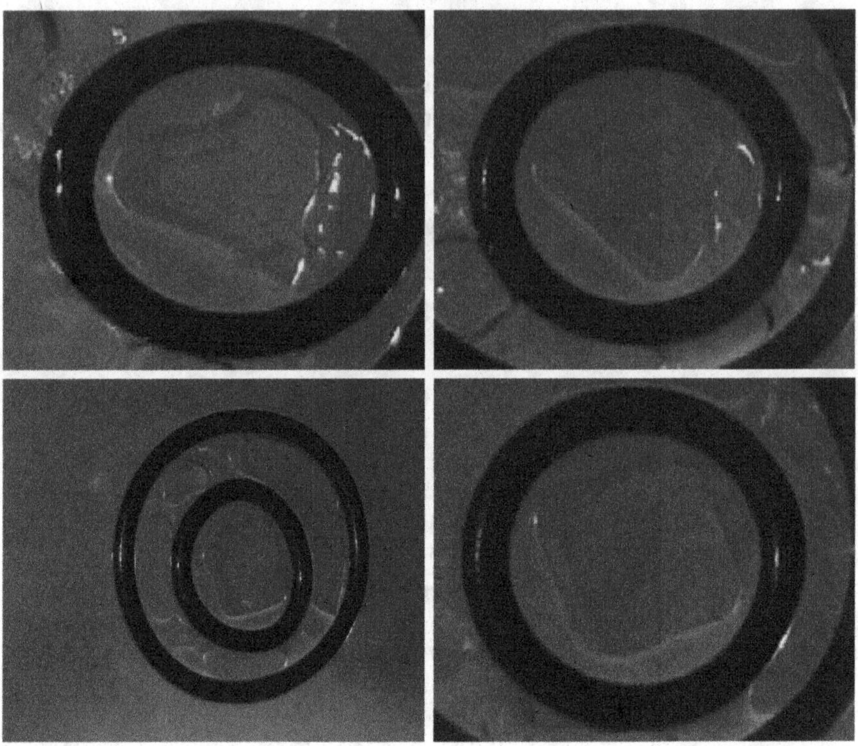

Fig. 12.12 Internal nonlinear surface waves in a two-dimensional shallow circular layer of liquid nitrogen around a rigid core (dark ring). Triangular and square modes with amplitude in the range 1–5 mm and rotating with a circular speed of 20–40 cm s^{-1} are obtained. In the figure the core has a diameter of 3 cm. The waves are stable for about 4 s, and then brake up because of evaporation and volume loss

If the droplets are surrounded by rigid contours, they will perform a wide range of motions because of the fast evaporation process (18.2 g evaporated per hour per Watt of surface thermal energy). Studies of small droplets of liquid nitrogen have been performed and several types of waves and patterns have been detected.[1] Moreover, because during the experiment the mass and the volume of the droplets or layer continuously decrease by evaporation, one can watch in real time succession of resonant modes and circular traveling waves corresponding to those dimensions of the system. Basically, on the free contour of the drop we notice initially the existence of high modes with $l = 20$–30. Later on, through evaporation, the high modes decay, and lower modes become more stable. At $l = 6$ we notice a special long time stability. After couple of seconds the $l = 6$ modes transform into lower

[1]We acknowledge Mrs. Tamika Thomas (NSU and JOVE) who performed the experiments and obtain the pictures with precision and accuracy.

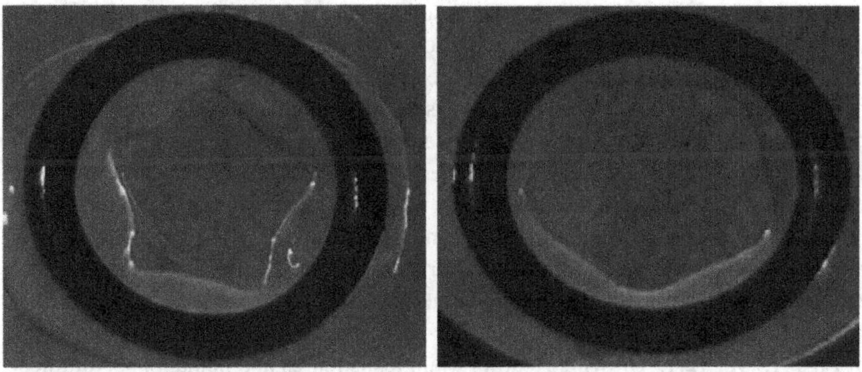

Fig. 12.13 Pentagonal internal waves inside rigid liquid nitrogen contours

Fig. 12.14 Traveling and decaying of cnoidal waves on the external contour taken at three different moments of time

Fig. 12.15 Soliton traveling on the external free surface of a two-dimensional layer of liquid nitrogen

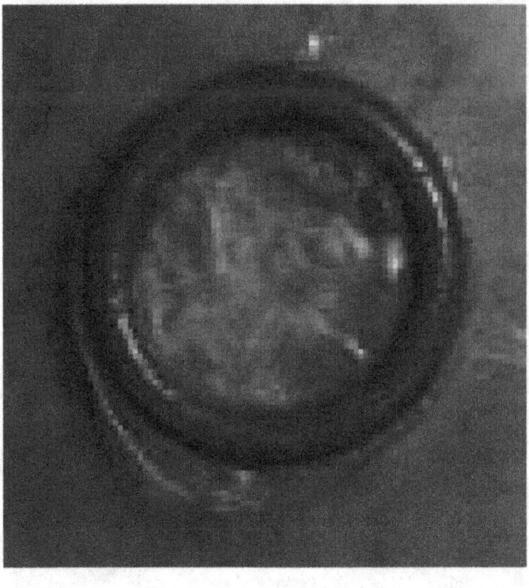

modes, $l = 4, 3$, ending up into a fast oscillating dipole which eventually freezes with water vapors. In the case of a rigid core and very shallow layer ($R = 15$ mm, $h = 4$ mm), the modes are more stable. We notice a dynamical regime of transitions between modes because of the loss of mass. The change of modes is accompanied by change of direction of rotation, in between stable modes, one can notice a sort of turbulent regime. Also, the localized waves tend to breakup or decay after a couple of rotations. In addition to the free surface modes, or the modes around a rigid core, internal modes inside of a hollow ring can be measured. These waves have slower modes $l = 4$–8, and the $l = 6$ mode is very stable. Usually, the waves are powered by the bubble from evaporation, and sometimes the surface waves travel together with a trapped bubble in their area. The most interesting patterns are presented in Figs. 12.12, 12.13 showing rotating triangles, squares and pentagons. Nonlinear waves, Fig. 12.14, that fit pretty well a cnoidal wave pattern can be noticed on the external region, and occasionally one can notice the occurrence of a soliton, like in Fig. 12.15.

Chapter 13
Nonlinear Surface Waves in Three Dimensions

The study of shape oscillations of drops has a wide variety of applications at different space and time scales. At microscopic scales this includes the liquid drop models of nuclei, especially heavy nuclei, super- and hyperdeformed nuclei, nuclear breakup and fission, where the surface energy plays an important role. They also play a role in the modeling of atomic clusters and clouds of electrons in high magnetic fields. Zooming out from the Angström scale, the study of drops is important in the study of motion and swimming of motile cells, and cellular division in biological systems. Bubble sonoluminescence represents a recent application of bubbles and droplets formed inside the bubbles [326, 328]. At lab scale there is a huge spectrum of applications, including container-less liquid processing in space, rheological and surfactant theory, pharmaceutical industry, mixture of fluids in droplet form, behavior of long wavelet jets emitted from noncircular orifices, coalescence of liquid drops [219], and surface oscillations of liquid drops, bubbles, and shells in combination with surfactants. At larger scales drops are important for calculations of the radar cross-section of rain clouds, modeling of impacts between stellar objects and neutron star tides, important for the gravitational waves emitted by such oscillations.

Nonlinear terms from Navier–Stokes equations and from the boundary conditions usually introduce couplings between modes of oscillations, even between modes of different nature, like radial and shear ones. Nonlinear terms coming from the geometry of curved, eventually closed, surfaces provide additional coupling. One general nonlinear phenomenon introduced by such couplings is the interrelation between kinematics and shape. For example, in the case of one-dimensional solitary waves, the dependence of the amplitude and the width on the group velocity is well known. Similarly, in the case of drops, bubbles, and shells, couplings induce interesting behavior.

For example, it is known that in the linear case [167], the core of the drop tends to have potential flow, while next to the free boundary the flow is rather vortical. This artificial fact is generated by the linear approximations. It can be explained by the existence of surface singularities of the spherical Bessel functions when the

A. Ludu, *Nonlinear Waves and Solitons on Contours and Closed Surfaces*, Springer Series in Synergetics, DOI 10.1007/978-3-642-22895-7_13,

damping constant becomes imaginary and the oscillation modes become weakly dissipative and even conservative. When we introduce the nonlinear terms in the model, because of the coupling between the vorticity and shape, the singularity is removed and the vorticity field is controlled by the local shape, i.e., surface vorticity is enhanced in regions with large curvature values.

Another example of coupling effects is the physical difference in the behavior of a flat fluid surface and a curved one. In the plane case the rate of local expansion of the surface Σ is given by the surface divergence of the tangent velocity, i.e., $\nabla_\Sigma \cdot \boldsymbol{V}_\parallel$, and usually this term is involved in nonlinear terms in the Navier–Stokes equations. So, surface elements can have radial displacement without producing local expansion. That implies radial oscillations to involve no tangent motion at the surface, and so shear deformation can be absent. This further implies that if we investigate interfaces with very high coefficient of surface dilatation (elastic and/or viscous), $D \to \infty$, but small and finite coefficient of surface shear S (also elastic and/or viscous), the motion is not frozen, and still small oscillations can occur. This situation happens even if both coefficients have very large values, like in the case of cellular membranes which are practically inextensible. So, even in the limit $S, D \to \infty$ the plane surface can radially oscillate.

However, in the case of curved interfaces, the rate of local expansion contains two terms, like, for example, in the spherical case

$$D\left(\nabla_\Sigma \cdot \boldsymbol{V}_\parallel + \frac{2}{R} V_r\right).$$

Consequently, a dilatation-rigid curved surface (high values for D) will have the rate of local expansion zero only if it either does not oscillate at all, or it performs radial oscillations, but these are coupled with tangent motion. So, for an inextensible curved surfaces, radial oscillations should be accompanied by sliding at the interface. This tangent sliding involves shear deformation, which is controlled by S, the coefficient of surface shear. Consequently, it is impossible to have large values for S, because such values will forbid tangent motions. In conclusion, in curved fluid interfaces it is impossible to have simultaneously high values for the coefficient of surface dilatation and the coefficient of surface shear. So, in the case of closed surfaces it is impossible to have motion under both shear (S) and expansion (D) resistance.

Another particularity of free surface oscillations and waves for (nonlinear) viscous drops is given by the fundamental parabolic nature of the equations (Navier–Stokes) [263]. That is, given the distribution of the energy balance between the vorticity terms and the velocity terms, the dynamics of the system is history dependent, hence can be correctly described only through integrodifferential equations. Indeed, if in the beginning the vorticity is zero, dissipation arises only through the velocity, i.e., through a term of the form

$$\iint_\Sigma (\boldsymbol{V} \cdot \nabla)\boldsymbol{V} \cdot \boldsymbol{N} dA.$$

As the motion develops, vorticity is created at the free surface and dissipates toward the inside of the drop, introducing another channel of dissipation through the term

$$\iiint_D \omega \cdot \omega dV.$$

Consequently, the energy dissipation depends on the vorticity, hence on the past history of the flow, and so the mathematical description should be integrodifferential.

There are many distinct features between the nonlinear drops and the linear ones: mode coupling, large amplitude oscillations, frequency shift, cubic or higher resonances, quasiperiodic motions, surface solitary waves, etc. In the following we present the Navier–Stokes normal mode approach, and the Lagrangian approach, for both linear and nonlinear three-dimensional drops with axial symmetry. This constraint does not introduce too much loss of generality concerning nonlinear effects, and it does not change the final theoretical expressions for frequencies. A short history of the models for linear toward nonlinear drops is presented in the introduction of Basaran [13].

13.1 Oscillations of Inviscid Drops: The Linear Model

In this section we study linear surface oscillations of an isolated (no gravitation, inert atmosphere) three-dimensional liquid drop with surface tension liquid surface under three simplifying hypotheses: the flow is inviscid (viscosity coefficients are zero), incompressible (density ρ_0 = constant), and irrotational ($\nabla \times V = 0$). These conditions, together with the Euler equation, form a system of seven partial differential equations (PDEs) for seven unknown functions of three variables: velocity V, velocity potential Φ, density ρ_0, pressure P, and the shape function $\xi(\theta, \varphi, t)$ of the free surface of the drop $r(\theta, \varphi, t) = R_0 + \xi(\theta, \varphi, t)$. To have a unique solution we add to this system boundary and initial conditions. The expression for the surface tension occurs for the first time within the boundary conditions. For drop without core and for bubbles the boundary condition is taken only on one closed surface, the free surface of the fluid. For drops with core or liquid shells we take into account two or more surfaces in the boundary conditions. In the following we use the spherical coordinates (r, θ, φ), so for example $V = (V_r, V_\theta, V_\varphi)$.

The flow inside the drop is potential and incompressible, and so the velocity potential $V = \nabla\Phi$ fulfills the Laplace equation $\Delta\Phi = 0$. In the absence of any external force field, Euler equation reduces to the Bernoulli equation. In the linear approximation Bernoulli equation has the form

$$\frac{\partial \Phi}{\partial t} = -\frac{1}{2}(\nabla\Phi)^2 - \frac{1}{\rho_0} P \simeq -\frac{1}{\rho_0} P, \tag{13.1}$$

where ρ_0 is the constant density. In spherical coordinates and in the same linear approximation, the kinematic condition for the free surface (9.30) reads in spherical coordinates

$$\left.\frac{\partial \Phi}{\partial r}\right|_S = V_r|_S = \frac{\partial \xi}{\partial t} + \frac{\partial \Phi}{\partial \theta} V_\theta + \frac{\partial \xi}{\partial \varphi} V_\varphi \simeq \frac{\partial \xi}{\partial t}, \tag{13.2}$$

where the free surface of the drop was defined as $r(\theta, \varphi) = R + \xi(\theta, \varphi)$ and R is the equilibrium radius of the stationary drop. The general approach of solving such linear problems is to expand the potential in a convenient series of orthogonal functions (e.g., (13.5) for spherical symmetry), then solve Laplace equation for the potential in the corresponding boundary conditions, and then plug the coefficients of the potential in the free surface equation (13.2) to find the shape ξ.

The surface pressure is

$$P_S = P_0 + \sigma(\kappa_1 + \kappa_2) = P_0 + 2\sigma H,$$

according to (10.35), where P_0 is the constant pressure outside the drop. From Sect. 10.4.6 we have the expression of the mean curvature H in spherical coordinates. According to the hierarchy of orders of smallness in ξ/R performed there, we will use for our linear case orders up to $\mathcal{O}(2)$ in (10.60)

$$2H = \frac{\rho}{R^2} + \frac{1}{2R^2}\left(\rho_{\theta\theta} + \cot\theta \rho_\theta + \frac{\rho_{\varphi\varphi}}{\sin^2\theta}\right). \tag{13.3}$$

The order zero term $1/R^2$ in the mean curvature was absorbed in P_0 and the sign of the mean curvature is chosen according to the convention that a positive surface pressure is directed toward inside the drop. If we differentiate with respect to time (13.2) and substitute ξ with Φ, we can write

$$\left.\frac{\partial^2 \Phi}{\partial t^2}\right|_S \simeq \frac{\sigma}{\rho_0 R^2}\left(2\frac{\partial \Phi}{\partial r} + \frac{\partial}{\partial r}\Delta_\Omega \Phi\right)_S. \tag{13.4}$$

Since the potential is a harmonic function, it can be written as a series of spherical harmonics Y_{lm} , and from the uniqueness warranted by the Cauchy condition through (13.2) we can determine its time-dependent coefficients.

$$\Phi(r, \theta, \varphi, t) = \sum_{l \geq 0, |m| \geq l} f_{lm}(r) Y_{lm}(\theta, \varphi) \sin(\omega_{lm} t + \varphi_{lm}). \tag{13.5}$$

From the Laplace equation we have

$$f_{lm} = \text{const.} r^l + \frac{\text{const.}}{r^{l+1}}, \tag{13.6}$$

and we introduce this form of potential in (13.4). By using $\Delta_\Omega Y_{lm} = -l(l+1) y_{lm}$, after identification of the coefficients of the spherical harmonics, and in the linear approximation $(R+\xi)^l \to R^l$, we obtain the normal frequencies all linear modes of this type of oscillation

$$\omega_{lm}^2 = \omega_l^2 = \frac{\sigma(l+2)(l+1)}{R^3\rho_0}\frac{l A_{lm} R^{2l+1} - B_{lm}(l+1)}{A_{lm} R^{2l+1} + B_{lm}}. \tag{13.7}$$

In the first case, for drops and bubbles, we have no core, and just one free surface. The fluid domain contains the origin of the coordinate axes, and to have differentiable solutions, we have to cancel the B_{lm} coefficients. The resulting normal modes for simple drops are

$$\omega_l^2 = \frac{\sigma}{R^3\rho_0} l(l+2)(l-1). \tag{13.8}$$

The modes $l = 0, 1$ are eliminated by the center of mass position conservation and by the incompressibility hypothesis, respectively.

13.1.1 Drop Immersed in Another Fluid

The second case we investigate is the case of a liquid drop of density ρ_{int} surrounded by infinite liquid of density ρ_{ext}, both in inviscid potential flow. We define the velocity potential in two distinct regions, inside ($r < \xi$) and outside ($r > \xi$) the drop, and we match these two functions according to physical continuity conditions (13.10) and (13.12). The potential in each zone is harmonic, and according to (13.5) and (13.6), we can write the two expressions by eliminating those terms that become singular in each zone

$$\begin{aligned} \Phi_{int} &= \textstyle\sum_{l,m} A_{lm} r^l Y_{lm} \cos(\omega_l t + \varphi_{lm}) \\ \Phi_{ext} &= \textstyle\sum_{l,m} \frac{B_{lm}}{r^{l+1}} Y_{lm} \cos(\omega_l t + \varphi_{lm}). \end{aligned} \tag{13.9}$$

In the case of two fluids, the linearized free surface condition (13.2) becomes a continuity condition for the radial component of the velocity v_r

$$\left.\frac{\partial \Phi_{int}}{\partial r}\right|_S = \left.\frac{\partial \Phi_{ext}}{\partial r}\right|_{r=\xi} = \left.\frac{\partial \xi}{\partial t}\right|_{r=\xi}, \tag{13.10}$$

where the condition S for surface is again realized by the relation $r = \xi$. From the first part of (13.10) we have

$$B_{lm} = -\frac{l A_{lm} R^{2l+1}}{l+1}. \tag{13.11}$$

The second matching condition is given by equating the pressures P at the free surface. We have

$$\frac{\partial \Phi_{int}}{\partial t} = -\frac{1}{\rho_{int}}(P + \sigma(\kappa_1 + \kappa_2))$$
$$\frac{\partial \Phi_{ext}}{\partial t} = -\frac{1}{\rho_{ext}}(P - \sigma(\kappa_1 + \kappa_2)). \tag{13.12}$$

and we have

$$-\rho_{int}\frac{\partial \Phi_{int}}{\partial t} + -\rho_{ext}\frac{\partial \Phi_{ext}}{\partial t} = 2\sigma(\kappa_1 + \kappa_2) = \sigma\left(-\frac{2\xi}{R^2} - \frac{1}{R^2}\Delta_{\Omega}\xi\right). \tag{13.13}$$

From (13.10) and (13.13) we obtain, by equating the terms with the same l, m, the expression of the normal modes frequencies of the two fluids case

$$\omega_l^2 = \frac{\sigma l(l+1)(l+2)(l-1)}{[\rho_{int}(l+1) + \rho_{ext} l]R^3}. \tag{13.14}$$

This expression was obtained first time by Lamb in Article 275 of [167]. We notice the absence of the first two modes ($l = 0, 1$) because of incompressibility and momentum conservation conditions, respectively.

In the limit $\rho' - >\rightarrow 0$, (13.14) approaches the ideal case of (13.8) for oscillations of a liquid drop in vacuum (linearized results). In Fig. 13.1, we present the variation of the frequencies of normal oscillations of such a drop in the linearized approach, for nine values of l, vs. the ratio of the density of medium over the density of the drop. Around zero we have the frequencies of free oscillations of

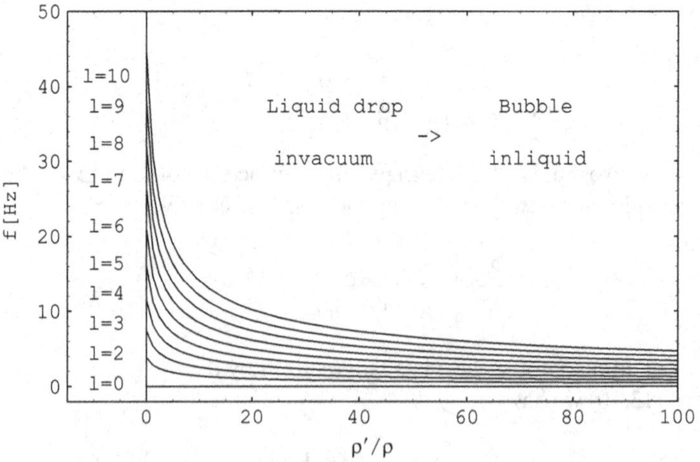

Fig. 13.1 Frequency of normal modes for a drop of density ρ submerged in a fluid of density ρ', in the linear approximation, vs. the ratio of the densities ρ'/ρ for $R = 1$ cm

drops in vacuum, while moving toward the right we increase the ambient density. For $\rho'/\rho = 10^3$ we have almost the oscillations of an air bubble in water

$$\omega^2_{l,bubble} = \frac{\sigma(l+1)(l+2)(l-1)}{\rho_{ext} R^3}. \tag{13.15}$$

We notice that in principle there is a "zero" radial mode ($l = 0$) for bubbles.

13.1.2 Drop with Rigid Core

The third case is a liquid drop containing a spherical rigid core of radius $a < R$. Such experimental configurations are easy to obtain for two- dimensional drops, but rather complicated for three-dimensional drops. However, such a model helps understanding the dynamics of heavy nuclei, where the external nuclear shells cover a stable (or even a double) magic number nucleus. They can also be used in motile cell investigations, where the cell nucleus can play the role of the rigid core. Also, in some neutron star models, the main dynamic part of the system is a deformable crust oscillating around a rigid core. For a rigid core the second boundary condition is the cancellation of the normal velocity at the core surface, $(\partial\Phi/\partial r)_{r=a} = v_r|_{r=a} = 0$. Again from the continuity conditions we have

$$B_{lm} = \frac{la^{2l+1}}{l+1} A_{lm}, \tag{13.16}$$

and consequently, the normal modes frequencies of drop plus rigid core in the linear approximation read

$$\omega_l^2 = \frac{\sigma l(l-1)(l+2)\left[1 - (\frac{a}{R})^{2l+1}\right]}{\rho_0 R^3 \left[1 + \frac{l}{l+1}(\frac{a}{R})^{2l+1}\right]}. \tag{13.17}$$

In Fig. 13.2, we present the frequency of the normal linear modes for a liquid drop with rigid core. The frequencies are practically equidistant when the core is small, and approach the modes without core, but tend to decrease to smaller values when the radius of the core increases. In the limit $a- > \to 0$, (13.17) approaches the ideal case of (13.8) for oscillations of a liquid drop in vacuum (linearized results).

In the following we calculate the velocity and pressure field within the oscillating drop. We begin with the coreless drop, and so we put $a = 0$, $B = 0$ in (13.16). From (13.5) we have

$$\Phi(r,\theta,\varphi,t) = \sum_{l\geq 0,|m|\geq l} A_{lm} r^l Y_{l,m}(\theta,\varphi) \sin\left(\sqrt{\frac{\sigma l(l+2)(l-1)}{R^3 \rho_0}}t + \varphi_{lm}\right). \tag{13.18}$$

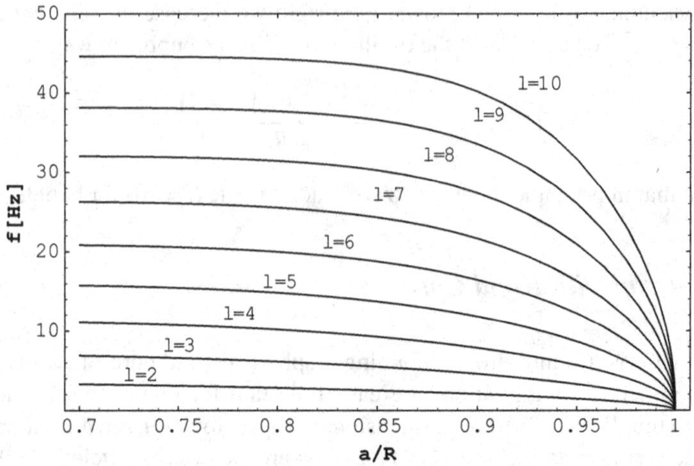

Fig. 13.2 Frequency of normal modes for a water drop with rigid core, in the linear approximation, vs. the ratio between the core radius and the drop radius, a/R. $R = 1$ cm

The velocity is given by

$$v = \nabla\Phi = \left(\frac{\partial\Phi}{\partial r}, \frac{1}{r}\frac{\partial\Phi}{\partial\theta}, \frac{1}{r\sin\theta}\frac{\partial\Phi}{\partial\varphi}\right). \tag{13.19}$$

From (13.2) in the linear approximation $\partial\xi/\partial t \simeq (\partial\Phi/\partial r)_\Sigma$, by integrating once with respect to time, we obtain the expression of the shape in terms of the A_{lm} coefficients of the potential

$$\xi(\theta,\varphi,t) = -\sum_{l\geq 0,|m|\geq l} \frac{l A_{lm} R^{l-1}}{\omega_l} Y_{l,m}(\theta,\varphi)\cos\left(\sqrt{\frac{\sigma l(l+2)(l-1)}{R^3\rho_0}}t + \varphi_{lm}\right). \tag{13.20}$$

We consider the shape known at the initial moment of time, and given by

$$\xi(\theta,\varphi)|_{t=0} = \sum_{l\geq 0,|m|\geq l} C_{lm} Y_{l,m}(\theta,\varphi), \tag{13.21}$$

where we choose $\varphi_{lm} = 0$, and C_{lm} are given. By identifying (13.20) at $t = 0$ with (13.21) we obtain

$$A_{lm} = -\frac{\omega_l C_{lm}}{l R^{l-1}}, \tag{13.22}$$

where ω_l is given by the core free frequency's formula (13.8). By introducing (13.22) in (13.20) and in (13.18) and (13.19), we determined the shape and velocity field inside the drop at any moment of time. We mention a technical calculation detail needed to adjust the form of the coefficients, since the spherical harmonics are complex functions and the shape and velocity must be real functions. Instead of (13.20) we use

$$\xi = \sum_{l\geq 2}\left(C_{l0}Y_{l,0}\cos(\omega_l t) + \sum_{m=1}^{l}(B_{lm}\cos m\varphi + D_{lm}\sin m\varphi)\theta_{lm}\cos(\omega_l t)\right),$$

where θ_{lm} are the Legendre generalized functions ($Y_{lm} = \theta_{lm}(\theta)e^{im\varphi}$) and the new coefficients are related to the old ones by

$$C_{l,\pm m} = \frac{(\pm 1)^m}{2}(B_{lm} \pm iD_{lm}),\ m > 2.$$

Then, the velocity potential may be written as

$$\Phi = -\sum_{l\geq 2}\frac{\omega_l C_{l0}}{lR^{l-1}}r^l Y_{l0}\sin(\omega_l t) - \sum_{l\geq 2}\frac{\omega_l r^l}{lR^{l-1}}\sum_{m=1}^{l}\sin(\omega_l t)(B_{lm}\cos(m\varphi)$$
$$+ D_{lm}\sin(m\varphi))\theta_{lm}.$$

In Fig. 13.3, we present some frames during the oscillation of such a liquid drop, starting from a given octupole shape, as an application of (13.20). In Fig. 13.4, we

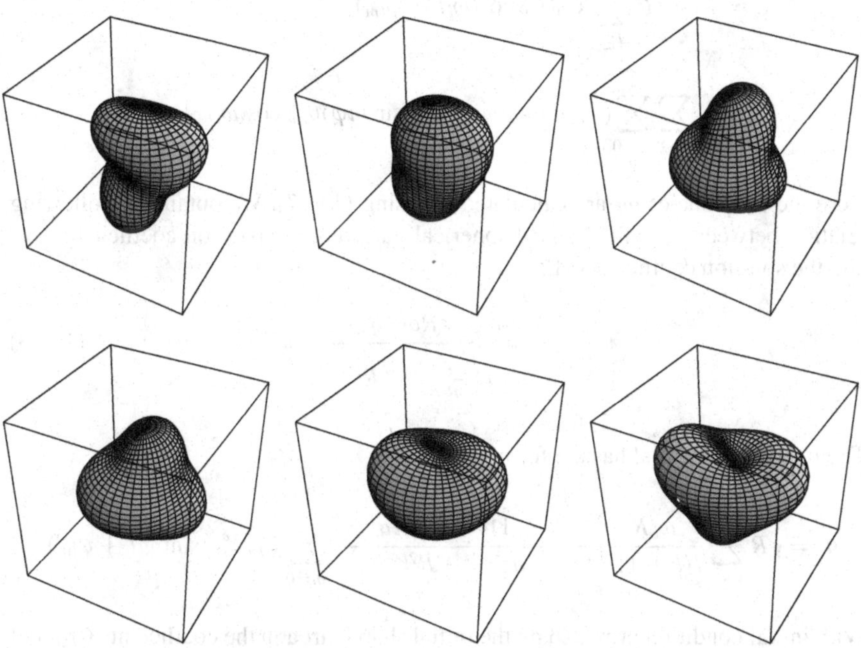

Fig. 13.3 Oscillations of an incompressible irrotational liquid coreless drop calculated from a given initial octupole shape

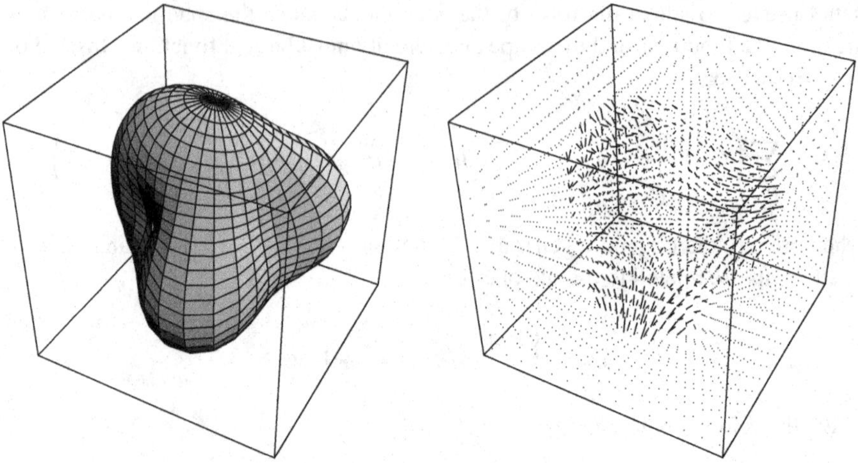

Fig. 13.4 Incompressible irrotational liquid drop: shape and velocity field

present the shape and the velocity field at a certain moment of time, with velocity calculated through (13.18) and (13.19).

In the case of liquid drops with rigid core we use for the shape a similar equation as (13.20), except we need to make sure that $|\xi| > a$ at all times

$$\xi = R + \epsilon R(\sum_{l\geq 2} C_{l0} Y_{l0} \cos(\omega_l t + \varphi_{lm})$$

$$+ \sum_{l\geq 2} \sum_{m=0}^{l} (A_{lm} \cos m\varphi + B_{lm} \sin m\varphi) \theta_{lm} \cos(\omega_l t + \varphi_{lm})).$$

Here the frequencies ω_l are calculated by using (13.17). We obtain the following relation between the initial shape spherical harmonics expansion coefficients C_{lm} and the solution coefficients A_{lm}

$$A_{lm} = \frac{\epsilon R \omega_l C_{lm}}{l\left(\frac{a^{2l+1}}{R^{l+2}} - R^{l-1}\right)}. \tag{13.23}$$

The resulting potential has the form

$$\Phi = \epsilon R \sum_{l\geq 2} \frac{\omega_l R^{l+2}}{l(l+1)r^{l+1}} \frac{(l+1)r^{2l+1} + l a^{2l+1}}{a^{2l+1} - R^{2l+1}} \sum_{m=0}^{l} C_{lm} Y_{lm} \sin(\omega_l t + \varphi_{lm}).$$

With initial condition provided by the initial shape through the coefficients C_{lm} and φ_{lm}, and by using (13.17) and (13.23) we obtain the velocity field and the shape at any moment of time. In Figs. 13.5 and 13.6, we present several snapshots of exact calculation of the shape of the drop linear oscillations plus core.

In Figs. 13.7 and 13.8, we present cross-sections in oscillating drops for two different core radii. One can notice the effect of the linearization of the free surface (13.2): oscillations happen only along the normal direction to the surface.

In Figs. 13.9 and 13.10, we present cross-sections and velocity field of oscillating drops for different core radii and different initial shapes, in irrotational incompressible flow.

Now it is easy to calculate the pressure distribution in the drop, by using (13.1)

$$P(r, \theta, \varphi, t) = -\rho \frac{\partial \Phi}{\partial t}.$$

In Figs. 13.11 and 13.12, we present the pressure field for two oscillating drops from three different orthogonal cross-sections. We notice that the pressure contour lines are always perpendicular on the boundaries. Close to the regions of the free surface where the shape is convex, we remark that the higher pressure contour lines extend more toward inside. This behavior can trigger different types of instabilities, or formation of inner jets of higher pressure like in the case of bubble sonoluminescence, for example.

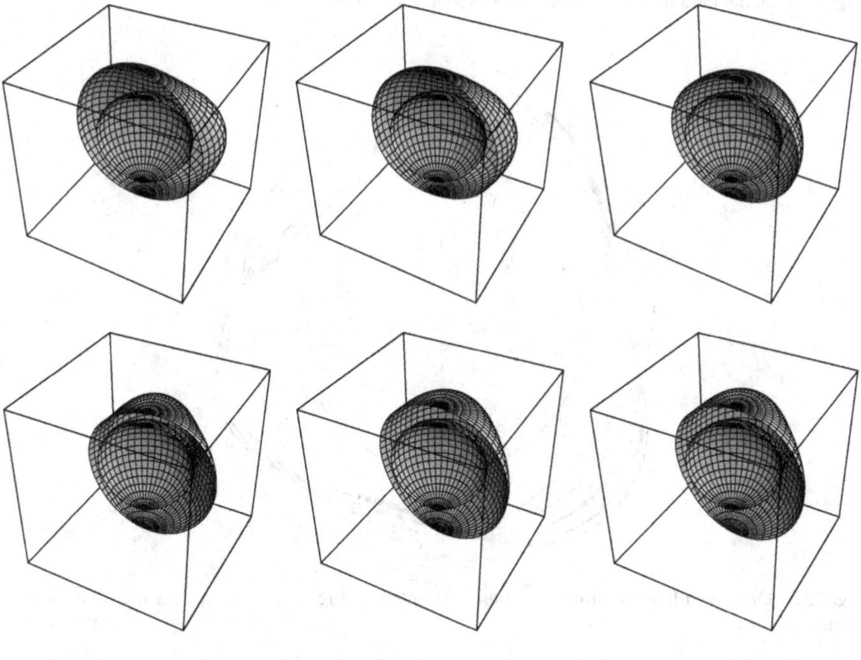

Fig. 13.5 Linear oscillations of a water drop of equilibrium radius $R = 10$ mm with a rigid core of radius $r = 7$ mm taken at intervals of 0.25 s. The smallness parameter was chosen $\epsilon = 0.12$, and we have $\rho = 10^3$ kg m^{-3} and $\sigma = 0.0728$ N m^{-1}. For these parameters we have $\omega_2 = 0.145$ s, $\omega_3 = 0.301$ s

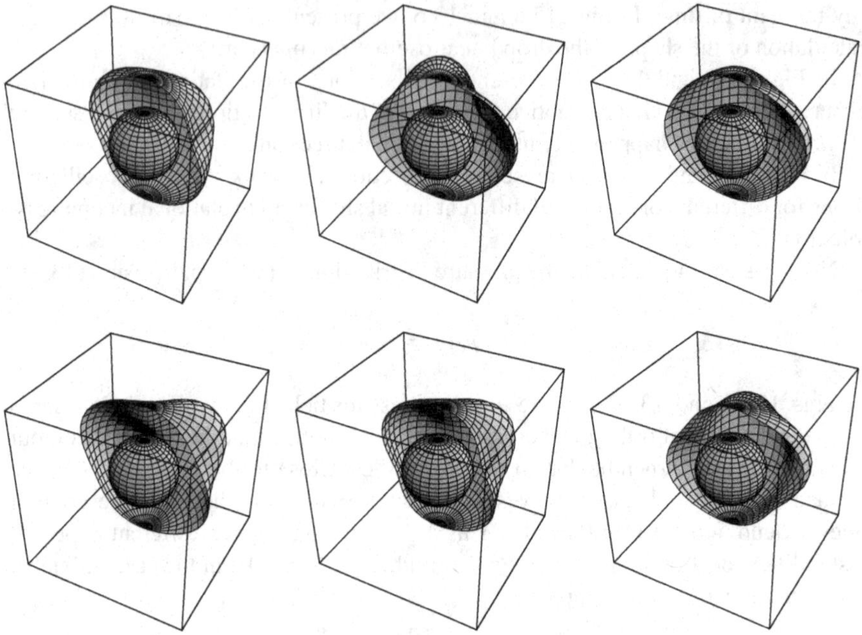

Fig. 13.6 Same parameters as in Fig. 13.5 except $r = 4\,\text{mm}$

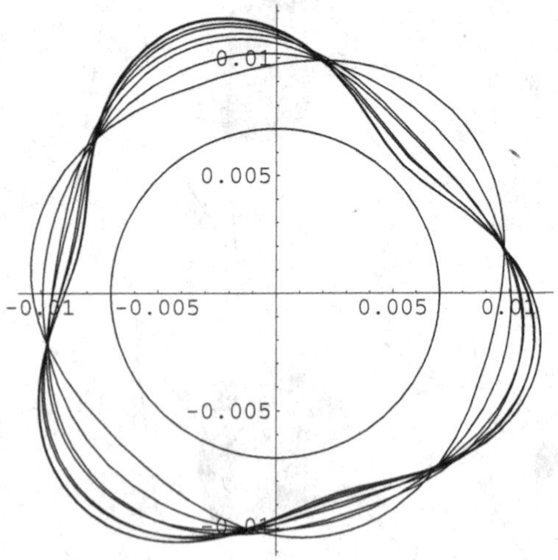

Fig. 13.7 Drop oscillations similar to those presented in Fig. 13.5, shown in a meridian cross-section

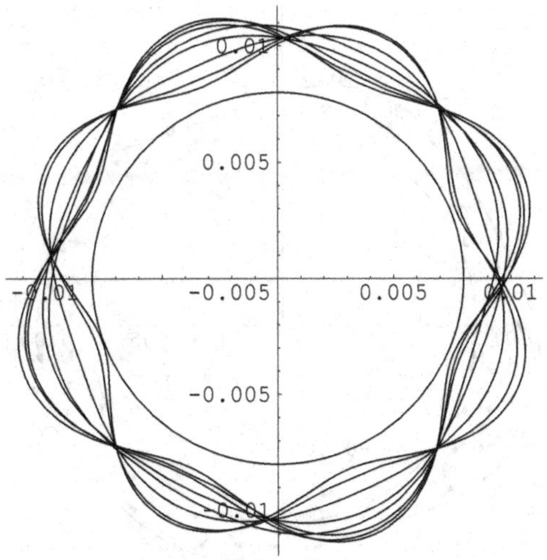

Fig. 13.8 Same as Fig. 13.7, but for a larger core

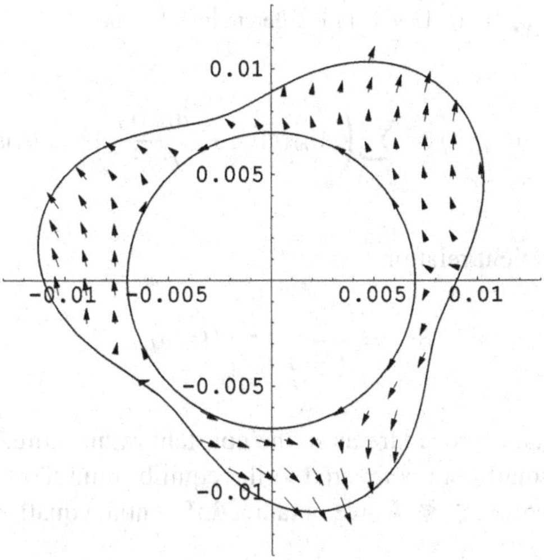

Fig. 13.9 Irrotational incompressible flow for a liquid drop with core

13.1.3 Moving Core

Another interesting situation occurs if we impose a certain type of motion to the core, $a = f(t)$. The inner boundary condition becomes $V_r(r = f(t)) = 0$. By plugging this boundary condition in a general potential of the form

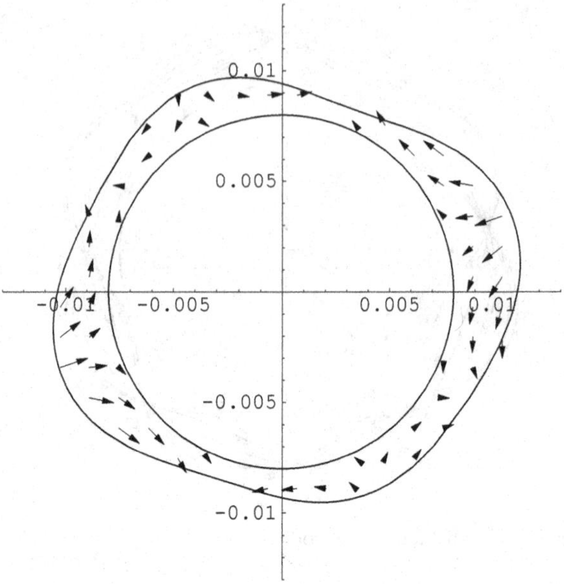

Fig. 13.10 Same drop as in Fig. 13.9, but for different initial shape

$$\Phi(r,\theta,\varphi,t) = \sum_{l,m}\left(A_{lm}(t)r^l + \frac{B_{lm}(t)}{r^{l+1}}\right)Y_{lm}(\theta,\varphi) \tag{13.24}$$

we obtain the coefficient relation

$$B_{lm} = \frac{l}{l+1} f^{2l+1}(t) A_{lm}. \tag{13.25}$$

By following the same procedure as in the constant radius core, and by using the approximation of small core compared to the equilibrium radius, $f(t) \ll R$, and the linear approximation $\xi \ll R$, we obtain a differential equation in time for each coefficient $A_{lm}(t)$

$$A''_{lm}R^l + \frac{2l(2l+1)}{(l+1)R^{l+1}}A'_{lm}f^{2l}f' + \frac{l(2l+1)}{(l+1)R^{l+1}}A_{lm}(f^{2l}f')'$$
$$= \frac{\sigma(l+2)(l-1)}{\rho}R^{l-3}A_{lm}. \tag{13.26}$$

This ODE is difficult to be solved exactly in the general case. For an exponential core motion, like, for example, the expansion of gas bubbles in a fluid $f(t) = be^{ct}$, with b, c constants, we have

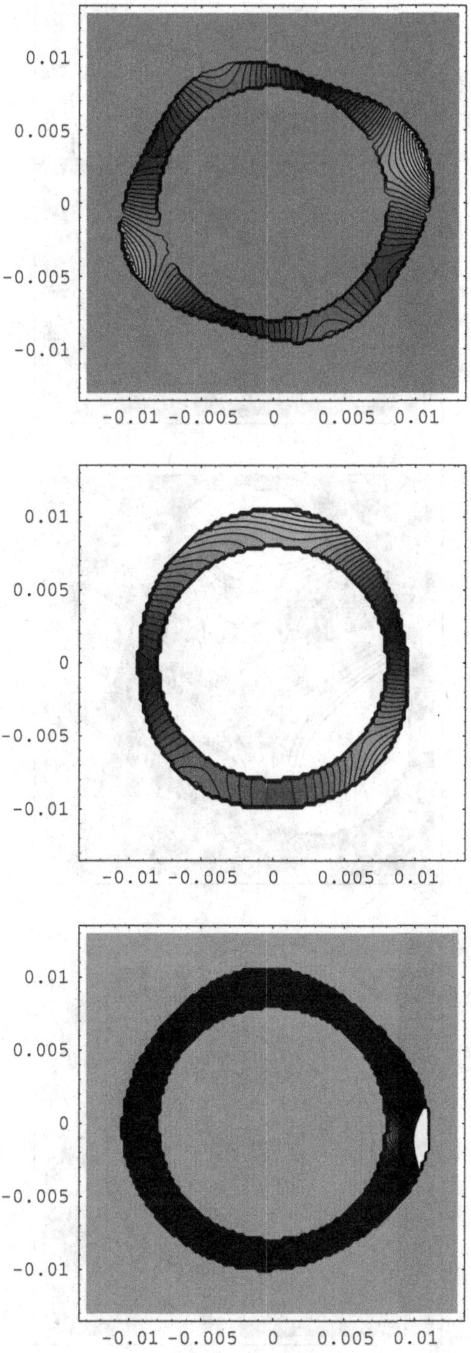

Fig. 13.11 Pressure contour lines for an incompressible irrotational flow in a liquid drop with rigid core, taken simultaneously in three orthogonal cross-sections

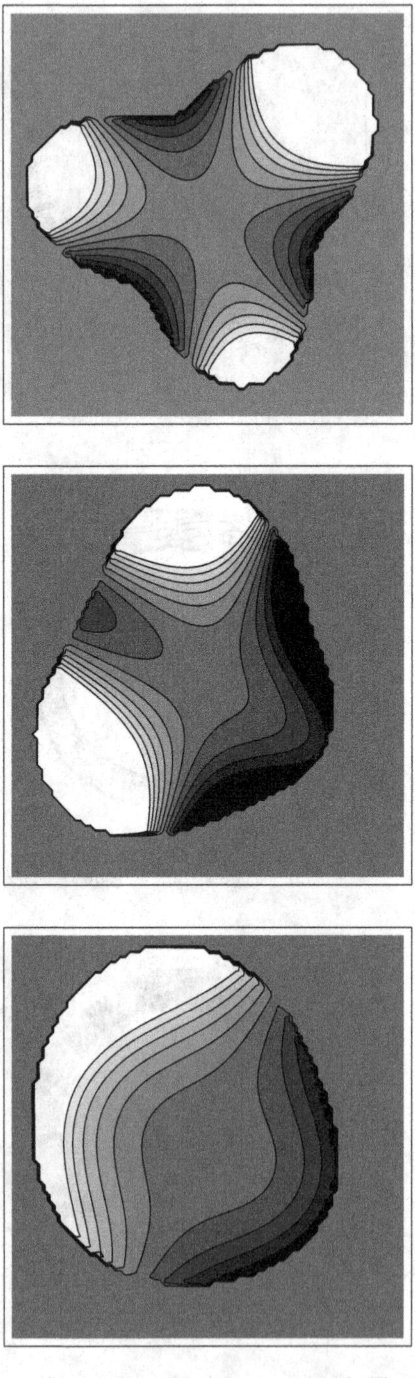

Fig. 13.12 Pressure contour lines similar with those presented in Fig. 13.11 except for different initial data of the flow

$$A^{\exp}_{lm}(t) = A_{lm,0} \exp\left(\frac{l b^{2l+1}}{(l+1)R^{2l+1}} e^{c(2l+1)t} \right)$$
$$\times I_n \left(\frac{\sqrt{(l+2)(l-1)\sigma}}{c(2l+1)\sqrt{\rho R^3}}, \frac{l b^{2l+1}}{(l+1)R^{2l+1}} e^{(2l+1)ct} \right), \tag{13.27}$$

where $I_n(\alpha, \beta)$ is the modified Bessel function of the first kind. This expression for the coefficients is plugged in (13.25), and then back in the potential (13.24) to obtain the flow. Such a solution can model situations like submarine explosions, see for example in Thomson [315, Sect. 16.21]. In this section Milne–Thomson supposes that a spherical cavity containing gas begins to expand rapidly in surrounding unbounded liquid, such that the gravity can be neglected. The potential can be approximated with the first singular term in the series (13.24), namely $\Phi \simeq 1/r$, and introduced in the Euler equation (13.1) it gives

$$\frac{P}{\rho} + \frac{1}{2}\left(\frac{f^2 f'}{r^2} \right)^2 - \frac{f^2 f'' + 2 f f'^2}{r} = C(t).$$

The arbitrary function of time $C(t)$ can be taken zero if the pressure is negligibly far away from the free surface.

13.1.4 Drop Volume

None of the above calculations guaranties the drop volume conservation. A correct treatment would request writing the Lagrangian of the drop and imposing volume conservation as a Lagrange multiplier. Obviously, the volume obtained from ξ is not conserved, but we can make estimations about the range of error in time for the volume conservation. In general we have

$$V = \int_0^{2\pi} d\varphi \int_0^{\pi} \sin\theta d\theta \int_a^{R+\epsilon\xi(\theta\phi t)} r^2 dr. \tag{13.28}$$

Without too much loss of generalization we expand (13.28) in the situation without core and we have

$$V = V_0 + \epsilon R^3 \int_0^{2\pi} d\varphi \int_0^{\pi} \xi \sin\theta d\theta + \epsilon^2 R^3 \int_0^{2\pi} d\varphi \int_0^{\pi} \xi^2 \sin\theta d\theta$$
$$+ \frac{\epsilon^3 R^3}{3} \int_0^{2\pi} d\varphi \int_0^{\pi} \xi^3 \sin\theta d\theta, \tag{13.29}$$

with

$$V_0 = \frac{4\pi R^3}{3}, \quad \xi = \sum_{l \geq 2, |m| \leq l} C_{lm} Y_{lm} \cos(\omega_l t + \chi_l).$$

The first term on the RHS of (13.29) is zero because all terms inside it have multipoles larger than 2, which are orthogonal on $\sin\theta$. The order 2 in ϵ term contributes only with those products of spherical harmonics $Y_{lm}Y_{l'm'}$ that fulfill the conditions $l = l'$ and $m = -m'$. The order ϵ^3 contains even less nonzero terms, for example, only those terms fulfilling $m_1 + m_2 + m_3 = 0$. These triple products of spherical harmonics are determined by the Wigner $3j$-symbols

$$\int_0^{2\pi} d\varphi \int_0^{\pi} Y_{l_1m_1}Y_{l_2m_2}Y_{l_3m_3} \sin\theta d\theta = \sqrt{\frac{(2l_1+1)(2l_2+2)(2l_3+1)}{4\pi}}$$

$$\times \begin{pmatrix} l_1\, l_2\, l_3 \\ 0\ 0\ 0 \end{pmatrix} \begin{pmatrix} l_1 & l_2 & l_3 \\ m_1 & m_2 & m_3 \end{pmatrix}. \qquad (13.30)$$

In general, the higher the order l, the "less" nonzero terms we have in the summations, compared to the total "number" of terms. We mention this in the sense of the measure theory applied to the "number" of terms in the series expansion. Consequently, the higher corrections are smaller and smaller on the top of the decrease produced by higher powers of ϵ. In Fig. 13.13, we show the ratio V/V_0 for a $l = 4$ mode vs. time. To have an estimation of the error we provide an example in the quadratic order in ϵ. We plot the relative change in volume $|V - V_0|/V_0$ vs. the maximum distance to the center of the free fluid surface, in a certain amount of time, over R. To estimate this we consider the shape function known, and given in terms of some arbitrary coefficients C_{lm} of ξ, namely

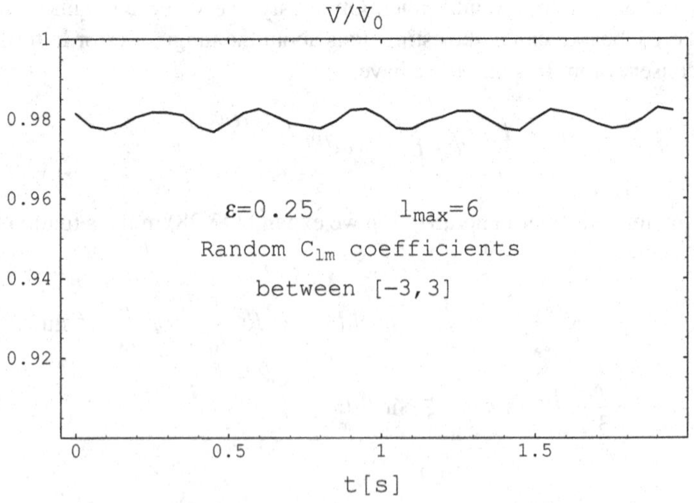

Fig. 13.13 Oscillations in the volume of the drop compared to the initial one in time

$$\frac{|r_{max} - R|}{R} \leq \frac{1}{R} \sum_{l\geq 2, |m|\leq l} |C_{lm}||Y_{lm}|_{max}$$

$$\leq \frac{\sqrt{\pi}}{R} \sum_{l\geq 2, |m|\leq l} |C_{lm}| \frac{2^m}{\Sigma\left(\frac{l-m}{2}+1\right)\Sigma\left(\frac{1-l-m}{2}\right)}, \tag{13.31}$$

where we use the upper bound of the maximum value taken by a spherical function. The quadratic term in ϵ normalized by the initial volume has the form

$$\left|\frac{V(O(\epsilon^2))}{V_0}\right| \leq V_0 + \frac{3\epsilon^2}{4\pi} \sum_{l\geq 2, |m|\leq l} |C_{lm}|^2 \frac{4\pi}{2l+1} \frac{(l+m)!}{(l-m)!}, \tag{13.32}$$

where we use the well-known norms of the spherical harmonics. The ratio between the two numerical series in (13.31) and (13.32) provides a numerical criterion about the errors in volume estimations compared to the deformations.

13.2 Oscillations of Viscous Drops: The Linear Model

In this section we study oscillations of three-dimensional liquid drops with surface tension and viscosity, embedded into a viscous fluid. Rayleigh described for the first time the small oscillations of a drop of liquid about the spherical form oscillation in air in Rayleigh [268]. In Article 275 in [167] Lamb slightly generalized the question by supposing that the liquid globule, of density ρ, is surrounded by an infinite mass of other liquid of density ρ'. Recent treatment of the same problem can be found in monographes like [171, Sect. 61], [50, Chap. VI], [147, 332], or in articles like [35, 61, 156, 157, 223, 263–265, 269, 316, 321, 346]. In all these approaches one takes the center of the stationary incompressible inviscid drop of initial (before oscillations) radius R as the origin of a spherical coordinate system, and describes the shape of the drop by the function $r(\theta, \varphi, t) = R(1 + f(\theta, \varphi, t))$. In the linear approach, the velocity, vorticity, pressure, and the shape of the drop are expanded in modes. That is series of orthogonal functions: spherical harmonics $Y_{lm}(\theta, \varphi)$ for the angular variables, spherical Bessel functions $j_l(\omega r), n_l(\omega r)$ for the radial variable, and trigonometric functions of time, $e^{i\beta_l t}$. We have shown in Sect. 13.1 that the frequencies of linear inviscid isolated oscillations are

$$\omega_l^2 = \frac{\sigma}{\rho R^3} l(l-1)(l+2) \tag{13.33}$$

where σ is the surface tension coefficient. The lowest two modes ($l = 0, 1$) are eliminated by the mass and momentum conservation, since radial oscillations are forbidden by incompressibility, and translation are not interesting. The influence of the external fluid and of the viscosity generate variations of this basic equation.

13.2.1 Model 1

If viscosity is taken into account, the standard frequency spectrum of the drop changes (even in the linear approximation), and in addition, the damping of oscillations occur. Miller and Scriven [223] calculated such oscillations for a three-dimensional incompressible, Newtonian drop immersed into another fluid, with viscosity. This type of dynamics of fluid drops occurs in many physical systems like transfer of one fluid immersed in another fluid, dispersed in small droplets and offering a large interfacial contact, in emulsions or biological cells. The space is Euclidean (x^i) and so all components of vectors will be considered contravariant by default, and on the tangent space to the compact surface of the deformed drop we can use the natural frame given by the outer normal ($\boldsymbol{r}$) and the tangent spherical coordinates. To neglect gravity we consider

$$\frac{gR^2\Delta\rho}{\sigma} \ll 1, \tag{13.34}$$

where g is the gravitational acceleration and $\Delta\rho$ is the difference between drop density and exterior medium density. The smallness parameter that controls the nonlinear effects is

$$\frac{\Delta r}{\lambda} \ll 1. \tag{13.35}$$

That is, if the radial displacement Δr (one can take for example $2\pi r$ for the order of magnitude) is small compared to the wavelength of the oscillations along the surface we are in the linear approximation, and we can neglect nonlinear terms in the Navier–Stokes equation

$$\frac{\partial \boldsymbol{V}}{\partial t} = -\frac{1}{\rho}\nabla P + \nu\Delta \boldsymbol{V}, \quad \nabla \cdot \boldsymbol{V} = 0, \tag{13.36}$$

with ν the viscosity of the drop fluid and Δ is the Laplacian. The density is denoted ρ, and we also denote by $\rho_{i,e}$ the density of the fluid inside the drop and outside it. The general approach to solve the problem is to first eliminate pressure from the Navier–Stokes equations by using vorticity, then we decouple the radial part from the angular part in the unknown functions. Because of the Laplace type of equations we can take profit of representation formulas in Sect. 10.6.2 and calculate velocity, vorticity, and pressure only from the radial components. Then, by including the boundary conditions we can write the whole algebraic system of equations to determine the coefficients of the spherical harmonic expansions. The determinant of this system will provide the damped modes exponents.

To eliminate the pressure P we apply a curl operator on (13.36) and we introduce the *vorticity* $\boldsymbol{\omega} = (\omega_r \boldsymbol{e}_r, \omega_\theta \boldsymbol{e}_\theta, \omega_\varphi \boldsymbol{e}_\varphi)$ in spherical coordinates,

$$\frac{\partial \boldsymbol{\omega}}{\partial t} - \nu \Delta \boldsymbol{\omega} = 0, \tag{13.37}$$

where obviously $\nabla \times \boldsymbol{\omega} = 0$.

Equations (13.36) and (13.37) can be further reduced to two scalar equations for the radial components of the velocity and vorticity. This is possible because $\boldsymbol{V}$ is a divergence-free poloidal field. Once we obtained the radial components it is easy to calculate the whole vectors by using the representation theorem from (10.81) and Sani [286]. We have

$$\boldsymbol{V} = \boldsymbol{e}_r V_r + \frac{r^2}{l(l+1)} \left[\nabla_\Sigma \left(\frac{1}{r^2} \frac{\partial r^2 V_r}{\partial r} \right) - \boldsymbol{e}_r \times \nabla_\Sigma \omega_r \right], \tag{13.38}$$

where ∇_Σ is the surface gradient operator (Sect. 6.5.1).

To decouple the radial and tangent components in the equations we can use the relation

$$\Delta \left(\sum_{i=1}^{3} x^i \omega^i \right) = \Delta (r \omega_r).$$

Consequently, we obtain from (13.37) the system

$$\left(\frac{\partial}{\partial t} - \nu \Delta \right) r \omega_r = 0, \tag{13.39}$$

$$\Delta \left(\frac{\partial}{\partial t} - \nu \Delta \right) r V_r = 0. \tag{13.40}$$

We assume that there is no external excitation to maintain the oscillations, so that the only physical regime will be exponential damping in time. We expand all quantities in spherical harmonics

$$(r\omega_r)(r, \Omega, t) = \sum_{l,m} e^{-\beta_l t} W_{lm}(r) Y_{lm}(\Omega),$$

$$(rV_r)(r, \Omega, t) = \sum_{l,m} e^{-\beta_l t} V_{lm}(r) Y_{lm}(\Omega), \tag{13.41}$$

as well as the pressure itself

$$P = \sum_{lm} P_{lm}(r) e^{-\beta_l t} Y_{lm}(\Omega), \tag{13.42}$$

where we denoted the angular spherical coordinates by $\Omega = (\theta, \varphi)$. From (13.36) and (13.42) we obtain for the pressure coefficients that depend only on the radial components of the velocity. We have the form

$$P_{lm} = \frac{\rho\nu}{l(l+1)} \frac{\partial}{\partial r}\left[r\left(\frac{\beta_l}{\nu} + \Delta(rV_r) \right) \right] e^{-\beta_l t}. \tag{13.43}$$

or the form

$$P_{lm}(r) = \frac{\nu\rho}{l(l+1)} \frac{\partial}{\partial r}\left[r\left(\frac{\beta_l}{\nu} + \frac{1}{r^2}\frac{\partial}{\partial r} r^2 \frac{\partial}{\partial r} - \frac{l(l+1)}{r^2} \right) rV_r \right], \tag{13.44}$$

and then use (13.42)

$$P(r, \Omega) = \sum_{lm} \frac{\rho\nu}{l(l+1)} \frac{\partial}{\partial r}\left[r\left(\frac{\beta_l}{\nu} + \Delta \right) rV_r \right] e^{-\beta_l t} Y_{lm}. \tag{13.45}$$

By introducing (13.41) in (13.39) we obtain for the radial functions W_{lm} a spherical Bessel functions differential equation

$$r^2 W_{lm}'' + 2rW_{lm}' + \left(\frac{\beta_l}{\nu} r^2 - l(l+1) \right) W_{lm} = 0, \tag{13.46}$$

with general solution

$$W_{lm} = a_l j_l\left(\sqrt{\frac{\beta_l}{\nu}} r \right) + b_l n_l\left(\sqrt{\frac{\beta_l}{\nu}} r \right) + h_{lm}, \tag{13.47}$$

where j_l, n_l are the spherical Bessel functions and h is a harmonic function, which in this case reduces to $h_{lm}(r) = a_{1l} r^l + a_{2l} r^{-l-1}$. For the radial velocity we obtain from (13.40) the differential radial equation

$$\left(\frac{\partial^2}{\partial^2} + \frac{2}{r}\frac{\partial}{\partial r} - \frac{l(l+1)}{r^2} \right)\left(-\beta_l V_{lm} - \nu V_{lm}'' - \frac{2\nu}{r} V_{lm}' + \frac{\nu l(l+1)}{r^2} V_{lm} \right) = 0, \tag{13.48}$$

with the solution

$$V_l = C_1 r^l + C_2 r^{-l-1} + C_3 j_l\left(\sqrt{\frac{\beta_l}{\nu}} \right). \tag{13.49}$$

We present more details about this solution in Exercise 1 at the end of this chapter. Solutions of types (13.47) and (13.49) have a polynomial part responsible for the inviscid type of flow and the Bessel part responsible for viscous flow.

From (13.47) and (13.49) we can write the final explicit form of the radial components of the velocity and vorticity, and the shape function. First we mention that we have two types of solutions: external and internal with respect to the drop and its exterior environment, labeled by subscripts e, i. In all the following equations the labels lm are suppressed, but all quantities actually contain them. We will make a note when we come back to explicit writing of the labels. We denote $\chi_{e,i} = \sqrt{\beta/\nu_{e,i}}$. The free coefficients $a_1, \dots, a_4$ and $b_0, \dots, b_2$ are dimensionless and B is the speed. We have

$$\begin{aligned}
V_{ri} &= B\left(a_1 \frac{r^{l-1}}{R^{l-2}} + a_3 R^2 \frac{j_l(\chi_i r)}{r}\right) e^{-\beta t} Y, \\
V_{re} &= B\left(a_2 \frac{R^{l+3}}{r^{l+2}} + a_4 R^2 \frac{n_l(\chi_e r)}{r}\right) e^{-\beta t} Y, \\
\omega_{ri} &= BRb_1 \frac{j_l(\chi_i r)}{r} e^{-\beta t} Y, \\
\omega_{re} &= BRb_2 \frac{n_l(\chi_e r)}{r} e^{-\beta t} Y, \\
r &= b_0 R e^{-\beta t} Y.
\end{aligned} \tag{13.50}$$

We recall that $j_l(\xi), n_l(\xi)$ are the Bessel spherical functions. For properties and relations the author can use any of the books [5, 129, 238, 281, 284].

The last step is to include the boundary conditions for the surface Σ of the drop. Miller and Scriven [223] use seven special boundary conditions, namely free surface (linearized) kinematic condition (9.5), (9.29), and (9.30)

$$V_{ri}|_\Sigma = \left.\frac{dr}{dt}\right|_\Sigma \rightarrow \beta b_0 + a_1 B + a_3 B j(\chi_i R) = 0, \tag{13.51}$$

continuity of the radial velocity

$$V_{ri}|_\Sigma = V_{re}|_\Sigma \rightarrow a_1 + a_3 j(\chi_i R) = a_2 + a_4 n(\chi_e R), \tag{13.52}$$

continuity of the radial vorticity

$$\omega_{ri}|_\Sigma = \omega_{re}|_\Sigma \rightarrow b_1 j(\chi_i R) = b_2 n(\chi_e R), \tag{13.53}$$

and continuity of the surface divergence of the velocity

$$\begin{aligned}
\nabla_\Sigma V_i = \nabla_\Sigma V_e \rightarrow a_1(l-1) + a_3[(l-1)j(\chi_i R) - \chi_i R j_{l+1}(\chi_i R)] \\
= -a_2(l+2) + a_4[(l-1)n(\chi_e R) - \chi_e R n_{l+1}(\chi_e R)].
\end{aligned} \tag{13.54}$$

We note that j, n without a subscript means j_l, n_l, but where it is the case we wrote explicitly j_{l+1}, etc. For the surface differential operators in (13.54), and in the following equations, we refer to Sect. 6.5 or Weatherburn [338] and Sani [286].

Next boundary conditions refer to balance of forces at the interface. Instead of using the continuity of the three components of the Euclidean forces, it is more convenient (in the spherical symmetry case) to use other three quantities: radial component of Euclidean force (F_r), surface divergence ($\nabla_\Sigma \cdot \boldsymbol{F}$), and radial part of the surface curl ($\nabla_\Sigma \times \boldsymbol{F}$) of the surface force. To write these boundary conditions we need to introduce some physical parameters specific to fluid interface physics. For reference the reader can consult [35,223]. We denote like before by σ the *coefficient of surface tension*, we introduce $\Gamma_l = \rho_e l + \rho_i(l+1)$ but we shall skip the subscript l in Γ, k is the *coefficient of interfacial dilatational viscosity*, ϵ is the *coefficient of interfacial shear viscosity*, Λ is the *coefficient of interfacial dilatational elasticity*, and M is the *coefficient of interfacial shear elasticity*. If the interface is clean and simple, the coefficients of interfacial viscosity and elasticity vanish, i.e., $K = 0, \Lambda = 0, D = 0$. On the contrary, very large values for D describe an inextensible interface like in the case of biological membranes [104]. Also, if $S \ll D$ we have an interface where the viscous dissipation of energy is mainly due to the boundary layer flow in the underlying bulk fluid, and much less due to shearing deformation. The densities and kinematic viscosities of the fluid inside and outside the drop are $\rho_{i,e}, \nu_{i,e}$, respectively. Based on these coefficients, it is useful to use the symbols

$$S_l = \frac{\epsilon}{R} - \frac{M}{\beta_l R}, \quad D_l = \frac{k}{R} - \frac{\Lambda}{\beta_l R},$$

namely S is the combined coefficient of surface shear elastic and viscosity and D is the combined coefficient of surface dilatational elastic and viscosity.

The boundary condition for the radial component of the surface force can be obtained from the Navier–Stokes (10.13) in radial coordinates (r, θ, φ) [171]

$$F^r = \sigma^{rr} = -P + 2\eta^* \frac{\partial v_r}{\partial r}, \tag{13.55}$$

where it is usual to introduce a correction in the viscosity by taking into account the interfacial dilatational elasticity since the interface may have elastic properties, too. Forces of elastic nature depend on the interfacial strain in the same manner that viscous forces depend on the interfacial rate of strain [223]. The correction is

$$\eta \to \eta^* = \eta - D = \eta - \frac{k}{R} + \frac{\Lambda}{\beta R}.$$

From (13.50) and (13.55), and the expression of surface tension (13.43) we can write for the pressure as the contribution of the internal, external, and surface terms (where again we skip writing the l subscript for β, etc.)

$$P = P_\sigma + P_{in} + P_{ext} = -\frac{\sigma b_0(l-1)(l+2)}{R} + \frac{\rho_i \beta B a_1 R^2}{l}$$
$$+ \frac{\rho_i \beta B a_3 R^2}{l(l+1)}\left[\left(l+\frac{3}{2}\right) j_l\left(\sqrt{\frac{\beta}{\nu_i}}R\right) - \sqrt{\frac{\beta}{\nu_i}} R j_{l+1}\left(\sqrt{\frac{\beta}{\nu_i}}R\right)\right]$$
$$+ \frac{\rho_e \beta B a_4 R^2}{l(l+1)}\left[\left(l+\frac{3}{2}\right) n_l\left(\sqrt{\frac{\beta}{\nu_e}}R\right) - \sqrt{\frac{\beta}{\nu_e}} R n_{l+1}\left(\sqrt{\frac{\beta}{\nu_e}}R\right)\right], \quad (13.56)$$

where j_l, n_l are the Bessel and von Neumann functions and the first term is the surface tension obtained from the linear fluid drop model in Sect. 13.1, or from literature, for example in Article 274 from [167], [50, Chap. VI], or [171, Chap. VII]. The derivative of the velocity in (13.55) can be calculated from (13.50) and (13.55)

$$2(\eta^* - D)\frac{\partial v_{r,i}}{\partial r} = 2B(\eta^* - D)\left[a_1(l-1) + a_3\left((l-1) j_l\left(\sqrt{\frac{\beta}{\nu_i}}R\right)\right.\right.$$
$$\left.\left. - \sqrt{\frac{\beta}{\nu_i}}R\right) j_{l+1}\left(\sqrt{\frac{\beta}{\nu_i}}R\right)\right]. \quad (13.57)$$

From (13.56) and (13.57) we have the next boundary condition for (13.55)

$$\frac{\sigma b_0(l-1)(l+2)}{R} - a_1\frac{\rho_i B\beta R^2}{l} - a_3\frac{\rho_i \beta B R^2}{l(l+1)}\left[\left(l+\frac{3}{2}\right) j_l\left(\sqrt{\frac{\beta}{\nu_i}}R\right)\right.$$
$$\left. - \sqrt{\frac{\beta}{\nu_i}} R j_{l+1}\left(\sqrt{\frac{\beta}{\nu_i}}R\right)\right] + 2a_1 B(\eta_i - D)(l-1)$$
$$+ 2a_3 B(\eta_i - D)\left((l-1) j_l\left(\sqrt{\frac{\beta}{\nu_i}}R\right) - \sqrt{\frac{\beta}{\nu_i}} R j_{l+1}\left(\sqrt{\frac{\beta}{\nu_i}}R\right)\right) - \frac{\rho_e \beta B a_2 R^2}{l+1}$$
$$= 2B(\eta_e - D)\left[-a_2(l+2) + a_4\left((l-1) n_l\left(\sqrt{\frac{\beta}{\nu_e}}R\right)\right.\right.$$
$$\left.\left. - \sqrt{\frac{\beta}{\nu_e}} R n_{l+1}\left(\sqrt{\frac{\beta}{\nu_e}}R\right)\right)\right] - \frac{B\rho_e \beta a_4 R^2}{l(l+1)}\left[\left(l+\frac{3}{2}\right) n_l\left(\sqrt{\frac{\beta}{\nu_e}}R\right)\right.$$
$$\left. - \sqrt{\frac{\beta}{\nu_e}} R n_{l+1}\left(\sqrt{\frac{\beta}{\nu_e}}R\right)\right]. \quad (13.58)$$

Similar equations can be written for the surface divergence and radial component of the surface curl of the surface force [223], where the surface differential operators are defined in Sects. 6.5.2 and 6.5.4. Finally, we have sets of seven equations from the seven boundary conditions in seven unknowns: $b_0, \dots, b_2$ and $a_1, \dots, a_4$, each set for one value of l. Once we obtained these series coefficients (13.47) and (13.49) for the radial parts of the velocity and vorticity, it is easy to calculate the full $\boldsymbol{V}, \boldsymbol{\omega}$ vectors by using the representation formula (13.38).

For each l, the system of seven equations splits into two systems, $S_{2\times 2}$ in b_1, b_2 and $S_{5\times 5}$ in $b_0, a_1, \dots, a_4$, where the first one is responsible for the vorticity coefficients only. The compatibility of these systems is provided by vanishing of the corresponding determinants, and this determines the β_l coefficients. This decomposition in $2+5$ equations induces two types of solutions corresponding to two types of waves.

If we choose solutions with $\det S_{2\times 2} = 0$ and $\det S_{5\times 5} \neq 0$, the second condition implies that we have no radial motion $v_r = 0$, and so the wave generated by these equations are *shear waves* or purely rotational waves without any oscillations involved. The first condition provides radial component of the vorticity and hence, by (13.38) we have only tangent components for the velocity. These waves always decay in time without oscillations because the corresponding coefficients β are pure real [35, 223, 293].

In Fig. 13.14, we present a numerical check of this fact for air, water, and oil. We used $\nu_{water} = 10^{-6}$ P, $\nu_{air} = 1.82 \times 10^{-3}$ P, $\nu_{oil} = 1.5 \times 10^{-4}$ P,

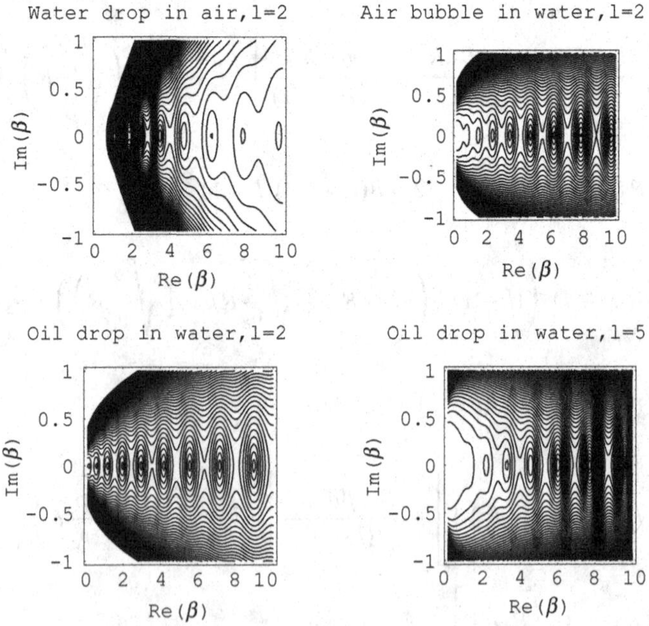

Fig. 13.14 The contour plots of the determinant of the system of equations $S_{2\times 2} = 0$ vs. the real and imaginary part of β, for two values of l, and different types of fluids. It is easy to see that the only zeros (closed contours) are along the $\mathrm{Im}\beta = 0$ axis

$\rho_{water} = 10^3\,\mathrm{kg\,m^{-3}}$, $\rho_{air} = 1\,\mathrm{kg\,m^{-3}}$, and $\rho_{oil} = 750\,\mathrm{kg\,m^{-3}}$ for $l = 2, 5$. In all these examples the result does not change with the value of the combined coefficient of surface shear in the range $S = 0 \to 500\,\mathrm{kg\,s^{-1}}$.

If we choose $\det S_{2\times 2} \neq 0$ and $\det S_{5\times 5} = 0$ we obtain solutions with radial velocity, but zero radial vorticity. The condition of zero determinant provides complex values for β, hence we have both oscillations and damping. In Fig. 13.15, we present numerical calculation of the roots of the 5×5 determinant to check the occurrence of both real and imaginary parts for β.

Figures 13.14 and 13.15 provide a numerical estimation of the evolution of the roots β. To have a better understanding on the influence of physical parameters on oscillating and damping regimes of the drop, we analyze the exact expression of β in some special cases. We use the same convention of subscript, i.e., (i, e) for inner and outer part of the drop.

In the case of low viscosities, for the droplet configuration, i.e., $\rho_e \ll \rho_i$, we can write the following expressions

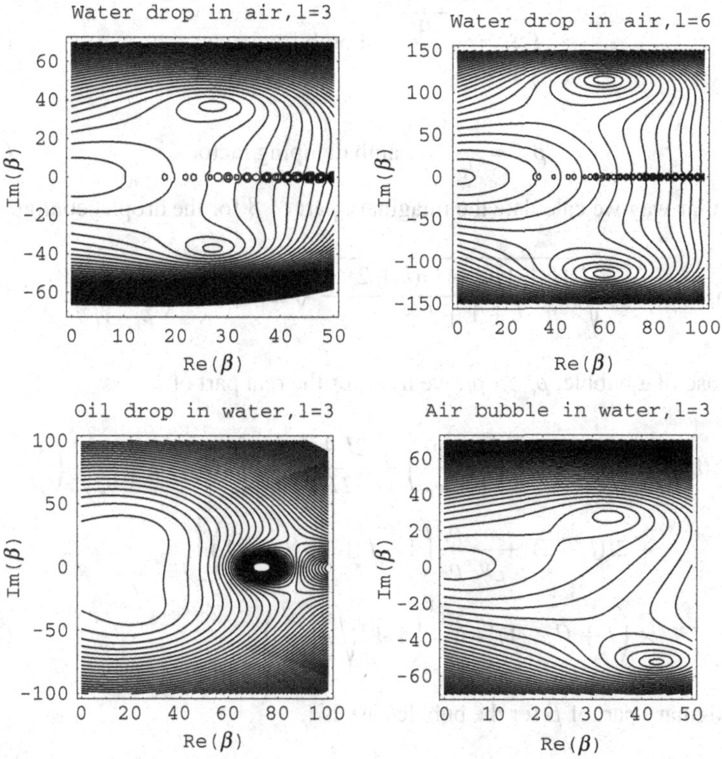

Fig. 13.15 The contour plots of the determinant of the system of equations $S_{5\times 5} = 0$ vs. the real and imaginary part of β, for two values of l, and different types of fluids. It is easy to see that the zeros (closed contours) involve both real and imaginary parts of $\beta = 0$

$$\mathrm{Re}\beta_l = \sqrt{\Omega_{L,i}\beta_{L,e}}\, F\left(\frac{\rho_e}{\rho_i}, \frac{\nu_e}{\nu_i}\right) + \frac{2l+1}{2R^2}\beta_{L,i}\left[2(l^2-1)\right.$$
$$+2l(l+2)\left(\frac{\nu_e}{\nu_i}\right)^2\left(\frac{\rho_e}{\rho_i}\right)^3 + \frac{\nu_e}{\nu_i}\frac{\rho_e}{\rho_i}\left.\left(l+2-\frac{\rho_e}{\rho_i}(l-1)\right)\right]$$
$$\times\left(l+1+l\frac{\rho_e}{\rho_i}\right)^{-1}\left(1+\sqrt{\frac{\nu_e}{\nu_i}\frac{\rho_e}{\rho_i}}\right)^{-2}, \tag{13.59}$$

where

$$F = \frac{(2l+1)^2}{2\sqrt{2}}(l(l+1)(l-1)(l+2))^{\frac{1}{4}}\frac{\rho_e}{\rho_i}\left\{\left[l\left(1+\frac{\rho_e}{\rho_i}\right)+1\right]^{\frac{5}{4}}\left[1+\frac{\rho_e}{\rho_i}\sqrt{\frac{\nu_e}{\nu_i}}\right]\right\}^{-1}$$

Here we choose to write the ratio of exterior parameters over the inner parameters to have a formula available for series expansion. The new symbols introduced are

$$\Omega_L = \sqrt{\frac{\sigma}{\rho R^3}}, \quad \text{Lamb frequency,} \tag{13.60}$$

and

$$\beta_L = \frac{\nu}{R^2}, \quad \text{Lamb damping factor.} \tag{13.61}$$

In a similar way we calculate the imaginary part of β for the droplet configuration

$$\mathrm{Im}\beta_l = \Omega_{L,i}\sqrt{\frac{l(l+1)(l-1)(l+2)}{l+1+l\frac{\rho_e}{\rho_i}}} - \sqrt{\Omega_{L,i}\beta_{L,e}}\,F\left(\frac{\rho_e}{\rho_i}, \frac{\nu_e}{\nu_i}\right). \tag{13.62}$$

In the case of a bubble, $\rho_e \gg \rho_i$, we have for the real part of β

$$\mathrm{Re}\beta_l = \sqrt{\Omega_{L,e}\beta_{L,i}}\, F\left(\frac{\rho_i}{\rho_e}, \frac{\nu_i}{\nu_e}\right) + \frac{2l+1}{2R^2}\beta_{L,e}\left[2(l^2-1)\left(\frac{\nu_i}{\nu_e}\right)^2\left(\frac{\rho_i}{\rho_e}\right)^3\right.$$
$$+\, 2l(l+2) + \frac{\nu_i}{\nu_e}\frac{\rho_i}{\rho_e}\left.\left(1-l+\frac{\rho_i}{\rho_e}(l+2)\right)\right]$$
$$\times\left(l+(l+1)\frac{\rho_i}{\rho_e}\right)^{-1}\left(1+\sqrt{\frac{\nu_i}{\nu_e}\frac{\rho_i}{\rho_e}}\right)^{-2}. \tag{13.63}$$

The imaginary part of β for the bubble case is

$$\mathrm{Im}\beta_l = \Omega_{L,e}\sqrt{\frac{l(l+1)(l-1)(l+2)}{l+(l+1)\frac{\rho_i}{\rho_e}}} - \sqrt{\Omega_{L,e}\beta_{L,i}}\,F\left(\frac{\rho_i}{\rho_e}, \frac{\nu_i}{\nu_e}\right). \tag{13.64}$$

From the general behavior of (13.59–13.64) we note that no matter if the system is droplet or bubble, the damping (real part) depends on both $\beta_{L,i}, \beta_{L,e}$, while the oscillations (imaginary part) depend only on Ω_L of the denser medium. Moreover, we can write

$$\frac{\Omega^2_{drop}}{\Omega^2_{bubble}} = \left(\frac{\Omega_{L,i}}{\Omega_{L,e}}\right)^2 \frac{l}{l+1} = \frac{\rho_{bubble,e}}{\rho_{drop,i}} \frac{l}{l+1} = \frac{l}{l+1}, \tag{13.65}$$

meaning that the higher modes, large l, namely the modes with more complicated shapes, have same frequencies no matter if they are drops or bubbles, but for lower modes, the droplet system is slower in oscillations.

To figure out how do (13.59), (13.62)–(13.64) work in a case study, we choose a drop of water of radius $R = 1$ cm, $\sigma = 73.4 \times 10^{-3}\,\mathrm{N\,m^{-1}}$, and we plot Re$\beta$ vs. the mode l and the external density ρ_{ext} in Fig. 13.16.

From this figure we infer that in the range $\nu_e/\nu_i = 0.1 \rightarrow 10$ the aspect of the Re$\beta(l, \rho_e)$ does not change qualitatively. For viscous droplet Reβ has a maximum when $\rho_e \le \rho_i$ and decreases when the two densities become more and more different. The highest dissipation happens when the densities are equal. If the exterior viscosity is higher than the internal one, dissipation increases with the

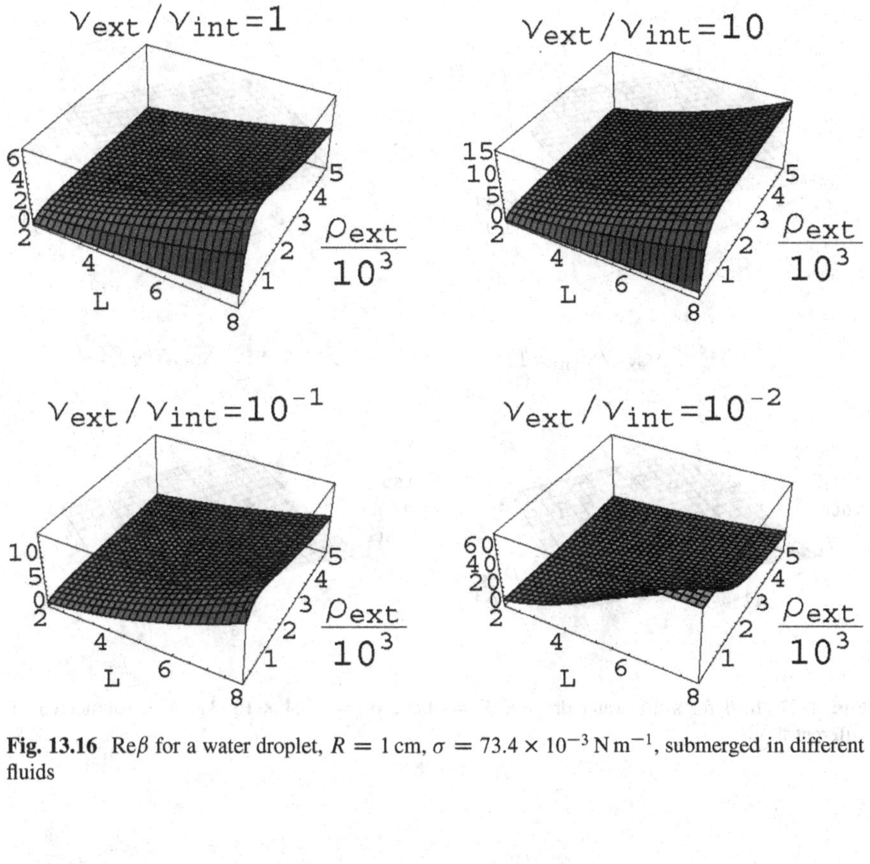

Fig. 13.16 Reβ for a water droplet, $R = 1$ cm, $\sigma = 73.4 \times 10^{-3}\,\mathrm{N\,m^{-1}}$, submerged in different fluids

density of the exterior fluid. For very viscous drops the dissipation increases if the exterior fluid is less dense.

In Fig. 13.17, we present the imaginary part of β as function of the same parameters and variables. We notice that the frequency of oscillations decreases with the density of the exterior fluid and increases with l. However, there is no significant variation of the frequencies with ν_e/ν_i. This happens because in the expression of Imβ (13.62) and (13.64), the first term (that one independent of ν_e) is always much larger than the second one, and there is no way to increase the second term for any range of R, σ, ρ_i, ν_i. Even if we approach $\nu_e \to \infty$ still there is no significant change in frequencies because this term has a horizontal asymptote in this limit. If ν_i increases very much we meet a new qualitative behavior. Im$\beta \to 0$ and the frequency of oscillations decreases to zero (especially if ρ_e has large values) until some oscillation modes completely vanish. This is shown by the gap in the right lower corner of Fig. 13.17. This occurrence of an aperiodic mode on behalf of annihilation of an oscillating mode when viscosity increases was noticed first time in Willson [346].

Other limiting situations. For inviscid fluids, $\nu_{i,e} = 0$ we have the well-known Lamb frequencies denoted β in literature [167], [50, (280) and (283), Sect. 98], and

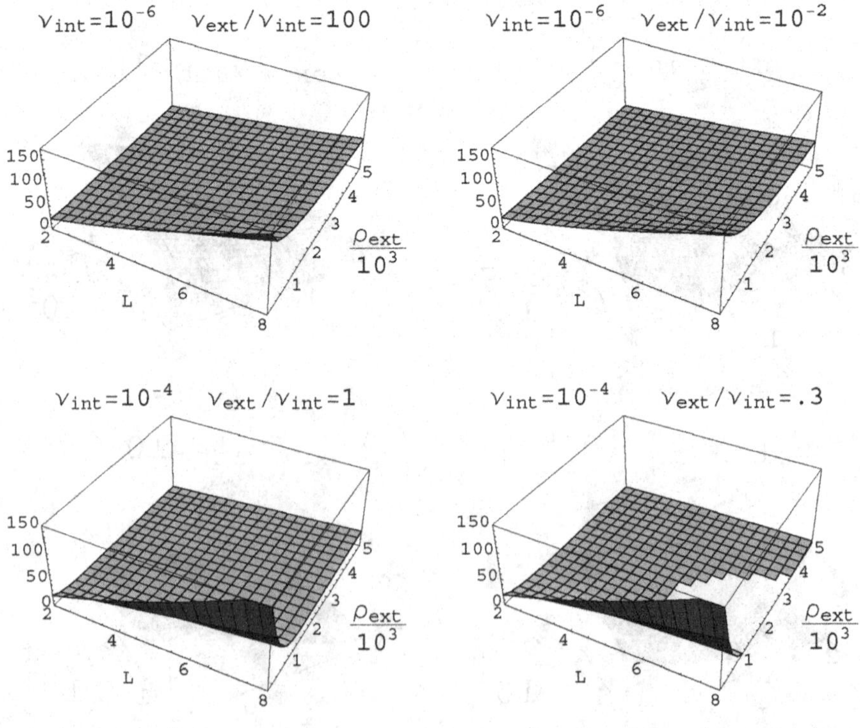

Fig. 13.17 Imβ for same water droplet, $R = 1\,\text{cm}$, $\sigma = 73.4 \times 10^{-3}\,\text{N}\,\text{m}^{-1}$, submerged in different fluids

[269], with

$$\beta = i\sqrt{\frac{\sigma l(l-1)(l+1)(l+2)}{R^3[\rho_e l + (l+1)\rho_i]}}, \tag{13.66}$$

which becomes

$$\beta_{bubbles} = i\sqrt{\frac{\sigma(l-1)(l+1)(l+2)}{R^3\rho_e}}, \tag{13.67}$$

for bubbles, and

$$\beta_{drops} = i\sqrt{\frac{\sigma l(l-1)(l+2)}{R^3\rho_i}}, \tag{13.68}$$

for droplets.

For small viscosities, $\nu_{i,e} \approx 0$, it is easy to verify the occurrence of the *slip effect* between the exterior and interior fluid layers. In this case the solutions are dominated by the terms expressed in terms of rational functions, while the Bessel function terms become negligible. The coefficients $a_{3,4}$ vanish, which cancels a whole column in the determinant of $S_{2\times 2}$. Consequently, β becomes pure imaginary, i.e.,

$$\beta = \pm i\Omega_L\sqrt{\frac{l(l-1)(l+1)(l+2)}{l+(l+1)\frac{\rho_i}{\rho_e}}}.$$

From (13.50), by neglecting j_l, n_l, we obtain $a_3 = a_4 = b_1 = b_2 = 0$. By plugging this result in (13.38) we obtain a simple relation between the tangent velocities at the fluid interface

$$\left.\frac{V_{\parallel,i}}{V_{\parallel,e}}\right|_\Sigma = -\frac{l+1}{l},$$

which put into evidence the strong slip effect for this situation.

If the viscosity of the exterior fluid is very large, the imaginary part of β approaches zero and the real part approaches infinity (Fig. 13.18). Consequently, the drop enters in a very rapidly decaying mode. For a bubble, $\rho_i, \nu_i \approx 0$, in a viscous fluid we obtain

$$\beta = \Omega_{L,e}\sqrt{\frac{(l-1)(l+1)(2l+1)R^2}{2(2l^2+1)\nu_e}},$$

while for a bubble in an inviscid fluid ($\nu_e = 0$) we have

$$\beta = \frac{(2l+1)(l+2)\nu_e}{R^2} \pm i\Omega_{L,e}\sqrt{(l-1)(l+1)(l+2)}.$$

Finally, in the limit of inextensible surface (controlled by large values of D compared to S) we have large values for Reβ (see Fig. 13.18). This effect happens because of the enhancement of the boundary layer flow next to the surface, pretty much like in the case of flat interfaces.

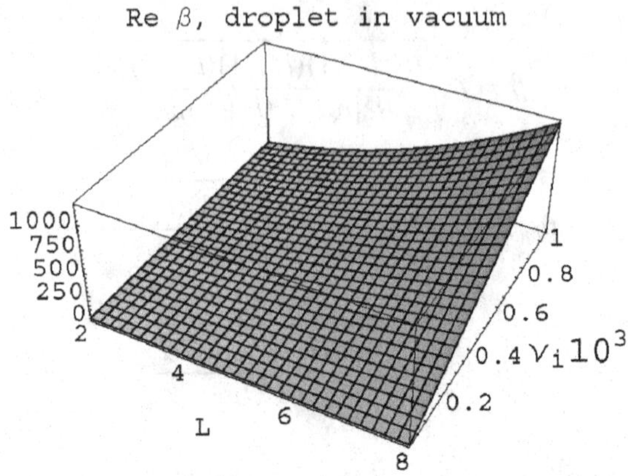

Fig. 13.18 The damping coefficient Reβ for $R = 1$ cm water drop in vacuum

A comprehensive analysis of small-amplitude axisymmetric shape oscillations of an isolated viscoelastic drop is performed in the paper [156]. The authors investigate the characteristic equation for the complex frequency and find exact solutions in several regimes: high-viscosity limit, viscoelastic drop (for different ranges of elasticities), low-viscosity limit, and quadrupole oscillations. The same authors contribute in Kishmatullin and Nadim [157] to the applications of the same model to the radial oscillations of gas microtubule encapsulated by a viscoelastic solid shell and surrounded by slightly compressible viscous liquid. These calculations are useful for research in the medical field, for example, the description of pulsations of such encapsulated bubbles in the blood flow for ultrasound diagnosis.

As an alternate approach to the drop and bubble shape oscillations in the small-amplitude viscous case, we mention the work of Prosperetti. Instead of using the traditional expansion of the potential, velocity, and pressure in spherical harmonics, in articles [61, 263–265] the author uses a decomposition in terms of poloidal and toroidal normal modes (Sect. 10.6.3) inspired by Chandrasekhar [50]. For example, we can represent the vorticity as $\boldsymbol{\omega} = \nabla \times (\boldsymbol{A} + \nabla \times \boldsymbol{B})$ with

$$\boldsymbol{A} = T(r) Y_{lm}(\theta, \varphi) e^{\lambda t} \boldsymbol{e}_r,$$

$$\boldsymbol{B} = S(r) Y_{lm}(\theta, \varphi) e^{\lambda t} \boldsymbol{e}_r,$$

where the functions T and S describe the toroidal and poloidal modes, respectively. Next, we can express the velocity

$$\boldsymbol{V} = \boldsymbol{A} + \nabla \times \boldsymbol{B} + \nabla \phi,$$

with

$$\Delta \phi = -\nabla \cdot \boldsymbol{A}.$$

Basically, such decomposition still uses the spherical harmonics, but combines them in more useful way to handle the curl, and curl(curl) operators occurring in the vorticity equations. We have

$$\boldsymbol{V} = \sum_{lm} (V_{lm}^{(1)} \boldsymbol{T}_{lm} + V_{lm}^{(2)} \boldsymbol{S}_{lm})$$

where

$$\boldsymbol{S}_{lm} = \nabla \times \nabla \times [S_{lm}(r,t) Y_{lm} \boldsymbol{e}_r], \quad \boldsymbol{T}_{lm} = \nabla \times [T_{lm}(r,t) Y_{lm} \boldsymbol{e}_r].$$

The algebra of these modes is described in Sect. 10.6.3. The radial dependence is not anymore controlled by spherical Bessel functions, like in the previous section, but by the Hankel $H_k^{(1,2)}$ and Bessel functions J_n. The resulting equation is related to the Plesset equation, which lately raised interest in bubble sonoluminescence problems [328]. The big advantage is that the toroidal and poloidal normal modes are effectively decoupled. The T modes, where $S = 0$, describes shape oscillations of the drop, and one can find the same results that have been obtained by the previously presented formalism. The S modes ($T = 0$) describe a motion in which different shells of fluid rotate about the center, i.e., shear waves or purely rotational waves. Since there is no tangent restoring force for these modes they will be aperiodically damped. In Prosperetti [264] an extended analysis on these modes is presented, for drops and bubbles embedded in fluids of different viscosities.

13.3 Nonlinear Three-Dimensional Oscillations of Axisymmetric Drops

Like in the case of viscous linear models, nonlinear viscous droplets oscillations are investigated by solving the Navier–Stokes equations in the incompressible fluid approximation, by using the same mode expansion. Several theoretical models describing the nonlinear drop dynamics were developed in the last three decades. For example, in [237, 321] inviscid nonlinear drops are investigated, and in [13,203,253,283] the analysis is extended to viscous droplet, but only numerically. The boundary integral method [203] and the Galerkin-finite element method [13] give good results in principle, but cannot model drops with viscosities in the physical range of interest. Also, the finite element methods have been used but limited to low viscosities, since higher Reynolds numbers require long computational times and a very fine discretization mesh. Another approach [16] still uses the modes expansion method for axisymmetric drops, but handles the resulting differential equations by the variational principle of Gauss.

In the following we describe a theoretical model for the nonlinear axisymmetric oscillations of viscous drops, which provided a good agreement with experimental data and also offers several predictions [16]. We consider a drop of viscous incompressible fluid, uniform surface tension coefficient, $\nu = \nu_i, \rho = \rho_i, \sigma$ constant, freely oscillating in a fluid of negligible density, and viscosity, $\rho_e = \nu_e = 0$. We study the case of an axisymmetric drop, with the symmetry axis along Oz. We use polar coordinates (θ, z) so that the interface is parametrized by the shape function

$$r(\theta, \varphi, t) = r(\theta, t) = R_0[A_0(A_2, A_3, \dots) + \sum_{l \geq 2} A_l(t) P_l(\cos\theta)], \qquad (13.69)$$

where P_l are the Legendre polynomials and $A_0 < 1$, $\theta \in [0, \pi]$. The dependence of the free term A_0 on the other coefficients fulfills the constraint of preserving constant volume (Sect. 13.1.4).

By momentum conservation the center of mass of the drops moves along the Oz axis with a displacement $s(t)$. The law of motion of the center of mass is given by

$$s(A_2, A_3, \dots) = \frac{3}{8} R_0 \int_{-1}^{1} \cos\theta \left[A_0 + \sum_{l \geq 2} A_l P_l(\cos\theta) \right]^4 d(\cos\theta). \qquad (13.70)$$

In the (noninertial) frame of the center of mass the Navier–Stokes equation (10.13)

$$\frac{\partial \boldsymbol{V}}{\partial t} + (\boldsymbol{V} \cdot \nabla)\boldsymbol{V} - \xi \boldsymbol{e}_z = -\frac{1}{\rho} \nabla P - \nu \nabla \times \nabla \times \boldsymbol{V},$$

$$\nabla \cdot \boldsymbol{V} = 0, \qquad (13.71)$$

where $\xi = s_{tt}$, the acceleration of the center of mass, and $\boldsymbol{e}_z$ is the unit vector in the direction of the Oz axis. By applying $\nabla\times$ on (13.71) we obtain

$$\frac{\partial \boldsymbol{\omega}}{\partial t} + \nabla \times (\boldsymbol{V} \cdot \nabla)\boldsymbol{V} = -\nu \nabla \times \nabla \times \boldsymbol{\omega}, \qquad (13.72)$$

where $\boldsymbol{\omega} = \nabla \times \boldsymbol{V}$ is the vorticity. The kinematic condition for the free fluid surface (see (9.5), (9.29), and (9.30)) reads

$$\boldsymbol{V} \cdot (r\boldsymbol{e}_r - r_\theta \boldsymbol{e}_\theta) = r \frac{\partial r}{\partial t}, \qquad (13.73)$$

where $\boldsymbol{e}_\theta$ is the tangent unit vector (Sect. 4.12) and subscripts like r_θ, ξ_{tt} mean differentiation. The driving force of the oscillations is the surface tension, which always acts normal to the surface (see (8.53) and (8.55)), while the tangent stress here is zero. Consequently, we can write the boundary conditions in the form (8.55)

$$\sigma^{ij} N_j t_{\theta,i} = 0, \quad \sigma^{ij} N_j t_{\varphi,i} = 0$$

$$\sigma^{ij} N_j N_i = 2\sigma H, \tag{13.74}$$

where we use for the stress tensor its three-dimensional Euclidean components, $\boldsymbol{N}$ is the unit normal to the surface, $\boldsymbol{t}_{\theta,\varphi}$ is the spherical coordinates basis of the tangent space to the surface, and H is the mean curvature of the surface (6.9).

To solve the system of nonlinear equations (13.72)–(13.74) we use the ansatz inspired by Becker et al. [16] and Brosa [31] and based on the representation Theorem 30 applied to the linearized version of the (incompressible) Navier–Stokes equation in an inertial frame

$$\frac{\partial \boldsymbol{V}}{\partial t} = -\frac{1}{\rho}\nabla P - \nu \nabla \times \nabla \times \boldsymbol{V}, \quad \nabla \cdot \boldsymbol{V} = 0. \tag{13.75}$$

The procedure is a sort of method of variation of constants doubled by an implicit substitution. Namely, we first build solutions for the linearized version of the Navier–Stokes equation, and write the velocity, pressure, and shape as series of spherical Bessel functions in r, Legendre polynomials in θ, and exponential in t, with constant coefficients. The linear coefficients of these series are calculated at $r = R_0$, which is again a linear approximation. To move to the nonlinear solution we couple the coefficients in the velocity and pressure series with the coefficients in the shape (13.69). In that, we assume that the linear constant coefficients depend actually on a_i. Also, where ever we have R_0 in the solutions we substitute it with $r(\theta, t)$. Finally, with these implicit equations at hand, we can run a numerical code.

From Theorem 30 we know that

$$\boldsymbol{V} = \nabla \times (\boldsymbol{Q}\beta) + \nabla \times \nabla \times (\boldsymbol{Q}b) + \nabla c, \quad P = -\rho \frac{\partial c}{\partial t}, \tag{13.76}$$

form a solution of (13.75) if β and b fulfill the diffusion equation

$$\nu \Delta \beta = \frac{\partial \beta}{\partial t}, \quad \nu \Delta b = \frac{\partial b}{\partial t},$$

where c is harmonic function $\Delta c = 0$ and $\boldsymbol{Q} = \boldsymbol{r}\cdot$const. From the above conditions we can build solutions for β and c

$$\beta, b \sim e^{-\lambda t} j_l\left(\sqrt{\frac{\lambda}{\nu}} r\right) Y_{lm}$$

$$c \sim e^{-\lambda t} \left(\frac{r}{R_0}\right)^l Y_{lm}, \tag{13.77}$$

where we eliminated the second type of spherical Bessel function from the solution, n_l, as being singular in $r = 0$. We obtain

$$\nabla \times (\boldsymbol{r}\beta) = \frac{1}{\sin\theta}\beta_\varphi \boldsymbol{e}_\theta - \beta_\theta \boldsymbol{e}_\varphi,$$

$$\nabla \times (\nabla \times (\boldsymbol{r}b)) = \frac{l(l+1)}{r} b\boldsymbol{e}_r + \left(b_{\theta r} + \frac{b_\theta}{r}\right)\boldsymbol{e}_\theta + \left(\frac{b_{\varphi r}}{\sin\theta} + \frac{b_\varphi}{r\sin\theta}\right)\boldsymbol{e}_\varphi,$$

$$\nabla c = \frac{l}{r} c\boldsymbol{e}_r + \frac{c_\theta}{r}\boldsymbol{e}_\theta + \frac{c_\varphi}{r\sin\theta}\boldsymbol{e}_\varphi,$$

where subscripts denote differentiation and $\{\boldsymbol{e}_r, \boldsymbol{e}_\theta, \boldsymbol{e}_\varphi\}$ is the orthonormal basis in spherical coordinates (Sect. 4.12). The velocity field (13.76) becomes

$$\boldsymbol{V} = \boldsymbol{e}_r \frac{l(l+1)b + lc}{r} + \boldsymbol{e}_\theta\left(\frac{\beta_\varphi}{\sin\theta} + b_{r\theta} + \frac{b_\theta + c_\theta}{r}\right)$$
$$+ \boldsymbol{e}_\varphi\left[-\beta_\theta + \frac{1}{\sin\theta}\left(b_{r\varphi} + \frac{b_\varphi + c_\varphi}{r}\right)\right]. \tag{13.78}$$

Because the fluid flow is divergence free, the divergence-free Newtonian tress tensor can be written in the form

$$\sigma^{ij} = \eta\left(\frac{\partial V^i}{\partial x^k} + \frac{\partial V^k}{\partial x^i}\right). \tag{13.79}$$

The boundary conditions (13.74) in spherical components are

$$\sigma^{ij} N_i N_j = \sigma^{rr},$$

$$\sigma^{ij} N_i t_{\theta,j} = \sigma^{r\theta}, \quad \sigma^{ij} N_i t_{\varphi,j} = \sigma^{r\varphi}. \tag{13.80}$$

By using (13.79) and (13.80), we obtain the components of σ^{ij} in spherical coordinates. For reference, these components can be also found in literature, like for example in Landau and Lifchitz [171].

$$\sigma^{r\theta} = \eta\left(\frac{1}{r}\frac{\partial V_r}{\partial\varphi} + \frac{\partial V_\theta}{\partial r} - \frac{V_\theta}{r}\right),$$

$$\sigma^{r\varphi} = \eta\left(\frac{1}{r\sin\theta}\frac{\partial V_r}{\partial\varphi} + \frac{\partial V_\varphi}{\partial r} - \frac{V_\varphi}{r}\right),$$

$$\sigma^{rr} = -P + 2\eta\frac{\partial V_r}{\partial r}. \tag{13.81}$$

If we want to eliminate the interface slip (that is not to take into account this phenomenon in our present solutions) we need to equate $\sigma^{r\varphi} = \sigma^{r\theta} = 0$. To do this, the only possibility is to choose $\beta = 0$, otherwise β and b have always Y_{lm} terms of different orders, and it is impossible to balance the tangent stresses. Using this ansatz and taking profit of the cylindrical symmetry of the present model we have

$$V_l = e^{-\lambda t}\left[b_l^0 \nabla \times \nabla \times \left(\mathbf{r}\, j_l\left(\sqrt{\frac{\lambda}{\nu}} r \right) P_l(\cos\theta) \right) + c_l^0 \nabla \left(\frac{r}{R_0} \right)^l P_l(\cos\theta) \right],$$

$$P_l = \rho\lambda e^{-\lambda t} c_l^0 \left(\frac{r}{R_0} \right)^l P_l(\cos\theta), \tag{13.82}$$

where b_l^0 and c_l^0 (in $\mathrm{m^2\,s^{-1}}$) are so far arbitrary initial conditions for the coefficients. The coefficients b_l describe the vortex flow and the coefficients c_l describe the potential flow.

It is natural to introduce now the hypothesis that the coefficients $a_l(t)$ of the shape function (13.69) have the same type of time dependence, to fulfill the kinematic surface condition (13.73)

$$a_l(t) = a_l^0 e^{-\lambda t}. \tag{13.83}$$

We plug (13.81–13.83) in the kinematic condition for the free interface (13.73), and in the normal and tangent stress (13.74), we obtain a 3×3 set of linear homogenous systems of equations, one for each l, in the unknowns a_l, b_l, c_l. We need to make some dimension adjustments to have in the end dimensionless determinants for the systems. Where ever it occurs, we substitute $j_{l,rr}$ with

$$j_{l,rr}\left(\sqrt{\frac{\lambda}{\nu}} r \right) = \frac{l(l+1)}{r^2} j_l - \frac{\lambda}{\nu} j_l - \frac{2}{r} j_{l,r},$$

from the corresponding spherical Bessel [5, 238, 284]. From now on we will denote the differentiation with respect to r with a prime, a second, etc. For example $j_{l,r} = j_l'$. Also, we introduce an arbitrary constant B of dimensions $\mathrm{m^2\,s^{-1}}$ and rescale the coefficients $\tilde{b}_l = b_l/B$ and $\tilde{c}_l = c_l/B$. With all these, we can write the equation for the determinant of the system of order l, taken at $r = R_0$ as linear approximation. This equation gives the compatibility condition for the systems and it results in the admissible values for λ. With the equations of the system in the order (from above) (13.73), (13.74) tangent, and (13.74) normal

$$\det \begin{pmatrix} \lambda R_0 & \frac{l(l+1) j_l B}{R_0} & \frac{lB}{R_0} \\ 0 & B\left(\frac{2l(l+1)}{R_0^2} - \frac{2}{R_0^2} - \frac{\lambda}{\nu} \right) j_l - \frac{2Bj_l'}{R_0} & \frac{2(l-1)B}{R_0^2} \\ -\frac{\sigma(l+2)(l-1)}{R_0} & \frac{2B\eta l(l+1)}{R_0}\left(j_l' - \frac{j_l}{R_0} \right) & B\left(-\lambda\rho + \frac{2\eta l(l-1)}{R_0^2} \right) \end{pmatrix} = 0. \tag{13.84}$$

With the notations

$$X = R_0 \sqrt{\frac{\lambda}{\nu}}, \quad \alpha = \left(-\frac{\sigma l(l-1)(l+2) R_0}{\rho\nu^2} \right)^{\frac{1}{4}}, \tag{13.85}$$

the determinant (13.84) becomes

$$j_l(X)\left[-4l^2(l-1)(l+2) - 2l\frac{\alpha^4}{X^2} + 2(2l^2-1)X^2 - X^4 + \alpha^4\right]$$

$$+ Xj_l'(X)\left[-2X^2 + 4l(l^2+l-2) - \frac{2\alpha^4}{X^2}\right] = 0. \tag{13.86}$$

This equation was obtained, for example, in Becker et al. [16] and same equation, in a different notation, is noted by Chandrasekhar in (280) of Article 98 in [50]. Equation (13.86) is a transcendental equation in λ, which allows only numerical solutions. As a check, we will expand it in the asymptotical limit $\nu \to 0$, i.e., $x \to \infty$. For the spherical Bessel functions we use the asymptotic formulas

$$j_l(X) \to \frac{1}{X}\sin\left(X - \frac{\pi l}{2}\right),$$

$$Xj_l' \to \cos\left(X - \frac{l\pi}{2}\right) - \frac{1}{2X}\sin\left(X - \frac{l\pi}{2}\right),$$

and we obtain

$$\lambda^2 \to \omega\frac{\sigma l(l-1)(l+2)}{\rho R_0^3}, \tag{13.87}$$

which is exactly the linear limit for the inviscid droplet oscillations (13.33), (13.68), and also [50, 167, 171]. In the approximation of small viscosity, (13.86) reduces to

$$X^4 - 2(2l+1)(l-1)X^2 - \alpha^4 = 0$$

with exact solutions

$$\lambda_{i,l} = \frac{(2l+1)(l-1)\nu}{R_0^2} \pm \sqrt{\frac{\sigma l(l-1)(l+2)}{\rho R_0^3} - \left(\frac{(2l+1)(l-1)\nu}{R_0^2}\right)^2}, \tag{13.88}$$

where i, called the radial wave label, counts the solutions of the polynomial equations, and l is called here the polar wave label. In this form, the solution for the damping and oscillating modes was obtained in [16, 167, 264, 265]. In Fig. 13.19, we present some numerical results for the general equation for λ (13.86). We plot the value of the determinant (LHS in (13.86)) function of $X = R_0\sqrt{\lambda/\nu} > 0$ parameter for several values of l and α. The real roots are numerically obtained and are represented by vertical bars in the figures. These roots form an almost periodic countable set and they are responsible for the damping or aperiodic modes. The real roots have a rather weak dependence on α (upper frames in Fig. 13.19), even if α runs in the range 5–10^7. This means that the aperiodic modes, especially the *strongly dissipative modes* for high values of real λ, are not very much influenced by the actual surface of the drop. These are modes of internal velocity fields that leave the drop surface at rest [16]. However, the dependence on l for fixed α is stronger (lower frames in Fig. 13.19), i.e., the roots are slightly shifted. This calculation also predicts existence of dissipative modes for $l = 1$, when the shape is not deformed.

The solutions of (13.86) depend on two labels i and l, where i labels solutions for a given l. We plug these solutions for X_{li} into the systems of equations, and we calculate the series coefficients for the velocity, pressure, and shape. We introduce a different notation for the initial values of the coefficients a, b, c, namely $A(0), B(0), C(0)$. We have

$$V(r,\theta,t) = \sum_l \sum_i e^{-\lambda_{li}t}[B_{li}(0)\boldsymbol{b}_{li}(r,\theta) + C_l(0)\boldsymbol{c}_l(r,\theta)], \tag{13.89}$$

where we defined

$$\boldsymbol{b}_{li}(r,\theta) = \frac{l(l+1)}{r} j_L\left(X_{li}\frac{r}{R_0}\right)P_l(\cos\theta)\boldsymbol{e}_r - \left[\frac{X_{li}}{R_0}j_l'\left(X_{li}\frac{r}{R_0}\right) + \frac{j_l\left(X_{li}\frac{r}{R_0}\right)}{r}\right]P_l'(\cos\theta)\sin\theta\,\boldsymbol{e}_\theta, \tag{13.90}$$

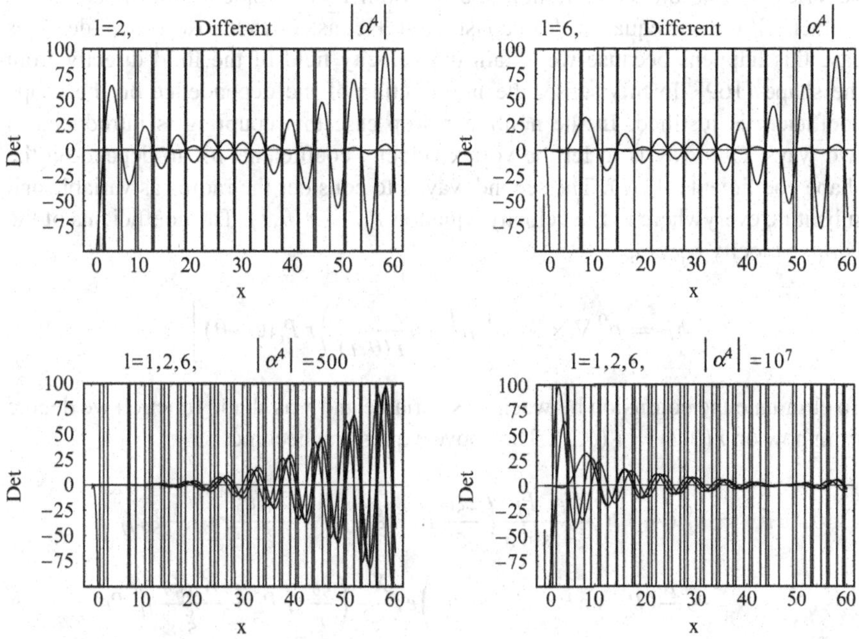

Fig. 13.19 Determinant in (13.86) plotted against X parameter showing real roots (*the vertical bars*) responsible for dissipative modes. The *upper frames* show a weak dependence of the real roots on α, but the *lower frames* show some shift in the roots induced by different l

$$c_l(r,\theta) = \frac{l r^{l-1}}{R_0^l} P_l(\cos\theta) \boldsymbol{e}_r - \frac{r^{l-1}}{R_0^l} P_l'(\cos\theta) \sin\theta \boldsymbol{e}_\theta. \tag{13.91}$$

$$r(\theta,t) = R\left(A_0 + \sum_l \sum_i A_{li}(0) e^{-\lambda_{li} t} P_l(\cos\theta)\right). \tag{13.92}$$

To introduce the contribution of nonlinearity, we generalize (13.89) to the form

$$V(r,\theta,t) = \sum_l \sum_i B_{li}(t) \boldsymbol{b}_{li}(r,\theta; A_k) + \sum_l C_l(t) \boldsymbol{c}_l(r,\theta). \tag{13.93}$$

This nonlinear ansatz consists of two main ideas. On the one hand it is breaking the fixed coupling between the time evolution of the vortex flow and the potential flow. The new coefficients $B_{li}(t)$ and $C_l(t)$ are independent, compare new (13.93) with the previous linear (and exponential time dependence) equation (13.89). This linear coupling assures that the tangent stress of any mode vanishes at the undeformed drop surface.

On the other hand, the nonlinear ansatz introduces the implicit dependence between the geometry and the dynamics. This is the typical shape-velocity coupling for nonlinear systems. For example, in the case of a one-dimensional Korteweg–de Vries soliton, the shape (amplitude A, width L) is coupled with the dynamics (velocity V) in one equation $L \sim \text{const.}/\sqrt{A} \pm \text{const.} V$. In the Korteweg–de Vries case this happens because we obtain the velocity field of the fluid directly from the shape [169]. In this sense, the introduction of the dependence on the shape coefficient is justified. In the nonlinear drop case this coupling is introduced in two ways. One way is to let the vortex velocity coefficients $\boldsymbol{b}_{li}$ to depend on the shape coefficients $A_{li}(t)$. The second way is to consider the radius as variable and substitute everywhere in the velocity equation $R_0 \to r(\theta,t)$. The coefficients of the vortex velocity become

$$\boldsymbol{b}_{li} = b_{li}^0 \nabla \times \nabla \times \left[j_l\left(X_{li} \frac{r}{r(\theta,t)}\right) \boldsymbol{r} P_l(\cos\theta)\right].$$

To eliminate the confusion between r as variable and r as shape function we denote from now on $r(\theta,t) = \xi(\theta,t)$. The above expression becomes

$$\boldsymbol{b}_{li} = b_{li}^0(A_k)\Bigg[\left(\frac{X_{li}}{\xi^2}\frac{P_l}{\sin\theta}\left(\frac{2\xi_\theta}{\xi} j_l' - \xi_{\theta\theta} j_l'' - \frac{X_{li}\xi_\theta^2}{\xi^2} r j_l'' - \xi_\theta j_l'\right)\right.$$
$$\left. + \frac{j_l}{r}(P_l'' \sin\theta - P_l' \cot\theta)\right)\boldsymbol{e}_r \frac{X_{li}}{\xi^2}\left(\frac{\xi_\theta}{r} j_l' P_l + \frac{X_{li}\xi_\theta}{\xi} j_l'' P_l\right.$$
$$\left. + \frac{\xi \sin\theta}{r} j_l' P_l'\right)\boldsymbol{e}_\theta\Bigg]. \tag{13.94}$$

With the $\boldsymbol{b}$ coefficients from (13.94), and the $\boldsymbol{c}$ coefficients from (13.91) plugged in (13.93) we have the velocity in explicit form, depending on the A_k coefficients. The vorticity can be calculated in a similar way

$$\boldsymbol{\omega} = \sum_l \sum_i b_{li}^0(A_k) \nabla \times \nabla \times \nabla \times \left[r j_l \left(\chi_{li} \frac{r}{\xi} \right) P_l \right] \boldsymbol{e}_\theta . \tag{13.95}$$

The final step in solving the nonlinear drop dynamics is to plug (13.69), (13.93), and (13.95) in the kinematical and dynamical (13.71–13.73), and minimize numerically the mean square errors.

A first consequence of taking into account the nonlinearities by the coupling between shape and vorticity is the generation of a more realistic dependence of vorticity on the distance to the center of the drop. In the linear case, the drops experience a singular concentration of vorticity in a thin layer below the drop surface for the weakly damped modes. For such modes λ is dominantly imaginary and spherical Bessel functions of imaginary argument have exponential growth. In the nonlinear case, because of the dependence $\boldsymbol{b}_{li}(A_k)$, the vorticity depends strongly on the shape. Numerical simulations show [16] an increase of vorticity below the surface where this has larger curvature and a diminishing of the vorticity under neighborhoods with low curvature. The flow in the boundary later becomes dominant when the Reynolds number, $R = (\sigma R_0/\rho)^{1/3}\nu^{-1}$, exceeds 1,000 [13]. Still, the asymptotic behavior of the spherical Bessel functions next to the surface is in effect, but is controlled by the coefficients A_k. This thin exterior layer of finite vorticity effect, also noticed [167], is again a direct consequence of the introduction of nonlinearities. Like in the one-dimensional soliton case (which we use here like a *Guinea pig* for comparison) nonlinear waves tend to occur rather in thin layers than in deep layers. Consequently, in numerical models based on the above calculations, one can split the drop in a thin exterior boundary layer where (13.94) and (13.95) are used for calculation of velocities and vorticity, and the nonlinear effects are dominant, and an inner core of spherical shape where the flow is dominantly potential [16, 203]. This approach is also used when the nonlinearities becomes stronger, like we will present in Sect. 13.3.1.

Following numerical minimization of the mean square errors of the above solutions, one can note the occurrence of specific features of the nonlinearity. The most important result is probably the fact that linear predictions are not anymore valid for modes with $l > 3$ and/or for Reynolds number larger than 100. For such higher modes the nonlinear shapes become less symmetrical and the time scale changes. Lower modes oscillate more slowly than higher modes, and higher modes decay faster than lower modes, and do not reach their linear solutions. Another typical nonlinear effect is the coupling between different modes. Through the dependence $\boldsymbol{b}_{li}(A_k)$ higher modes, of lower energy, can be generated by strong nonlinear coupling with lower modes. This effect can be detected if the coupling between two modes is time persistent and it does not depend on the initial conditions of the drop motion. For example, if we set up the initial condition with a certain shape described by a multipole of order l_0 (this can be experimentally done by applying ultracoustic waves or variable electric field on levitated droplets), after a while, new modes are excited (the new modes appear to have always the same parity as l_0) and an amplitude-dependent shift in the frequency of the initial mode is noticed. The higher modes dissipate faster than the lower ones because the mode coupling is inhibited by increase in viscosity, and so higher modes have no energy reserves to survive and die out. The coupling between modes can be detected either by checking for the coincidence in time of the extrema and zeros of different modes (Fig. 13.20), or by plotting the shape coefficients A_k of different modes in a *phase space*, i.e., plot one coefficient vs. another one in time.

Another effect induced by the nonlinearity is changing the relations between the frequency, viscosity, and the amplitude of oscillation. For small amplitude oscillations the frequency decreases monotonically with the increase of the viscosity [264, 265]. In the nonlinear case the frequency has a maximum at a value different from $\nu = 0$ [13, 318]. Not only the frequency is affected by nonlinear couplings, but also the periodicity of oscillations. For drops undergoing $l = 2$ modes there is slight tendency to spend more time in the prolate shape than in the oblate one (about 60–70% more) [321]. This asymmetric type of oscillation is a sign for the occurrence of nonlinear surface waves, like, for example, cnoidal waves. Such nonlinear waves can trigger the occurrence of solitary waves on the surface of the drop, and for stronger oscillations, can even initiate the breakup of the drop. In Sect. 13.3.1, we will show how the resonant approach will clarify the existence of such time asymmetric oscillations. Extensive examples of numerical simulations of shapes of nonlinear drops can be found in [13, 203, 321].

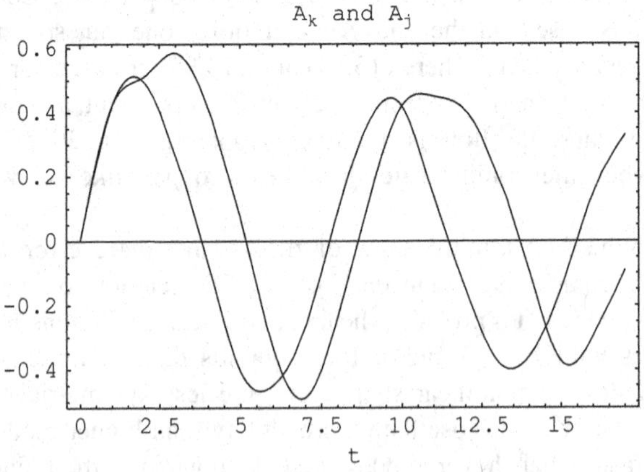

Fig. 13.20 An example of how the coupling between two different modes, k and j, can be detected by checking the simultaneous occurrence of their zeros and extrema in time

13.3.1 Nonlinear Resonances in Drop Oscillation

The approach toward analysis of nonlinear oscillations of drops presented above is based on substitution of amplitude-dependent corrections in the linear solutions. In this process, nonlinear terms that may have the same spatial dependence and frequency as some linear terms (secular terms) can alter linear oscillations in an unexpected way, or can build blowup solutions. Such solutions grow in time enough fast (usually polynomial law) to disturb the perturbational structure of the system. A nonlinear mechanism responsible for such situations is the existence of *resonant terms*. By definition, resonance involving two or more linear normal modes is possible when the frequencies of these modes are commensurate, i.e., if a linear combination of the frequencies with integer coefficients is zero. For the inviscid drop oscillations, for example (13.8), the typical low modes resonances are $\omega_4 \pm 3\omega_2 = 0, \omega_8 \pm 2\omega_5 = 0$, and $\omega_{16} \pm 2\omega_{10} = 0$, in general

$$\sum_{j=1}^{N} k_j \omega_{l_j} = 0, \quad k_j, l_j \in \mathbf{Z}. \tag{13.96}$$

Such resonances occur usually in the third order of approximation in the amplitude (smallness parameter being in this case ϵ, the ratio between the amplitude of the oscillations and the radius of the drop in equilibrium) either by cubic self-interaction of the linear modes or by interaction between the linear modes and second-order harmonics [236, 237, 321]. There is one more interest in studying resonances. They produce couplings between modes that allow transfer of energy and angular momentum between these modes in addition to the usual amplitude dependence frequency shifts discussed in Sect. 13.3.

For an inviscid linear drop the frequencies are given by (13.8). In Fig. 13.21, we present all possible resonances between modes up to $l = 100$, in comparison with the possible resonances for a bubble in linear oscillations in an inviscid fluid (13.14). The interacting modes are denoted with n, m and the resonances are denoted by symbols. The above mentioned resonances for drops are presented in this figure. We notice that the resonances for drops differ from those for the bubbles. In Fig. 13.22, we present the evolution of possible resonances for an inviscid linear drop with rigid core of radius $a = \epsilon R_0$ (13.17), function of the radius of the core. In this figure each sector of circle represents a resonance, and the angle of this sector is related to ϵ. For example, no core is represented by a black sector lying between 0 and $\pi/6$, an $\epsilon = 0.1$ core resonance is represented by a black sector lying between $\pi/6$ and $2\pi/6$, etc. In this way we can identify the figure resonances that persists when the core grows or resonances vanish. Of course $l = 100$ is nonrealistic, but it just gives an idea about the distribution of the resonances in this discrete phase space. For example, the traditional resonances at $n = 5, m = 8, n = 10, m = 16, n = 11$, and $m = 96$ are pretty stable no matter of the radius of the core, while lower modes resonances vanish when the fluid layer becomes thinner. The dependence of the

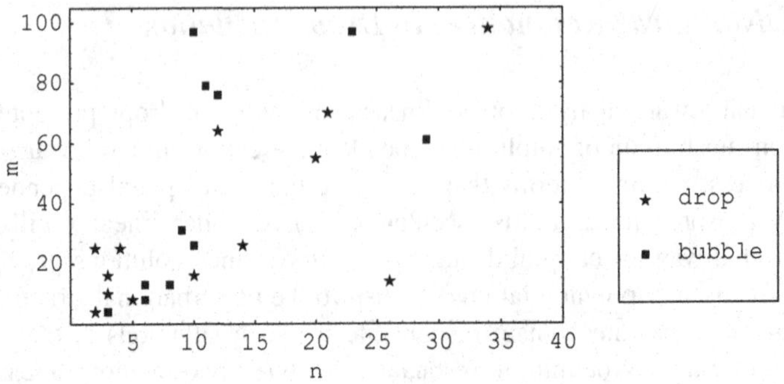

Fig. 13.21 Comparison between the resonances of linear oscillating modes ω_n and ω_m $(n < m)$ for a three-dimensional inviscid drop (stars) and a bubble (squares)

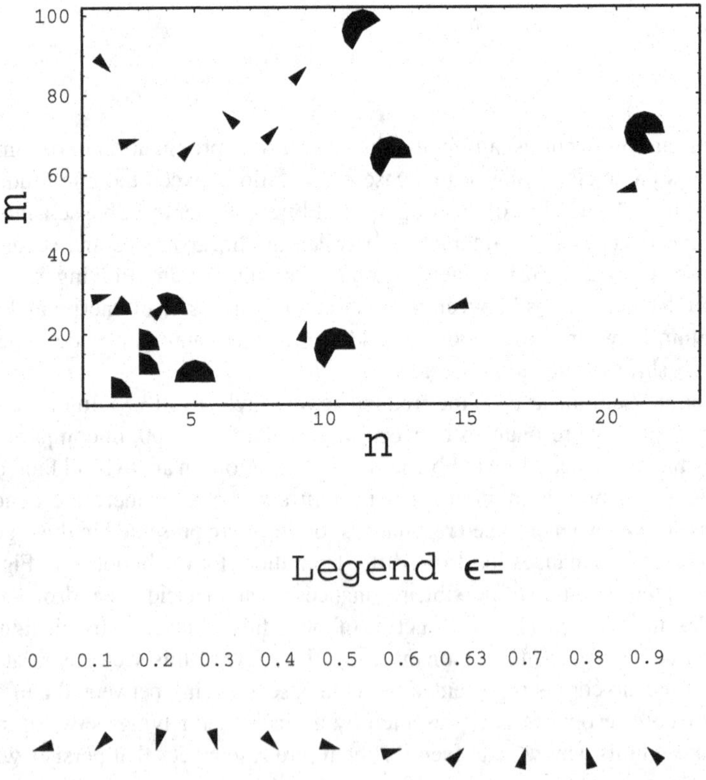

Fig. 13.22 Each sector of circle represents a resonance between two linear oscillating modes ω_n and ω_m $(n < m)$ for a three-dimensional inviscid drop with rigid core. ϵ is the radii ratio core/drop. While the core extends (in the figure the black sector rotates CCW) some resonances vanish, some new occur, and some are stable

resonance pairs from the core is not a simple or smooth function because it is given actually as a solution of a two-dimensional diophantine equation with parameter. Unexpected new resonances can become abundant next to a situation where there are no resonances. For example, in Fig. 13.22, we have for $\epsilon \in [0.4, 0.8]$ in between four and six resonant pairs, but for $\epsilon = 0.63$ we have nine resonant pairs. A numerical estimation of the density of resonances in this discrete phase space function of the radius of the core can be performed by applying the Rouche theorem for complex functions [303] to the function

$$f(n, m, \epsilon) = \sin\left(P\pi \frac{\omega_n(\epsilon)}{\omega_n(\epsilon)} \right).$$

For a given core (i.e., ϵ), when the two frequencies are commensurate, the function $f(n, m, \epsilon)$ becomes zero if the integer P is chosen sufficiently large in the domain of definition of n, m. Then we can estimate the number of zeros of f for fixed n and ϵ as a function of m, i.e., the number of possible resonances with a given n, in a given range, by the Rouche formula

$$N_{zeros} = \frac{1}{2\pi i} \oint_{\Gamma} \frac{F'(\hat{m})}{F(\hat{m})} d\hat{m},$$

where Γ is a contour surrounding the real domain of definition for n, m, $F(z) = f(n, z, \epsilon)$, and $\hat{m} \in \mathbb{C}$ is prolongation of m in the complex plane. In Fig. 13.23, we present such an estimation for $\epsilon = 0.7$ and $n, m \in [2, 100]$. The possible resonances can be found by looking for closed contours in the figure.

An efficient tool for the resonances analysis is the Lagrangian approach [202, 236, 301, 342], if the hypotheses of the flow allow its existence. For an inviscid isolated incompressible drop in potential flow we define its Lagrangian in spherical coordinates as the functional $L[\Phi, \Phi_\theta, \Phi_\varphi, \Phi_r, \Phi_t, \xi, \xi_\theta, \xi_\varphi, \xi_t, \delta P]$, where Φ is the velocity potential, ξ is the shape function defined here in the form $r|_\Sigma = \tilde{r}(\theta, \varphi, t) = R_0(1 + \epsilon\xi)$, and δP is the difference between the ambient pressure and the pressure for the spherical equilibrium shape. The action is the integral of the Lagrangian taken between two fixed moments of time. The Lagrangian depends also on the derivatives with respect to the coordinates and time. The Lagrangian density contains a term responsible for the kinetic energy density of the drop $\rho V^2/2$, one for the surface tension potential energy σdA and a Lagrange multiplier term for the volume conservation $\mathbf{V} = 4\pi R^3/3$. So the Lagrangian reads

$$L = \iiint_{\mathbf{V}} \frac{\rho(\nabla\Phi)^2}{2} d\mathbf{V} + \iint_{\Sigma} \left[\sigma + \delta P \left(\frac{\tilde{r}^3}{3} - \frac{\mathbf{V}}{4\pi} \right) \right] dA, \tag{13.97}$$

where $\mathbf{V}$ is the volume of the drop, Σ is the boundary of the drop, ρ is the density, and dA is the spherical area element. The parameter δP works here as a Lagrangian multiplier. In spherical coordinates $\boldsymbol{r} = (r(\theta, \varphi, t) \sin\theta \cos\varphi,$

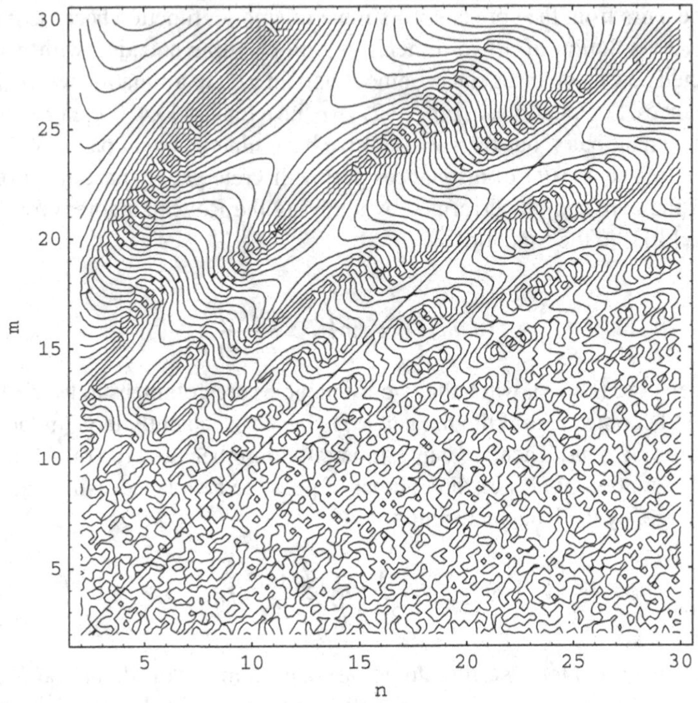

Fig. 13.23 Structure of possible resonant pairs for an inviscid drop with core

$r(\theta,\varphi,t)\sin\theta\cos\varphi, r(\theta,\varphi,t)\cos\theta)$, we have the first fundamental form coefficients (Sect. 6), $E = r^2 + r_\theta^2$, $G = r^2\sin^2\theta + r_\varphi^2$, and $F = r_\theta r_\varphi$. The area element is

$$dA = \sqrt{EG - F^2}d\theta d\varphi = r^2\sqrt{1 + \frac{r_\theta^2}{r^2} + \frac{r_\varphi^2}{r^2\sin^2\theta}}\sin\theta d\theta d\varphi.$$

With these notations the Lagrangian reads

$$L = \frac{\rho}{2}\int_0^{2\pi}\int_0^{\pi}\int_0^{\tilde{r}}\left(\Phi_r^2 + \frac{\Phi_\theta^2}{r^2} + \frac{\Phi_\varphi^2}{r^2\sin^2\theta} - 2\Phi_t\right)r^2 dr \sin\theta d\theta d\varphi$$
$$+ \int_0^{2\pi}\int_0^{\pi}\left[\sigma\hat{r}^2\left(1 + \frac{\xi_\theta^2}{\xi^2} + \frac{\xi_\varphi^2}{\xi^2\sin^2\theta}\right)^{1/2}\right.$$
$$\left. + \delta P\left(\frac{\hat{r}^3}{3} - \frac{V}{4\pi}\right)\right]\sin\theta d\theta d\varphi. \tag{13.98}$$

Next step is to expand the velocity potential and the shape function in series of orthogonal functions

$$\xi(\theta,\varphi,t) = \sum_l \Xi_l(t) Y_{lm}(\theta,\varphi),$$

$$\Phi(r,\theta,\varphi,t) = \sum_l C_l(t) r^l Y_{lm}(\theta,\varphi), \tag{13.99}$$

where the r dependence is imposed by the constraint that any term of the sum (13.99) should fulfill Laplace equation and should also be so regular in origin [236, 237]. We plug (13.99) in the Lagrangian equation (13.98) and write the corresponding Euler–Lagrange equations

$$\frac{d}{dt}\frac{\partial L}{\partial \dot{\Xi}_l} - \frac{\partial L}{\partial \Xi_l} = 0,$$

$$\frac{d}{dt}\frac{\partial L}{\partial \dot{C}_l} - \frac{\partial L}{\partial C_l} = 0, \tag{13.100}$$

where the dot represents differentiation with respect to time. The general analysis of these equations is a difficult algebraic task. For this reason the Euler–Lagrange equations are expanded themselves in series with respect to the smallness parameter ϵ. Order zero is always identical zero, and so the main analysis is concentrated on the second and third orders in this formalism. The time variation of the physical quantities is divided into two disparate time scales: the fast time scale of the primary oscillations (usually linear oscillations excited from initial conditions) and the slow time scale on which the amplitude and frequency are modulated because of the nonlinear coupling. The fast time scale is the time parameter t itself, while the slow time scale is taken as an independent coordinate $t_1 = \epsilon t$.

In the second order in ϵ, the Euler–Lagrange equations are still linear so that the time dependence of Ξ_l and C_l is exponential, i.e., $\Xi_l, C_l \sim e^{i\omega_l t}$. Moreover, the linear structure of the differential equations in (13.100) in order ϵ^2 allows us to reduce "half" of the system. Namely, we will obtain linear relations between the coefficients of the potential and shape function series expansions [236]

$$C_l = C_l^0(\omega_l, l)\Xi_l, \tag{13.101}$$

where C_l^0 are obtained directly. For example, in the case of three-dimensional inviscid isolated incompressible irrotational drop, we obtain in this order $C_l^0 = -i\omega_l/l$.

The coefficients in front of this exponential are not considered constant, like in the linear theory, but they are allowed to depend on the slow time scales to account for the resonant modulation of the amplitudes and the frequencies of the primary oscillations. This next step could be approached either by numerical procedures or by focusing on certain modes and trying to find the behavior of resonance modes. Under these approximations we chose to limit the t-time dependence of the potential and shape only through a finite number N of frequencies, namely those fulfilling a resonance condition of the type (13.96). From (13.101), for an N-coupling, we have

$$\xi(\theta,\varphi,t) = \sum_{j=1}^{N} \Xi_j(\theta,\varphi,t_1)e^{i\omega_{l_j}t}$$

$$\Phi(r,\theta,\varphi,t) = \sum_{j=1}^{N} r^{l_j} C_{l_j}^{0}(\omega_{l_j},l_j)\Xi_j(\theta,\varphi,t_1)e^{i\omega_{l_j}t}. \tag{13.102}$$

This substitution reduces the infinite number of equations in (13.100) to a finite number, reducing hence the dynamical problem to a description of the interaction of N resonant modes. Next, we plug (13.102) in L and we average L over the most rapid time scale. The procedure works if this fast scale is small compared to the other slow modulation scales. We assume, without loss of generality, that we can average with respect to the first frequency, ω_{l_1}. Through this procedure the averaged Lagrangian density becomes a functional depending only on $L_{ave}[l_j,\omega_{l_j},\Xi_j(\theta,\varphi,t1)]$. When we plug the finite sums (13.102) in the quadratic terms in (13.98), we obtain the *coupling terms*, as quadratic products of coefficients $\Xi_{l_j}\Xi_{l_k}$. We expand again the N selected Ξ coefficients in terms of spherical harmonics over the shape degeneracy of the frequencies

$$\Xi_j(\theta,\varphi,t_1) = \sum_{m_j=-l_j}^{l_j} \Xi_{l_j,m_j}^{0}(t_1)Y_{l_j,m_j}(\theta,\varphi), \tag{13.103}$$

and write again a new set of N Euler–Lagrange equations for $\Xi_{l_j,m_j}^{0}(t_1)$ that emerges in the form

$$\frac{d\Xi_{l_i,m_i}^{0}}{dt_1} = \sum_{k=1}^{N}\sum_{p=1}^{N}\sum_{m_k=-l_k}^{l_k}\sum_{m_p=-l_p}^{l_p} E_{l_k,m_k,l_p,m_p}^{l_i,m_i}\Xi_{l_k,m_k}^{0*}\Xi_{l_p,m_p}^{0*}, \quad i=1,\dots,N, \tag{13.104}$$

where $*$ represents complex conjugation. Equation (13.104) is a nonlinear ODE system of $N+2\sum_{j=1}^{N} l_j$ equations. The coefficient matrix E is not symmetric. Equation (13.104) represent a system of generalized Riccati-type equations (Sect. 18.2), and consequently, we expect it to have first integrals. For quadratic coupling the two first integrals are the total energy and the angular momentum [236]. In general, for higher orders of nonlinear coupling it is rather the exception than the rule to find $N+2\sum_{j=1}^{N} l_j$ first integrals, unless the system is integrable, and it has an infinite number of invariants and hence *soliton* solutions. Otherwise, the system behaves stochastically. The system (13.104) has always trivial stationary solutions of the form $\Xi_{l_j}^{0} = \delta_{l_j}^{l_{j0}}$const. for some $j0 = j_1,\dots,j_N$. This is an oscillation with frequency $\omega_{l_{j0}}$ corresponding to a unique mode, with shape degeneracy of the order $2l_{j0}+1$.

The interesting solutions are the time periodic ones. To accomplish an exact calculation we choose a quadratic resonance $N = 2$ similar to those presented in the beginning of the section, $k_1\omega_{l_1} + k_2\omega_{l_2} = 0, k_{1,2} \in \mathbf{Z}$. In this case the system (13.104) reduces to two ordinary differential equations in t_1, in the two Ξ^0 functions. We mention that, because we integrate the initial Lagrangian over the period of one of the two resonant frequencies, say ω_{l_1}, the averaged Lagrangian is not symmetric in the two frequencies or in the two shape functions Ξ^0. We have

$$\frac{d\Xi^0_{l_1,m_1}}{dt_1} = \sum_{m=-l_2}^{l_2} \sum_{n=-l_2}^{l_2} A^{l_1,m_1}_{l_2,l_2,m,n} \Xi^{0*}_{l_2,m} \Xi^{0*}_{l_2,n}$$

$$\frac{d\Xi^0_{l_2,m_2}}{dt_1} = \sum_{m=-l_1}^{l_1} \sum_{n=-l_2}^{l_2} A^{l_2,m_2}_{l_1,l_2,m,n} \Xi^{0*}_{l_1,m} \Xi^{0*}_{l_2,n}, \tag{13.105}$$

where the coefficients A are obtained directly from (13.104), and in general are represented by products of $3 - j$ symbols and rational functions of $l_{1,2}$ [236, 237]. To work a simple example we assume that we confine all the energy of the drop in one single component of each of the $l_{1,2}$ modes. This reduces the summations in (13.105) to one single term in the RHS of each equation, and for compatibility reasons this is $m = m_1 = -2m_2 = -2n$

$$\frac{d\Xi^0_{l_1,m_1}}{dt_1} = A_1(\Xi^{0*}_{l_2,m})^2$$

$$\frac{d\Xi^0_{l_2,m_2}}{dt_1} = A_2 \Xi^{0*}_{l_1,m_1} \Xi^{0*}_{l_2,m_2}, \tag{13.106}$$

where $A_{1,2}$ is just a simplified way of writing the coefficients from (13.105) in the no-summation case. In the following, we use a procedure similar to those used in the nonlinear Schrödinger equation, namely break the functions in a magnitude and a complex phase

$$\Xi^0_{l_j,m_j} = R_j e^{i\theta_j}, \quad R_j(t_1), \theta_j(t_1) \in \mathbb{R}. \tag{13.107}$$

If we plug these forms in (13.106) and separate the real and imaginary parts, we form a system of four real differential equations for R_i, θ_i. By multiplying the real and imaginary parts of the first equation in (13.106) and then we subtract them, we have

$$A_1 R_2^2 \sin(\theta_1 + \theta_2) + A_1 R_1 \frac{d\theta_1}{dt_1} = 0. \tag{13.108}$$

By substituting this derivative of θ_1 back in the real part of the first equation in (13.106) we obtain

$$\frac{dR_1}{dt_1} = \frac{\cos 2\theta_2 - \sin\theta_1 \sin(\theta_1 + 2\theta_2)}{\cos\theta_1} A_1 R_2^2. \tag{13.109}$$

We use another simplification hypothesis, namely we choose that the phases do not depend on time, i.e., the derivatives of $\theta_{1,2}$ are zero, which implies from (13.108) the constraint $\theta_1 = -2\theta_2$. In this situation (13.109) reduces to

$$\frac{dR_1}{dt_1} = CR_2^2, \tag{13.110}$$

where C is the abbreviation for the constant resulting from (13.109). From the real and imaginary part of the second equation in (13.106) we have

$$\cos(\theta_1 + \theta_2)A_2R_1R_2 = \cos\theta_2 \frac{dR_2}{dt_1},$$

$$\sin(\theta_1 + \theta_2)A_2R_1R_2 = -\sin\theta_2 \frac{dR_2}{dt_1}, \tag{13.111}$$

and it results

$$\frac{dR_2}{dt_1} = A_2R_1R_2\frac{\cos(\theta_1 + \theta_2)}{\cos\theta_2},$$

or simply denoted

$$\frac{dR_2}{dt_1} = DR_1R_2. \tag{13.112}$$

From the (13.110) and (13.112) we obtain the relation

$$R_2^2 = \left(\frac{D}{C}\right)^2 R_1^2 + E, \tag{13.113}$$

with E arbitrary constant of integration. If we plug the invariant (13.113) in the system (13.110) and (13.112) the equations for $R_{1,2}$ decouple and the resulting equation for say R_1 becomes

$$\frac{d^2R_1}{dt_1^2} = 2CDER_1 + 2D^2R_1^3. \tag{13.114}$$

But this last equation is just the differential equation for the *Jacobi elliptic functions* (18.3). One solution for (13.114) is the well-known *cnoidal* cos function

$$R_1(t_1) = R_1^0 \mathrm{cn}(\lambda t_1 + t_1^0|m), \tag{13.115}$$

with the relation between its amplitude R_1^0 and scaling coefficient λ given by $D^2(R_1^0)^2 = \lambda^2$ and the *modulus* m given by $\lambda^2(m+2) = -2CDE$. The number t_1^0 is an arbitrary constant. The modulus m is a real number between 0 and 1 and is related to the period T of the cnoidal cos function $\mathrm{cn}(\lambda t_1 + T|m) = \mathrm{cn}(\lambda t_1|m)$, namely $T = 4K(m)$, where $K(m)$ is the *complete elliptic integral of the first kind*

(Sect. 18.3). For $m = 0$ the cn function is precisely the regular cos. For $m \in (0, 1)$ the function is still periodic and oscillating and the period increases with m. In the limit $m = 1$ $\text{cn}(\lambda t_1|1) \to \text{sech}(\lambda t_1)$. The other mode has the amplitude

$$R_2(t_1) = \sqrt{E + \frac{D}{C}(R_1^0)^2 \text{cn}^2(\lambda t_1 + t_1^0|m)}. \qquad (13.116)$$

It results that the motion of the drop in the quadratic resonance, under the above simplifications, is a nonlinear oscillation whose period and amplitude are strongly dependent on the initial conditions. In some specific limit the motion becomes aperiodic and slows down toward an asymptotic approach toward an equilibrium position, i.e., the profile of a soliton in time (in the slow time scale). However, this is just radial oscillations, and no actual solitary wave travels on the surface of the drop. This is because we neglected from the beginning the vorticity of the velocity.

The aspect of the cnoidal solution for values of m close to 1 suggests an explanation for the different amounts of time that the drop spends in different shapes, contrary to the case of a linear oscillation. In Fig. 13.24, we present the graphics of $R_{1,2}(t_1)$ for two values of m. We note the odd distribution of different amplitudes in time. We also note the coupling between the two modes, since they oscillate in phase. Example of drop shapes for the quadratic resonance for $l_1 = 5$ and $l_2 = 8$ are given in Fig. 13.25.

Some cnoidal oscillations with same resonance $l_1 = 5$ and $l_2 = 8$ are presented in Fig. 13.26 in a cross-section in the vertical yz-plane ($\varphi = \pi/2$). In this case we choose $m_1 = m_2 = 0$ (axial symmetry), $\Xi_1^0 = \Xi_2^0$ (equal contribution of both modes), the modulus of the cnoidal oscillation $m = 0.9$, and the perturbation $\epsilon = 0.3$. The period of the oscillation is $T = 4K(m) = 10$ s. We note how the energy is transferred back and forth from the $l = 5$ mode to the $l = 8$ mode. From upper left to lower right, in frame 1 we have a mixture of 5+8 modes, then in frame 2 we have a pure $l = 5$ mode, then in the next two frames we have $l = 8$. In the

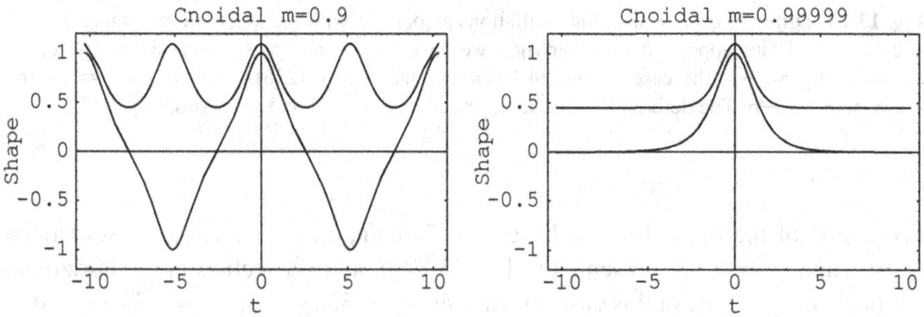

Fig. 13.24 The two amplitudes vs. time for a quadratic resonance in a nonlinear drop. These are a cnoidal oscillation, $R_1(t_1)$ (the larger amplitude oscillation), and the oscillation of $R_2(t_1)$ from (13.116)

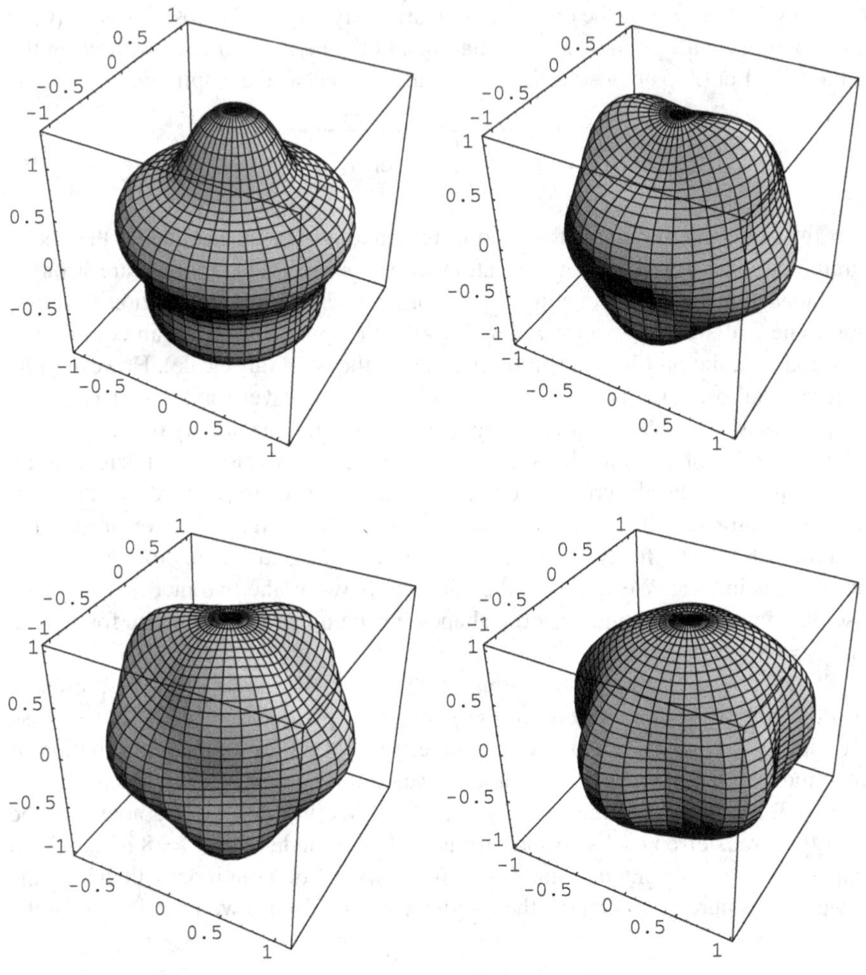

Fig. 13.25 Different drop shapes for oscillations associated with the quadratic resonance $l_1 = 5$ and $l_2 = 8$. From upper left to lower right we have $m_1 = m_2 = 0$ (axial symmetric case), $m_1 = 2, m_2 = -1$ (the case studied in Natarajan and Brown [236]), $m_1 = 3, m_2 = 5$, and $m_1 = 5, m_2 = -6$. The deformation is characterized by $\epsilon = 0.3$, and we choose $\Xi_1^0 = \Xi_2^0$

first frame of the lower line we have $l = 5$ again, etc. More cnoidal oscillations $l_1 = 5$ and $l_2 = 8$ are presented in Fig. 13.27 in a cross-section in the horizontal xy-plane ($\theta = \pi/2$). In this case we choose $m_1 = 3$,$m_2 = 8$, $\Xi_1^0 = \Xi_2^0$, $m = 0.9$, and the perturbation $\epsilon = 0.3$. We note again energy transfer and coupling between the modes: From upper left to lower right, in frame 1 we have a $l = 3$ mode, then in the next four frames we have $l = 8$ modes, and in the last two frames we have a mixture $8 + 3$, and back toward a $l = 3$ mode.

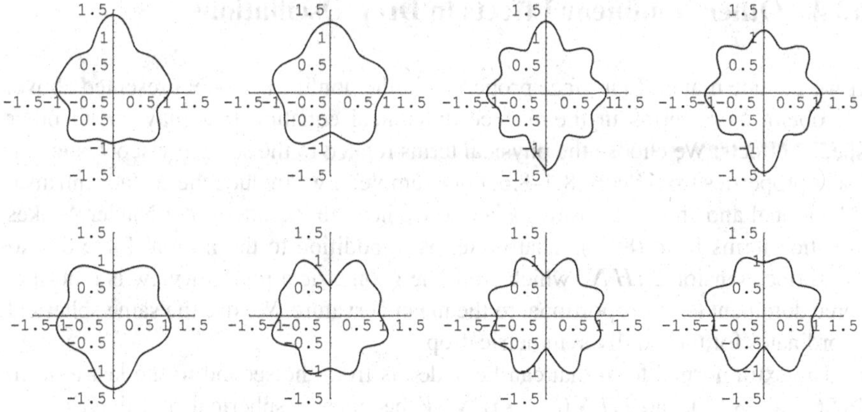

Fig. 13.26 Cnoidal oscillations $l_1 = 5$, $l_2 = 8$ in the yz-plane ($\varphi = \pi/2$) for axial symmetry $m_1 = m_2 = 0$, $\Xi_1^0 = \Xi_2^0$, $m = 0.9$, $\epsilon = 0.3$

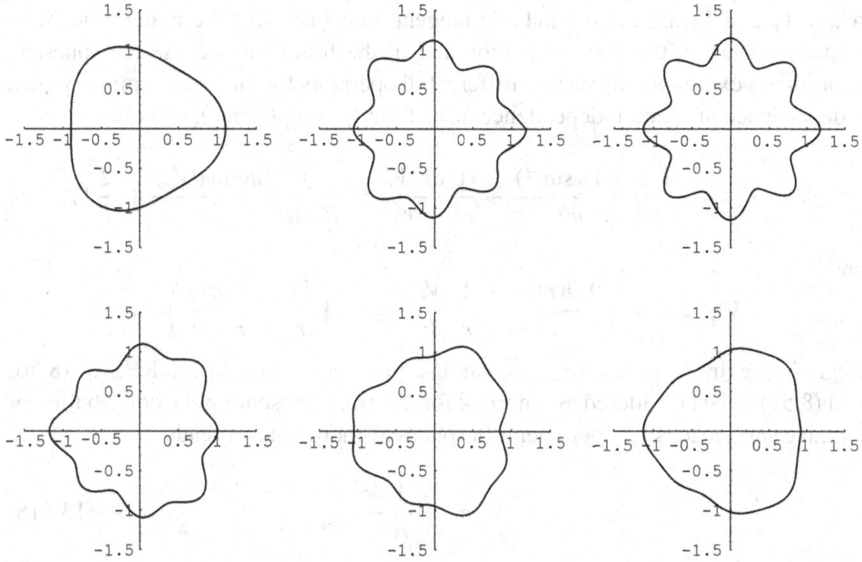

Fig. 13.27 Cnoidal oscillations $l_1 = 5$, $l_2 = 8$ in the xy-plane ($\theta = \pi/2$) for $m_3, m_2 = 8$, $\Xi_1^0 = \Xi_2^0$, $m = 0.9$, $\epsilon = 0.3$

More complex oscillations can be described if we choose to keep all the terms in the summations in (13.105). Numerical calculations show, however, that axisymmetric drop oscillations are unstable to nonaxial symmetric perturbations. Also, in this section we have omitted the effect of cubic or higher-order resonances which will complicate the interactions.

13.4 Other Nonlinear Effects in Drop Oscillations

If we include material interface properties in the nonlinear model presented above, we obtain extra terms in the surface dynamical equation that may create other special effects. We choose the physical terms related to the surface viscoelastic and shear properties from Sects. 8.4–8.6. For example, if we include the surface intrinsic dilatational and shear viscosities k and ϵ we need to include in the Navier–Stokes equation terms from (8.57). That is, terms in addition to the normal force due to the surface tension $2\sigma H N$, which from the geometrical point of view is a normal force dominant term proportional to the mean curvature. We use the same spherical coordinates for the axially symmetric drop.

The extra normal term that can be added is from the second to the last term in (8.56) and (8.57), and $2HN(k+\epsilon)\hat{B}\nabla_\Sigma V$ becomes in spherical coordinates

$$2HN(k+\epsilon)\hat{B}\nabla_\Sigma V \to \frac{2\epsilon}{R}\left(\frac{1}{R\sin\theta}\frac{\partial(V_r)\sin\theta}{\partial\theta}+\frac{2V_r}{R}\right), \tag{13.117}$$

where V_r and V_θ are the normal and tangent components of the material velocity on the surface. In the above equation and in the following we use the spherical coordinates expression of surface differential operators for the axial symmetry (i.e., independence of φ and independence of r of the V components), namely

$$\nabla_\Sigma \cdot V = \frac{1}{r\sin\theta}\frac{\partial(V_\theta\sin\theta)}{\partial\theta}+\frac{1}{r^2}\frac{\partial r^2 V_r}{\partial r} = \frac{1}{R\sin\theta}\frac{\partial(\sin\theta V_\theta)}{\partial\theta}+\frac{2}{R}V_r,$$

and

$$\nabla_\Sigma \times V = \left(\frac{1}{r}\frac{\partial(rV_\theta)}{\partial r}-\frac{1}{r}\frac{\partial V_r}{\partial\theta}\right)e_\varphi = \left(\frac{V_\theta}{r}-\frac{1}{R}\frac{\partial V_r}{\partial\theta}\right)e_\varphi.$$

Regarding extra tangent terms, we can use the second term in the RHS of (8.56) and (8.57) to be considered as a normal force term, representing the contribution of variable surface tension coefficient. In spherical coordinates it reads

$$\hat{B}\nabla_\Sigma\sigma \to \frac{1}{R}\frac{\partial\sigma}{\partial\theta}. \tag{13.118}$$

Also, from the term $(k+\epsilon)\hat{B}\nabla_\Sigma(\hat{B}\nabla_\Sigma\cdot V)$ we have in spherical coordinates

$$(k+\epsilon)\hat{B}\nabla_\Sigma(\hat{B}\nabla_\Sigma\cdot V) \to \frac{k+\epsilon}{R}\frac{\partial}{\partial\theta}\left(\frac{1}{R\sin\theta}\frac{\partial(\sin\theta V_\theta)}{\partial\theta}\right). \tag{13.119}$$

Another tangent term can occur from the double curl operator

$$-\frac{\epsilon}{g}\hat{B}\nabla_\Sigma\times(\nabla_{Sigma}\times\hat{B}V) \to \frac{2\epsilon V_\theta}{R^2}+\frac{2\epsilon}{R^2}\frac{\partial V_r}{\partial\theta}. \tag{13.120}$$

A comprehensive study of the effects of these surface viscoelastic terms was performed numerically in Tian et al.[316]. In the same spirit of Sects. 13.2 and 13.3 and (13.84) and (13.117), we calculate characteristic equations, i.e., the determinant of the linear system of equations for the series expansion coefficients as a function of the complex oscillation frequencies λ. From the Navier–Stokes equation (13.71) plus all the viscoelastic terms from (13.117) to (13.120) we obtain an equation similar to the condition (13.86)

$$\alpha^2 + X_k - 12X_\epsilon + \frac{1}{\alpha^2} + R_0\sqrt{\frac{\lambda}{\nu}}\frac{j_{l+1}\left(R_0\sqrt{\frac{\lambda}{\nu}}\right)}{j_l\left(R_0\sqrt{\frac{\lambda}{\nu}}\right)}$$

$$\times\left[\frac{16X_kX_\epsilon}{\alpha^2} - (1+\alpha^4)\frac{3X_k - 4X_\epsilon}{\alpha^4}\right] = 0, \tag{13.121}$$

where

$$\alpha^2 = \frac{\lambda}{\rho\omega_L}, \quad X_\epsilon = \frac{\nu R_0 + \epsilon}{\rho R_0^3\omega_L},$$

$$X_k = \frac{\sigma_s(c^*) - \sigma_0}{\sigma_s(c^*)} + \frac{2(\nu R_0 - k)}{\rho R_0^3}, \quad \omega_L^2 = \frac{\sigma_0 l(l-1)(l+2)}{\rho R_0^3}.$$

In these equations we take into account the variation of the surface tension coefficient with concentration of the surfactant, c^*, namely $\sigma_s(c^*), \sigma_s(0) = \sigma_0$.

The results of numerical calculation of the roots of (13.121) [316] show that, in addition to the roots we found from previous equations, there is one additional one generated by the term responsible for the surface elasticity, namely the first term in X_k. This new root is equivalent to the occurrence of a new type of longitudinal surface waves. These waves are strongly damped, unless there is a nonzero tangent gradient of surface tension (Marangoni effect). These new modes can be excited by applying an external tangent stress along the droplet surface or by the nonlinear coupling between the shape oscillating modes. In Sect. 13.5 we show that such longitudinal modes can be modeled with nonlinear equations of modified Korteweg–de Vries type, having for solutions cnoidal waves or their limiting solution, solitary waves.

A very good review on experimental results, and some theoretical trends about liquid drops, breaking-up, and collision is done in Eggers [84].

13.5 Solitons on the Surface of Liquid Drops

Several experiments and numerical tests [77, 154, 182, 236, 237, 298] performed on droplets suggest the existence of standing traveling waves on the surface [77,154]. In this section we introduce a slightly different nonlinear liquid drop model, compared to the models treated in the previous sections of this chapter. The differences consist first in retaining higher-order nonlinear terms in the dynamical equations, and second, by searching especially for traveling surface oscillations, instead of combined radial and transverse modes. The result is that we obtain surface waves in the form of cnoidal functions that approach in limiting cases solitary waves [189–192]. In the following we present two parallel approaches: the traditional Euler equation approach and a Hamiltonian approach, both leading to the same result. The same model adapted for microscopic systems is considered again in Sect. 16.3. Another particular feature of this model is that instead of the traditional series expansion in terms of spherical harmonics, we use other types of localized functions defined on the sphere surface.

We restrict our model to inviscid irrotational flow; therefore, we have a velocity potential governed by the Laplace equation $\Delta\Phi = 0$, and the dynamics is described by Euler's equation,

$$\rho\left(\frac{\partial}{\partial t}\boldsymbol{v} + (\boldsymbol{v}\cdot\nabla)\boldsymbol{v}\right) = -\nabla P + \boldsymbol{f}, \tag{13.122}$$

where P is pressure. If the density of the external force field is also potential, $\boldsymbol{f} = -\nabla\Psi$, where Ψ is proportional to the potential (gravitational, electrostatic, etc.), then (13.122) reduces to Bernoulli's scalar equation. We apply two types of boundary conditions: one on the external free surface of the drop, Σ_1, and one on an inner rigid core surface of radius a, Σ_2. These types of boundary conditions are also used in literature [169, 182, 236, 237, 298]. We can express the boundary conditions in the form

$$\left.\frac{dr}{dt}\right|_{\Sigma 1} = \left(\frac{\partial r}{\partial t} + \frac{\partial r}{\partial\theta}\frac{\partial\theta}{\partial t} + \frac{\partial r}{\partial\varphi}\frac{\partial\varphi}{\partial t}\right)_{\Sigma_1}, \quad \left.\frac{\partial r}{\partial t}\right|_{\Sigma_2} = 0,$$

respectively. The radial velocity and tangential velocities are, respectively,

$$\frac{\partial\Phi}{\partial r} = \frac{\partial r}{\partial t}, \quad \frac{\partial\Phi}{\partial\theta} = r^2\frac{\partial\theta}{\partial t}, \quad \frac{\partial\Phi}{\partial\varphi} = r^2\sin\theta\frac{\partial\varphi}{\partial t}.$$

The second boundary condition is applied only if the drop has some rigid core inside or in the case of liquid shells. An interesting situation which, to our present knowledge, was not yet studied experimentally is when the liquid layer is bounded from outside by a rigid circumference and the free surface is toward inside. For example, a shallow layer of liquid adhering on the inner surface of a hollow sphere. A convenient geometry places the origin at the center of mass of the distribution and, according to our previous hypothesis concerning the traveling waves, the shape is described by

$$r(\theta,\varphi,t) = R_0[1 + \xi(\theta,\varphi,t)] = R_0[1 + g(\theta)\eta(\varphi - Vt)].$$

$\xi = g\eta$ is dimensionless function. Here R_0 is the radius of the undeformed spherical drop and V is the tangential velocity of the traveling solution ξ moving in the φ direction and having a constant transversal profile $g(\theta)$ in the θ direction. We mention that the linearized form of the first boundary condition

$$\frac{\partial r}{\partial t}_{\Sigma_1} = \frac{dr}{dt}|_{\Sigma 1},$$

allows only radial vibrations and no tangential motion of the fluid on $\Sigma 1$, [169, 182, 236, 237, 298], and so nonlinearity is mandatory for the existence of this tangent traveling modes. The second boundary condition restricts the radial flow to a spherical layer of depth $h(\theta)$ by requiring $\Phi_r|_{r=R_0-h} = 0$. This condition stratifies the flow in two layers: the surface layer, $R_0 - h \le r \le R_0(1+\xi)$, and the liquid bulk, $r \le R_0 - h$. This is again a typical situation in nonlinear, irrotational, or viscous flow. Usually, inside compact domains of flow, the external layer develops irrotational flow, while the inside bulk is potential, and they separate in a natural way [50, 91, 167, 171]. In what follows the flow in the bulk will be considered negligible compared to the flow in the surface layer. This condition does not restrict the generality of the argument because $h \in [0, R_0]$ is still arbitrary at this stage. Nonetheless, keeping $h < R_0$ opens possibilities for the investigation of more complex fluids, e.g., superfluid, flow over a rigid core, multilayered systems [216, 236, 237, 319] or multiphasic, etc. Instead of an expansion of Φ in term of spherical harmonics, consider the following form

$$\Phi(r,\theta,\varphi,t) = \sum_{n=0}^{\infty}\left(\frac{r}{R_0} - 1\right)^n f_n(\theta,\varphi,t). \tag{13.123}$$

The convergence of the series is controlled by the value of the small quantity $\epsilon = max|\frac{r-R_0}{R_0}|$ [169]. The condition $max|h/R_0| \simeq \epsilon$ is also assumed to hold in the following development. Laplace's equation introduces a system of recursion relations for the functions f_n, namely

$$f_n = [(-1)^{n-1}(n-1)\Delta_\Omega f_0 - 2(n-1)f_{n-1} + \sum_{k=1}^{n-2}(-1)^{n-k}\frac{(2k-(n-k-1)\Delta_\Omega f_k)}{n(n-1)}, \quad n > 2, \tag{13.124}$$

where

$$\Delta_\Omega = \frac{1}{\sin\theta}\frac{\partial}{\partial\theta}\left(\sin\theta\frac{\partial}{\partial\theta}\right) + \frac{1}{\sin^2\theta}\frac{\partial}{\partial\varphi}$$

is the angular Laplacian operator in spherical coordinates. Equation (13.124) reduces the unknown functions to only two, $\Delta_\Omega f_0$ and f_1:

$$f_2 = -\frac{1}{2}(\Delta_\Omega f_0 + 2f_1),$$

$$f_3 = \frac{1}{6}(4\Delta_\Omega f_0 - 4\Delta_\Omega f_1 + 4f_1 + 2),$$

$$f_4 = \frac{1}{24}(\Delta_\Omega^2 f_0 - 14\Delta_\Omega f_0 + 8\Delta_\Omega f_1 - 8f_1) \quad \ldots . \tag{13.125}$$

If f_0 is harmonic on the sphere surface, still the series does not reduce to spherical harmonics, because in the second order we have again Laplacian of f_1. In a special case when all f_n are harmonic, the series is determined by f_1 only. If we choose the independent functions $\Delta_\Omega f_0$ and f_1 to be smooth on the sphere, they must be bounded together with all the f_ns (these being linear combinations of higher derivatives of f_0 and f_1) and hence the convergence of the series in (13.123) is controlled by these two functions only.

The second boundary condition plus the condition of having a traveling wave along φ only: $\xi_\varphi = -V\xi_t$, yield, up to second order in ϵ,

$$f_{0,\varphi} = VR_0^3 \sin^2\theta \xi(1 + 2\xi)/h + \mathcal{O}_3(\xi), \tag{13.126}$$

i.e., a connection between the flow potential and the shape, which is typical of nonlinear systems. Equation (13.126) together with the relations

$$f_1 \simeq R_0^2 \xi_t \simeq \frac{2h}{R_0} f_2 \simeq -\frac{h\Delta_\Omega f_0}{R_0 + 2h}, \tag{13.127}$$

which follow from the boundary condition and recursion, characterize the flow as a function of the surface geometry. The balance of the dynamic and capillary pressure across the surface $\Sigma 1$ follows by expanding up to third order in ξ the square root of the surface energy of the drop

$$U_S = \sigma R_0^2 \int_{\Sigma 1} (1 + \xi)\sqrt{(1 + \xi)^2 + \xi_\theta^2 + \xi_\varphi^2/\sin^2\theta}\, d\Sigma, \tag{13.128}$$

and by equating its first variation with the local mean curvature of $\Sigma 1$ under the restriction of the volume conservation. The surface pressure, in third order, reads

$$P|_{\Sigma 1} = \frac{\sigma}{R_0}(-2\xi - 4\xi^2 - \Delta_\Omega \xi + 3\xi\xi_\theta^2 ctg\theta), \tag{13.129}$$

where σ is the surface pressure coefficient. Equation (13.129) was obtained in a general frame in Sect. 10.4, too.

In the above equation, and subsequently, we consider that for all the surface wave and perturbations studied with this model, the relative amplitude of the deformation ϵ is smaller than the angular half-width L, i.e.,

$$\xi_{\varphi\varphi} \sim \xi_{\theta\varphi} \sim \xi_{\theta\theta} \sim \epsilon^2/L^2 \ll 1, \tag{13.130}$$

as most of the experiments [130, 154, 216, 318, 319] concerning traveling surface patterns show. This is a typical request in shallow water soliton deduction, too [2, 169]. Consequently, we can neglect the terms $\xi_{\varphi,\theta}$, $\xi_{\varphi,\varphi}$, and $\xi_{\theta,\theta}$ in this approximation. We comment here that, after solving the dynamical equations for the surface traveling waves and obtaining cnoidal and solitary solutions, we plugged these solutions back in the dynamical equation to compare the orders of magnitude of different terms (Fig. 13.28). The comparison of these terms appears to be in good agreement with the approximations in (13.130).

Equation (13.126) plus the boundary conditions yield, to second order in ϵ,

$$\Phi_t|_{\Sigma 1} + \frac{V^2 R_0^4 \sin^2\theta}{2h^2}\xi^2 = \frac{\sigma}{\rho R_0}(2\xi + 4\xi^2 + \Delta_\Omega \xi$$
$$-3\xi^2 \xi_\theta \mathrm{cotan}\theta). \tag{13.131}$$

The linearized version of (13.131) together with the linearized boundary condition, $\Phi_r|_{\Sigma 1} = R_0 \xi_t$, yield a limiting case of the model, namely, the normal modes of oscillation of a liquid drop with spherical harmonic solutions [182, 298]. Differentiation of (13.127) and (13.131) with respect to φ yields the dynamical equation for the evolution of the shape function $\eta(\varphi - Vt)$:

$$A\eta_t + B\eta_\varphi + Cg\eta\eta_\varphi + D\eta_{\varphi\varphi\varphi} = 0, \tag{13.132}$$

which is the Korteweg–de Vries (KdV) equation with coefficients depending parametrically on θ

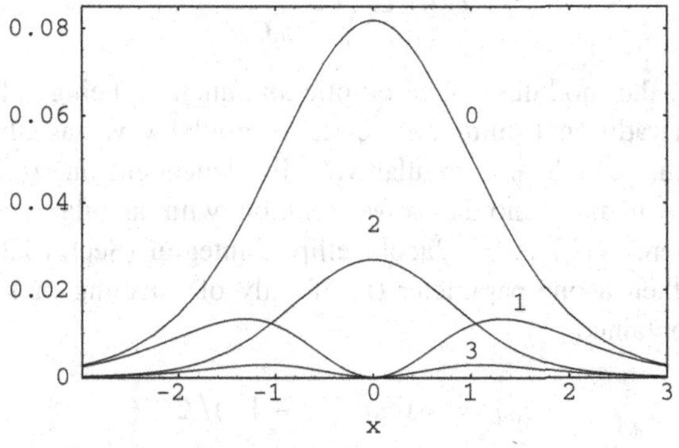

Fig. 13.28 Plot of different terms of different orders of magnitude in (13.131), after we found the solutions, as a general check of the expansions. It is easy to check that the approximations performed were appropriate

$$A = V\frac{R_0^2(R_0 + 2h)\sin^2\theta}{h}, \quad B = -\frac{\sigma}{\rho R_0}\frac{(2g + \Delta_\Omega g)}{g},$$
$$C = 8\left(\frac{V^2 R_0^4 \sin^4\theta}{8h^2} - \frac{\sigma}{\rho R_0}\right), \quad D = -\frac{\sigma}{\rho R_0 \sin^2\theta}. \tag{13.133}$$

In the case of a two-dimensional liquid drop, the coefficients in (13.133) are all constant. Equation (13.132) has traveling wave solutions in the φ direction if $Cg/(B-AV)$ and $D/(B-AV)$ do not depend on θ. These two conditions introduce two differential equations for $g(\theta)$ and $h(\theta)$, which can be solved with the boundary conditions $g = h = 0$ for $\theta = 0, \pi$. For example, $h_1 = R_0 \sin^2\theta$ and $g_1 = P_2^2(\theta)$ is a particular solution that is valid for $h \ll R_0$. It represents a soliton with a quadrupole transverse profile, being in good agreement with [236, 237, 318]. We mention that the next higher-order term in (13.131), $-3\xi^2\xi_\theta ctg\theta$, introduces a $\eta^2\eta_\varphi$ nonlinear term into the dynamics and transforms the KdV equation into the MKdV equation. The traveling wave solutions of (13.132) are then described by the Jacobi elliptic function (Sect. 18.3)

$$\eta(\theta, \varphi, t) = \eta_1 + \eta_0 \,\mathrm{sn}^2\left(\frac{\varphi - Vt}{L}\middle| k\right), \tag{13.134}$$

where the η_0 and η_1 are the constants of integration introduced through (13.132) and are related to half-width and the velocity (Sect. 18.4) by

$$V(\theta) = \frac{B}{A} + \frac{Cg}{3A}\left[\eta_0\left(1 + \frac{1}{k}\right) + \eta_1\right]$$

and

$$L(\theta) = \sqrt{-\frac{12kD}{\eta_0 Cg}}$$

with $k \in [0, 1]$, the modulus of the elliptic sn function, being a free parameter. Different from a traditional soliton, this circular cnoidal wave has all its parameters, amplitude, width, period, and angular velocity dependent on θ. This result for (13.134) is known as a cnoidal wave solution with angular period $T(\theta) = 4K[k]L(\theta)$, where $K(k)$ is the Jacobi elliptic integral (Sect. 18.3). If $m \to 1$ and $T \to \infty$ then a one-parameter (η_0) family of traveling pulses (solitons or antisolitons) is obtained,

$$\eta_{sol} = \eta_0 sech^2[(\varphi - Vt)/L], \tag{13.135}$$

with velocity

$$V = \frac{b}{A} + \frac{Cg}{3A}\left(\eta_0 + 3\eta_1\right),$$

and angular half-width $L = \sqrt{-12D/Cg\eta_0}$. Taking for the coefficients A to D the values given in (13.133) for $\theta = \pi/2$ (the equatorial cross-section), one can calculate numerical values of the parameters of any cnoidal excitation, function of the constants η_0, η_1, k and the structure functions $g(\theta)$, $h(\theta)$. The solitary waves, among other wave patterns, have a special shape–kinematic dependence $\eta_0 \simeq V \simeq 1/L$; a larger amplitude perturbation is narrower and travels faster. This relation can be used to experimentally distinguish solitons from other modes or turbulence. When a layer thins ($h \to 0$) the coefficient C in (13.133) approaches zero on average, producing a break in the traveling wave solution (L becomes singular) because of the change of sign under the square root (13.134). Such wave turbulence from capillary waves on thin shells was first observed in Holt and Trinh [130]. For the water shells described there, (13.133) gives $h(\mu m) \leq 20\nu/k$, i.e., $h = 15$–$25\,\mu$ m at $V = 2.1$–$2.5\,\text{ms}^{-1}$ for the onset of wave turbulence, in good agreement with the abrupt transition experimentally noticed (ν is the kinematic viscosity). The cnoidal solutions provide the nonlinear wave interaction and the transition from competing linear wave modes ($C \leq 0$) to turbulence ($C \simeq 0$). In the KdV (18.8), the nonlinear interaction balances or even dominates the linear damping and the cnoidal (roton) mode occurs as a bend mode (h small and coherent traveling profile). The condition for the existence of a positive amplitude soliton is $gCD \geq 0$ which, for $g \leq 0$, limits the velocity from below to the value $V \geq h\omega_2/R_0$, where ω_2 is the Lamb frequency for the $l = 2$ linear mode.

This inequality can be related to the “independent running wave” described in [318], which lies close to the $l = 2$ mode. We stress that here we describe the equatorial modes, i.e., standing traveling profiles in the φ direction, and so the Legendre polynomials $P_l(\cos\varphi)$ we talk about are defined on φ. The periodic limit of the cnoidal wave is reached for $k \simeq 0$, and the shape is characterized by harmonic oscillations ($sn \to \sin$ in (13.134)) which realize the quadrupole mode of a linear theory P_2 limit [182, 236, 237, 298] or the oscillations of Legendre polynomials (Fig. 13.29). In Fig. 13.30, we present a cross-section in two solitary waves traveling along the equator.

The NLD model introduced in this paper yields a smooth transition from linear oscillations to solitary traveling solutions (“rotons”) as a function of the parameters η_0, η_1, k; namely, a transition from periodic to nonperiodic shape oscillations. In between these limits the surface is described by nonlinear cnoidal waves. In Fig. 13.29, some configurations from this transition from a periodic limit to a solitary wave are shown, in comparison with the corresponding normal modes that can initiate such cnoidal nonlinear behavior. This situation is similar to the transformation of the flow field from periodic modes at small amplitude to traveling waves at larger amplitude. The solution goes into a final form if the volume conservation restriction is enforced: $\int_\Sigma (1 + g(\theta)\eta(\varphi,t))^3 d\Omega = 4\pi$ and requires $\eta(\varphi,t)$ to be periodic. The periodicity condition

$$2n\pi = K(k)L,$$

for any positive integer n, is only fulfilled for a finite number of n values, and hence a finite number of corresponding cnoidal modes. In the roton limit the periodicity condition becomes a quasiperiodic one because the amplitude decays rapidly. This approach could be extended to describe elastic modes of surface as well as their nonlinear coupling to capillary waves. The double-periodic structure of the elliptic solutions [169] could describe the new family of normal wave modes predicted in Tian et al. [316].

Because the Euler equations reduced to an integrable equation, we expect that the system should have a Hamiltonian attached to it, at least in some order of approximation. In Natarajan and Brown [237] the drop has associated a Lagrangian with volume conservation condition being a Lagrange multiplier. In the third order of smallness the dynamical equation inferred from hydrodynamics becomes a KdV infinite-dimensional Hamiltonian system described by a nonlinear Hamiltonian function $H = \int_0^{2\pi} \mathcal{H} d\varphi$. In the linear approximation, the system has a linear wave Hamiltonian. If terms depending on θ are absorbed into definite integrals (becoming

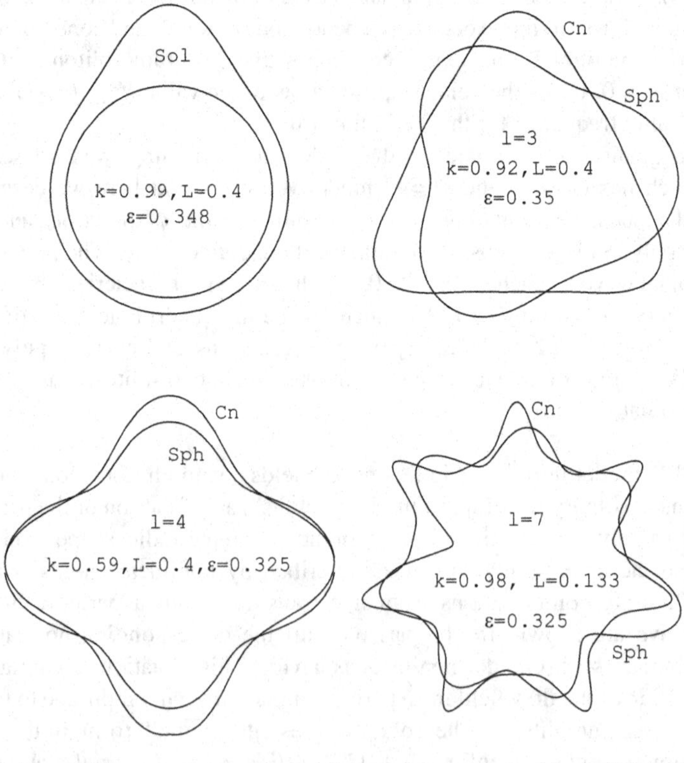

Fig. 13.29 Equatorial cross-sections ($\theta = 0$) in a drop excited with cnoidal surface waves (13.134). The soliton limit plus rigid core and a 3-, 4-, and 7-mode solution are shown, together with the closest matching Legendre functions for each cnoidal wave for comparison. The labels l for the corresponding Legendre polynomials $P_l(\cos\varphi)$ and the parameters k, L, and $\epsilon = \eta_0/R_0$, of the corresponding cnoidal solution, are given

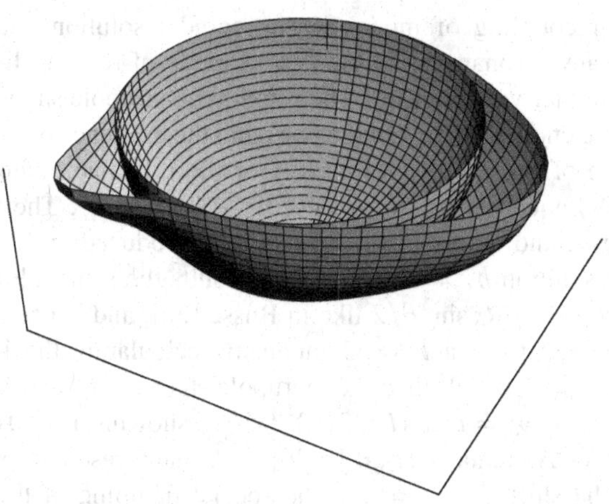

Fig. 13.30 Cross-section of the droplet excited by two solitary waves traveling along the equator

parameters) the total energy is a function of η only. Taking the kinetic energy from Natarajan and Brown [237], Φ from (13.123), and using the boundary conditions, the dependence of the kinetic energy on the tangential velocity along θ direction, Φ_θ, becomes negligible and the kinetic energy can be expressed as a $T[\eta]$ functional. For traveling wave solutions $\partial_t = -V\partial_\varphi$, to third order in ϵ, after a tedious but feasible calculus, the total energy is

$$E = \int_0^{2\pi} (C_1\eta + C_2\eta^2 + C_3\eta^3 + C_4\eta_\varphi^2)d\varphi, \tag{13.136}$$

where $C_1 = 2\sigma R_0^2 S_{1,0}^{1,0}$, $C_2 = \sigma R_0^2(S_{1,0}^{1,0} + S_{0,1}^{1,0}/2) + R_0^6\rho V^2 C_{2,-1}^{3,-1}/2$, $C_3 = \sigma R_0^2 S_{1,2}^{1,0}/2 + R_0^6\rho V^2(2S_{-1,2}^{3,-1}R_0 + S_{-2,3}^{5,-2} + R_0 S_{-2,3}^{6,-2})/2$, $C_4 = \sigma R_0^2 S_{2,0}^{-1,0}/2$, with $S_{i,j}^{k,l} = R_0^{-l}\int_0^\pi h^l g^i g_\theta^j \sin^k\theta d\theta$. Terms proportional to $\eta\eta_\varphi^2$ can be neglected since they introduce a factor η_0^3/L^2, which is small compared to η_0^3, i.e., it is in the third order in ϵ. If (13.136) is taken to be a Hamiltonian, $E \to H[\eta]$, then the Hamilton equation for the dynamical variable η, taking the usual form of the Poisson bracket, gives

$$\int_0^{2\pi} \eta_t d\varphi = \int_0^{2\pi} (2C_2\eta_\varphi + 6C_3\eta\eta_\varphi - 2C_4\eta_{\varphi\varphi\varphi})d\varphi. \tag{13.137}$$

For the function $\eta(\varphi - Vt)$ the LHS of (13.137) is zero. Consequently, KdV solitons $\eta(\varphi)$, with appropriate choice of parameters, are allowed solutions, since they cancel the integral on the RHS, too. Hence, the energy of the NLD model, in the third order, is interpreted as a Hamiltonian of the KdV equation. This is in full agreement with the result finalized by (13.132) for an appropriate choice of the parameters and the Cauchy conditions for g and h.

The nonlinear coupling of modes in the cnoidal solution could explain the occurrence of many resonances for the $l = 2$ mode of rotating liquid drops at a given (higher) angular velocity [36]. The rotating quadrupole shape is close to the soliton limit of the cnoidal wave. On the one hand the existence of many resonances is a consequence of the multivalley profile of the effective potential energy for the KdV (MKdV) equation: $\eta_x^2 = a\eta + b\eta^2 + c\eta^3 + d\eta^4$. The frequency shift predicted by Busse and others in [8, 36] can be reproduced in the present theory by choosing the solution $h_1 = R_0 \sin\theta/2$. It results the same additional pressure drop in the form of $V^2\rho R_0^2 \sin^2\theta/2$ like in Busse [36], and hence a similar result. For a roton emerged from a $l = 2$ mode, by calculating the half-width (L_2) and amplitude ($\eta_{max,2}$) which fit the quadrupole shape, it results in a law for the frequency shift: $\Delta\omega_2/\omega_2 = (1 \pm 4L^2/3R_0)^{-1} V/\omega_2$, showing a good agreement with the observations of Annamalai et al. [8, 36], i.e., many resonances and nonlinear dependence of the shift on $\Omega = V$. The special damping of the $l = 2$ mode for rotating drops could also be a consequence of the existence of the cnoidal solution. An increase in the velocity V produces a modification of the balance of the coefficients C/D, which is equivalent with an increasing in dispersion.

13.6 Problems

1. Comments on (13.40). The solution used in the text is appropriate for the analysis of drop oscillations, but this equation has a richer spectrum of solutions. If we substitute $rV_R = \phi(r)$ and write the equation like $\Delta D\phi = D\Delta\phi = 0$, with the operator D being the parenthesis in (13.40) we have two classes of solutions. Solutions with property $D\phi = 0$ belong to the class represented in (13.49). Another possibility is to have $\Delta\phi = \psi \neq 0$. Then, it is convenient to solve $D\psi = 0$, since we know the solutions from (13.40). With ψ such a solution we obtain ϕ as the integral representation of the radial Laplace equation.
2. For the cnoidal solution defined on the sphere in (13.134) calculate the total angular momentum of the flow in the drop. Since the initial hypothesis was irrotational motion, we expect this angular momentum to be zero for this solution.
3. Improve the model presented in Sect. 13.5 by introducing the relative motion of the center of mass. Solutions should be considered at least in pairs to have the position of the center of mass unchanged.
4. Verify if cnoidal waves or solitary waves exist on the surface of a drop by plugging a solitary wave type of test solution for the surface directly in the Euler equations for the drop.
5. Find a property of the nonlinear tangential surface wave from Sect. 13.5, and implications on the surface velocity field based on the *hairy ball theorem*, i.e., there is no zero smooth, regular tangent vector field on the two-sphere.

Chapter 14
Other Special Nonlinear Compact Systems

In this chapter we present an interesting back up of the previous chapters devoted to solitons on closed free surfaces, like drops. Namely, one can predict the possibility of existence of such exotic shapes from some first geometric principles. In the frame of geometric collective models, for example, it can be shown that these types of shapes can be created through the formalism of nonlinear symmetry groups. We conclude the chapter by presenting an example of Hamiltonian structure for systems with free closed boundaries.

14.1 Nonlinear Compact Shapes and Collective Motion

In Sect. 13.5 we introduced a special nonlinear mode of oscillation of a liquid droplet in terms of cnoidal functions and solitary waves. Similar nonlinear compact shapes can be obtained by using a different geometrical approach, namely by an integrable nonlinear theory of a many-body system. The theory was applied in the geometric and Riemannian ellipsoidal models for large amplitude collective modes of oscillations in heavy nuclei [277, 279, 320].

A geometrical model of collective motion is defined by a group of transformations, called the *motion group*, of the three-dimensional Euclidean space. The motion group acts on the Euclidean space and, among other things, transforms surfaces into other surfaces. For example, the rotation group $SO(3)$ is the linear group of motion for the rigid bodies in mechanics, and it is also the adiabatic rotational model in nuclear physics. Another example is the real general linear three-dimensional group $GL(3,\mathbb{R})$, which is the group for the Riemannian ellipsoidal models in fluid dynamics or elasticity and also yields the microscopic extension of the Bohr–Mottelson nuclear model [49, 87]. However, such traditional models have limitations imposed by the linear character of the transformation. The classes of shapes generated by these linear groups can never include exotic shapes like hour-glass, breakup droplet shapes, fissionable shapes, toroidal shapes, etc.

A. Ludu, *Nonlinear Waves and Solitons on Contours and Closed Surfaces*, Springer Series in Synergetics, DOI 10.1007/978-3-642-22895-7_14,

A nonlinear geometrical model, if algebraically closed under commutation, can construct collective models compatible with such nonlinear shapes. Such a collective model will be integrable on behalf of the closeness property. It could be applied to many-body collective motion problems in astronomy (nonelliptical galaxies, tides in neutron stars, cosmic object collisions), in plasma physics, nuclear physics of heavy ions and superheavy elements, mean field theory, and geometric quantization.

To construct a nonlinear motion group we need first a Hilbert space H of wave functions. We recall that a Hilbert space is a Euclidean space E of vectors over the complex numbers which is complete in the norm. The norm is defined in the Euclidean space as a function $||\cdot|| : E \to \mathbb{R}^+$, which assigns a positive length or size to all vectors in the space, other than the zero vector. For example, we can introduce the space of the square integrable functions as the space of complex integrable functions defined on E having finite value for the integral of the square over the whole space

$$\int_E |f(x)|^2 dx < \infty,$$

denoted $L_2(E)$. Consequently, the above integral is the norm of the $L_2(E)$ space. A norm is complete if any Cauchy sequence of functions from the space has a limit in this space. Not any Euclidean space is Hilbert, but the good news are that any Euclidean space can be densely embedded in some Hilbert space. The Hilbert space of wave functions we need in the following construction of a nonlinear collective model is $L_2(\mathbb{R}^3)$. Its vectors are functions $\Psi(\boldsymbol{r},t) : \mathbb{R}^3 \to \mathbb{C}$. For more details about the construction of this space, as well as for more details about the operators acting in it, the reader can consult one of the best books in axiomatic quantum mechanics, namely [266].

Next object we need for the geometric model is a nonlinear Lie algebra (Definition 19) of operators acting on the Hilbert space of wave functions. We consider for any real number Λ the following nine differential operators acting on $L_2(\mathbb{R}^3)$

$$N_{jk} = x_j p_k - i\delta_{jk}\frac{\hbar}{2} + \frac{\Lambda x_j x_k}{r^5}\boldsymbol{r}\cdot\boldsymbol{p}, \tag{14.1}$$

with $j,k = 1,\dots,3$, x_k are Euclidean coordinates in $\mathbb{R}^3$ (so it is not important if they carry covariant or contravariant indices) and $\boldsymbol{p} = -i\hbar\nabla$ is the momentum operator. The Planck constant $\hbar$ has the common meaning from quantum mechanics. All these nine operators are Hermitian operators, i.e., $\forall\Phi, \Psi \in H$ the equality holds

$$\int_{\mathbb{R}^3} \Phi^* N_{jk}\Psi d^3\boldsymbol{r} = \left(\int_{\mathbb{R}^3} \Psi^* N_{jk}\Phi d^3\boldsymbol{r}\right)^*,$$

which guaranties that N_{jk} represent physical observables. The most important fact about the N_{jk} operators is that they are closed under commutation relations, i.e., for any two N_{jk} and N_{il} we have $[N_{jk}, N_{il}] = c_{jkilmp}N_{mp}$, with c_{jkilmp} being complex constants. Let us denote N the set of all possible linear combinations of the N_{jk}

operators with real coefficients. This structure is a Lie algebroid [300]. However, we can consider it a Atiyah algebra which is a generalization of a Lie algebra.

For any 3×3 matrix $X = (X_{ij})$ with real entries we can build the mapping $\sigma : M_3(\mathbb{R}) \to N$ given by $\sigma(X) = (i/\hbar)X_{jk}N_{jk}$. Such a mapping is a linear representation of the Lie algebra $M_3(\mathbb{R})$ in the Lie algebra N. For example, we can introduce a representation defined by the following operators

$$\begin{aligned}
L_l &= \epsilon_{jkl} N_{jk} = x_j p_k - x_k p_j, \\
T_{jk} &= N_{jk} + N_{kj} - \frac{2}{3}\delta_{jk}\,\mathrm{Tr}(N) \\
&= \left(x_j p_k + x_k p_j - \frac{2}{3}\delta_{jk} \boldsymbol{r} \cdot \boldsymbol{p} \right) + \frac{2\Lambda}{r^5}\left(x_j x_k - \frac{1}{3}\delta_{jk} r^2 \right) \boldsymbol{r} \cdot \boldsymbol{p}, \\
S &= \mathrm{Tr}(N) = \left(1 + \frac{\Lambda}{r^3} \right) \boldsymbol{r} \cdot \boldsymbol{p} - \frac{3}{2} i\hbar.
\end{aligned} \tag{14.2}$$

The first three operators are closed under commutation and generate the rotation group Lie algebra $so(3)$ of the angular momentum $\boldsymbol{L}$. These operators together with the next ones, called the quadrupole vibration operators, T_{jk}, are also closed and form a Lie algebra isomorphic with $sl(3, \mathbb{R})$ of traceless matrices from $M_3(\mathbb{R})$. The name comes from the fact that their average value over a wave function provides the third order term in a spherical harmonics expansion. Finally, all the operators in (14.1) and (14.2), including the nonlinear operator S, are closed under commutation, and so they generate the Lie algebra N.

To involve the geometry we map these operators into a system of differential vector fields defined on $\mathbb{R}^3$

$$V_{jk} = \frac{i}{\hbar} N_{jk} - \frac{\delta_{jk}}{2},$$

which reads in the Euclidean coordinates

$$V_{jk} = \left(x_j \delta_{lk} + \frac{\Lambda x_j x_k x_l}{r^5} \right) \frac{\partial}{\partial x_l}. \tag{14.3}$$

Some of these vector fields are divergence free, i.e., $\nabla \cdot V_{jk} = \delta_{jk}$, and so they generate transformations that conserve the volume. If $\Lambda = 0$ these vector fields become linear, and they generate the six-dimensional Lie algebra of rotations and dilations (when we make $\Lambda = 0$, only six generators remain independent, while the other three reduce to Casimir elements). The nine vector fields in (14.3) generate a nine-dimensional Lie algebra, and the exponential of these vector field (i.e., infinitesimal generators) form the associate local Lie motion group. This structure is only a local Lie group because for some values of Λ the vector fields are not complete (Sect. 4.4), and their exponential is integrable only locally. In the following we are looking for the classes of Euclidean compact surfaces that are left invariant by the nonlinear motion local group elements. In the original article, Rosensteel uses the adjoint representation of this Lie algebra to find invariant surfaces. An alternate possibility to find the invariants is to use the method of the symmetry group of differential equations [242]. According to Sect. 4.7, a smooth real function $F : \mathbb{R}^3 \to \mathbb{R}$ is invariant to the action of the motion local Lie group if

$$V_{jk}(F) = 0 \text{ for all } j,k = 1,\ldots,3. \tag{14.4}$$

To construct the invariant functions we use (14.3) and the associate characteristic system of equations becomes

$$\frac{dx_1}{x_j\delta_{k1} + \Lambda \cdot \frac{x_j x_k x_1}{r^5}} = \frac{dx_2}{x_j\delta_{k2} + \Lambda \cdot \frac{x_j x_k x_2}{r^5}} = \frac{dx_3}{x_j\delta_{k3} + \Lambda \cdot \frac{x_j x_k x_3}{r^5}}, \tag{14.5}$$

for $j,k = 1,\ldots,3$. The general solution of system (14.5) is given by the six symmetric functions

$$Q_{jk} = \left(1 + \frac{\Lambda}{r^3}\right)^{\frac{2}{3}} x_j x_k, \tag{14.6}$$

plus some arbitrary constants. These functions are the linear independent invariant functions of the motion local Lie group, and any linear combination of them is also an invariant function. In Rosensteel and Troupe [277] it is proved that these functions also generate a six-dimensional Lie algebra, which in semidirect product with the Lie algebra generated by the vector fields V_{jk} form a 15-dimensional Lie algebra called $gcm(3)$, i.e., the Lie algebra of the *nonlinear motion group*. Its corresponding local Lie group is $GCM(3)$. For different values of Λ these algebras are isomorphic, but their physical interpretation varies. The surfaces parametrized by the implicit equation

$$\sum_{jk} C_{jk}\left(1 + \frac{\Lambda}{r^3}\right)^{\frac{2}{3}} x_j x_k = C_0, \tag{14.7}$$

are invariant surfaces to the local Lie group $GCM(3)$ for any combination of constants $C_{jk}, C_0, j,k = 1,\ldots,3$. Of course, only the symmetric sets of constants count. In other words, if for a given choice of the constants C_{jk} and C_0 we generate a surface by (14.7), this surface will be left unchanged by the action of any of the group transformations, i.e., the $GCM(3)$ local group transforms a drop surface in another allowable drop surface of the model. The surfaces described by (14.7) are compact (Sect. 6.4) if the C_{jk} are a real positive-definite symmetric matrix. Actually, if $\Lambda = 0$, these functions reduce to ellipsoids of different semiaxes and orientation in space. Since C_{jk} are symmetric they can be diagonalized, and actually, only the diagonal elements count for different surfaces. Since the $gcm(3)$ Lie algebra contains the infinitesimal rotations $so(3)$ as a subalgebra, the nondiagonal coefficients C_{jk} generate same surfaces like the diagonal ones,

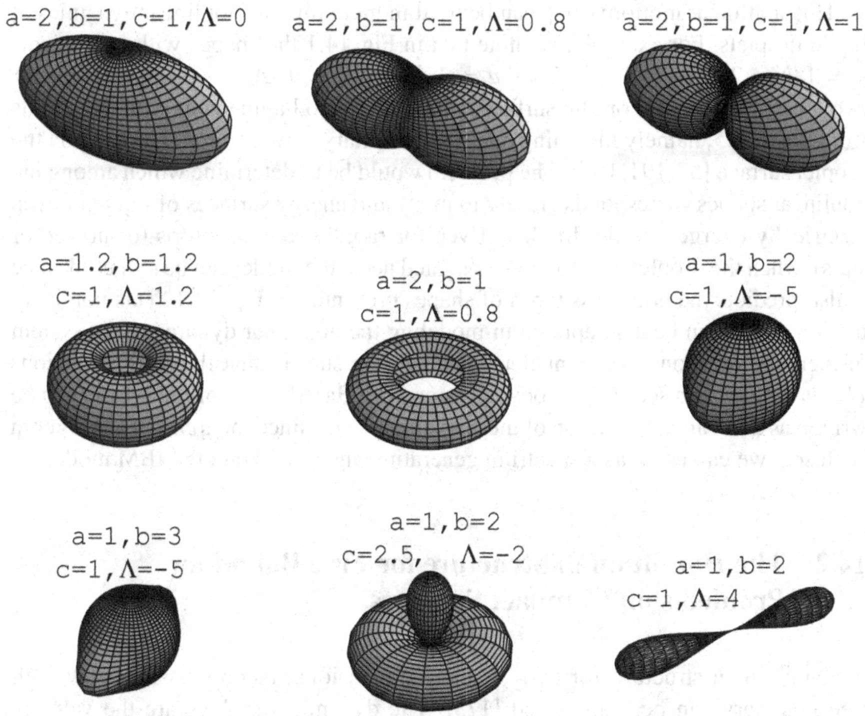

Fig. 14.1 Examples of compact surfaces invariant to the nonlinear motion group $GCM(3)$. The values of the three diagonal parameters and the nonlinear one are given next to each surface

except they are rotated. Compact nonlinear surfaces are generated by diagonal elements (C_{11}, C_{22}, C_{33}) with positive signature. From dimensional analysis we note that these coefficients are m^{-2} units, and so a better physical notation for them is (a^{-2}, b^{-2}, c^{-2}). Actually, the numbers a, b, c represent the semiaxes of the ellipsoidal surfaces generated by (14.7) for $\Lambda = 0$. In Fig. 14.1, we present some typical nonlinear surfaces obtained through (14.7). To plot these surfaces we just write (14.7) in polar coordinates

$$r(\theta, \varphi) = \left[-\Lambda \pm \left(\frac{C_0}{\frac{\sin^2\theta\cos^2\varphi}{a^2} + \frac{\sin^2\theta\sin^2\varphi}{b^2} + \frac{\cos^2\theta}{c^2}} \right)^{\frac{3}{2}} \right]^{\frac{1}{3}}.$$

For positive values of Λ the invariants Q_{jk} are well defined in all the points, except the origin. For axially symmetric solutions, the deformed droplets are surfaces of revolutions with a central neck. When $\Lambda = a^3, b^3$, or c^3 the neck reduces to zero diameter and the drop breaks-up in two symmetric parts, like in a fission process. If $\Lambda < 0$ the invariant functions are not defined all over the space, and so this negative Λ motion local group could model droplets with missing parts, i.e., smaller droplets emission, fusion, exotic bubble shapes, and two-fluid models.

This nonlinear motion group can be used in modeling the nonlinear dynamics of liquid droplets. For example, we note that in Fig. 14.1 the shapes with coefficients $a = 1, b = 2, c = 1, \Lambda = -5$ and $a = 1, b = 3, c = 1, \Lambda = -5$ represent one or two localized bump(s) on the surface, which is in good agreement with the results from Sect. 13.5, namely modeling one or two solitary waves (*rotons*) moving on the droplet surface [57, 191, 192]. The problem would be to determine which among the nonlinear shapes corresponds closely to minimum energy surfaces of liquid or even electrically charged liquid droplets. Even for rapidly rotating drops (or nuclei, or stars), when the droplet develops an elongated neck, the model can be still used since it also predicts the hourglass types of shapes presented in Fig. 14.1. This nonlinear motion group can be also applied in modeling the nonlinear dynamics of a system of identical fermions, like a nucleus or a neutron star. In that the set of functions Q_{jk} is defined as a set of one-body operators.The Hamiltonian of the system can be written as a linear combination of these operators, and since the $gcm(3)$ Lie algebra is closed, we can use it as a spectrum generating algebra, like in the IBM model.

14.2 The Hamiltonian Structure for Free Boundary Problems on Compact Surfaces

A Hamiltonian structure for two- or three-dimensional incompressible flow with free boundary can be constructed [173]. The dynamic variables are the velocity field $\boldsymbol{V}$ and the compact surface Σ that surrounds the fluid domain D_Σ. These two entities form the basic phase space $N = \{(|vec V, \Sigma)\}$ for the representation of the canonical bracket. Incompressibility condition assures $\nabla \cdot \boldsymbol{V} = 0$. According to the representation formulas in Ebin and Marsden [82] we can write the velocity as $\boldsymbol{V} = \boldsymbol{V}_{\parallel} + \nabla \Phi$, where $\boldsymbol{V}_{\parallel}$ is both divergence free and tangent to Σ. The potential is determined modulo an additive constant by

$$\Delta \Phi = 0, \quad \frac{\partial \Phi}{\partial N} = \boldsymbol{V} \cdot \boldsymbol{N}, \tag{14.8}$$

where $\boldsymbol{N}$ is the normal to Σ. We introduce three types of formal derivatives for functions $F : N \to \mathbb{R}$

$$\frac{\delta F}{\delta \boldsymbol{V}}, \text{ defined by } \left(\frac{\partial F}{\partial \boldsymbol{V}}\right)_\Sigma (\boldsymbol{V}, \Sigma) \cdot \delta \boldsymbol{V} = \int_{D_\Sigma} \frac{\delta F}{\delta \boldsymbol{V}} \cdot \delta \boldsymbol{V} d^3 x, \tag{14.9}$$

where $(\partial F / \partial vec V)_\Sigma = d_c F / d\boldsymbol{V}$ is the convective derivative with respect to $\boldsymbol{V}$ (see (9.16) in Sect. 9.2.6).

$$\frac{\delta F}{\delta \Phi} = \frac{\delta F}{\delta \boldsymbol{V}} \cdot \boldsymbol{N}, \tag{14.10}$$

$$\frac{\delta F}{\delta \Sigma}, \text{ defined by } \left(\frac{\partial F}{\partial \Sigma}\right)_V (\boldsymbol{V}, \Sigma) \cdot \delta \Sigma = \int_\Sigma \frac{\delta F}{\delta \Sigma} \delta \Sigma d^3 x. \tag{14.11}$$

With these three derivatives we can introduce the Poisson bracket of F, G in the form

$$\{F, G\} = \int_{D_\Sigma} \omega \cdot \left(\frac{\delta F}{\delta V} \times \frac{\delta G}{\delta V} \right) d^3x + \int_\Sigma \left(\frac{\delta F}{\delta \Sigma} \frac{\delta G}{\delta \Phi} - \frac{\delta G}{\delta \Sigma} \frac{\delta F}{\delta \Phi} \right) dA, \tag{14.12}$$

where $\omega = \nabla \times V$ is the vorticity. This bracket makes the phase space N into a Poisson manifold, satisfies Jacobi's identity, is real bilinear, antisymmetric, and it is a derivation in F and G. For irrotational flow (14.12) reduces to a canonical bracket in Φ and Σ. The authors in Lewis et al. [173] provide an interesting example of application of this Hamiltonian system to the dynamics of an incompressible (we choose $\rho = 1$) inviscid liquid drop with free boundary and surface tension. We recall the dynamical equation in this case: Euler, boundary, incompressibility, and surface tension balance, namely

$$V_t + (V \cdot \nabla)V = -\nabla P, \quad \Sigma_t = V \cdot N, \quad \nabla \cdot V = 0, \quad P_\Sigma = 2\sigma H, \tag{14.13}$$

with H the mean curvature of Σ. The Hamiltonian is

$$H = \frac{1}{2} \int_{D_\Sigma} V^2 d^3x + \sigma \int_\Sigma dA. \tag{14.14}$$

We have the following theorem. Equation (14.13) is equivalent to

$$F_t = \{F, H\}.$$

The proof is by direct calculation and it can be found in Lewis et al. [173].

This Hamiltonian approach can be applied directly to some nonlinear compact systems. There are situations, for example, on spheres, when the solutions of the dynamical system can be expressed as spherical harmonics plus small corrections, and these solutions retain this property for a long time, i.e., they are *near-monochromatic*. Such a situation is provided by a free surface potential flow of a fluid layer surrounding a gravitating sphere. The dynamical equations for traveling or standing water waves are obtained in a weakly nonlinear gravitational interaction on a sphere. Some numerical and classical perturbation theory studies [178, 252] proved that these solutions possess Hamiltonian structure.

We consider a spherical fluid layer of depth h surrounding a sphere of radius b in spherical coordinates (r, θ, φ). The outer free surface has the equation

$$r(\theta, \varphi, t) = b + h + \eta(\theta, \varphi, t),$$

and we assume that the flow inside the layer is potential. The Euler equation for the free surface potential flow takes the form

$$\frac{\partial \eta}{\partial t} = \frac{\partial \Phi}{\partial \theta} - \frac{1}{r^2} \frac{\partial \Phi}{\partial \theta} \frac{\partial \eta}{\partial \theta} - \frac{1}{r^2 \sin^2 \theta} \frac{\partial \Phi}{\partial \phi} \frac{\partial \eta}{\partial \phi},$$

and at the free surface we have

$$\frac{\partial \Phi}{\partial t} = -\frac{1}{2}|\nabla \Phi|^2 + \frac{1}{b+h+\eta}.$$

In the region occupied by the fluid we have $\Delta\Phi = 0$ and at the bottom $r = b$ we have

$$\frac{\partial \Phi}{\partial r} = 0.$$

The wave amplitude $\eta(\theta,\varphi)$ and the surface potential $\Phi(\theta,\varphi) = \Phi(\theta,\varphi,b+h+\eta(\theta,\varphi))$ determine uniquely the hydrodynamic potential Φ inside the layer at any moment of time t. The above equations can be written as a Hamiltonian system where the canonical variables are η and Φ at the surface. The kinetic energy term in the Hamiltonian can be formally expanded in powers of the wave amplitude η. We can write

$$H = \sum_{j=0}^{\infty} H_j,$$

with the first two terms in the series

$$H_0 = \frac{(h+b)^2}{2}\sum_{\gamma}\left(\frac{u'_\gamma(h+b)}{u_\gamma(h+b)}\Phi_\gamma\Phi^*_\gamma + \eta_\gamma\eta^*_\gamma\right) \tag{14.15}$$

and

$$H_1 = \sum_{\gamma_1,\gamma_2,\gamma_3} I_{\gamma_1,\gamma_2,\gamma_3}\Phi_{\gamma_1}\Phi_{\gamma_2}\Phi_{\gamma_3}, \tag{14.16}$$

with

$$\begin{aligned} I_{\gamma_1,\gamma_2,\gamma_3} = {} & \frac{(h+b)^2}{2}\left(\frac{u''_{\gamma_2}(h+b)}{u_{\gamma_2}(h+b)} - \frac{u'_{\gamma_1}(h+b)\,u'_{\gamma_3}(h+b)}{u_{\gamma_1}(h+b)\,u_{\gamma_3}(h+b)}\right)\int_{S_2} Y_{\gamma_1}Y_{\gamma_2}Y_{\gamma_3} \\ & + \frac{1}{2}\int_{S_2} Y_{\gamma_1}\nabla Y_{\gamma_2}\cdot\nabla Y_{\gamma_3}. \end{aligned} \tag{14.17}$$

In the above relations we use the notations

$$\eta = \sum_{\gamma}\eta_\gamma Y_\gamma, \quad \Phi = \sum_{\gamma}\Phi_\gamma Y_\gamma,$$

with $\gamma = (l,m)$, and

$$u_\gamma(r) = (l+1)\left(\frac{r}{b}\right)^l + l\left(\frac{b}{r}\right)^{l+1}.$$

The Hamilton equations read

$$\dot{\eta}_\gamma = \frac{\partial H}{\partial \Phi_\gamma^*}, \quad \dot{\Phi}_\gamma = -\frac{\partial H}{\partial \eta_\gamma^*}. \tag{14.18}$$

To solve numerically the initial value problem the authors in [178, 252] used a Galerkin truncation of the Hamilton equations (14.18).

Part III
Physical Nonlinear Systems at Different Scales

This last part is devoted to applications of solitons on closed or bounded systems at different physical scales, from elementary particles to neutron stars. We devote a whole chapter to the dynamics of free shape one-dimensional nonlinear systems like filaments, vortex filaments and polymer chains. Application of soliton dynamics are given at microscopic scale (heavy nuclei, quantum Hall effect) as well as at macroscopic scale (plasma and MHD systems, elastic spheres and neutron stars).

Chapter 15
Filaments, Chains, and Solitons

One of the most successful applications of the theory of nonlinear integrable systems on free one-dimensional systems is related to the existence of solitons on filaments. In the following we describe such systems from the hydrodynamic perspective and obtain the vortex filament equation, also called the binormal equation. Next, we describe a gas dynamical model which has an equivalent dynamics, and we obtain several soliton solutions and corresponding shapes. One interesting special feature of vortex filaments, namely by representing a unifying model for the Riccati and the NLS equations, is also presented. There are many applications of these nonlinear-geometric models, extending from nuclear physics to severe weather and astrophysics. In solid state physics filament solitons occur through the complex cubic Ginzburg-Landau equation (CCGLE) which resembles a magnetic Schrödinger equation, [197], similar to the cubic NLS equation. The nonlinear solutions of interest are strings and tubes with quantized angular momentum, namely vortex structures, their stability and their interactions [51, 258].

15.1 Vortex Filaments

Riemann–Christoffel tensor Rotational or vortex motion was first investigated by Kelvin [153], Helmholtz[125], and Thomson [315]. In absence of viscosity an isentropic fluid is described by the Euler equation (10.15), which in an Eulerian frame, takes the form

$$\frac{\partial}{\partial t}\boldsymbol{\omega} = \nabla \times (\boldsymbol{v} \times \boldsymbol{\omega}), \tag{15.1}$$

where $\boldsymbol{\omega}(\boldsymbol{r}, t) = \nabla \times \boldsymbol{v}(\boldsymbol{r}, t)$ is the vorticity field. Because it is a solenoidal vector field $\nabla \cdot \boldsymbol{\omega} = 0$, vorticity has some interesting properties, like, for example, it has zero flux on surfaces represented by tubes of flow (see (10.32)). This is just a geometric property and has not to do with the specific type of fluid. In the case of a perfect fluid (inviscid and isentropic) in potential force fields, this property

A. Ludu, *Nonlinear Waves and Solitons on Contours and Closed Surfaces*, Springer Series in Synergetics, DOI 10.1007/978-3-642-22895-7_15,

of vorticity yields the invariant circulation theorem (Theorems 25 and 26, see Sect. 10.3), which states that the circulation of the velocity of such a fluid, along a closed particle contour, is constant in time. If the fluid has nonzero vorticity, localized on a material surface, then the integral of the vorticity on this surface (the *strength of the vortex*) is constant during the motion of the material surface and it is also constant along the vorticity field lines. The tube of flow generated by the motion of such a material surface carries in time a constant amount of vorticity. This is the physical background for the introduction of *vortex tubes* or simply *vortex*. Such a vortex tube contains the vorticity field perpendicular on each of its cross-sections and oriented along the generator of the tube. An intuitive description is given, for example, in Article 145 of [167]. If such a vortex has an almost constant cross-section area along the vorticity lines, and if its diameter is much smaller than its length, we call it a vortex filament. If we can set the initial conditions such that the vorticity is almost negligible outside of a vortex filament, following the theorem of invariance of the circulation, we find out that this vortex filament is a stable structure and has a dynamics of its own. Moreover, a vortex filament will support shape solitary waves traveling along it.

To analyze the existence of solitons on vortex filaments we follow an approach presented in Lamb's book [169], originally introduced in Hasimoto [123] and Batchelor [14]. We consider an isolated vortex filament described by a *tube* (or *tubular neighborhood*) of constant radius r_0 (see Definition 60) around a simple regular differentiable curve Γ of finite length (length L_Γ). We denote by κ and τ the curvature and torsion of Γ, and we investigate the vortex filament under three approximating hypotheses:

1. The fluid is considered to be incompressible.
2. The filament is "narrow," i.e., the ratio $\epsilon = r_0/L_\Gamma \ll 1$ is one smallness parameter of the problem.
3. We also consider that the filament is not excessively bent or twisted compared to its length. We introduce a second smallness parameter $\eta = \kappa L_\Gamma$, i.e., the radius of curvature of the filament is much larger than its length.

Let $\boldsymbol{r}(s)$ be the equation of the curve Γ in the arc-length parameterization. We assume that inside of this tube of constant cross-section πr_0^2, the vorticity is constant and uniform in magnitude $|\boldsymbol{\omega}| = \omega_0$ =const., oriented along the tangent to Γ, and it is zero outside of the tube. From $\nabla \cdot \boldsymbol{v} = 0$ and $\boldsymbol{\omega} = \nabla \times \boldsymbol{v}$ we can always define a solenoidal vector potential $\boldsymbol{B}(\boldsymbol{r}, t)$, $\nabla \cdot \boldsymbol{B} = 0$, such that $\boldsymbol{v} = \nabla \times \boldsymbol{B}$ and $\Delta \boldsymbol{B} = 0$. We calculate the velocity at a point of the tube $\boldsymbol{r}_1 \notin \Gamma$ by using the fundamental solution of the Poisson equation

$$\boldsymbol{v}(\boldsymbol{r}_1) = -\frac{1}{4\pi} \int_{\boldsymbol{r} \in \Gamma} \frac{(\boldsymbol{r}_1 - \boldsymbol{r}) \times \boldsymbol{\omega}(\boldsymbol{r})}{|\boldsymbol{r}_1 - \boldsymbol{r}|^3} dV, \tag{15.2}$$

where $dV = \pi r_0^2 ds$ is the volume element, and we did not write explicitly the time dependence. From the hypotheses we have $\boldsymbol{\omega}(\boldsymbol{r}) = \omega_0 \boldsymbol{t}(s)$, where $\boldsymbol{t}$ is the unit tangent to Γ. Equation (15.2) can be written as

$$v(r_1) = -\frac{C}{4\pi}\int_0^{L_\Gamma} \frac{(r_1 - r(s)) \times t(s)}{|r_1 - r(s)|^3} ds, \tag{15.3}$$

where

$$C = \oint_{\partial A} v \cdot dl = \int_A (\nabla \times v) \cdot dA = \int_A \omega \cdot dA = \omega_0 \pi r_0^2, \tag{15.4}$$

is the circulation of v along any circle surrounding the tube and A is the cross-section circular area of the tube.

We choose a point $r_1 \notin \Gamma$ placed on the surface of the filament at a distance r_0 from its axis. We choose a reference point $s = s_0$ on Γ as the closest point to r_1, and we expand $r(s)$ in Taylor series with respect to $\delta s = s - s_0$ around $s = s_0$. By using the Serret–Frenet equations (5.3) for the derivatives with respect to the arc-length, we obtain

$$r(s) = r(s_0) + t(s_0)\delta s + \frac{\kappa(s_0)}{2} n(s_0)\delta s^2 + \frac{1}{6}\Big(\kappa'(s_0)n(s_0) + \kappa(s_0)\tau(s_0)b(s_0) - \kappa^2(s_0)t(s_0)\Big) + \mathcal{O}(\delta s^3), \tag{15.5}$$

with $\delta s \in [-l_\Gamma, l_\Gamma]$. By the definition of s_0 we have $t(s_0) \cdot (r_1 - r(s_0) = 0$ and $|r_1 - r(s_0)| = r_0$, and so we can assume that $r_1 - r(s_0) = r_0(\alpha n(s_0) + \beta b(s_0))$ with $\alpha^2 + \beta^2 = 1$. We have

$(r_1 - r(s)) \times t(s) =$

$$r_0\left[(\beta n - \alpha b) - \kappa_0 \beta \delta s t + \frac{\delta s^2}{2}((\alpha\kappa\tau - \beta\kappa_s)t - \beta\kappa^2 n + \alpha\kappa^2 b) - \frac{\kappa \delta s^2}{2} b\right]_{s_0} + \mathcal{O}(\eta^3). \tag{15.6}$$

The orders of smallness of the terms in the RHS of (15.6) are

$$r_0, \quad r_0\kappa(s_0)\delta s < r_0\mathcal{O}(\eta), \quad r_0(\kappa(s_0)\delta s)^2 < r_0\mathcal{O}(\eta^2), \quad \kappa(s_0)\delta s^2, \quad \ldots,$$

where $\kappa(s_0)\delta s < \kappa l_\Gamma = \eta \ll 1$ is from hypothesis (3) introduced above. We can write $\kappa(s_0)\delta s^2 = r_0\eta\delta s / r_0 > r_0\eta$ so the last term in RHS of (15.6) has its order larger than the second term in η. We can approximate (15.6) with

$$(r_1 - r(s)) \times t(s) \approx -r_0\left(\alpha + \frac{\kappa(s_0)\delta s^2}{2}\right) b(s_0) + r_0\beta n(s_0). \tag{15.7}$$

The denominator of (15.2) has the form

$$|r_1 - r(s)|^3 = \delta s^3\left(1 + \frac{r_0^2}{\delta s^2} + \frac{(\kappa(s_0)\delta s)^2}{4} + \mathcal{O}(\eta^3)\right)^{\frac{3}{2}}, \tag{15.8}$$

and now we need to compare the contribution of the two terms inside the parenthesis. The term $r_0/\delta s$ is lower bounded by $\epsilon = r_0/l_\Gamma$ according to hypothesis (2). This fact does not help too much in the comparison with the third term in the parenthesis of (15.8), because δs runs between zero and l_Γ. In a very interesting way, the topology of the filament vortex shape will help us here. We have to compare the terms

$$\kappa(s_0)\delta s \lesssim \frac{r_0}{\delta s},$$

and we can write this expression as

$$\frac{\kappa(s_0)}{r_0} \lesssim \frac{1}{\delta s^2}.$$

The LHS of the above inequality is actually the Gaussian curvature of the tube surface at r_0, $K \approx \kappa / r_0$, i.e., the product of the principal curvature $\kappa(s_0)$ along the generator of the tube and the principal curvature of the base circle $1/r_0$. According to the Bonnet Theorem 21 if there is a positive number δ_0 such that the Gaussian curvature of a complete surface $K \geq \delta_0$, then the surface is compact. In the case of the vortex filament this is false, because the surface is a long cylinder, and so the Gaussian curvature can be arbitrarily small. As a consequence we have

$$\kappa(s_0)\delta s < \eta \ll \frac{r_0}{\delta s},$$

and it results that the dominant term in (15.8) is $r_0/\delta s$. From (15.3), (15.7), and (15.8) we obtain the velocity of an arbitrary point on the surface of the vortex filament obtained in the order η^2 of smallness

$$\boldsymbol{v}(\boldsymbol{r}_1) \approx \frac{r_0 C}{4\pi} \int_\Gamma \frac{\left(\alpha + \frac{\kappa(s)s^2}{2}\right)\boldsymbol{b}(s) - \beta \boldsymbol{n}(s)}{s^3\left(1 + \frac{r_0^2}{s^2}\right)^{\frac{3}{2}}} ds, \tag{15.9}$$

where r_0 is the radius of the filament, the limits of integration for s depend on the specific position of the chosen point $\boldsymbol{r}_1$ along Γ, α and β describe the position of $\boldsymbol{r}_1$ on the base circumference of the tube, and C is the circulation of the velocity around this circumference, supposed constant.

The equation of motion of the vortex filament (15.9) can be simplified more [14, 123, 169] if we consider very narrow filaments $\alpha \approx \beta \approx 0$

$$\boldsymbol{v}(\boldsymbol{r}_1) \approx \text{cst.} \int_\Gamma \kappa(s)\boldsymbol{b}(s) \frac{s^2}{(s^2 + r_0^2)^{\frac{3}{2}}} ds, \tag{15.10}$$

and we notice that a part of the integrand

$$\varphi(s, r_0) \equiv \frac{s^2}{(s^2 + r_0^2)^{\frac{3}{2}}}$$

is actually a sequence of functions weakly converging toward the δ-Dirac distribution when $s \simeq 0$ [274]

$$\lim_{s \leftarrow 0} \varphi(s, r_0) = \delta(r_0).$$

Consequently, we can approximate (15.10) and obtain the most simplified version of the dynamical equation for a long and narrow vortex filament of incompressible fluid

$$\boldsymbol{v}(\boldsymbol{r}_1) \approx \text{cst.}\, \kappa(s_0)\boldsymbol{b}(s_0). \tag{15.11}$$

The constant term on the RHS can be eliminated by a special choice for the velocity vector. Equation (15.11) represents the well-known *vortex filament equation* first introduced in Hasimoto [123], and later on investigated in many books or articles among we mention [14, 169], [11, Chap. VI], [12, 45, 117, 139, 166, 172, 232, 240, 243, 250, 257, 272, 285, 289].

In the following we confine the discussion to the investigation of the filaments governed by (15.11) in its simplest form

$$\frac{d\boldsymbol{r}}{dt}(s,t) = \dot{\boldsymbol{r}} = \kappa \boldsymbol{b}, \tag{15.12}$$

where for the notation in the following we use $\dot{\boldsymbol{r}}$ for time derivative and $\boldsymbol{r}'$ for $d\boldsymbol{r}/ds$. That is, we neglect the filament width and consider it just a (time dependent) regular arc-length parameterized curve $\boldsymbol{r}(s,t)$. Then (15.12) is equivalent to

$$\dot{\boldsymbol{r}} = \boldsymbol{r}' \times \boldsymbol{r}'', \tag{15.13}$$

as we can obtain from Serret–Frenet equations (5.3), (5.4), and (5.11) from Chap. 5. From (15.13) we have

$$\frac{\partial \kappa^2}{\partial t} = \frac{\partial}{\partial t}(\boldsymbol{r}'' \cdot \boldsymbol{r}'') = 2\left(\frac{\partial^2 \dot{\boldsymbol{r}}}{\partial s^2}\right) \cdot \boldsymbol{r}'',$$

where we used $\boldsymbol{t}' = \boldsymbol{r}'' = \kappa \boldsymbol{n}$ and $\boldsymbol{r}'' \cdot \boldsymbol{r}'' = \kappa^2$. In the following, using (5.3) and (15.12) their consequences $\boldsymbol{r}' \times \boldsymbol{r}'' = \boldsymbol{t} \times \kappa \boldsymbol{n} = \kappa(\boldsymbol{t} \times \boldsymbol{n}) = \kappa \boldsymbol{b}$, and $\boldsymbol{b}' = -\tau \boldsymbol{n}$, we have

$$2\left(\frac{\partial^2 \dot{\boldsymbol{r}}}{\partial s^2}\right) \cdot \boldsymbol{r}'' = 2\left(\frac{\partial^2 \kappa \boldsymbol{b}}{\partial s^2}\right) \cdot \kappa \boldsymbol{n} = -2(\kappa^2 \tau)',$$

where κ and τ are the curvature and the torsion of the filament. It results a sort of continuity equation for the curvature and the torsion of the filament

$$\frac{\partial}{\partial t}\left(\frac{\kappa^2}{2}\right) + \frac{\partial}{\partial s}(\kappa^2 \tau) = 0. \tag{15.14}$$

The same result can be obtained for arbitrary parameterization of the filament curve.

Next, we want to obtain a similar relation for the time derivative of the torsion. We begin from the time derivative of $\tau^2 = \boldsymbol{b}' \cdot \boldsymbol{b}'$ and, by using again Serret–Frenet, and (15.12) and (15.13) we obtain

$$\dot{\tau} = -\frac{\partial^2 (\boldsymbol{t} \times \boldsymbol{n})}{\partial s \partial t} \cdot \boldsymbol{n} = -(\dot{\boldsymbol{t}} \times \boldsymbol{n} + \boldsymbol{t} \times \dot{\boldsymbol{n}})' \cdot \boldsymbol{n}. \tag{15.15}$$

The expression in the RHS parenthesis can be expanded by taking into account the orthonormality of the Serret–Frenet trihedron and the relation $\boldsymbol{r}''' \times \boldsymbol{n} = \kappa$. We obtain

$$\begin{aligned} \dot{\boldsymbol{t}} \times \boldsymbol{n} + \boldsymbol{t} \times \dot{\boldsymbol{n}} &= (\boldsymbol{r}' \times \boldsymbol{r}'')' \times \boldsymbol{n} + \boldsymbol{t} \times \frac{\dot{\boldsymbol{r}}''\kappa - \boldsymbol{r}''\dot{\kappa}}{\kappa^2} \\ &= -\kappa' \boldsymbol{t} - \frac{\dot{\kappa}}{\kappa}\boldsymbol{b} + \frac{1}{\kappa}(\boldsymbol{t} \times \dot{\boldsymbol{r}}''), \end{aligned} \tag{15.16}$$

with

$$\dot{\boldsymbol{r}}'' = \kappa^2 \tau \boldsymbol{t} - (2\kappa'\tau + \kappa\tau')\boldsymbol{n} + (\kappa'' - \kappa\tau^2)\boldsymbol{b}.$$

By combining the last two equations we have

$$\dot{\tau} = -\left[\left(-\kappa' \boldsymbol{t} - \frac{\dot{\kappa} + 2\kappa'\tau + \kappa\tau'}{\kappa}\boldsymbol{b} + \frac{\kappa\tau^2 - \kappa''}{\kappa}\boldsymbol{n}\right)'\right] \cdot \boldsymbol{n}, \tag{15.17}$$

and in the end

$$\dot{\tau} = \kappa\kappa' - \frac{\dot{\kappa}\tau}{\kappa} - 2\frac{\kappa'\tau^2}{\kappa} - 3\tau\tau' + \frac{\kappa\kappa'''}{\kappa^2} - \frac{\kappa'\kappa''}{\kappa^2}. \tag{15.18}$$

To process (15.18) we need (15.14) for the value of $\dot{\kappa} = -2\kappa'\tau - \kappa\tau'$. We obtain

$$\dot{\tau} + 2\tau\tau' = \left(\frac{\kappa^2}{2} + \frac{\kappa''}{\kappa}\right)'. \tag{15.19}$$

It is interesting to write (15.14) and (15.19) with the substitution

$$\rho(s,t) = \frac{\kappa^2}{4}, \quad u(s,t) = 2\tau, \tag{15.20}$$

where usually $\kappa^2/2$ is called the *energy density* of the filament curve. We have

$$\begin{aligned} &\frac{\partial \rho}{\partial t} + \frac{\partial}{\partial s}(\rho u) = 0 \\ &\frac{\partial u}{\partial t} + u\frac{\partial u}{\partial s} = \frac{\partial}{\partial s}\left(4\rho + \frac{2}{\sqrt{\rho}}\frac{\partial^2 \sqrt{\rho}}{\partial s^2}\right). \end{aligned} \tag{15.21}$$

Equations (15.21) represent the so-called *gas dynamics* model of the filament because they describe the filament (15.14) and (15.19), in terms of the velocity u and density ρ fields for a one-dimensional fluid. In Arnold and Khesin [11, Chap. VI], these equations are described as a Marsden–Weinstein Hamiltonian structure. Different approaches of the filament problem include the Hasimoto model of the filament equation, derived from the nonlinear Schrödinger equation [123], which will be analyzed in the next sections.

15.1.1 Gas Dynamics Filament Model and Solitons

In this section we discuss some particular traveling solutions of the gas model for the filament equation (15.21), or equivalently (15.14) and (15.19). Obviously plane filaments ($\tau = 0$) do not exist. We are looking for traveling solutions in the form

$$\rho(s,t) = R(s - Vt) = R(x)$$
$$u(s,t) = U(s - Vt) = U(x),$$

with V an arbitrary constant. By integrating the first of (15.21) we obtain

$$U(x) = V - \frac{C_1}{R(x)}, \tag{15.22}$$

where C_1 is a constant of integration. The resulting equation for R reads

$$\frac{V^2}{2} + 4R - \frac{C_1^2}{2R^2} + 2\frac{\sqrt{R}\,''}{\sqrt{R}} + C_2 = 0, \tag{15.23}$$

or in terms of the curvature

$$\kappa'\kappa = \pm\sqrt{-C_1^2 - \frac{\kappa^6}{4} - \left(\frac{V^2}{4} + \frac{C_2}{2}\right)\kappa^4 + C_3\kappa^2}. \tag{15.24}$$

By substituting in (15.24) $\kappa\kappa' = 2dR/dx$, we integrate (15.23), and we obtain

$$\int_{C_4}^{R} \frac{dR'}{\sqrt{\mathcal{P}_3(R')}} = \pm x + C_5, \tag{15.25}$$

where $\mathcal{P}_3(R) = -4(R - R_1)(R - R_2)(R - R_3)$ is a third order polynomial in $R(x)$ and $C_i, i = 1, \ldots, 5$ are integration constants. The three roots of $\mathcal{P}_3$ depend on C_1, C_2, C_3, V, from (15.22) to (15.24). The structure of the roots determine the structure of the solutions $R(x)$. Let us study some examples:

1. *Three real solutions*, $R_i \in \mathbb{R}, i = 1, 2, 3$.
   In this case the solution reads

$$\frac{1}{\sqrt{R_3 - R_1}} F\left(\arcsin \sqrt{\frac{R - R_3}{R_2 - R_3}} \,\middle|\, \frac{R_3 - R_2}{R_3 - R_1} \right) = \pm x + C_4, \tag{15.26}$$

where F is an elliptic integral of the first kind

$$F(\alpha|m) = \int_0^\alpha \frac{d\theta}{\sqrt{1 - m^2 \sin^2 \theta}}.$$

By inverting (15.26) we have

$$R(x) = R_3 + (R_2 - R_3) sn^2 \left(\pm \sqrt{R_3 - R_1}, x + C_4 \,\middle|\, \frac{R_3 - R_2}{R_3 - R_1} \right), \tag{15.27}$$

where $sn(\alpha|m) = \sin am(\alpha|m)$ is the cnoidal sine Jacobi function obtained from the Jacobi amplitude am for the Jacobi elliptic functions. As a consequence, the filament is described by the intrinsic equations

$$\kappa(s,t) = 2\sqrt{R(s - Vt)} \quad \kappa(s,t) = \frac{1}{2} U(s - Vt),$$

with $R(x)$ given in (15.27) and U from (15.22). This solution is a cnoidal wave which can approach a trigonometric function or a solitary wave when $m \in [0, 1]$.

$$\kappa(x,t) = 2\sqrt{R_3 + (R_2 - R_3) sn^2 (\pm\sqrt{R_3 - R_1}(x - Vt) + C_4|m},$$

$$\tau(x,t) = \frac{V}{2} - \frac{C_1}{2[R_3 + (R_2 - R_3) sn^2 (\pm\sqrt{R_3 - R_1}(x - VT) + C_4|m)]}, \tag{15.28}$$

with

$$m = \frac{(R_3 - R_2)}{(R_3 - R_1)}. \tag{15.29}$$

The solution for the filament curvature in (15.28) is similar with the solution given in Lamb [169, (7.2.25)]. For example, to obtain a soliton in curvature we need $m = 1$ in (15.29) and also $C_1 = C_4 = 0$. To convert the cnoidal sine Jacobi elliptical function into a hyperbolic tangent we also need $(R_3 - R_2) = R_3$. By using these constraints we obtain $R_1 = R_2 = 0$ and

$$\kappa(s,t) = 2\sqrt{R_3} sech(\sqrt{R_3}(s - Vt)), \quad \tau(s,t) = \frac{V}{2} = \tau_0 \tag{15.30}$$

which is a single-soliton solution of the cubic nonlinear Schrödinger equation (NLS) or of the modified Korteweg–de Vries (mKdV) equation. This filament is a constant torsion helix with a traveling localized soliton-like disturbance in curvature. Moreover, for such a soliton solution it is easier to integrate the corresponding Serret–Frenet equations, by mapping them into a Riccati differential equation, and then finding the shape of the filament.

Since the cnoidal sine is a periodic function, it is interesting to verify if (15.28) can support closed filaments as parameterized loops. Finding the criterium for a curve to be closed in terms of a differential equations is still an open problem [106]. There are no simple conditions on curvature and torsion which would force a curve to close up. For planar curves, on the other hand, where one is concerned only with curvature, it is known that any positive periodic function with at least four extremum points may be realized as the curvature of some closed planar curve [107]. However, there is no simple condition on curvature that would guarantee the existence of a closed planar curve parameterized by arc-length. We can test this behavior by integrating the Serret–Frenet equations with κ and τ given in (15.28). For example, in Fig. 15.1, we notice that a periodic structure for curvature and torsion generates a strongly oscillating filament, yet still open.

2. *Two distinct real solutions*, $R_1, R_2 = R_3 \in \mathbb{R}$.
   We have $\mathcal{P}_3 = -4(R - R_1)(R - R_2)^2$ and by integration we obtain

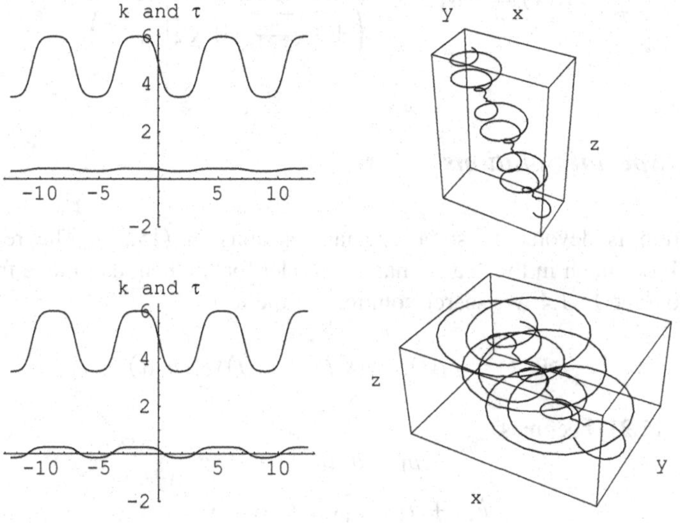

Fig. 15.1 *Left*: curvature (the upper curve) and torsion (the lower curve) from (15.28) for $R_1 = 2.9, R_2 = 3, R_3 = 6, C_4 = 0, V = 1$, and $C_1 = 1$ for the upper part and $C_1 = 4$ for the lower part. *Right*: corresponding filament shapes

$$R(x) = R_1 + (R_2 - R_1)\tanh^2\left(\pm\sqrt{R_1 - R_2}\frac{x}{2} + C_4\right), \tag{15.31}$$

which is a propagating kink, similar to a nontopological solitary wave.

3. *One real root* $R_1 = R_2 = R_3$.

   The solution can be directly integrated and we obtain

$$R(x) = R_1 - \frac{4}{(\pm x + C_4)^2}.$$

4. *One real root* R_1 *and other two complex conjugated roots.*

   The polynomial has the form

$$\mathcal{P}_3 = -4(R - R_1)(R^2 + a^2).$$

The solution can be written again in the form of the Jacobi elliptic integral

$$-2F\left(i \operatorname{arcsinh}\sqrt{\frac{R_1 + ia}{R + R_1}} \,\middle|\, \frac{iR_1 + a}{iR_1 - a}\right) = (\pm x + C_4)\sqrt{R_1 + ia},$$

and we have

$$R(x) = -R_1 - \frac{R_1 + ia}{sn^2\left(\pm\frac{\sqrt{R_1+ia}}{2}x + C_4 \,\middle|\, \frac{iR_1+a}{iR_1-a}\right)}.$$

15.1.2 Special Solutions

This section is devoted to some special solutions of (15.21). The reader not interested too much in the "gas dynamic" model for filament can move from here directly to Sect. 15.1.4. We search solutions of the form

$$\rho(s,t) = \rho_1(t), \quad u(s,t) = u_1(t)s + u_2(t).$$

Equation (15.21) becomes

$$\dot{\rho}_1 + \rho_1 u_1 = 0,$$
$$\dot{u}_1 s + \dot{u}_2 + u_1^2 s + u_1 u_2 = 0.$$

We have an "exploding" type of solution

$$\kappa(s,t) = 2\sqrt{\frac{C_1}{t - C_1}},$$

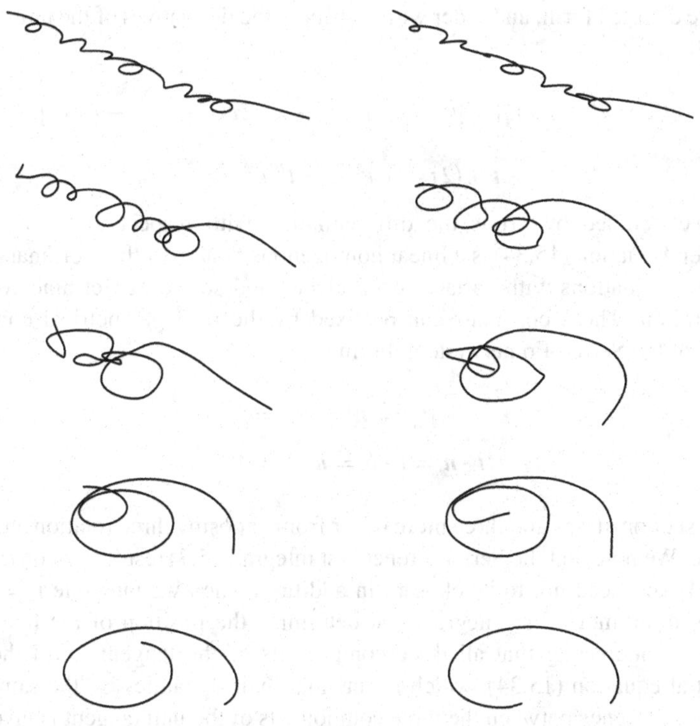

Fig. 15.2 Filament shapes obtained by numerical integration of the intrinsic equations for κ and τ given in (15.32). The two columns represent two different values for the integration constants, while the time evolution is from top to bottom

$$\tau(s,t) = \frac{s + C_1}{2(t - C_1)}, \qquad (15.32)$$

and a rigid helix type of solution with $\kappa, \tau = \text{const}$. We present these solutions in Fig. 15.2.

15.1.3 Integration of Serret–Frenet Equations for Filaments

With curvature and torsion determined by a certain filament model we need to integrate the Serret–Frenet equations (5.3) and (5.4) to have the filament shape. A direct integration can be performed, for example, starting with the first two equations in (5.3) and obtaining

$$\left\{ \frac{1}{\tau} \left[\left(\frac{1}{\kappa} t' \right)' + \kappa t \right] \right\}' + \frac{\tau}{\kappa} t' = 0. \qquad (15.33)$$

In a more detailed form, and order with respect to the derivatives of the unit tangent, (15.33) reads

$$t\kappa^3(\tau\kappa' - \tau'\kappa) + t'[\kappa^2\tau(\kappa^2 + \tau^2) + \kappa'(2\tau\kappa' + \kappa\tau') - \kappa\tau\kappa'']$$

$$- t''\kappa(2\tau\kappa' + \kappa\tau') + t'''\kappa^2\tau = 0, \qquad (15.34)$$

where we denoted by prime the differentiation with respect to the arc-length parameter. Equation (15.34) is a linear homogenous system of three ordinary vector differential equations with variable coefficients, and so we expect nine constants of integration. These constants can be fixed by the nine geometrical conditions imposed to the Serret–Frenet system. From

$$|t| = |n| = |b| = 1 \qquad (15.35)$$

$$t \cdot n = t \cdot b = n \cdot b = 0,$$

we have six constrains and three more occur from choosing three rotation angles for the curve. We note that the Serret–Frenet first integrals (5.7) result as a consequence of (15.34) and need not to be chosen. In addition, when we integrate $r = \int t ds$ we bring three more first integrals that determine the position of the filament in space. It is interesting that all three components of the tangent fulfill the same differential equation (15.34), which means that their dynamics is "the same" in a way. The difference between the three components of the unit tangent is given only by the choice of initial conditions. More specific, we can map any given solution of (15.34) into another solution of the same equation by using the symmetry group of transformations [242]. That is, we can map any component of the tangent to the curve into another component of the tangent by using the symmetries of the Serret–Frenet system of equations.

Let us present some particular cases. If $\tau = 0$ and $\kappa = \kappa_0 =$ const. we choose $b = 0$ and we have

$$t'' = \kappa_0 n' = -\kappa_0^2 t,$$

which results in the general solution

$$t = (t_0^1 \sin(\kappa_0 s + s_0^1), t_0^2 \sin(\kappa_0 s + s_0^2), t_0^3 \sin(\kappa_0 s + s_0^3))$$

$$n = (t_0^1 \cos(\kappa_0 s + s_0^1), t_0^2 \cos(\kappa_0 s + s_0^2), t_0^3 \cos(\kappa_0 s + s_0^3))$$

and $b = 0$. From (15.35) we can choose $t_0^3 = 0$, $s_0^3 = 0$, $s_0^1 = 0$, $s_0^2 = \pi/2$, and $t_0^1 = t_0^2 = \pm 1/\sqrt{2}$, so in the end we obtain the solution of circular shape like it should be.

In the following we give some examples of filament shapes obtained by numerical integration of the Serret–Frenet relations by using (15.34). For example, by choosing the solution in (15.28) for the curvature and constant torsion, we obtain a periodic structure in the filament. The period is given by $T = 4K(m)$ with m

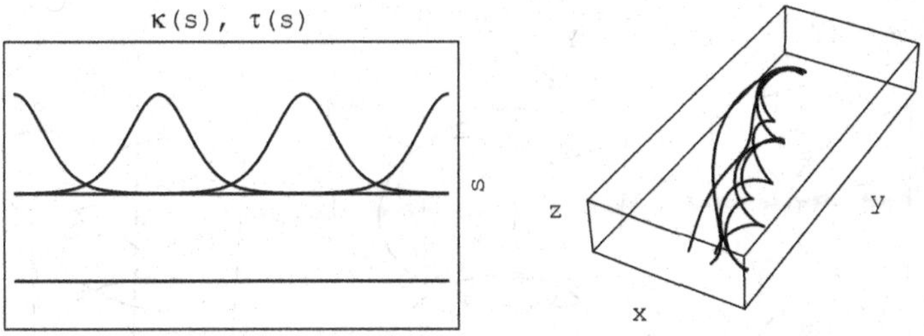

Fig. 15.3 *Left*: periodic solutions in curvature Upper curves) and torsion (lower curve, almost a constant) from (15.28) for $R_1 = R_2 = 2$, $R_3 = 6$, $C_1 = 0$, $C_4 = 0$, $V = 2$, and $\tau = 1, m = 1$ at times $t = 1, 2, 3$, and 4. *Right*: the corresponding numerically integrated filaments

given by (15.29), and

$$K(m) = F\left(\frac{\pi}{2}|m\right) = \int_0^{\pi/2} \frac{d\theta}{\sqrt{1 - m\sin^2\theta}},$$

being the complete elliptic integral of the first kind. For a soliton solution we have $m = 1$ and hence $T = \infty$. A plot of a traveling soliton in curvature along a very elongated helix, at different moments of time, is presented in Fig. 15.3.

To visualize the effect of a localized perturbation in curvature on a filament, we use a single-soliton solution (pretty much like the one in (15.30)). This specific curvature can be obtained from (15.28) with $m \lesssim 1$, for an appropriate choice of the parameters. We add this perturbation to a constant curvature, constant torsion helix, and present the numerical integration of the Serret–Frenet equations in Fig. 15.4. The soliton-like perturbation is propagating along the filament in the positive z direction. A wider soliton (left in Fig. 15.4) produces a longer arc-length change in the filament shape and shrinks it toward smaller radii. A soliton with the half-width comparable with the helix pitch produces a little wiggle (Fig. 15.4, center) in the helix and little deformations in the rest. A narrow soliton (Fig. 15.4, right) produces a sort of global bent in the helix. Also, in Fig. 15.5, we show the propagation of a soliton in curvature $\kappa(s) = \kappa_0 + \kappa_1 \text{sech}(8(s - Vt))$.

15.1.4 The Riccati Form of the Serret–Frenet Equations

In this section we present a specific procedure to integrate the Serret–Frenet equations by reducing them to the Riccati differential equation. We work the case of vortex filaments, especially when soliton solutions are investigated. Such an example is worked out in detail in Lamb [169] and Hasimoto [123], while the differential geometry details are provided in Eisenhart [88] and Struik [310], for example. We begin by using the Serret–Frenet equations written in components (5.6). From the

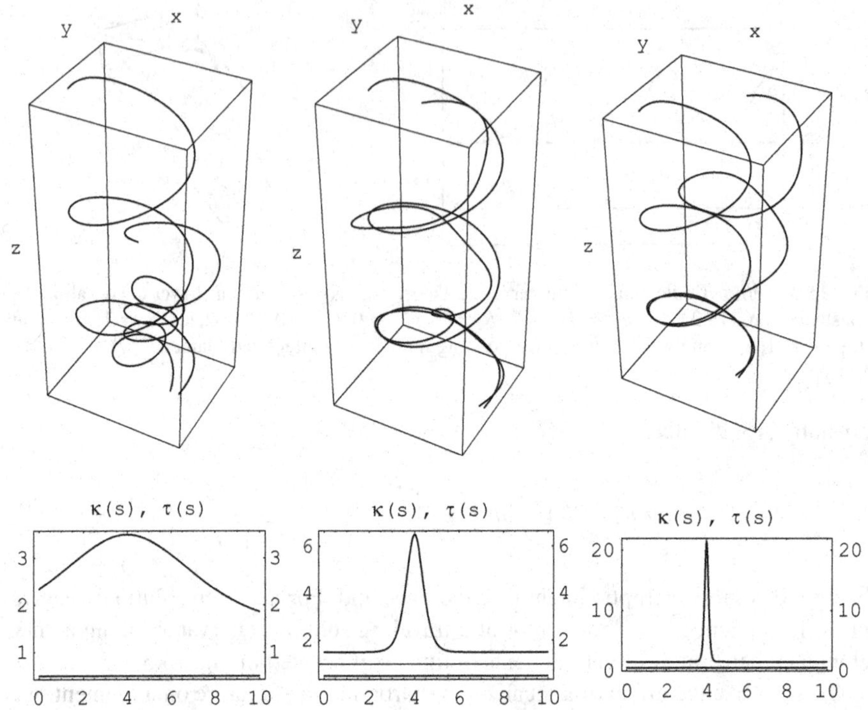

Fig. 15.4 Effects of localized perturbations in curvature (upper curves) $\kappa(s) = 1.5 + \kappa_1$ sech (s/L) on a helix of curvature $\kappa = 1.5$ and constant torsion $\tau = 0.5$ (the lower constant curves). Each frame overlaps the un-deformed and the deformed helices for three different values of the soliton half-width L and amplitude κ_1

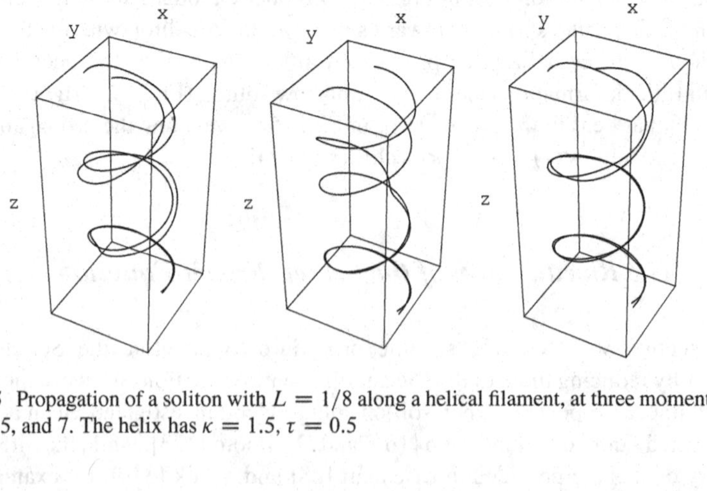

Fig. 15.5 Propagation of a soliton with $L = 1/8$ along a helical filament, at three moment of time $Vt = 3, 5$, and 7. The helix has $\kappa = 1.5$, $\tau = 0.5$

three first integrals of motion in (5.7) we can define two test vector functions φ^j and χ^j

$$\varphi^j = \frac{t^j + in^j}{1 - b^j} = \frac{1 + b^j}{t^j - in^j}, \tag{15.36}$$

and

$$-\frac{1}{\chi^j} = \varphi^{*j} = \frac{t^j - in^j}{1 - b^j} = \frac{1 + b^j}{t^j + in^j}, \tag{15.37}$$

where $j = 1, 2, 3$ and * means complex conjugated. We have

$$t^j = \frac{\varphi^j \chi^j - 1}{\chi^j - \varphi^j},\ n^j = -i\frac{\varphi^j \chi^j + 1}{\chi^j - \varphi^j},\ b^j = \frac{\varphi^j + \chi^j}{\varphi^j - \chi^j}. \tag{15.38}$$

Now we can calculate the derivative of the φ test function

$$\frac{d\varphi^j}{ds} = -i\kappa\varphi^j + \tau\frac{ib^j - \varphi^j n^j}{1 - b^j}. \tag{15.39}$$

From the two expressions of φ in (15.36) we have

$$it^j = i\varphi^j(1 - b^j) + n^j \text{ and } n^j\varphi^j = -\varphi^j(it^j) + i + ib^j, \tag{15.40}$$

respectively. By substituting the left of (15.40) into the right one we obtain $2n^j\varphi^j = i(1 + b^j - (\varphi^j)^2(1 - b^j))$ and by substituting this result in the derivative of φ (15.39) we have

$$\frac{d\varphi^j}{ds} - \frac{i\tau}{2}(\varphi^j)^2 + i\kappa\varphi^j + \frac{i\tau}{2} = 0. \tag{15.41}$$

Equation (15.41) is the resulting Riccati equation for any of the three components of the test function $\varphi^j(s)$. Similarly, we obtain another Riccati equation in χ. Indeed, by coupling the derivative

$$\frac{d\chi^j}{ds} = \frac{-\kappa n^j - i\tau b^j + i\kappa t^j + \tau n^j \chi^j}{1 + b^j},$$

with the two expressions for χ from (15.37), we have a Riccati equation in χ^j of the same form

$$\frac{d\chi^j}{ds} - \frac{i\tau}{2}(\chi^j)^2 + i\kappa\chi^j + \frac{i\tau}{2} = 0. \tag{15.42}$$

We present some basic facts about the Riccati differential equation in Sect. 18.2. To find the shape of the vortex filament we need to choose the curvature and the torsion expressions from the physical model. We used similar procedure in the *gas dynamics* filament model (Sect. 15.1.1). Usually, the curvature and torsion are related to local interactions between the filament and the surrounding medium. Once we choose a model for κ and τ, we plug them in (15.41) and (15.42) and solve for the auxiliary functions φ and χ. Finally, the last step is to introduce these values of φ, χ in (15.38) and to obtain the unit tangent vector field $\boldsymbol{t}(s)$, hence the shape of the filament by one more integration. The general solution of each Riccati equation depends on six arbitrary constants of integration (ODE of order 1, complex solution, three components) so all in all we need the same number of 12 arbitrary constants of integration like in the case of the Serret–Frenet equation (15.34).

15.2 Soliton Solutions on the Vortex Filament

To find the motion of such an isolated vortex filament the next step forward from the previous section is to integrate the Riccati equations (15.41) and (15.42) for a given model for the curvature and torsion (i.e., the functions $\kappa(s), \tau(s)$). To find an analytic solution for the filament shape in general, by this integration, is not a straightforward task. There is no general procedure to integrate the Riccati equation, unless we know some of its particular solutions [120, 133]. However, there are some interesting particular situations when we can obtain analytic solution, for example, when we choose a solitary wave profile in the curvature while keeping the torsion constant. This is again a nice match between the theory of motion of curves and nonlinear dynamics. The fact that the one-soliton solution of the cubic nonlinear Schrödinger equation (NLS3) allows the Riccati equation to be integrated exactly is rather an exception than the rule. Details of the following calculations can be found in Lamb [169], Eisenhart [88], and Struik [310].

15.2.1 Constant Torsion Vortex Filaments

We work this example for constant torsion vortex filaments $\tau = \tau_0$, which restricts the vortex filament class to helix-like curves. The Riccati equations (15.41) and (15.42) for φ^j and χ^j can be written in a generic form for the unknown function $\varphi(s)$ in the form

$$\varphi' - \frac{i\tau_0}{2}\varphi^2 + i\kappa\varphi + \frac{i\tau_0}{2} = 0. \tag{15.43}$$

We choose the following form for the curvature

$$\kappa = \kappa_0 \mathrm{sech}(\alpha s + \beta t). \tag{15.44}$$

The specific choice of this form (which is a NLS3 single-soliton solution) will be explained in more detail on a physical background in Sect. 15.2.2. For the moment we take it as a working example. We substitute $\varphi(s) = \Phi(\alpha s + \beta t) = \Phi(z)$ and (15.43) becomes

$$\frac{2\alpha}{\kappa_0}\Phi' - i\tau_0\Phi^2 + 2i\ \mathrm{sech} z\ \Phi + i\tau_0 = 0, \tag{15.45}$$

and becomes integrable if we choose $\kappa_0 = 2\alpha$. Of course, this is a restrictive choice, but fortunately the NLS3 single-soliton solution can fulfill such a condition. Equation (15.45) reads

$$\Phi' + 2i \ \text{sech} z \ \Phi + i\tau_0(1 - \Phi^2) = 0. \tag{15.46}$$

To integrate this equation we make one more substitution by

$$\Phi = \frac{i}{\tau_0}\frac{\Psi'}{\Psi},$$

and transform (15.46) into

$$\Psi'' + 2i \ \text{sech} z \ \Psi' + \tau_0^2 \Psi = 0. \tag{15.47}$$

We introduce a new substitution

$$\Psi(z) = \theta(z) e^{-i\int^z \text{sech} z' dz'}, \tag{15.48}$$

and (15.47) becomes

$$\theta'' + (\tau_0^2 + \text{sech}^2 z + i \ \text{sech} z \ \tanh z)\theta = 0. \tag{15.49}$$

Equation (15.49) can be mapped in a "harmonic oscillator plus Pöschl–Teller potential" equation

$$\frac{d^2\tilde{\theta}}{d\omega^2} + (4\tau_0^2 + 2\text{sech}^2 \omega)\tilde{\theta} = 0, \tag{15.50}$$

with $4\omega = -2z + i\pi$ and $\tilde{\theta}(\omega) = \tilde{\theta}(-z/2 + i\pi/4) = \theta(z)$. The advantage of this series of substitutions is that (15.50) can be easily integrated and we have its general solution in the form

$$\theta(\omega) = \theta_1 e^{2i\tau_0\omega}(2i\tau_0 - \tanh \omega) + \theta_2 e^{-2i\tau_0\omega}(2i\tau_0 + \tanh \omega), \tag{15.51}$$

with $\theta_{1,2}$ constants of integration. Equation (15.47) becomes

$$\Psi(z) = \theta(z)\frac{1 - ie^z}{1 + ie^z}, \tag{15.52}$$

and then we have

$$\frac{\Psi'}{\Psi} = \frac{\theta'}{\theta} - \frac{2ie^z}{1 + e^{2z}}, \tag{15.53}$$

or

$$\varphi(z) = \frac{i}{\tau_0}\frac{\theta'}{\theta} + \frac{1}{\tau_0} \text{sech} z. \tag{15.54}$$

We can draw the conclusion of these series of substitution by Proposition 11.

Proposition 11. *The general solution of the Riccati equation* (15.46) *for the vortex filament dynamics in* (15.41) *and* (15.42)*, with single-soliton perturbation in curvature* (15.44) *has the form*

$$\varphi(z) = \left[C\left((1+2\tau_0)\cosh\frac{z}{2} + i(2\tau_0-1)\sinh\frac{z}{2}\right) + (1-2\tau_0)e^{(\pi+2iz)\tau_0}\cosh\frac{z}{2} - i(1+2\tau_0)e^{\pi+2iz}\sinh\frac{z}{2}\right]\left(\cosh\frac{z}{2} + i\sinh\frac{z}{2}\right)^{-1} \times\left[C\left(2\tau_0 - \tan\frac{\pi+2iz}{4}\right) + e^{(\pi+2iz)\tau_0}\left(2\tau_0 + \tan\frac{\pi+2iz}{4}\right)\right]^{-1}, \tag{15.55}$$

where C is an arbitrary constant of integration.

Actually there are three such equations for each component φ^j, and we can denote them C_j. With τ_0 and C_j chosen we have φ^j, and we plug them in (15.37) to obtain χ^j. Then we plug both φ^j and χ^j in (15.38), integrate the unit tangent field, and obtain the filament shapes function of the parameters $C_j, \kappa_0 = 2\alpha, \beta$ and τ_0. In Fig. 15.6, we present some typical helical (τ = const.) shapes obtained through Proposition 11 for different values of the constants of integration. Such shapes are also described in Lamb [169, Chap. 7]. These filaments twist locally around their asymptotic direction over an arc-length equal to the width of the single-soliton perturbation in curvature, i.e., $2/\kappa_0$. The localized loop travels along the vortex filament with the soliton velocity $-2\beta/\kappa_0$.

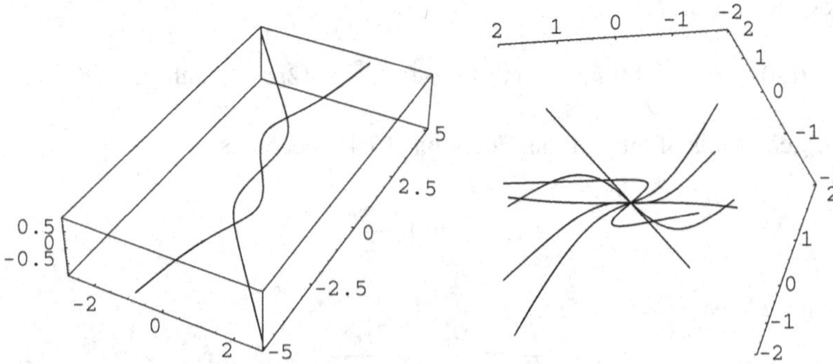

Fig. 15.6 Vortex filaments generated by (15.55). *Left*: $C_1 = 1, \tau_0 = 0.1$ and $C_1 = 5, \tau_0 = -0.1$ for the two intertwined curves and $C_2 = 1, C_3 = -2$, for both of them. *Right*: different vortex filaments intersecting at the same origin for several values for $-10 < C, \tau_0 < 10$. The three axes show that the filaments are twisted in full 3-D space.

15.2.2 Vortex Filaments and the Nonlinear Schrödinger Equation

In Sect. 15.2.1, to integrate the vortex filament equation, we used an example of localized perturbation in the curvature, in the form of a NLS single-soliton (15.44). The fact that precisely this type of soliton profile is an exact solution for the Riccati version of the Serret–Frenet equations for the vortex filament is more than a coincidence. The dynamics of the vortex filament is actually related to the dynamics of the cubic NLS through all its solutions, not only through traveling solutions. We noted already in Sect. 15.1.1 a connection between solutions of the vortex filament equation (15.12) and solitons. In the following, following the line introduced in Hasimoto [123], we present the connection between the motion of vortex filaments and the cubic NLS equation. Details of calculations could be found also in Lamb [169].

For a given smooth parametrized by arc-length curve we can introduce the complex normal and the complex curvature in the form

$$\boldsymbol{N} = (\boldsymbol{n} + i\boldsymbol{b})e^{i\int^s \tau(s')ds'}$$
$$\Psi = \kappa e^{i\int^s \tau(s')ds'}, \tag{15.56}$$

where all quantities depend on arc-length and time. From (15.56) and by using again the Serret–Frenet relations (5.3) and (5.4), we can write

$$\boldsymbol{N}' = -\Psi \boldsymbol{t},$$
$$\dot{\boldsymbol{t}} = \frac{i}{2}(\Psi' \boldsymbol{N}^* - \Psi'^* \boldsymbol{N}),$$
$$\boldsymbol{t}' = \frac{1}{2}(\Psi^* \boldsymbol{N} + \Psi \boldsymbol{N}^*), \tag{15.57}$$

where the prime means differentiation with respect to s, the dot means differentiation with respect to time, and * is complex conjugation. The time derivative of $\boldsymbol{N}$ needs more attention, but in the end we can have it in the form

$$\dot{\boldsymbol{N}} = \frac{i}{2}(|\Psi|^2 + A(t))\boldsymbol{N} - i\Psi' \boldsymbol{t}, \tag{15.58}$$

where $A(t)$ is an integration term. The equation fulfilled by $\Psi(s,t)$ reads

$$i\dot{\Psi} + \Psi'' + \frac{1}{2}(|\Psi|^2 + A(t))\Psi = 0, \tag{15.59}$$

and by using the substitution

$$u(s,t) = \frac{\Psi}{2} e^{-\frac{i}{2}\int^t A(t')dt'},$$

we reduce (15.59) to the cubic NLS equation

$$i\dot{u} + u'' + 2|u|^2 u = 0. \tag{15.60}$$

In conclusion, the procedure to determine the motion of the vortex filament is the following. We choose a solution $u(s,t)$ of the cubic NLS (15.60) and an arbitrary function $A(t)$, plug them into

$$\Psi = 2ue^{\frac{i}{2}\int^t A(t')dt'},$$

and then identify the relations

$$Re\,\Psi = \kappa \cos \int^s \tau(s',t)ds',$$

$$Im\,\Psi = \kappa \sin \int^s \tau(s',t)ds'. \tag{15.61}$$

After solving (15.61) with respect to κ and τ we can integrate the equation of motion of the curve. The previous single-soliton perturbation in the curvature can now be easily obtained following this procedure. Moreover, the soliton described by (15.28) can approach the soliton solutions of the cubic NLS equation described here, if we make $C_1 = 0$ in (15.28).

A direct example of using (15.60) and (15.61) for other vortex filament shapes is to look for many-soliton solutions. A two-soliton solution of the cubic NLS equation (15.60) can be constructed in the center of mass frame of the two solitons, which are moving with relative velocity $2v$ and amplitude a [72]

$$u(s,t) = 2a \exp\left[it\left(a^2 - \frac{v^2}{4}\right)\right](\phi(s,t) - \phi^*(s,-t))$$
$$\times\left[1 + 2e^{-2as}\left(\cosh(2avt) - 4a^2 \mathrm{Re}\left(\frac{e^{ivs}}{(v+2ia)^2}\right)\right) + \frac{v^4}{(v^2+4a^2)^2}e^{-4as}\right]^{-1}, \tag{15.62}$$

where

$$\phi(s,t) = e^{\frac{ivs}{2}}\left(e^{-as+avt} + \frac{v^2}{(v-2ia)^2}e^{-3as-avt}\right).$$

In (15.62) a, v are free parameters and the same for both solitons, and the initial phases and initial positions of the solitons are set equal to zero. In Fig. 15.7, we present the case of two NLS solitons departing from opposite initial positions, with equal phases. Consequently, they wind in the same direction. In Fig. 15.8, we present two solitons also departing from opposite initial positions, but having a phase shift of π. Consequently, the loops along the vortex filament changes its chirality in time.

As Lamb [169] points out, the relation between the vortex filament dynamics and the NLS equation is provided by the special binormal equation of motion (15.12).

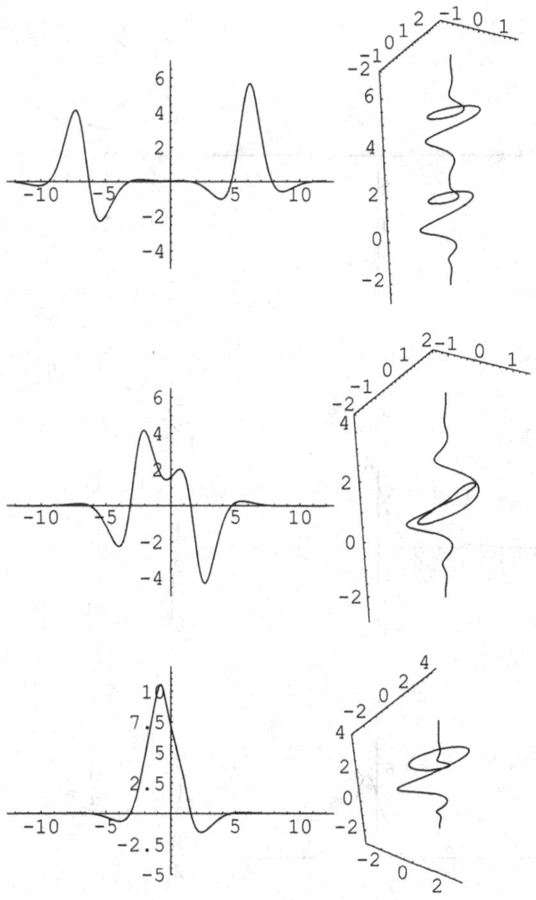

Fig. 15.7 *Left*: Curvature versus arc-length parameter at three different moments of time for the two-soliton solution (15.62) of the cubic NLS equation. *Right*: the corresponding vortex filament shape obtained with (15.61) at the corresponding three moments of time. The solitons have same phase, and so the localized helices wind in the same direction

This type of motion leads to a special orientation of the rate of change in time of the unit tangent to the filament, i.e., $\dot{\boldsymbol{t}}$ belongs to the normal plane of the curve. In general, if $\dot{\boldsymbol{t}}$ has an arbitrary orientation, the equation governing the function u in (15.60) becomes more general than the cubic NLS. For example, one can relate the motion of twisted curves with the Hirota equation [128]

$$\dot{u} + 3A|u|^2u' + iB|u|^2u + iCu'' + Du''' = 0,$$

or, for curves of constant curvature, with the sine–Gordon equation

$$\dot{u}' + \kappa_0 \sin u = 0.$$

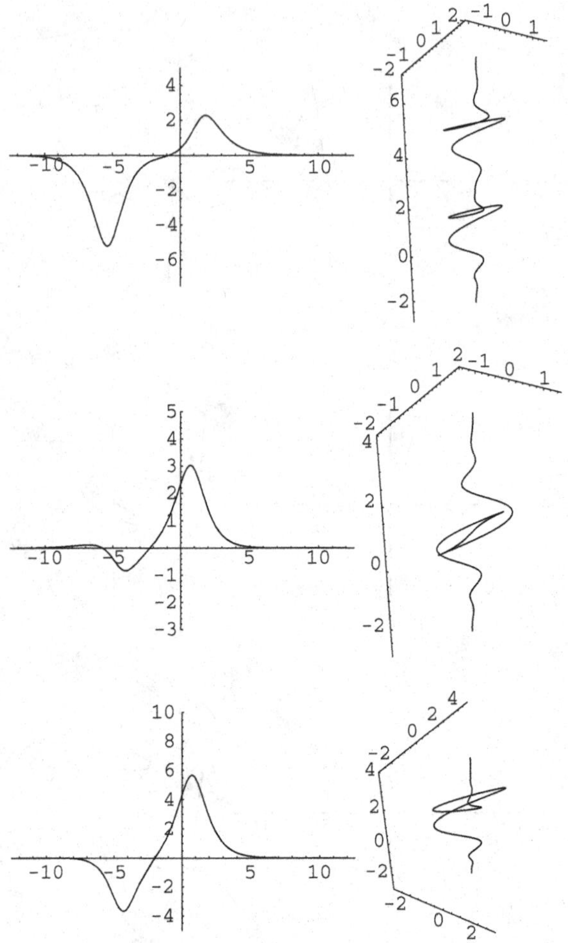

Fig. 15.8 *Left*: Curvature versus arc-length parameter at three different moments of time for the two-soliton NLS solution with opposed phases. *Right*: the corresponding vortex filament shapes at the three moments of time, respectively, showing change in chirality

15.3 Closed Curves Solitons

The filament flow can be regarded as a completely integrable PDE, having an infinite number of conserved integrals

$$\int \tau ds, \int \kappa^2 ds, \int \kappa^2 \tau ds, \int \left[(\kappa')^2 + \kappa^2 \tau^2 - \frac{1}{4}\kappa^4 \right] ds, \dots,$$

and the associated commuting flows

$$t, \kappa b, \frac{1}{2}\kappa^2 t + \kappa' n + \kappa\tau b, \ldots$$

Some closed curve solutions for the vortex filament equation can be obtained from special, periodic or non-periodic NLS potential solutions [40, 41]. It is known now that the position vector of a moving curve representing a vortex filament can be obtained (Sym-Pohlmeyer reconstruction formula) from the fundamental solution matrix $\Psi(s,\lambda)$

$$r(s,t) = \Psi^{-1}(s,\lambda)\frac{d\Psi(s,\lambda)}{d\lambda},$$

of a Zakharov-Shabat spectral problem for the NLS equation with real eigenvalues λ [40]

$$\Psi_s = \begin{pmatrix} i\lambda & q(s) \\ -\bar{q}(s) & -i\lambda \end{pmatrix}.$$

The NLS complex-valued potential $q(s)$ is related through the Hasimoto map to the curvature of the filament $\kappa = |q|$, while the torsion is given by $\tau = \arg[q(s)]' - 2\lambda$. For example, for plane wave solutions of the NLS equation the resulting curve solution is closed and represented by either a multiply-covered circle or a helix. It has been proved, [140] (and references herein), that the algebraic properties of the eigenvalue spectrum λ are related to the geometry and topology of such closed vortex filaments. The Floquet spectrum of the Zakharov-Shabat problem for the NLS equation, for a given potential q, is the set of λ which generate bounded NLS equation solutions Ψ. This spectrum is symmetric under complex conjugation and is constant in time evolution. It has a discrete part responsible for NLS periodic solutions, and several continuous components responsible for aperiodic bounded NLS soliton solutions. Typically, the continuous spectrum consists in the real axis reunited with finite length open curves in $\mathbb{C} = \mathbb{R}^2$ (called "spine-branches"), each intersecting maximum at one point of the real axis $\text{Im}\lambda = 0$. The discrete part of the spectrum contains isolated points placed on these curves, or at their ends. The algebraic properties (like multiplicity) of the discrete points of the Floquet spectrum is in relation to the topology of the corresponding filament curves. The NLS solutions corresponding to isolated points are not necessary periodic, but they can be quasi-periodic. Namely, they are periodic up to a multiplicative factor called the Floquet multiplier. Among various types of spectra, the so-called *finite-gap* spectra (and consequently the corresponding finite-gap potentials q) have only a finite number of isolated points of multiplicity one (simple points).

The main result of this theory, [40, 41, 140], shows that the curve associated to the vortex filament is *smoothly closed* for a certain real eigenvalue λ if that eigenvalue is a real double point (algebraic multiplicity two), and the derivative of the corresponding Floquet multiplier with respect to λ is zero for this eigenvalue. Typically, these points are the double multiplicity points on the real axis where those continuous finite curves intersect it. When the isolated eigenvalue is placed at the intersection with one spine-branch the resulting filament is a deformation of a circle. If the isolated point belongs to a multiple intersection of spine-branches the curves become knotted curves (like the trefoil, cable knot, or torus knots), or star-shaped allowing KdV types of flow [39].

15.4 Nonlinear Dynamics of Stiff Chains

We consider a one-dimensional deformable system (neglecting the width) characterized by finite length L, inextensibility, and elastic bending rigidity ϵ, moving in a very viscous fluid (kinematic viscosity ν or friction coefficient between the polymer and the fluid ζ). At the macromolecular space scale ($L \sim 100\,\mu\text{m}$, $V = 10$–$100\,\mu\,\text{s}^{-1}$, $\nu = 10^{-6}\,\text{m}^2\,\text{s}^{-1}$) the flow is dominated by zero Reynolds number

$$\text{Re} = \frac{VL}{\nu} \sim 0.$$

Because of its geometry (elastic potential energy depending on the square of the curvature) the stiff polymer problem is strongly nonlinear. There are several models in literature about the dynamics of such thin rigid systems, and we mention here just a few of them: DNA molecules [290], actin filaments and motile cells flagella [30, 126, 176, 193, 195, 231, 324, 334], polymeric liquid crystals and stiff polymers in general [66, 109, 290, 345], etc. In the following we present an interesting nonlinear geometrical model based on a Lagrangian approach [109]. The Euler–Lagrange equation for a system described by a smooth parametrized curve $\boldsymbol{r}(\alpha, t)$ with $\alpha \in [0, 1]$ for convenience is given by

$$\frac{d}{dt}\frac{\partial L}{\partial \boldsymbol{r}_t} - \frac{\partial L}{\partial \boldsymbol{r}} = -\frac{\partial R}{\partial \boldsymbol{r}}, \tag{15.63}$$

where R is defined as the Rayleigh dissipation function and it measures the rate of energy dissipation by viscous forces

$$R = \frac{\zeta}{2}\int_0^L |\boldsymbol{r}_t|^2 ds, \tag{15.64}$$

where we used here the arc-length parametrization of the curve. Because of zero Reynolds number we can neglect the first term in (15.63), i.e., the inertia of the system [38, 122, 231, 260, 261, 297, 307]. Consequently the Lagrangian reduces to the minus potential energy of the system. In addition, the unstretching condition enters into the equations in two places: on the one hand as a Lagrange multiplier, and on the other hand as the condition for the metrics g to be independent of time. For a quadratic elastic potential energy (Euler–Bernoulli energy functional for macroscopic systems, or monomer pair interaction for microscopical ones) the dynamical equation reads

$$\frac{\partial}{\partial \boldsymbol{r}}\left[\epsilon \int_0^L (\kappa(s)-\kappa_0(s))^2 ds + \zeta \int_0^L |\boldsymbol{r}_t|^2 ds - \int_0^L \Lambda(s) ds\right] = 0, \qquad (15.65)$$

where we assume that the polymer chain has an equilibrium shape of curvature $\kappa_0(s)$, and $\Lambda(s)$ is the linear tension in the polymer [66], which secures through the last integral (functional Lagrange multiplier) in (15.65), the condition of constant length. We decompose the local forces acting on the polymer along the Serret–Frenet frame

$$\boldsymbol{F} = \zeta \boldsymbol{r}_t = T\boldsymbol{t} + N\boldsymbol{n} + B\boldsymbol{b}. \qquad (15.66)$$

By applying to (15.65) the variational approach, we obtain the following dynamical equations

$$\zeta\kappa_t = \left(\frac{\partial^2}{\partial s^2} + \kappa^2 - \tau^2\right)N - \left(2\tau\frac{\partial}{\partial s} + \tau_s\right)B + \kappa_s T,$$

$$\zeta\tau_t = \frac{\partial}{\partial s}\left[\frac{1}{\kappa}\left(\left(\frac{\partial^2}{\partial s^2} + \kappa^2 - \tau^2\right)B + \left(2\tau\frac{\partial}{\partial s} + \tau_s\right)N\right)\right] + 2\kappa\tau N - \kappa_s B + \tau_s T,$$

$$\frac{\partial T}{\partial s} + \kappa N = 0, \qquad (15.67)$$

where κ, τ are the curvature and torsion and the last equation comes from the condition of time independent metric. The equations are complicated and, except numerical simulations, it is difficult to sense the contribution of the nonlinear terms. By using the Hasimoto transformation (7.56)

$$\Psi(s,t) = \kappa(s,t)\exp\left(i\int^s \tau(s',t)ds'\right), \qquad (15.68)$$

in (15.67) the problem is simplified a lot. With the complex notation

$$\Gamma = (N + iB)\exp\left(i\int^s \tau(s',t)ds'\right), \qquad (15.69)$$

the system (15.67) reduces, like in the case of moving curves (7.57), to an integrodifferential equation

$$\zeta\Psi_t = \left(\frac{\partial^2}{\partial s^2} + |\Psi|^2\right)\Gamma + \Psi\,\mathrm{Im}\int^s \Psi_s\Gamma^* ds' + \Psi_s T. \qquad (15.70)$$

The dynamics of the stiff polymer should be related to the dynamics of elastic beams (Euler's *elastica* theory [170]). In this theory the inflexion points ($\kappa = 0$) of the beam are important because at these points the net torque is zero. For (15.70), in the plane case $V = 0$ (to simplify the equations) we have the following condition holding at inflexion points

$$\zeta\kappa_t = N_{ss} + \kappa_s T.$$

Considering the tangential components to be irrelevant (just a reparametrization of the curve) this further reduces to a very simple condition

$$\zeta \kappa_t = N_{ss}.$$

From the nonlinear Schrödinger equation type of structure of the dynamical equations, we expect that some solutions in curvature to have nontopological soliton behavior. That would imply zero curvature along the polymer chain, except in some isolated points (many-soliton solutions) where the curvature increases drastically. Indeed, numerical simulations of the dynamical equations show such multiple *hairpin loop* shapes.

15.5 Problems

1. Prove that starting from (15.12) written in an arbitrary parametrization (not the arc-length one) we obtain the same continuity equation (15.14). Check if the same continuity equation is obtained if we start from the equation $\dot{\boldsymbol{r}} = \kappa \boldsymbol{r}' \times \boldsymbol{r}''$.
2. Show that any rigid helix ($\kappa, \tau =$ const.) is a solution of the filament equations (15.14) and (15.19).
3. By identifying the Navier–Stokes one-dimensional equation with the second equation in (15.21), find that the pressure associated with the gas dynamics filament model is
$$P = \frac{\rho'^2}{\rho} - \rho^2 - \rho''.$$
4. Solve (15.34) and (15.35) for simple examples of $\kappa(s)$ and $\tau(s)$. Consider that at the initial point $s = 0$ the Serret–Frenet trihedron has the orientation of the canonical frame of reference, i.e., $\boldsymbol{t}(0) = (1,0,0), \boldsymbol{n}(0) = (0,1,0)$, etc. and $\boldsymbol{r}(0) = 0$. Find the expression of $\boldsymbol{r}''(0)$ and $\boldsymbol{r}'''(0)$ in terms of $\kappa(0)$ and $\tau(0)$. Show that a better result (faster convergence) of numerically integrating (15.34) could be obtained if we use for initial conditions a configuration inspired by a helix:
$$\boldsymbol{r}(0) = \left(\frac{\kappa(0)}{\kappa^2(0) + \tau^2(0)}, 0, 0 \right),$$
$$\boldsymbol{r}'(0) = \left(0, \frac{\kappa(0)}{\sqrt{\kappa^2(0) + \tau^2(0)}}, \frac{\tau(0)}{\sqrt{\kappa^2(0) + \tau^2(0)}} \right),$$
$$\boldsymbol{r}''(0) = (-\kappa(0), 0, 0), \quad \boldsymbol{r}'''(0) = \left(0, -\kappa(0)\sqrt{\kappa^2(0) + \tau^2(0)}, 0\right).$$
5. Find the third-order differential equation fulfilled by $\boldsymbol{b}(s)$ in the case of constant torsion, from (15.34).
6. Prove that (15.9) and (15.10) contain a logarithmic divergence in ϵ^{-1}, hence these expressions for the velocity of the vortex filament are not valid in the limits $r_0, r_1 \to 0$.

Chapter 16
Solitons on the Boundaries of Microscopic Systems

In this chapter, we focus on some applications of soliton theory in microscopic compact systems with boundary, like nuclei or quantum Hall liquids. At this space scale, the solitons correspond to solutions of field equations with finite energy and with a localized, nondispersive energy density. Since the field theories describing many-body systems of elementary particles are quantum theories, one should perform the so-called quantization of solitons procedure. This is done in principle by using a semiclassical expansion to associate with a classical soliton solution both a quantum soliton-particle states, and a whole series of excited state by quantizing the fluctuations around the soliton. Since the soliton solutions are nonperturbative, their quantum versions are themselves nonperturbative [161, 267].

Soliton models have been successfully used to incorporate the quark structure of hadrons into nuclear physics, by using phenomenological quantum chromodynamics field theories. Examples of such simple models are the soliton bag model, chiral quark-meson models, and color dielectric model, which permit calculations of nucleon structure and interactions [23]. Soliton solutions are also involved in semimicroscopic nuclear models like the quasimolecular shapes model [282] and cluster model [115]. Dynamical calculations in these models are based on the use of coherent states to provide quantum states corresponding to these solitons. For example, some authors [116] consider the Glauber coherent states as possible candidates for the wave functions of the nucleons clustered in the α-particle before decay. Moreover, at mesoscopic scale, localized nonlinear field solutions can occur in superconductors as vortices [197, 199].

Coming back to our main topic, the description of compact many-body systems gives best results if performed in terms of collective modes, especially if the collective modes have lower excitation energies when compared with the single-particle excitations. By microscopic compact system, we understand a bounded system of particles having one or more closed boundaries. Collective modes are coherent, in-phase motion of the nuclear matter, as opposed to individual single-particle motions. We can divide the collective modes of excitation in two categories: bulk and boundary. For example, in a nucleus the bulk collective modes could be

A. Ludu, *Nonlinear Waves and Solitons on Contours and Closed Surfaces*, Springer Series in Synergetics, DOI 10.1007/978-3-642-22895-7_16,

rotations of deformed nuclei, the photonuclear giant resonance, while boundary, surface, or contour excitations could be the low-lying "rotation–vibration" surface modes, or nuclear fission [87]. More example of boundary modes are sound waves in solids [83, 180, 322], plasmons in charged systems [244], shape oscillations or surface waves in liquid drops [13, 16, 31, 35, 61, 156, 157, 189, 203, 223, 236, 237, 253, 263–265, 269, 283, 316, 318, 321, 346], vortex patches in ideal fluids [65, 329], atmospheric plasma clouds [248], pattern formation in ferromagnetic fluids [280], two-dimensional electron systems, tides in neutron stars [340, 341], etc. If the collective modes associated with both single-particle and collective bulk excitations are absent or reduced (gapped as atomic physics people would say), the system is referred as *incompressible* and the *boundary modes* take over. From the energy point of view, the boundary modes will be lower in energy and "softer," with lower frequencies than the bulk modes. In such situations, one has to study the dynamics of the boundaries or contours, which has the advantage of less calculations than the whole bulk: the system has lower dimension. Moreover, the global constraints like length, area, and volume conservation can be useful in the model. These conservation laws enter into the calculation as Lagrange multipliers and global conservation laws.

16.1 Solitons as Elementary Particles

In spite of the great progress been made by lattice quantum field theory combined with perturbative (Feynman diagram) methods, and experiments on particle spectra and particle scattering there is so far only a primitive understanding of the detailed geometry of the quark structure of nucleons [208]. Moreover, effective calculation of the properties of nuclei using QCD is currently too hard, and there is no agreed understanding of the force between nucleons at short range directly from QCD. Along these lines, the use of a soliton paradigm for elementary particles is an attractive direction: the same field equations can explain simultaneously the existence, structure and interaction of these soliton-like particles [21].

It is not an exception when, by solving the nonlinear field equations precisely, and deals with quantum aspects perturbatively, one finds solitons solutions that behave like particles, as we know from the case of pion (the nonlinear sigma model and Skyrmions), or gluon fields [209]. Such a different paradigm for an elementary particle travels uniformly with any relativistic velocity in a flat space and along a geodesic in a constantly curved space; it has mass, momentum, spin, and has localized field density of energy.

In the case of elementary particles field equations, the stability of the soliton against radiative dissipation is provided by its topological conservation laws. Pretty much like in the case of shallow water KdV soliton where the nonlinearity arise from the geometry through the surface tension (in addition to the intrinsic nonlinearity of the Navier–Stokes equations), in the case of elementary particles the nonlinearity arises from the non-trivial topological structure of the field itself. The interactions between soliton particles can be obtained exactly when the separation between solitons is large compared to their size (energy profile half-width), by superimposing the asymptotical (linearized) soliton fields. The amplitude of the linearized soliton field far from the soliton core is called the charge of the soliton particle.

There are many developments of the soliton particle field equations, mainly arising from examples of nonlinear integrable systems like sine-Gordon model, Skyrme model, Q-balls, [6, 208], etc. A generic example is provided by the Bogomolny equations, [208]. These equations can describe, for example, magnetic flux vortices in superconductors at the critical coupling which separates type I from type II superconductivity. In type II superconductors the vortex structure is described by the Ginzburg-Landau equation, and the vortices strongly repel and form Abrikosov lattices at low temperatures. Another possible application place of the Bogomolny equations is found in supersymmetry or grand unification theory of elementary particles at higher energy scale where monopoles become relevant. Other examples of Bogomolny equation occur for abelian Higgs vortices, pure Yang-Mills theory for four space dimensions where the solitons are called instantons, types of domain wall models, and in some string theories.

The Bogomolny field equations are obtained from a Yang-Mills-Higgs model with gauge group $SU(2)$, and with the fields potentials (A_i, Φ) taking values in the $su(2)$ Lie algebra, where the potential energy has the form

$$U = -\frac{1}{4}\int \Big(\mathrm{Tr}(B_k B_k) + \mathrm{Tr}(\nabla_k \Phi \nabla_k \Phi)\Big) d^3x,$$

where $B_k = -\epsilon ijk F_{ij}/2$ and $F_{ij} = \partial_i A_j - \partial_j A_i + [A_i, A_j]$ are the field intensities, and the operator ∇_k is the covariant derivative, e.g. $\nabla_k \Phi = \partial_k + [A_k, \Phi]$. By using the Bianchi identities, see (4.55),(4.56),(4.36), $\nabla_k B_k = 0$ one can re-write the potential energy in the form

$$U = -\frac{1}{4} \in \mathrm{Tr}(B_k \pm \nabla_k \Phi)(B_k \pm \nabla_k \Phi) d^3x \pm \frac{1}{2}\int \partial_k(\mathrm{Tr}(B_k \Phi)) d^3x,$$

such that the second term can be expressed as a surface integral at infinity. From the boundary conditions it results that the value of this integral is $2\pi \deg(\Phi) = 2\pi N$ with N integer, since the Higgs field Φ at infinity can be interpreted as a Gauss map from $S_2 \in \mathbf{R}^3$ to the unit 2-sphere in $su(2)$. From the expression of the potential energy it results that $U \geq 2\pi|N|$, where equality holds if the Higgs field satisfies the Bogomolny equation

$$B_k \pm \nabla\Phi = 0,$$

where the sign $\pm$ is chosen function of the signature of N, respectively. Solutions of the Bogomolny equation can be physically interpreted as static superposition of N magnetic monopoles (antimonopoles for $N < 0$) and describe the minima of the potential energy for a fixed value for N.

16.2 Quantization of Solitons on a Closed Contour and Instantons

Let us consider a real scalar field theory on a circle of radius R, described by the following Lagrangian density

$$\mathcal{L} = \frac{1}{2}\left\{\Phi_t^2 - (\Phi_x^2 - U(\Phi))^2\right\}, \tag{16.1}$$

where $x = R\theta$, $\theta \in [-\pi, \pi]$. The function U is left, for the moment arbitrary. We impose the boundary condition:

$$\Phi(t, -\pi R) = \Phi(t, \pi R). \tag{16.2}$$

The Euler–Lagrange equation is

$$\Phi_{tt} - \Phi_{xx} + U(\Phi)U'(\Phi) = 0. \tag{16.3}$$

Clearly, this Lagrangian is invariant under time translations, and the corresponding conserved quantity given by Noether's theorem (the energy) is:

$$\mathcal{E}[\Phi] = \int_{-\pi R}^{\pi R} dx \frac{1}{2}\left\{\Phi_t^2 + (\Phi_x - U(\Phi))^2\right\}. \tag{16.4}$$

The energy functional, given by (16.4) is obviously bounded from below. It attains its minimum ($\mathcal{E} = 0$) at a field configurations which satisfies

$$\begin{cases} \Phi_t = 0 \\ \Phi_x = U(\Phi) \end{cases} \tag{16.5}$$

We prove that a configuration that satisfies (16.5) also satisfies the equation of motion (16.3). The first equation in (16.5) means that a minimum-energy Φ is time independent. For time-independent Φ, (16.3) becomes

$$\Phi_{xx} - U(\Phi)U'(\Phi) = 0. \tag{16.6}$$

Now, taking the x-derivative of second equation in (16.5) and eliminating Φ_x with the same (16.5), we obtain (16.6). So, any minimum-energy configuration satisfies the equation of motion, i.e., it is a *vacuum* configuration. Let us take a look now at second ordinary differential equation (ODE) in (16.5). Clearly

$$\Phi(x) = K = constant, \quad \text{where} \quad U(K) = 0, \tag{16.7}$$

is (are) solution(s) of (16.5). In terms of ODE's language, these are "singular" solutions. Equation (16.5) has also the "regular" solution given by

$$\int \frac{d\Phi}{U(\Phi)} = x - x_0 \tag{16.8}$$

Depending on the explicit form of U, (16.5) could have, also, some singular, nonconstant solutions. The solution in (16.8) will be referred as the *bubble* solution. To be a vacuum, a solution of (16.5) should satisfy the boundary condition (16.2). This depends on the explicit form of U. We will assume that solution (16.8) do so. Obviously, a solution like (16.7) satisfies (16.2).

This model has also Hamiltonian structure. The action is

$$\mathcal{S}[\Phi] = \int dt L[\Phi], \tag{16.9}$$

where L is the Lagrangian

$$L[\Phi] = \frac{1}{2}\int_{-\pi R}^{\pi R} \{\Phi_t^2 - (\Phi_x - U(\Phi))^2\} = \mathcal{T} - \mathcal{U}\cdot$$

The canonical momenta are

$$\Pi(x,t) = \frac{\delta L}{\delta \Phi_t(x,t)} = \Phi_t(x,t)$$

and the Hamiltonian becomes

$$\mathcal{H} = \int_{-\pi R}^{\pi R} dx \Pi(x,t)\Phi_t(x,t) - L = \mathcal{T} + \mathcal{U}\cdot \tag{16.10}$$

The canonical equations of (16.10) are

$$\begin{cases} \frac{\partial \Pi(x,t)}{\partial t} = \Phi_{xx}(x,t) - U(\Phi)U'(\Phi) \\ \frac{\partial \Phi(x,t)}{\partial t} = \Pi(x,t) \end{cases}. \tag{16.11}$$

Clearly, the Hamiltonian (16.10) is the energy of the classical vacua. These solutions are static, so we may use the standard procedure [267] to find the low excited states associated with each vacuum solution. Of course, this procedure ignores the tunneling between the different vacua. The net effect of tunneling is to add an imaginary part to the energies. In case of the "false" vacua, this imaginary part is small compared with the real part (the energy obtained if the tunneling is ignored) (see [58, 59]). We expect this to be true for the low excited states also. According to this procedure, a "tower" of states is associated to each static solution. We will not discuss here the towers associated with the vacuum solutions. We will present only a "tower" that is not associated with a vacuum solution but it is somehow related to the "bubble" solution.

Let $\Phi^{(0)}(x)$ be the "bubble" solution. Because the static Euler–Lagrange equation (16.6) is of order 2 in x, $\Phi^{(1)}(x) = \Phi^{(0)}(-x)$ it is also a solution. Obviously, $\Phi^{(1)}$ it is not a vacuum solution and it has a nonvanishing energy. Consequently, the "tower" of states built around $\Phi^{(1)}$ will have the energies higher than the "tower" built around the "bubble" solution. Note that the shape of $\Phi^{(1)}$ is identical with the shape of $\Phi^{(0)}$. So, we have two classical configurations that are identical in shapes. One ($\Phi^{(0)}$) is a vacuum configuration; therefore, it might be attainable in the corresponding quantum theory by *spontaneous* transitions from other vacuum configurations (for example "normal" configurations), and the other is a classically excited configuration. This feature clearly supports the interpretation that the "bubble" is at least related with a separate object (the cluster) that may exist alone, separate from the object that created it. The "bubble" vacuum configuration ($\Phi^{(0)}$) would be, in this interpretation, the configuration that consists in a *preformed* cluster plus whatever remains if the cluster is emitted (the descendant), and the "bubble" excited configuration ($\Phi^{(1)}$) would be a bound-state configuration of descendant and cluster.

To study the tunneling between the classical vacua, we will use the standard instanton-based method [58, 59, 267]. Instantons are solutions of the Euclidean Euler–Lagrange equation

$$\Phi_{\tau\tau} + \Phi_{xx} - U(\Phi)U'(\Phi) = 0, \tag{16.12}$$

having a finite Euclidean action. The Euclidean action for our model is

$$S_E[\Phi] = \int d\tau \int_{-\pi R}^{\pi R} dx \frac{1}{2} \left\{\Phi_\tau^2 + (\Phi_x + U(\Phi))^2\right\}. \tag{16.13}$$

If x would range on the entire real axis, it could be proved that there are no finite-action, nontrivial solutions of (16.12). This is similar with Derrick's theorem (see [267]). We will not present the demonstration here. We note it to show the necessity for considering the model on a circle. Note that the missing of instantons does not mean that there is no tunneling, but the tunneling (if exists) cannot be revealed by the semiclassical instanton-based method. The desired feature is the presence of the normal vacuum disintegration. Note that in all models studied in [58,59], the false vacuum has a higher energy than the true vacuum. We will prove that in this model, even if the two classical vacua under study (the normal vacuum and the "bubble") are degenerate (i.e., we cannot call one of them "true vacuum" and the other "false vacuum") the normal (classical) vacuum is quantum unstable. Let $\Phi(x) = K$ be a constant vacuum (the normal vacuum) and $\Phi^0(x)$ a nonconstant one (the "bubble"). We will suppose that $\lim_{x\to\pm\infty} \Phi^0(x) = K$ and Φ^0 have only one local extremum. We are interested to find nonconstant (τ) solutions of (16.12), which satisfy the following boundary conditions

$$\Phi(x, \tau = \pm\infty) = K, \tag{16.14}$$

It can be seen that a satisfactory solution is

$$\Phi(x, \tau) = \Phi^0(\tau). \tag{16.15}$$

In Coleman's terminology, this is a *bounce*. The fact that a bounce exists, subjected to boundary conditions (16.14) is the first step to prove the quantum instability of the normal (classical) vacuum. The last step is to prove that the following operator has a negative eigenvalue (see [58, 59] for more details)

$$\mathcal{O} = -\frac{\partial^2}{\partial \tau^2} - \frac{\partial^2}{\partial x^2} + 2U'(\Phi^{(0)}(\tau))\frac{\partial}{\partial x} + \left(UU'' + (U')^2\right)\big|_{\Phi^{(0)}(\tau)} \tag{16.16}$$

Let us restrict the study of the eigenvalue problem of the operator (16.16) to the x-independent eigenfunctions. Of course, by doing this restriction we lose some eigenvalues. But we are interested only in proving that there is at least one negative eigenvalue. By this restriction, the eigenvalue problem become:

$$\left[-\frac{\partial^2}{\partial x^2} + \left(UU'' + (U')^2\right)\big|_{\Phi^{(0)}(\tau)}\right]\chi(\tau) = \epsilon\chi(\tau). \tag{16.17}$$

Note that (16.17) is a time-independent Schrödinger equation. This equation has the following particular solution

$$\chi_0(\tau) = \frac{d}{d\tau}\Phi^{(0)}(\tau). \tag{16.18}$$

This solution corresponds to $\epsilon = 0$ and is associated to τ-translation symmetry of the system. Clearly, χ_0 has a node where $\Phi^{(0)}$ has an extremum. By the balancing theorem there is at least one eigenfunction corresponding to an eigenvalue lower than $\epsilon = 0$. This proves the previous assertion. In principle, the instanton-based method may be used to compute the disintegration probability.

16.3 Clusters as Solitary Waves on the Nuclear Surface

We devote this section to the application of soliton models on compact shapes in the study of α or heavier cluster formation in heavy nuclei resulting in radioactive decay, α-cluster states in scattering processes, quasimolecular resonances in heavy ions, highly deformed exotic nuclear shapes and fission.

The liquid drop model, as a collective model of the nucleus, describes very well the spectra of spherical nuclei as small vibrations around the equilibrium shape. On the other hand, it is known that on the nuclear surface of heavy nuclei close to the magic nuclei (^{208}Pb, ^{100}Sn) a large enhancement of clusters (alpha, carbon, oxygen, neon, magnesium, silicon) exists, which leads to the emission of such clusters as natural decays [262]. Traditional collective models [113] are unable to give a complete explanation of such natural decays, i.e., they still did not completely answer the main physical question: why should nucleons join together and spontaneous form an isolated cluster on the nuclear surface? In the following, we present how soliton solutions in the nuclear (nonlinear) drop model plus shell corrections can give an answer in a positive way to this question [105, 183–186, 188].

We describe the surface Σ of a nucleus as a function of the polar angles θ and φ, by writing the nuclear radius in the form

$$r = R_0(1 + \xi(\theta, \varphi, t)), \tag{16.19}$$

where R_0 is the radius of the spherical nucleus. Without loss of generality we choose a special shape as a traveling perturbation (η) in the φ-direction, having a given transversal profile (g) in the θ-direction

$$\xi(\theta, \varphi, t) = g(\theta)\eta(\varphi - Vt) \tag{16.20}$$

with g an arbitrary bounded, nonvanishing continuous function, η a rapidly decreasing function, and V defining the tangential velocity of the traveling solution η on the surface. This choice is different from the traditional liquid drop model case where the shape function is expanded in spherical harmonics and we need ten multipoles to fit a soliton shape [183–185]. In the liquid drop model, we consider the nucleus as an inviscid incompressible fluid layer described by the irrotational field velocity $\boldsymbol{v}(r, \theta, \varphi, t)$ and by the constant mass density $\rho = \text{const}$. From the continuity equation and the irrotational condition, we have the Laplace equation

$$\boldsymbol{v} = \nabla\Phi, \quad \Delta\Phi = 0. \tag{16.21}$$

The dynamics of this perfect fluid is described by the Euler equation (10.15)

$$\frac{\partial \boldsymbol{v}}{\partial t} + (\boldsymbol{v} \cdot \nabla)\boldsymbol{v} = -\frac{1}{\rho}\nabla P + \frac{1}{\rho}\boldsymbol{f}, \tag{16.22}$$

where P is the pressure and $\boldsymbol{f}$ is the volume density of the Coulombian force, $\boldsymbol{f} = -\rho_{el}\nabla\Psi$, with Ψ the electrostatic potential and ρ_{el} the charge density, supposed to be constant, too. We have

$$\left(\Phi_t + \frac{1}{2}|\nabla\Phi|^2\right)\bigg|_\Sigma = -\frac{1}{\rho}P - \frac{\rho_{el}}{\rho}\Psi|_\Sigma. \tag{16.23}$$

To determine the functions Φ and ξ, we need in addition boundary conditions for the scalar harmonic field Φ, on two closed surfaces: the external free surface of the nucleus (9.30) and the inner surface (if it exists) of the fluid layer. The latter condition requests zero radial velocity of the flow on its inner surface.

$$\left.\frac{dr}{dt}\right|_{\Sigma} = \left.\left(\frac{\partial r}{\partial t} + \frac{\partial r}{\partial \theta}\frac{d\theta}{dt} + \frac{\partial r}{\partial \varphi}\frac{d\varphi}{dt}\right)\right|_{\Sigma}. \tag{16.24}$$

This equation allows general types of movements, including traveling and vibrational waves. Equation (16.24) reduces to the form $\left.\frac{dr}{dt}\right|_{\Sigma} = \left.\frac{\partial r}{\partial t}\right|_{\Sigma}$ in the linear approximation (the Bohr–Mottelson model). This linearization restricts the oscillations to only collective radial vibrations, and does not allow any motion along the tangential direction. Equation (16.24) can be written in terms of the derivatives of the potential of the flow and the shape function ξ

$$\left.\Phi_r\right|_{\Sigma} = R_0\left.\left(\xi_t + \frac{\xi_\theta}{r^2}\Phi_\theta + \frac{\xi_\varphi}{r^2\sin^2\theta}\Phi_\varphi\right)\right|_{\Sigma}. \tag{16.25}$$

where $\frac{\partial \Phi}{\partial r} = v_r = \dot{r}$ is the radial velocity and $\frac{1}{r}\frac{\partial \Phi}{\partial \theta} = v_\theta = r\dot{\theta}$, $\frac{1}{r\sin\theta}\frac{\partial \Phi}{\partial \varphi} = v_\varphi = r\dot{\varphi}\sin\theta$ are the tangential velocities. We denote here the partial differentiation by suffixes, $\partial\Phi/\partial\varphi = \Phi_\varphi$, etc. The existence of a rigid core of radius $R_0 - h(\theta) > 0$, $h(\theta) \ll R_0$, introduces the second boundary condition for the radial velocity on the surface of this core in the form

$$v_r|_{r=R_0-h} = \left.\frac{\partial \Phi}{\partial r}\right|_{r=R_0-h} = 0. \tag{16.26}$$

The motion of the fluid is described by the Laplace equation and by the two boundary conditions. We use for the potential of the flow the expansion

$$\Phi = \sum_{n=0}^{\infty}\left(\frac{r-R_0}{R_0}\right)^n f_n(\theta,\varphi,t), \tag{16.27}$$

where the functions f_n do not form in general a complete system on the sphere. The convergence of (16.27) is assured by the value of the small quantity $\frac{r-R_0}{R_0} \le max|\xi| = \epsilon$. From the Laplace equation (in spherical coordinates) and the expansions

$$\frac{1}{r^n} = \frac{1}{R_0^n}\sum_{k=0}^{\infty}(-1)^k((n-1)k+1)\xi^k, \quad k = 1,2, \tag{16.28}$$

we obtain a system of equations that result in the recurrence relations for the unknown functions f_n

$$f_n = [(-1)^{n-1}(n-1)\Delta_\Omega f_0 - 2(n-1)f_{n-1} + \sum_{k=1}^{n-2}(-1)^{n-k}(2k-(n-k-1)\Delta_\Omega f_k)]\frac{1}{n(n-1)}, \tag{16.29}$$

with $n \geq 2$ and where $\Delta_\Omega = \frac{1}{\sin\theta}\frac{\partial}{\partial\theta}\left(\sin\theta\frac{\partial}{\partial\theta}\right) + \frac{1}{\sin^2\theta}\frac{\partial}{\partial\varphi}$ is the angular part of the Laplacian operator in spherical coordinates. Equation (16.29) reduces the unknown functions to only two: $\Delta_\Omega f_0$ and f_1:

$$f_2 = -\frac{1}{2}(\Delta_\Omega f_0 + 2f_1),$$

$$f_3 = \frac{1}{6}(4\Delta_\Omega f_0 - 4\Delta_\Omega f_1 + 4f_1 + 2),$$

$$f_4 = \frac{1}{24}(\Delta_\Omega^2 f_0 - 14\Delta_\Omega f_0 + 8\Delta_\Omega f_1 - 8f_1) \quad \ldots. \tag{16.30}$$

If we choose the independent functions $\Delta_\Omega f_0$ and f_1 to be smooth on the sphere, they must be bounded together with all the f_ns (these being linear combinations of higher derivatives of f_0 and f_1) and hence the series in (16.27) is indeed controlled by the difference in the radii between the deformed and the spherical one. However, in the following we will use only truncated polynomials of these series. By introducing (16.29) and (16.30) in the second boundary condition (16.26), we obtain the condition

$$\sum_{n=1}^{\infty} n\left(-\frac{h}{R_0}\right)^{n-1} f_n = 0, \tag{16.31}$$

which reads, in the first order in h/R_0

$$f_1 = \frac{2h}{R_0} f_2. \tag{16.32}$$

From (16.30) and (16.32), the unknown function f_1 is obtained, in the smallest order in h/R_0

$$\Delta_\Omega f_0 = -\left(\frac{R_0}{h} + 2\right) f_1. \tag{16.33}$$

Concerning the free surface boundary condition, we need to calculate the derivatives of the potential of the flow on that surface

$$\Phi_r|_\Sigma = \sum_n n \frac{(r - R_0)_\Sigma^{n-1}}{R_0^n} f_n = \frac{f_1}{R_0} + \frac{2\xi f_2}{R_0} + \mathcal{O}_2(\xi),$$

$$\Phi_\varphi|_\Sigma = \sum_n \xi^n f_{n,\varphi} = f_{0,\varphi} + \xi f_{1,\varphi} + \mathcal{O}_2(\xi), \tag{16.34}$$

$$\Phi_\theta|_\Sigma = \sum_n \xi^n f_{n,\theta} = f_{0,\theta} + \xi f_{1,\theta} + \mathcal{O}_2(\xi).$$

By introducing the series (16.28) and (16.34) in (16.25) for the traveling wave solution (16.20), we have the equation

$$f_1 + 2\xi f_2 = R_0^2 \xi_t + \frac{\xi_\varphi(1 - 2\xi)}{\sin^2\theta}(f_{0,\varphi} + \xi f_{1,\varphi})$$

$$+ \xi_\theta(1 - 2\xi)(f_{0,\theta} + \xi f_{1,\theta}). \tag{16.35}$$

We keep the nonlinearity of the boundary conditions in the first order in the expression of f_0 and the second order in the expression of f_1. Consequently, to be consistent, it is enough to take the linear approximation of the solution for f_1 in (16.35), like in the case of the normal modes of vibrations

$$f_1 = R_0^2 \xi_t + \mathcal{O}_2(\xi). \tag{16.36}$$

Hence, by introducing the linear approximation for f_1 (16.36) in (16.35) we have

$$2\xi f_2 = \frac{1}{\sin^2\theta}\left(-\xi_\varphi f_{0,\varphi} + \xi\xi_\varphi(f_{1,\varphi} - 2f_{0,\varphi})\right) + \xi\xi_\theta(f_{1,\theta} - 2f_{0,\theta}), \tag{16.37}$$

and by taking the expression of f_2 from the recurrence relations (16.32) and $\Delta_\Omega f_0$ from (16.33), we obtain the form of f_0, in the second order in ξ

$$f_{0,\varphi} = -\frac{R_0^3 \sin^2\theta}{h}\frac{\xi\xi_t}{\xi_\varphi}(1 + 2\xi) - \frac{\xi_\theta f_{0,\theta}}{\xi_\varphi} + \mathcal{O}_3(\xi). \tag{16.38}$$

In the case of traveling wave profile of the form $\xi(\theta, \varphi, t) = g(\theta)\eta(\varphi - Vt)$, it occurs the restriction $\xi_\varphi = -V\xi_t$, and consequently the tangential velocity in the θ-direction becomes zero. Equation (16.38) reads

$$f_{0,\varphi} = \frac{V R_0^3 \sin^2\theta}{h}\xi(1 + 2\xi) + \mathcal{O}_3(\xi). \tag{16.39}$$

Equations (16.36), (16.38), and (16.39) describe, in the second order in ξ, the connection between the velocity potential, the shape function, and the boundary conditions. This fact is a typical feature of nonlinear systems. The dependence of

$\Phi|_\Sigma$ on the polar angles, in the second order in ξ, has the form of a quadrupole in the θ-direction and depends only on ξ and its derivatives in the φ-direction. For traveling wave profiles the tangential velocity in the direction of motion of the perturbation, $v_\varphi = \Phi_\varphi / r \sin\theta$ is proportional with ξ in the first order

$$v_\varphi = \frac{2VR_0 \sin\theta}{h}\xi + \mathcal{O}_2(\xi)\cdot \tag{16.40}$$

To obtain the dynamical equation for the surface Σ, we follow the formalism for the normal vibration of droplets described in Chap. 13. The surface pressure is obtained from the surface energy of the deformed nucleus, U_S, and according to Sect. 10.4 is given by

$$P|_\Sigma = 2\sigma H = \frac{\sigma}{R_0}(-2\xi - 4\xi^2 - \Delta_\Omega \xi + 3\xi\xi_\theta^2 \text{cotan}\,\theta) + \text{const.} \tag{16.41}$$

where H is the mean curvature of the fluid surface. The terms of order three in $\xi_{\varphi,\theta}, \xi_{\varphi,\varphi}$, and $\xi_{\theta,\theta}$, can be neglected in (16.41) because of the high localization of the solution (the relative amplitude of the deformation ϵ is smaller than its angular half-width L, $\xi\xi_{\varphi\varphi}/R_0^2 \simeq \epsilon^2/L^2 \ll 1$, etc.).

The Coulomb potential is given by a Poisson equation, $\Delta\Psi = \rho_{el}/\epsilon_0$, with ϵ_0 the vacuum dielectric constant. By using the same method like for Φ [183], we obtain in the second order for ξ, the form

$$\Psi|_\Sigma = \frac{\rho_{el} R_0^2}{3\epsilon_0}\left(1 - \xi - \frac{\xi^2}{6}\right). \tag{16.42}$$

To write the Euler equation we take the surface pressure from (16.41), the velocity potential from (16.27), (16.32), (16.36), (16.39), and the Coulomb potential from (16.42) and we write, in the second order in ξ, and in the first order in its derivatives the dynamic equation

$$\Phi_t|_\Sigma + \frac{V^2 R_0^4 \sin^2\theta}{2h^2}\xi^2 = \frac{\sigma}{\rho R_0}(2\xi + 4\xi^2 + \Delta_\Omega \xi - 3\xi^2 \xi_\theta ctg\theta) + \frac{\rho_{el}^2 R_0^2}{3\epsilon_0 \rho}\left(\xi + \frac{\xi^2}{6}\right) + \text{const.} \tag{16.43}$$

This is a nonlinear PDE in variables θ and φ. By differentiating it again with respect to φ, and by using (16.33) and (16.36) we obtain in the second order, after reordering the terms

$$A(\theta)\eta_t + B(\theta)\eta_\varphi + C(\theta)g(\theta)\eta\eta_\varphi + D(\theta)\eta_{\varphi\varphi\varphi} = 0, \tag{16.44}$$

which is a Korteweg–de Vries (KdV) equation with coefficients depending perimetrically on θ

$$A = \frac{VR_0^2(R_0 + 2h)\sin^2\theta}{h}; \quad B = -\frac{\sigma}{\rho R_0}\frac{(2g + \Delta g)}{g} - \frac{\rho_{el}^2 R_0^2}{3\epsilon_0 \rho};$$
$$C = 8\left(\frac{V^2 R_0^4 \sin^4\theta}{8h^2} - \frac{\sigma}{\rho R_0}\right) - \frac{\rho_{el}^2 R_0^2}{9\epsilon_0\rho}; \quad D = -\frac{\sigma}{\rho R_0 \sin^2\theta}, \tag{16.45}$$

It is obvious now that the depth of the fluid layer inside the nuclear surface should be considered as a function of θ, itself, $h = h(\theta)$. Same reasoning applies to $V, L \rightarrow V(\theta), L(\theta)$. It means that the nonlinear flow under the surface interacts with the core in a variable way, function of the azimuthal angle. The KdV (16.44) has cnoidal waves solutions (Sect. 11.2)

$$\eta(\varphi, t) = g(\theta)\mathrm{sn}^2\left(\frac{\varphi - V(\theta)t}{L(\theta)}\middle| k(\theta)\right) + \Upsilon(\theta), \tag{16.46}$$

depending on three arbitrary parametric functions $g(\theta), \Upsilon(\theta), k(\theta)$. The angular (poloidal) velocity and the angular half-width have the forms

$$V(\theta) = \frac{B}{A} + \frac{Cg}{3A}\left(1 + \frac{1}{k}\right) + \frac{Cg\Upsilon}{A},$$
$$L(\theta) = 2\sqrt{-\frac{3Dk}{Cg^2}} \tag{16.47}$$

where all symbols on the RHS are functions of θ as shown above. The periodicity condition on the closed path around a parallel circle reads

$$K(k(\theta))L(\theta) = \pi/2, \tag{16.48}$$

where $K(k)$ is complete elliptic integral of the first kind (Sect. 18.3). To have constant traveling waves along the φ-direction, the parameters of the soliton solution must have constant angular velocity V so as to keep the shape of the wave stable in time and along the equatorial motion. If we couple all these constraints with the periodicity condition (16.45), (16.47), and (16.48), we have the following dependence

$$V^2(\theta) = \frac{8h^2}{R_0^2 \sin^2\theta} \frac{\left[\beta + \left(\frac{g(k+1)}{3k} + \Upsilon\right)(\beta + 8\gamma) + \gamma\frac{2g+\Delta g}{g}\right]}{\left[8h(R_0 + 2h) - \left(\frac{g(k+1)}{3k} + \Upsilon\right)R_0^2 \sin^2\theta\right]}$$
$$L(\theta) = 2\sqrt{\frac{3\gamma k}{g^2\left[\gamma + \frac{V^2 R_0^4 \sin^4\theta}{8h^2}\right]\sin^2\theta}}, \tag{16.49}$$

where we denoted $\gamma = -\sigma/(\rho R_0)$ and $\beta = -\rho_{el}^2 R_0^2/(3\epsilon_0\rho)$. We have to further couple these equations with the expressions of the coefficients of the KdV equation

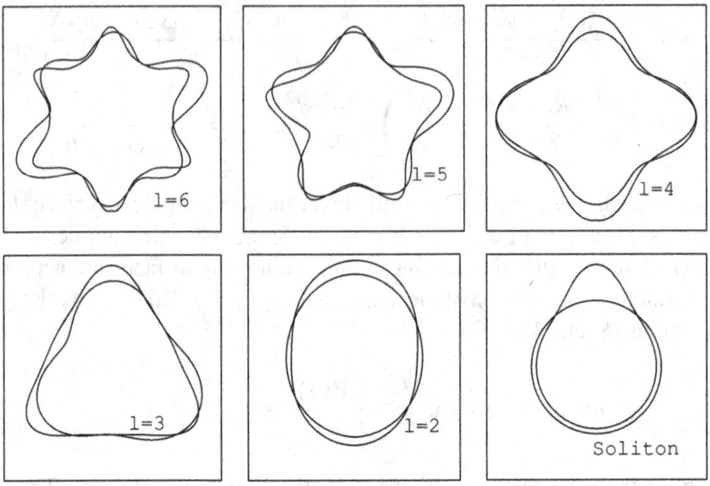

Fig. 16.1 Cnoidal waves excitation of the equatorial plane of the nuclear surface for different values of the modulus k of the cnoidal function, plotted together with the closest spherical harmonic combination that matches the nonlinear excitations

(16.45) and with the coefficients in the cnoidal solution (16.47), and of course with the periodicity condition. The remaining conditions of constancy of $V(\theta)$ and the periodicity condition (16.48), introduce two restrictions in the set of four arbitrary functions of θ: k, g, h and Υ, which provides the possible shapes, depths of the layers, amplitude, and velocities with a great deal of freedom. There are boundary conditions attached to these functions, namely at the poles, $\theta = 0, \pi$, all of them should be zero. In Figs. 16.1 and 16.2, we present some possible shapes of cnoidal excitations of the nuclear surface. In Fig. 16.1, the cnoidal solutions are plotted together with the closest possible match in terms of spherical harmonics. One can see that, with the exception of the solitary wave all other excitations are close to the linear modes. In the limit $k \to 1$, the cnoidal waves approach a solitary wave profile.

To verify the model we can estimate, in a very simplistic way, the spectroscopic factors of a certain cluster decay with the experimental results. The spectroscopic factor S is given by the penetrability of the quantum barrier associated with the process of preformation of the cluster from the parent nucleus. We can parametrize this process with the amplitude $\eta_0 = \max |g(\theta)|$ of the solitary wave, from spherical equilibrium shape $\eta_0 = 0$ to a certain maximum value. The quantum penetrability can be calculated with the formula [113]

$$S = \exp\left(-\frac{2}{\hbar} \int_0^{\eta_0} (A_{cluster} - E[\eta])^{1/2} d\eta \right), \tag{16.50}$$

where η_0 is the final amplitude of the soliton, $A_{cluster}$ is the nuclear mass of the preformed cluster of a certain α or other decay process, and E is the total nuclear

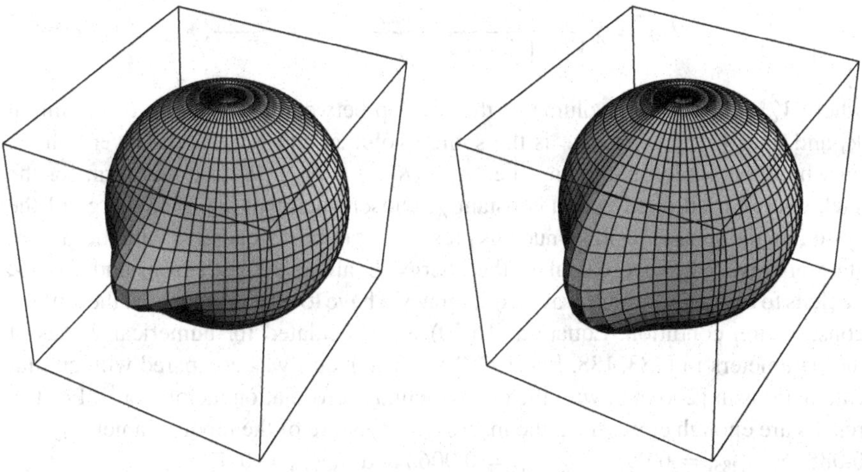

Fig. 16.2 Cnoidal waves excitation of nuclear surface in two cases. *Left*: the arbitrary functions $k(\theta), g(\theta), h(\theta)$ are chosen to have Gauss bell profiles of the same width with the solitary wave half-width. *Right*: the arbitrary functions k, h are chosen constant, and $g(\theta)$ is chosen to have a sech profile

energy calculated function of the amplitude of the excitation. This energy is actually defined on a multidimensional space of parameters, involving all types of energies in the model. The liquid drop mechanical energy consists in the sum of the kinetic energy

$$K = \frac{\rho}{2} \iiint_D (\nabla \Phi)^2 dV$$

where Φ can be calculated from (16.27), and potential energy of the surface

$$U_\Sigma = \sigma(A - A_0),$$

where A is the area of the deformed surface that can be calculated from (16.20). In addition to these two terms we have the Coulomb interaction energy

$$U_C[\eta] = \frac{\rho_e}{2} \int_{\mathcal{V}}' \int_{\mathcal{V}} \frac{1}{|\boldsymbol{r} - \boldsymbol{r}'|} d\mathcal{V} d\mathcal{V}'. \tag{16.51}$$

and the shell correction energy, which, in this model, takes care of the quantum effects. The shell energy is introduced by considering that the main contribution is from to the final nucleus, usually close to the double magic nucleus ^{208}Pb in α and heavier fragments decay. The spherical core $r \leq R_0 - h$ represents the final nucleus, which is also unexcited for the even–even case. We introduce the shell energy like a measure of the overlap between the core and the final nucleus, on one hand, and between the final emitted cluster and the bump, on the other hand

$$E_{sh} = \chi \frac{V_{over}}{V + [V_0 - (V_{cluster} + V_{layer})] - V_{over}}, \quad (16.52)$$

where V_{over} denotes the volume of the overlap between the volumes of the initial V_0 and final V nuclei, $V_{cluster}$ is the soliton volume, and V_{layer} is the layer volume on which the soliton is moving (i.e., $r \in [R_0 - h, R_0]$). We use this form for the shell energy multiplied with a constant χ, chosen such that the total energy of the system in the state of residual nucleus plus cluster to be degenerated with the ground state energy. When we calculate the energy E along the path from undeformed nucleus to a certain solitary wave excitation, we have to take into account the volume conservation condition. Equation (16.50) was calculated for numerical values of the parameters in [183, 188, 191, 192, 206]. The result was compared with similar calculations in [206] and with the experimental preformation factors for ^{208}Pb. The results are enough close given the macroscopic nature of the model, namely $S_{exp} = 0.085$, $S_{[281,285]} = 0.095$, $S_{[282,284]} = 0.0063$ and $S_{soliton} = 0.07$.

16.4 Solitons and Quasimolecular Structure

Since the soliton model for cluster preformation presented in Sect. 16.3 describes the dynamics of the nuclear surface, it is natural to search among possible experiments those in which the main contribution in the reactions is due – to some extent – to the surface. The soliton model could be proved or disproved more easily in the light of such measurements. A possible channel for such a goal is provided by the α-particle scattering with nuclei, in which the α-particle, being composite, interacts with nuclei in a more complex way than the nucleons namely, its high stability (high binding energy, zero spin, and isospin) restricts the interaction to a shallow surface layer region of the nucleus.

First interesting thing revealed by such complex interaction is the occurrence of a quasimolecular structure, i.e., states with structure polarized strongly into subunit nuclear clusters, which can be defined as molecule-like structures [114]. Specific features in all these experiments are a very high density of resonances, a good spin and parity assignment, irregular spacing of these spectra, and the relatively small moment of inertia of the $\alpha +$nucleus-systems. There are many theoretical attempts (microscopical and phenomenological models) to explain such resonances or the intermediate structure. Some of them were developed for α-cluster states [134, 287], or by anharmonic quadrupole surface vibrations analogues [56]. Other models include Morse-potential, quadrupole vibration–rotation model, coupled-channel calculations, two-center shell model, semimicroscopic algebraic models, and band-crossing models [1, 15, 89, 97, 99, 131, 132, 151].

In all these theoretical models, to explain the above mentioned features of these interactions, one has to introduce in the shell model a cluster-like component, or the many-body correlations. A more natural way to explain and/or predict these energy spectra is to consider that the α-particle interacts with the nucleus as a soliton or a breather. This is a new coexistence model consisting of the usual shell model and a cluster-like model describing a soliton moving on the nuclear surface. The energy spectrum is obtained from the quantum fluctuations around the classical soliton solutions by a nonperturbative weak-coupling procedure. The corresponding energy spectra are similar to a sum of nonlinear harmonic oscillators determined uniquely by the soliton geometry.

In the case of the resonances observed in the scattering of an α-particle on ^{28}Si, this surface soliton quantized model produces a surprising agreement between the predicted angular momentum states and energies and the measured ones [185, 186]. In addition, this model does not use any supplementary fragmentations of levels because of the neglected collective levels, like in the traditional models mentioned above. The spectrum obtained from the semiclassical quantization of the soliton state given in (16.46) reads [186, 267]

$$E_{n,I,N} = E_0 + \hbar\omega_1(n_1 + 1/2) + \hbar\omega_2(n_2 + 1/2) - B(n + 1/2)^2 + CJ(J + 1), \tag{16.53}$$

where J is the quantum number associated to the angular momentum and the corresponding constant $C = \hbar^2/2\mathcal{I}(R_0, L, h)$ is the reciprocal moment of inertia I, which is calculated from the soliton geometry by considering the rotation of the soliton plus the layer about the center of mass of the system. All the parameters in this term of rotation are obtained from physical considerations (and not numerical fit calculations): the daughter nucleus radius R_0, the width of the soliton L, and the depth of the fluid layer under the soliton h (Sect. 16.3; see (16.45) and (16.49)). The terms in $\hbar\omega_k$ are the excitations of the soliton state, and for the soliton whose geometry fits the α+^{28}Si reaction ($\eta_0 = 0.41, L = 0.546, h = 0.17R_0$) we obtain $\hbar\omega_1 = 0.23$ MeV, $\hbar\omega_2 = 0.801$ MeV, and $N = 0.015$ MeV. With these values obtained from the theory, the calculations reproduce about 190 observed experimental energies and spins of the intermediate states of α+^{28}Si (within errors of 2.5% or less), and predict positions and spins of other levels. Both even and odd parities are reproduced using the same parameters. This model explains that the odd–even parity splitting of the band members predicted by other models like RGM and OCM, etc., is not needed nor supported by the present experimental data or this nonlinear model. In support of this soliton-like model the experiments show that the even and odd states form mixed parity bands, which implies that the rotating mass has an asymmetric shape as it is natural for the α+^{28}Si-system, for example. If such a theoretical description is acceptable it implies that an alpha (or heavier) cluster modeled as a soliton orbiting on the nuclear surface could be viewed as another type of large amplitude nuclear collective deformation.

This quantum approach of cluster formation on the nuclear surface was applied to other resonances in the elastic scattering of alpha particles on ^{20}Ne [186]. For this lighter nucleus the constants in (16.53) are $E_0 = 9.465$ MeV, $h = 0.121\,R_{^{20}\mathrm{Ne}}$, $L = 0.636$, $\eta_0 = 0.62$, $C = 0.13052$, $\hbar\omega_1 = 0.533$ MeV, $\hbar\omega_2 = 0.1.0655$ MeV, and $N = 0.07878$ MeV. About 90 states and spins values are also predicted.

16.5 Soliton Model for Heavy Emitted Nuclear Clusters

In Sects. 16.3 and 16.4, we presented the nonlinear hydrodynamic model plus semimicroscopic corrections. To provide a more realistic description of large cluster formation on the nuclear surface, we have to add more detailed microscopic structure to the parent heavy nucleus and to the emitted cluster. The microscopic substructure further allows one to add shell corrections to the usual macroscopic liquid drop energy, and thus to give a complete description of the system, from the initial undeformed nucleus, to the parent nucleus with a shape deformation, and out to the cluster emission process. A straightforward way to accomplish this is to calculate shell effects obtained from the single-particle levels of an asymmetric shell model. Such nuclear asymmetric models are called "two-center" shell models, and allow a microscopic description of the nuclear evolution from one to two independent quantum systems. The procedure is presented in detail in [105], and involves calculating the total potential energy as the sum of the macroscopic (hydrodynamic) energy, and shell corrections, which is then minimized. This approach usually yields a potential energy barrier along the evolution of the parameter that describes the cluster formation. This barrier increases with the amplitude of the new formed cluster. We sketch here the soliton-model calculations for the nuclear reaction $^{248}\text{No} \rightarrow {}^{208}\text{Pb} + {}^{40}\text{Ca}$. Cluster emission processes are described by using soliton-like shapes on the nuclear surface of the heavy fragment like those developed in Sect. 16.3. For a given cluster geometry, the model calculates the corresponding soliton parameters (A, L, V) as functions of the separation parameter, i.e., along the static path of the cluster emission process.

The deformation energy, E_{def}, is calculated in a macroscopic–microscopic approach

$$E_{def} = E_C + E_{Y+E} + \delta E_{shell} + \delta P, \tag{16.54}$$

where E_C is the Coulomb energy and E_{Y+E} is the surface or nuclear energy calculated within the Yukawa-plus-exponential model [288]. The Coulomb energy of interaction is calculated by the double-volume integral

$$E_C = \frac{1}{2}\int_V\int_V \frac{\rho_e(r_1)\rho_e(r_2)d^3r_1d^3r_2}{r_{12}}. \tag{16.55}$$

The general form of the Yukawa-plus-exponential energy is:

$$E_{Y+E} = -\frac{a_2}{8\pi^2 r_0^2 a^4}\int_V\int_V \left(\frac{r_{12}}{a} - 2\right)\frac{\exp(-r_{12}/a)}{r_{12}/a} d^3r_1 d^3r_2, \tag{16.56}$$

where $r_{12} = |\mathbf{r}_1 - \mathbf{r}_2|$, $a = 0.68$ fm accounts for the finite range of nuclear forces, and $a_2 = a_s(1 - \kappa I^2)$. κ is the asymmetry energy constant, and the surface energy constant is $a_s = 21.13\,\text{MeV}$. In addition to the macroscopic energy, the model contains the energy corrections for the two-center shell model, where the two centers are taken in the center of the daughter nucleus, and in the center of mass of the emerging soliton. The energy corrections contain the energy of two coupled oscillators plus the spin–orbit interaction depending on the mass asymmetry in the final reaction products.

The level scheme of a soliton shape is used to obtain the shell corrections of the system. As the soliton is assimilated with an emerging fragment, it will provide the shell correction value of the independent nucleus of similar shape. Shell corrections are obtained by means of the Strutinsky procedure [309]. The relative velocity distribution V of the two presumed solitons along the minimum-energy path, together with the scaled values of the half-width L and the relative amplitude $a = A/R_1$, are plotted in Fig. 16.3. In the first stages, the tendency is that the amplitude and half-width increase with the elongation parameter, when the emitted cluster is emerging out from the parent nucleus. During the formation of the cluster the half-width remains practically constant, since the surface energy controls this stage. When the two nuclei are well separated, the soliton envelope hardly fits the two spheres, and in this limit, the half-width approaches zero value. This gives the limiting configuration for this soliton model. The velocity is increasing with the amplitude of the soliton, hence with the elongation of the cluster-like emission shape.

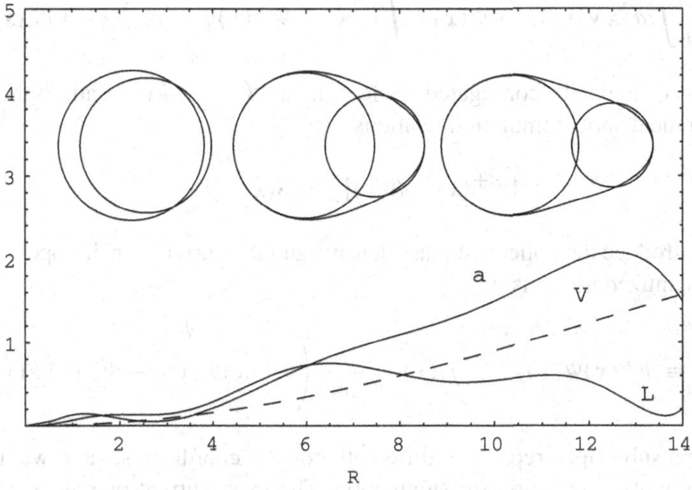

Fig. 16.3 The evolution of the soliton geometrical parameters $a = A/R_1$, L, and V in relative units vs. the elongation R in fm for the ^{40}Ca emission. The corresponding nuclear configurations (parent nucleus, daughter cluster, and embedding soliton shape) are plotted for the initial stage (when the emitted cluster is only slightly displaced off the common center), an intermediate stage, and the final stage when the two nuclei are almost separated. The oscillations in the soliton parameters are related to the shell corrections

16.5.1 Quintic Nonlinear Schrödinger Equation for Nuclear Cluster Decay

The soliton descriptions in the above sections are actually extensions of the traditional *geometric collective model* (Bohr–Mottelson), which allows not only the nuclear deformations leading to collective rotational and vibrational motions coupled with single-particle states, but also the creation of the bumps on the nuclear surface. The different approach followed by [149, 150] starts from the nonlinear irrotational hydrodynamic equations in a compact domain of space, with boundary, and introduces a Hamiltonian system in terms of collective mass and current densities, which satisfy the Euler and continuity equations. These equations reduce to nonlinear Schrödinger equation with a nonlocal "potential." By using the realistic effective Skyrme contact δ-interaction the "potential" becomes local polynomial in density [302]. This leads to a new quintic nonlinear Schrödinger equation whose highest order nonlinear term is essential for the Skyrme interaction, and describes well the main properties of real nuclei [149, 150].

In the second-quantization formalism, a system of A spin-less and isospin-less nucleons is described by a nonrelativistic Hamiltonian with a local two-body potential $U(\boldsymbol{x})$

$$\hat{H} = \frac{\hbar^2}{2m}\int d^3x \nabla\Psi^+(\boldsymbol{x})\nabla\Psi(\boldsymbol{x}) + \int d^3x d^3y \Psi^+(\boldsymbol{x})\Psi^+(\boldsymbol{y})U(\boldsymbol{x}-\boldsymbol{y})\Psi(\boldsymbol{x})\Psi(\boldsymbol{y}), \tag{16.57}$$

where the canonically conjugated nucleon fields $\Psi^+(\boldsymbol{x}), \Psi(\boldsymbol{x})$ satisfy the equal-time canonical anticommutation relations

$$\{\Psi^+(\boldsymbol{x}),\ \Psi(\boldsymbol{y})\}_+ = \delta(\boldsymbol{x}-\boldsymbol{y}).$$

We can introduce the collective mass density, and the current density operators in a second quantized formalism

$$\hat{\rho}(\boldsymbol{x}) \equiv \Psi^+(\boldsymbol{x})\Psi(\boldsymbol{x}), \qquad \hat{j}_k(\boldsymbol{x}) = \frac{\hbar}{2mi}\left(\Psi^+(\boldsymbol{x})\Psi_{,k}(\boldsymbol{x}) - \Psi^+_{,k}(\boldsymbol{x})\Psi(\boldsymbol{x})\right),$$

where the subscripts represent differentiation to coordinates, and we use the Einstein's mute convention for summation. The mass-current operators fulfill the following equations of motion

$$\begin{aligned} \hat{\rho}_t(\boldsymbol{x}) &= \frac{1}{i\hbar}[\hat{\rho}(\boldsymbol{x}), \hat{H}] = -\hat{j}_{k,k}(\boldsymbol{x}) \\ \hat{j}_{k,t}(\boldsymbol{x}) &= \frac{1}{i\hbar}[\hat{j}_k(\boldsymbol{x}), \hat{H}] = -\frac{\hbar^2}{2m^2}\left(\hat{T}_{nk,n}(\boldsymbol{x}) - \frac{1}{2}\hat{\rho}_{,knn}(\boldsymbol{x})\right) \\ &\quad -\frac{2}{m}\hat{\rho}(\boldsymbol{x})\left(\int d^3y U(\boldsymbol{x}-\boldsymbol{y})\hat{\rho}(\boldsymbol{y})\right)_{,k}. \end{aligned} \tag{16.58}$$

In the case of irrotational flow, the velocity operator can be defined through a potential operator $\hat{\varphi}(x)$ in the equation

$$\hat{j}_k(x) \equiv \frac{1}{2}\left\{\hat{\rho}(x), \hat{\varphi}, k(x)\right\}_+ . \tag{16.59}$$

Equations (16.58) and (16.59) provide a complete collective hydrodynamical description of the nuclear system. In the semiclassical limit (16.58) and (16.59) can be reduced for irrotational flow motion (16.59) to a nonlinear Schrödinger equation [148]

$$i\hbar\frac{\partial u}{\partial t} = -\frac{\hbar^2}{2m}\Delta u + \tilde{U}[|u|^2]u \tag{16.60}$$

where the local density and the velocity potential are given by

$$u(x,t) = \sqrt{\rho(x,t)}e^{\frac{im}{\hbar}\varphi(x,t)}. \tag{16.61}$$

Equation (16.60), (16.61) result in a natural way from the quantum field formalism if we think them in terms of Madelung's analogy between fluid dynamics and quantum mechanics.

In a case of a general two-body interaction $U(x)$, the potential $\tilde{U}[\rho]$ is a nonlocal one. The well-known effective Skyrme contact δ-interaction [302] leads to the following local nonlinear "potential" $\tilde{U}[\rho]$, after providing the following renormalization [148]

$$\int d^3xd^3y\rho(x)U(x-y)\rho(y) \Longrightarrow \int d^3x\left(\frac{3}{8}t_0\rho^2(x) + \frac{1}{16}t_3\rho^3(x)\right). \tag{16.62}$$

The introduction of the Skyrme force involves the following substitutions

$$m \to m^* = \frac{1}{\frac{1}{m} + (3t_1 + 5t_2)\frac{\rho_n}{8\hbar^2}},$$

$$\frac{\hbar^2}{8m} \to \frac{\hbar^2}{8m} + \frac{\rho_n}{64}(9t_1 - 5t_2),$$

where ρ_n is the nuclear matter density and t_i are parameters of the Skyrme forces. After a re-scaling, (16.60) becomes

$$i\frac{\partial\psi}{\partial\tau} = -\Delta\psi - 4\mid\psi\mid^2\psi + 3\mid\psi\mid^4\psi, \tag{16.63}$$

that is a nonlinear Schrödinger equation (NLS) with a quintic term in ψ. Such an equation is not completely integrable in the sense of the soliton theory. Also, the corresponding Bäklund transformation does not exist, and it is not possible to build exact N-soliton one-dimensional solutions. So we have to deal with the so-called quasisolitons, which are also under a constant intensive investigation [205]. However, there are methods to build N-soliton solutions of the one-dimensional cubic nonlinear Schrödinger equation, for example, the inverse scattering method [330], direct type method [127]. For the alpha and cluster decay we have the case of axial symmetric interaction of two small overlapping nonlinear waves. The both initially isolated waves (a large target and a small projectile) are solitary type spherically symmetric solutions. The general analysis of the collision of two three-dimensional initially localized nonlinear waves (nuclei) in the framework of nonlinear hydrodynamics can be made only numerically, where only density distributions and not the velocity fields are calculated. Example of numerical calculations for ^{208}Pb+^{20}Ne along the z-axes are presented in Fig. 16.4. One can see the transition from the two well-localized waves to the practically absorbed in the surface region. The angular dependence of the density distribution for ^{208}Pb +^{20} Ne is presented in the right frame of the same figure. The quintic Schrödinger equation (16.63) admits also antisoliton solutions. From the three-dimensional perspective, such a fast rotation antisoliton (Fig. 16.5), or rather an antisoliton pair where the two antisoliton are separated with π and travel along the same circle with the

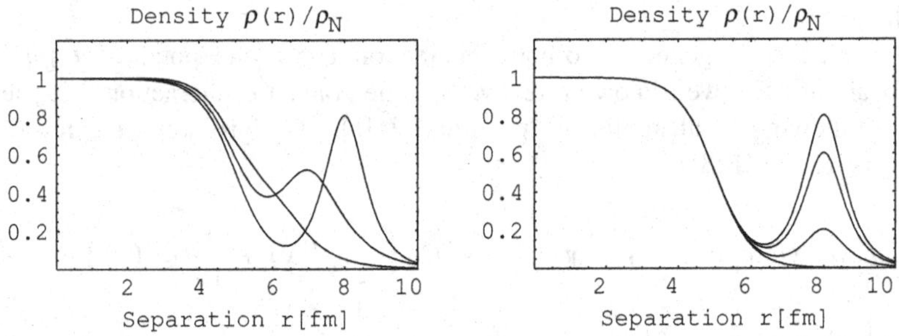

Fig. 16.4 *Left*: the density distribution for ^{208}Pb + ^{20}Ne at the angle $\theta = 0$ for three different separations. *Right*: the angular dependence of the density distribution for the maximum separation presented in the left frame, $\theta = 0°, 10°, 20°$

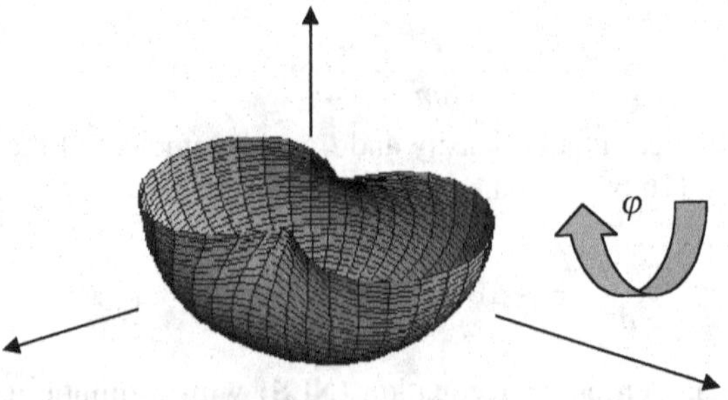

Fig. 16.5 A fast rotating antisoliton can cut a virtual channel in the nuclear shape, increasing so the probability for fission through that channel

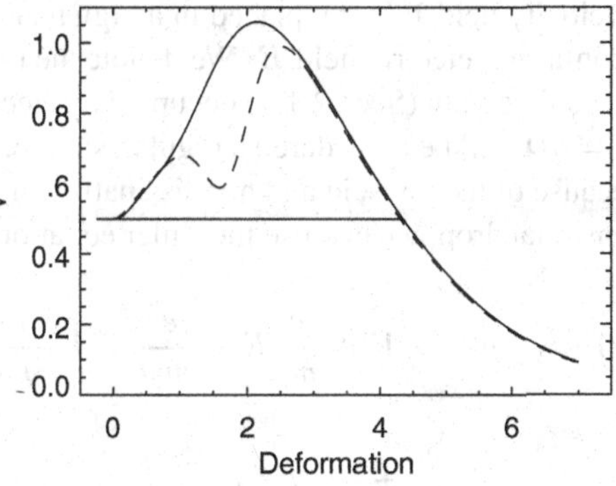

Fig. 16.6 Change in the fission barrier produced by the introduction of an antisoliton pair on the surface

same angular velocity can have interesting consequences on the probability of preformation of a spontaneous fission, or exotic radioactivity channel [76]. The rotating antisoliton can create a sort of virtual channel in the surface, so it can enhance the probability of breakup. Some preliminary numerical calculations show that the fission barrier (Fig. 16.6) can be lowered by the occurrence of such an antisoliton.

16.6 Contour Solitons in the Quantum Hall Liquid

An example of a nonlinear integrable system originating from the contour dynamics formalism at microscopic scale is provided by the excitations on the edge of a two-dimensional electron system in a perpendicular strong magnetic field. This practically two-dimensional system was theoretically investigated in [103] by using field-theoretical treatments of the edge excitations. Also, studies of edge channels in *quantum Hall* (QH) samples have shown the presence of nonlinear waves [333]. In this study, the origin of the nonlinearity is the variation of the intensity of the confining electrical field. From the contour dynamics point of view (theory of plane curve motion surrounding an incompressible inviscid fluid), this system was investigated by Wexler and Dorsey [340, 341]. In this study, the nonlinearity arises here from geometrical effects. These authors obtained explicit MKdV soliton solutions for the curvature of the contour, in agreement with the theory of motion of two-dimensional curves (Sect. 7.1), and with the results obtained in Sect. 13.5 [189].

In the following we present elements of this geometric nonlinear model. The boundary of a two-dimensional electron system, or a QH liquid can be investigated in a clean and controlled environment because the QH liquid is incompressible, there are no other low-lying excitations except the boundary ones so dissipative effects can be eliminated. We consider a bounded two-dimensional system of electrons of density $n(\boldsymbol{r},t)$ and velocity field $\boldsymbol{V}(\boldsymbol{r},t)$, placed in a high magnetic field $\boldsymbol{B} = B\boldsymbol{e}_z$, and a background (confining) electric field $\boldsymbol{E}$. We denote the two-dimensional connected, simply connected domain (Sect. 2.1.4) occupied by electrons with D and its moving boundary $\Gamma = \partial D$ will be considered a regular, simple, plane, parametrized curve (Sect. 5.1). Because of the inviscid and nondissipative character of the motion in the QH two-dimensional drop, we can use the Euler equation (10.15)

$$\frac{\partial \boldsymbol{V}}{\partial t} + (\boldsymbol{V}\cdot\nabla)\boldsymbol{V} = -\omega_c \boldsymbol{e}_z \times \boldsymbol{V} + \frac{e}{m_e}\boldsymbol{E} - \frac{e^2}{m_e \epsilon}\nabla \int_D \frac{n(\boldsymbol{r}')}{|\boldsymbol{r}-\boldsymbol{r}'|} dA' = 0,$$

$$\frac{\partial n}{\partial t} + \nabla\cdot(n\boldsymbol{V}) = 0\cdot \qquad (16.64)$$

In the second equation, $\omega_c = eB/m_e$ is the cyclotron frequency, m_e and e are the mass and charge of electron, and ϵ is the dielectric constant of the medium. The first two RHS terms are the Lorentz force, and the last one is the Coulomb interaction. The last equation is the continuity equation. If both velocity and density of electrons are supposed to oscillate around their equilibrium values $\boldsymbol{V}_e = 0, n_e$ with some frequency ω and wave number k, by using (16.64) we note that

$$\omega = \omega_c + \frac{2\pi n_e e^2}{m_e \epsilon} k.$$

A simple estimation shows that bulk excitations can be neglected since $\hbar\omega_c \sim 200\,\mathrm{K}$, while the equilibrium temperature is around 1 K. So, the QH system can be well modeled with nonlinear contour dynamics approach. This is equivalent with an incompressible inviscid two-dimensional liquid drop model with sharp contour, except here we have electromagnetic interactions in addition. The incompressibility condition introduces already a global conservation law, i.e., constant area of D. If we multiply in a crossproduct the first equation in (16.64) with $\boldsymbol{e}_z$, neglect the inertial terms and the drift terms produced by the parasite external electric field $\boldsymbol{e}_z \times \boldsymbol{E}$ [341], we obtain a simpler equation the velocity field of confined electrons

$$\boldsymbol{V}(\boldsymbol{r}) = -\frac{e^2}{\epsilon m_e \omega_c}\nabla \times \boldsymbol{e}_z \int_D \frac{n(\boldsymbol{r}')}{|\boldsymbol{r}-\boldsymbol{r}'|} dA. \qquad (16.65)$$

Moreover, because of the incompressibility we can pull outside of the integral the electron density, and by using Stoke's theorem we have

$$\boldsymbol{V}(\boldsymbol{r}) = \frac{n_e e^2}{\epsilon m_e \omega_c} \oint_\Gamma \frac{\boldsymbol{t}(s')}{|\boldsymbol{r}-\boldsymbol{r}(s')|} ds', \qquad (16.66)$$

where $\boldsymbol{t}$ is the unit tangent to the contour Γ and s is the arc-length (Sect. 5.1). Equation (16.66) is a nonlocal representation formula (Sect. 10.6) telling us that the motion of the boundary is determined by the flow of the electronic fluid at the surface. The motion of the boundary can be described in the formalism developed in Chap. 7.1. We parametrize the contour either with the arc-length s or with the azimuthal angle (it is a simple closed curve) φ, i.e., $r(\varphi, t)$. The contour has the following geometric parameters

$$\boldsymbol{t} = \frac{r\boldsymbol{e}_\varphi + \frac{\partial r}{\partial \varphi}\boldsymbol{e}_r}{\sqrt{r^2 + (\frac{\partial r}{\partial \varphi})^2}}, \quad g = \sqrt{r^2 + \left(\frac{\partial r}{\partial \varphi}\right)^2},$$

$$\boldsymbol{n} = \frac{1}{\sqrt{g}}\left(-r\boldsymbol{e}_r + \frac{\partial r}{\partial \varphi}\boldsymbol{e}_\varphi\right), \quad \kappa = \frac{\sqrt{r^2 + 2\left(\frac{\partial r}{\partial \varphi}\right)^2 - r\left(\frac{\partial^2 r}{\partial \varphi^2}\right)^2}}{\left[r^2 + \left(\frac{\partial r}{\partial \varphi}\right)^2\right]^{\frac{3}{2}}}$$

where we define a local orthogonal curvilinear basis attached to the contour $\{\boldsymbol{e}_r, \boldsymbol{e}_\varphi\}$, $\boldsymbol{e}_r = (\cos\varphi, \sin\varphi)$, $\boldsymbol{e}_\varphi = (-\sin\varphi, \cos\varphi)$. The plane velocity of the contour can be expressed by (7.1)

$$\boldsymbol{V}(s,t) = U(s,t)\boldsymbol{n}(s,t) + W(s,t)\boldsymbol{t}(s,t), \tag{16.67}$$

where (U, W) are the normal and tangential components of the velocity of the boundary.

Steady traveling contour waves move along the circumference as a perturbation. Consequently, we can write the parametric equation for the contour in the azimuthal angle parametrization, $\boldsymbol{r}(s,t) \to \boldsymbol{r}(\varphi, t)$,

$$r(\varphi, t) \to r(\varphi - \Omega t), \tag{16.68}$$

with Ω being the constant angular frequency of the boundary rotation. From $\boldsymbol{n} = \boldsymbol{t} \times \boldsymbol{e}_z$, and (16.67) and (16.68) we have a condition for the normal velocity

$$U = \boldsymbol{n} \cdot \boldsymbol{V}|_{\boldsymbol{r}\in\Sigma} = \Omega \boldsymbol{n} \cdot (\boldsymbol{e}_z \times \boldsymbol{r}). \tag{16.69}$$

The normal velocity can be obtained from the velocity field of the electron fluid (16.66) taken at the boundary $\boldsymbol{r}(\varphi, t)$

$$U(\varphi, t) = \frac{n_e e^2}{\epsilon m_e \omega_c}\int_0^{2\pi} \frac{\boldsymbol{n}(\varphi, t)\cdot \boldsymbol{t}(\varphi', t)}{|\boldsymbol{r}(\varphi, t) - \boldsymbol{r}(\varphi', t)|}\sqrt{g}\,d\varphi', \tag{16.70}$$

where g is the metric of the Γ curve (5.1).

16.6.1 Perturbative Approach

We expand the boundary curve in the azimuthal parameter in a Fourier series

$$r(\varphi, t) = R_0\left(1 + \sum_{n=-\infty}^{\infty} C_n e^{in\varphi}\right). \tag{16.71}$$

By using the Serret–Frenet relations for the expression of the unit tangent and principal normal of the curve in the φ parametrization in (16.69)–(16.71), we have

$$U(\varphi, t) = -\frac{i\Omega}{\sqrt{g}}\left(\sum_n nC_n e^{in\varphi} + \frac{1}{2}\sum_n\sum_m C_{n-m}C_m e^{in\varphi}\right). \tag{16.72}$$

Equations (16.70) and (16.72) form a nonlinear system of equations for the coefficients C_n and Ω. The solution of this system provides the nonlinear boundary standing traveling modes (dispersionless perturbations).

From the expansion (16.71), and by denoting $\varphi' = \varphi + \omega$, we can write the numerator of the integrand in (16.70) in the form

$$\begin{aligned}\boldsymbol{n}(\varphi)\cdot\boldsymbol{t}(\varphi+\omega) &= \frac{1}{g}\Bigg[-\sin\omega - 2\sum_n C_n e^{in\varphi}\left(\cos\frac{n\omega}{2}\sin\omega + n\cos\omega\frac{n\omega}{2}\right)\\ &+\sum_{n,m} C_{n-m}C_m e^{in\varphi}\Bigg((nm - m^2 - 1)\cos\frac{(n-2m)\omega}{2}\sin\omega,\\ &-(n-2m)\cos\omega\sin\frac{(n-2m)\omega}{2}\Bigg)\Bigg],\\ |\boldsymbol{r}(\varphi+\omega) - \boldsymbol{r}(\varphi)|^2 &= 4\sin^2\frac{\omega}{2}\Bigg[1 + 2\sum_n C_n e^{\frac{in\omega}{2}}\cos\frac{\omega}{2}\\ &-\sum_{n,m} C_{n-m}C_m e^{\frac{in\omega}{2}}\frac{\sin\frac{(n-m+1)\omega}{2}\sin\frac{(m-1)\omega}{2} + \sin\frac{(n-m-1)\omega}{2}\sin\frac{(m+1)\omega}{2}}{2\sin^2\frac{\omega}{2}}\Bigg].\end{aligned}$$

Next step is to expand the whole integrand of (16.70) according to the above sums, and then integrate over φ', i.e., over ω. After this integration, if we identify the coefficients of various products of C_n between (16.70) and (16.72) we obtain an infinite dimensional nonlinear system for C_n. By considering this system up to the fifth order in products of C_n coefficients we obtain the condition

$$\tilde{\Omega}\left(C_n + \frac{1}{2}\sum_{n_2} C_{n-n_2}C_{n_2}\right) = Q_n^{(1)}C_n$$

$$+\sum_{n_2} Q^{(2)}_{n,n_2} C_{n-n_2} C_{n_2} + \sum_{n_2,n_3} Q^{(3)}_{n_2,n_3} C_{n-n_2} C_{n_2-n_3} C_{n_3}$$

$$+ \sum_{n_2,n_3,n_4} Q^{(3)}_{n,n_2,n_3,n_4} C_{n-n_2} C_{n_2-n_3} C_{n_3-n_4} C_{n_4} + \dots. \quad (16.73)$$

Here we denoted $\tilde{\Omega} = \epsilon m_e \omega_c R_0 \Omega/(n_e e^2)$, and the tensors $Q^{(k)}$ have the form

$$Q^{(1)}_n = 8(\gamma + \ln 4) + \frac{1}{2}\Psi\left(n + \frac{1}{2}\right),$$

$$Q^{(2)}_{n,n_2} = \frac{1}{4}(Q^{(1)}_n - Q^{(1)}_{n-n_2} - Q^{(1)}_{n_2}) - 1,$$

$$Q^{(3)}_{n,n_2,n_3} = -\frac{5}{n}\left(\frac{n}{1-4n^2} + \frac{n_2}{1-4n_2^2} + \frac{n_3}{1-4n_3^2} + \frac{n-n_2}{1-4(n-n_2)^2}\right.$$
$$\left. + \frac{n-n_3}{1-4(n-n_3)^2} + \frac{n_2-n_3}{1-4(n_2-n_3)^2} + \frac{n-n_2+n_3}{1-4(n-n_2+n_3)^2}\right)$$
$$+ \frac{1}{48}[-(3+4n^2)(Q^{(1)}_n + 4) - (1+4n_2^2)(Q^{(1)}_{n_2} + 4) + (5+4n_3^2)(Q^{(1)}_{n_3} + 4)$$
$$+ (5+4(n-n_2)^2)(Q^{(1)}_{n-n_2} + 4) - (1+4(n-n_3)^2)(Q^{(1)}_{n-n_3} + 4)$$
$$+ (5+4(n_2-n_3)^2)(Q^{(1)}_{n_2-n_3} + 4) - (1+4(n-n_2+n_3)^2)(Q^{(1)}_{n-n_2+n_3} + 4)]$$

where $\gamma \sim 0.577216\dots$ is the Euler constant, and $\Psi(x) = \Gamma'(x)/\Gamma(x)$ is the digamma function, and $\Gamma(x)$ is the gamma function. In these equations above we used the relation [5, 284]

$$\sum_{n=1}^{N} \frac{1}{2n-1} = 2(\gamma + \ln 4) + \frac{1}{2}\Psi\left(n + \frac{1}{2}\right).$$

To evaluate the correct orders of smallness, we can expand the digamma function in a Bernoulli series

$$Q^{(1)}_n \sim 4\left(\frac{\gamma}{2} + \ln 2 + \frac{\ln n}{2} + \frac{B_2}{8n^2} + \frac{7B_4}{64n^4} + \dots\right),$$

where B_k are the Bernoulli numbers, i.e., $B_2 = 1/6, B_4 = -1/30, \dots$. The first terms on the RHS of the above series are in order $O(1)$, the term containing B_2 is in order $O(3)$, the next term is in order $O(5)$, so a pretty good approximation would be to approximate the series up to order $n, n_2, \dots \leq 5$. In Fig. 16.7, we present numerical estimation of the $Q^{(1)}_n$ sums vs. the order taken into account.

From the above conditions, the solutions for Ω are introduced in the system (16.70) and (16.72) allowing to calculate the coefficients C_n up to order five. The receipe used in [341] consists in choosing the largest coefficient $C_{max} = C_{n^*} =$

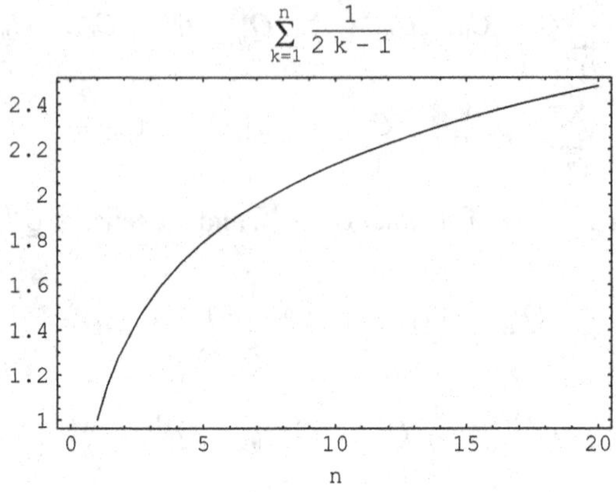

Fig. 16.7 The sums $\sum_{k=1}^{n}(2k-1)^{-1}$ plotted vs. n. Truncation of the sum up to the fifth term introduces a relative error of about 20%. The series is convergent even if in this figure is not obvious

$\max\{C_n\}_{n=1,\ldots,5}$ as being of order $O(1)$. Next, one needs to expand the remaining coefficients C_k, and the solution Ω of (16.73), in series of smaller and smaller orders, of the form $C_n = C_n^{(2)} + C_n^{(3)} + \ldots$. The linear approximation, i.e., the first-order term C_{n*}, provides the fundamental harmonic of the angular frequency

$$I(\tilde{\Omega}) = Q_{n*}^{(1)} + 4.$$

This result was previously obtained in [103]. Next orders obey the typical behavior of nonlinear oscillations of drops (Sect. 13.3) that is involving coupling between modes. The second-order mode in the C_{n*} expansion couples the fundamental harmonic with the second harmonic, the third-order term couples the fundamental mode to the first and third harmonics, the fourth-order couples the fundamental to the second and fourth harmonics, etc. Also, each next order brings additional corrections to the angular frequency Ω. The drop shapes obtained from these C_n coefficients are presented in [340, 341], and they include: ellipsoids of different eccentricities, elongated ellipsoids with neck, convex or concave triangular shapes, and convex or concave four-lobe shapes, with the contours going all the way to superdeformed ones like cruciform quartic curves, etc. These nonlinear shapes are in good agreement with the nonaxisymmetric shapes of liquid drops obtained through other theoretical approaches or experiments [8, 13, 96, 189, 319, 321]. To illustrate such types of shapes we generated typical examples in Fig. 16.8 by help of cnoidal sine functions, for different amplitudes and different values for the modulus k.

16.6.2 Geometric Approach

The two-dimensional incompressible inviscid model for the QH electron drop is susceptible for a geometric approach. We use the boundary velocity formula (16.66)

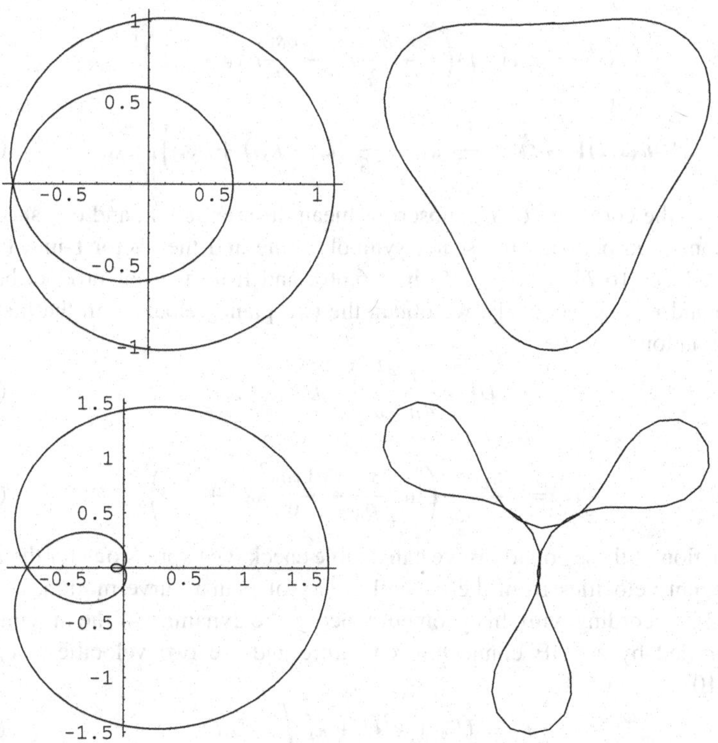

Fig. 16.8 *Left frames*: curvature κ in a polar representation along the loop (that is plotting the function $(x = (1+\kappa(s))\cos\varphi(s), y = (1+\kappa(s))\sin\varphi(s))$). *Right frames*: the nonlinear drop shapes generated by the cnoidal periodic solution in (16.84). For *upper figure* the coefficients are $A = -0.2, B = 0.3, D = 0.98, F = 10^{-3}$, and $m = 0.95$, and for the *lower figure* the coefficients are $A = -0.2, B = 0.9, D = 0.98, F = 10^{-3}$, and $m = 0.993$. Both loops represent octupole shapes. The lower one is an exaggerated case similar to a symmetric breakup or fission. The period, width, and angular velocity are given by (16.85)

and the formalism of plane curve motion developed in Sect. 7.1. The physics of the problem allows us to approximate the value of the loop integral (16.66) in $\boldsymbol{r}(s,t)$ with an integral along the contour Γ taken only in a neighborhood of s, i.e., integrated on $I = [s-\delta s/2, s+\delta s/2]$, where δs can be chosen relatively small when compared with the perimeter of Γ. This is possible because the dominant interaction is the Coulombian one which, in the plane case, decays as $1/r$. Consequently, the value of the integrand in s can be expanded in Taylor series on $s' \in I$. From the Serret–Frenet equations (5.3)–(5.5), (5.8), we have the series expansion in powers of δs

$$\boldsymbol{r}(s',t) = \boldsymbol{r}(s,t) + \boldsymbol{t}(s,t)\left(\delta s - \frac{\delta s^3}{6}\kappa^2 - \frac{\delta s^4}{8}\kappa\kappa_s + \ldots\right)\Bigg|_{(s,t)}$$
$$+\boldsymbol{n}(s,t)\left(-\frac{\delta s^2}{2}\kappa - \frac{\delta s^3}{6}\kappa_s + \frac{\delta s^4}{24}(\kappa^3 - \kappa_{ss})\ldots\right)\Bigg|_{(s,t)}, \qquad (16.74)$$

and

$$\boldsymbol{t}(s',t) = \boldsymbol{t}(s,t)\left(1 - \frac{\delta s^2}{2}\kappa^2 - \frac{\delta s^3}{2}\kappa\kappa_s + \ldots\right)\Big|_{(s,t)}$$

$$+ \boldsymbol{n}(s,t)\left(-\delta s\kappa - \frac{\delta s^2}{2}\kappa_s + \frac{\delta s^3}{6}(\kappa^3 - \kappa_{ss}) + \ldots\right)\Big|_{(s,t)}, \tag{16.75}$$

where κ is the curvature of Γ, subscripts mean differentiation, and one should not make confusion between the scalar symbol t-time and the vector $\boldsymbol{t}$-unit tangent. We introduce (16.74) and (16.75) in (16.66), and from the dot product between (16.67) and $\boldsymbol{t}, \boldsymbol{n}$, respectively, we obtain the two plane velocities in the first-order approximation

$$U = \frac{n_e e^2}{\epsilon m_e \omega_c}\frac{\delta s^2}{8}\kappa_s + \ldots, \tag{16.76}$$

and

$$W = \frac{n_e e^2}{\epsilon m_e \omega_c}\left(\ln\frac{\delta s^2}{2R_0} - \frac{11\delta s^2}{96}\kappa^2 + \ldots\right). \tag{16.77}$$

In deduction of these equations we can double check the expressions for the normal and tangent velocities from the general theory of planar curve motion, i.e., (7.4) and (7.7). According with this geometric theory, the dynamics of the moving curve is controlled by a PDE connecting curvature and the two velocities, i.e., (7.8) and (7.10)

$$\kappa_t = U_{ss} + \kappa^2 U + \kappa_s \int_0^s \kappa U ds'. \tag{16.78}$$

By introducing the expressions (16.76) and (16.77) in (16.78), we obtain exactly the modified Korteweg–de Vries equation (MKdV) for the curvature $\kappa(s,t)$ in the form

$$\kappa_t = -\frac{n_e e^2 \delta s^2}{8\epsilon m_e \omega_c}\left[\frac{3}{2}\kappa^2\kappa_s + \kappa_{sss} + \left(\frac{5}{12}\kappa^2(0) - \frac{8}{\delta s^2}\ln\frac{\delta s^2}{2R_0}\right)\kappa_s\right]. \tag{16.79}$$

The MKdV system is integrable and contains an infinite countable set of integrals of motion related strictly to the curvature and its derivatives with respect to s [2]. We need to make here a comment about these integrals of motion. For this model in particular as well as for two-dimensional incompressible traditional liquid drops, and actually even for a simple two-dimensional moving curve with same U, W as in (16.76) and (16.77), there are two conserved quantities that have nothing to do with this infinite series of conserving quantities of the MKdV hierarchy. Namely, we have constant perimeter of Γ and area of $\partial\Gamma$, and this is somehow expected to happen since the HQ liquid is incompressible and we did not associate any elasticity properties with the boundary. Indeed, by using the expressions of time variation of length L and area A of a plane curve, (7.34) and (7.37), (7.39), respectively, for the model velocities obtained in (16.76) and (16.77) we have

$$\frac{dL}{dt} = -\int_0^L \kappa U ds \sim \kappa^2|_0^L = 0,$$

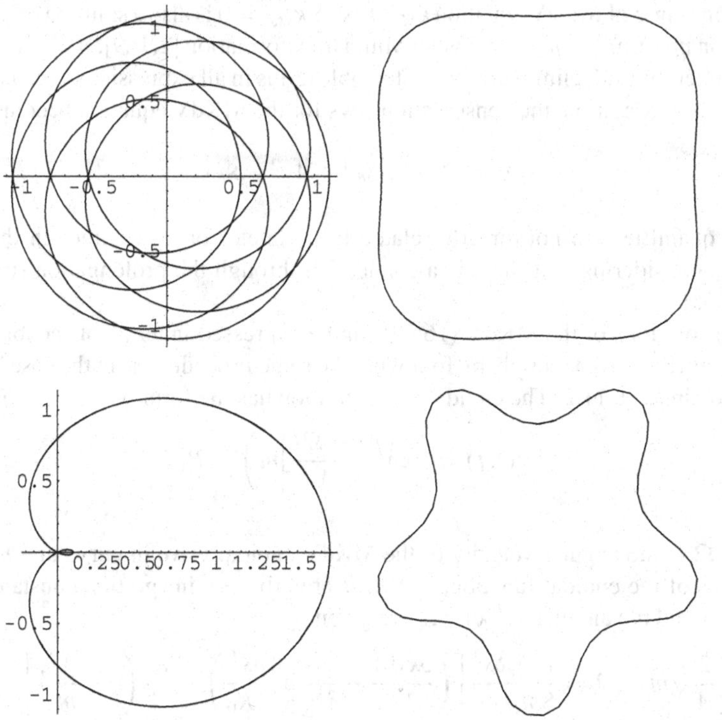

Fig. 16.9 Same as Fig. 16.8, but for higher-order multipoles. In the *upper frames* the coefficients are $A = -0.2$, $B = 0.3$, $D = 0.9$, $F = 5 \times 10^{-3}$, and $m = 0.825$, and in the *lower frames* the coefficients are $A = -0.3$, $B = 0.9$, $D = 0.9$, $F = 10^{-3}$, and $m = 0.992$

since the curve is closed. Some examples are presented in Fig. 16.9. The same conservation occurs for the area

$$\frac{dA}{dt} = -\int_0^L U ds \sim \kappa|_0^L = 0.$$

So, the perimeter and area conservation occur actually only because of the special form of the normal velocity of the contour. The infinite number conservation laws for the MKdV equation are actually integrals of polynomials of the curvature and its arc-length derivatives, so they "live in a higher space" (in the sense of lifting the problem of invariants to the tangent bundle over the equations of motion) than infinitesimal arc-length and area. We can check this easily, since the we know that the first conservation laws for the KdV equation in the function $\eta(s)$ are given by [2]

$$I_1 = \eta, \quad I_2 = \eta^2, \quad I_3 = \eta^3 - \frac{1}{2}\eta_s^2, \ldots$$

Any solution κ of MKdV equation $\kappa_t - 6\kappa^2\kappa_s + \kappa_{sss} = 0$ is also a solution of the KdV equation $\eta_t + 6\eta\eta_s + \eta_{sss} = 0$ by the Miura transformation [2,169], $\eta = -(\kappa^2 + \kappa_s)$. Consequently, after eliminating the integrable terms in all expressions because of the closed loop condition, the conservation laws for the MKdV equation become

$$J_1 \sim \kappa^2, \quad J_2 \sim -3\kappa^4 + 4\kappa_s^2 - 8\kappa\kappa_{ss}, \ldots$$

These quantities are not directly related to perimeter or area, although there are authors considering that there is a connection through the prolongation structures [242, 335].

The solutions of the MKdV (16.79) can be expressed in terms of Jacobi elliptic functions (Sect. 18.3) simply by following the same procedure as in the case of KdV equation in Sect. 11.2. The cnoidal wave solution has the form

$$\kappa(s,t) = A\mathrm{cn}\left(\frac{s - \Omega t}{\Lambda}\middle| m\right) + B, \tag{16.80}$$

where Ω is the angular velocity of the MKdV cnoidal wave in curvature, m is the modulus of the cnoidal function, and A, B are arbitrary integration constants. The width Λ and the angular velocity Ω are given by

$$\Lambda = \frac{2}{A}\sqrt{m}, \quad \Omega = \frac{e^2 n_e \delta s^2}{8 m_e \epsilon \omega_c}\left[\left(\frac{5\kappa(0)}{12} - \frac{8}{\delta s^2}\ln\frac{\delta s^2}{2R_0}\right) + \frac{A^2}{4}\left(2 - \frac{1}{m}\right)\right]. \tag{16.81}$$

This solution approaches the MKdV one-soliton solution in the limit $m \to 1$

$$\kappa_{sol} = A\mathrm{sech}\left(\frac{s - \Omega t}{\Lambda}\right) + B, \tag{16.82}$$

with

$$\Lambda = \frac{2}{A}, \quad \Omega = \frac{e^2 n_e \delta s^2}{8 m_e \epsilon \omega_c}\left[\left(\frac{5\kappa(0)}{12} - \frac{8}{\delta s^2}\ln\frac{\delta s^2}{2R_0}\right) + \frac{A^2}{4}\right]. \tag{16.83}$$

However, this solution is not appropriate for our closed contour problem. It is true that the curvature is a periodic function, and we can even request the tangent of the contour to be periodic. However, the curve itself obtained by the Fresnel integration of this curvature (5.15) is open. This is easy to observe: the curvature of a closed curve should be a constant plus a correction, to guarantee a perturbed closed circle. The KMdV soliton equation is always oscillating around zero, so the resulting curve is an oscillating open curve.

To provide the closure of the contour one needs to look for a different solution of (16.79), more related to a breather one. The authors in [340, 341] found the form

$$\kappa(s,t) = \frac{A + B\mathrm{cn}\left(\frac{s-\Omega t}{\Lambda}\middle| m\right)}{D + F\mathrm{cn}\left(\frac{s-\Omega t}{\Lambda}\middle| m\right)}. \tag{16.84}$$

If we plug this solution in the differential equation, we obtain the following form for the parameters of the solution

$$
\begin{aligned}
m &= -\frac{F(2BF^2 - BD^2 - ADF)}{2(AD - BF)(D^2 - F^2)},\\
\Lambda &= \sqrt{\frac{2DF(D^2 - F^2)}{[DF(A^2 + B^2) - AB(D^2 + F^2)]}},\\
\Omega &= \frac{e^2 n_e \delta s^2}{8 m_e \epsilon \omega_c}\left[\frac{5\kappa(0)}{12} - \frac{8}{\delta s^2}\ln\frac{\delta s^2}{2R_0} + \frac{AB}{3DF}\right.\\
&\quad \left. + 2\frac{A^2D^2 + B^2F^2 - 2ABDF}{3(D^2 - F^2)^2}\right].
\end{aligned}
\tag{16.85}
$$

Such a solution is periodic of period $4\Lambda K(m)$ (Sect. 18.3). In order for the contour to be a smooth loop, it needs to fulfill the condition of matching modulo 2π of the tangent at the ends, i.e.,

$$\int_0^L \kappa(s,t)ds = 2\pi.$$

This condition can be resolved for the solution in (16.84) and (16.85) and, by the Fresnel integration, one can obtain all the shapes presented in Fig. 16.8. Because it depends on four parameters, this solution for curvature generates a large variety of curves including self-intersecting curves, multifoils, etc., many of them very much related to the vortex filaments shapes (Sect. 15.1), since the two systems occur from the same type of nonlinear equation. Of course not all of them are appropriate for modeling a closed contour. The same type of MKdV dynamics was obtained for normal liquid drops in Sect. 13.3 by using a different approach.

We close this section with a note concerning the possibility of having rigid cores inside or outside such droplets, like for example in the experiments described in Sect. 12.6. Let us assume that a rigid boundary is placed at a radius a. The normal velocity for any point of coordinate φ_0 should cancel, so we need

$$U(a,\varphi_0) \sim \int_0^{2\pi} \frac{r_\varphi(\varphi)\sqrt{g(\varphi)}d\varphi}{(a\cos\varphi_0 - r(\varphi)\cos\varphi)^2 + (a\sin\varphi_0 - r(\varphi)\sin\varphi)^2} = 0, \tag{16.86}$$

where g is the metric of the contour. In principle, the equation of the contour (and its curvature) need to be expanded in cnoidal modes and then exploit the orthogonality relations between the Jacobi elliptic functions to cancel this integral, but this problem would be beyond the purpose of this book.

Chapter 17
Nonlinear Contour Dynamics in Macroscopic Systems

In this chapter we study several macroscopic applications of the closed contour dynamics problem by using theorems for differential geometry. A first application presented is the study of the geometry of trajectories of charged particles in magnetic fields. We present some closeness trajectories criteria based on Bonnet and Fenchel theorems. Another example is given by the application of the Gauss–Bonnet theorem to problems of trapping particles inside closed magnetic surfaces. At larger physical scales, we present the occurrence of very localized stable waves orbiting around elastic spheres, and we conclude the chapter with a description of nonlinear modes in neutron stars.

17.1 Plasma Vortex

17.1.1 Effective Surface Tension in Magnetohydrodynamics and Plasma Systems

In this section, we consider another situation where the geometry of the free surface controls the dynamics of the fluid inside. We shall consider the problem of confining some electrically conducting fluid by an external magnetic field configuration. This problem, part of a more general subject known under the name of *magnetohydrodynamics*, is important in hot and dense plasma systems, and in controlled thermonuclear fusion installations. To produce extreme pulses of neutrons through the initiation of a thermonuclear fusion reaction between helium, deuterium, and tritium for example, matter should be compressed and heated to ultrahigh densities, pressures, and temperatures for a long enough time. Under such conditions, matter becomes a dense and hot plasma namely a combination of positive ions, electrons, neutral particles, and electromagnetic radiation. Left to itself, a plasma – like a gas – will occupy all the geometrical space available

A. Ludu, *Nonlinear Waves and Solitons on Contours and Closed Surfaces*,
Springer Series in Synergetics, DOI 10.1007/978-3-642-22895-7_17,

because of the collisions between the particles. At these high energy densities, the plasma–wall interaction is enough intense to damage any type of material, so practically there is no type of material strong enough to keep such a plasma confined. Consequently, the only possibility for plasma confinement is through magnetic fields.

17.1.2 Trajectories in Magnetic Field Configurations

Magnetic fields can confine a plasma, because the electrically charged particles follow helical paths around the magnetic field lines. Indeed, let us assume that a charged particle moves in a region where there is a constant and uniform field of force, $\boldsymbol{F}_0$. This is the case of electric $\boldsymbol{E}$ and/or gravitational field $\boldsymbol{G}$, only. Let $\boldsymbol{r}(t)$ be the particle law of motion, as a three-dimensional curve parametrized by time. We have the metrics $g = \dot{\boldsymbol{r}} \cdot \dot{\boldsymbol{r}} = v^2$ and the arc-length $ds = \sqrt{g}dt = vdt$. The velocity is given by $\boldsymbol{v} = \dot{\boldsymbol{r}} = v\boldsymbol{t}$, where $\boldsymbol{t}, \boldsymbol{n}$, and $\boldsymbol{b}$ are the three Serret–Frenet unit vectors associated to the particle trajectory. The acceleration has the form

$$\ddot{\boldsymbol{r}} = \frac{ds}{dt}\frac{d\dot{\boldsymbol{r}}}{ds} = \left(\frac{v^2}{2}\right)_s \boldsymbol{t} + v^2\kappa\boldsymbol{n},$$

where κ, τ are the curvature and torsion of the trajectory, and subscript means differentiation. Newton's second law $\boldsymbol{F}_0 = m\boldsymbol{a}$ reads

$$\left(\frac{v^2}{2}\right)_s \boldsymbol{t} + v^2\kappa\boldsymbol{n} = \left(\frac{g}{2}\right)_s + g\kappa\boldsymbol{n} = \frac{\boldsymbol{F}_0}{m}. \tag{17.1}$$

It is easy to identify the geometrical meaning of the kinematics quantities: the linear acceleration $a = g_s/2$, and the centripetal acceleration $a_{cp} = g\kappa$. Because the force is constant, by differentiating (17.1) with respect to the arc-length, and by using the Serret–Frenet equations (5.3) we have

$$\left[\left(\frac{g}{2}\right)_{ss} - g\kappa^2\right]\boldsymbol{t} + \left[\left(\frac{g}{2}\right)_s \kappa + (g\kappa)_s\right]\boldsymbol{n} - g\kappa\tau\boldsymbol{b} = 0.$$

Since $v \neq 0, g \neq 0$ we have from the above equation

$$\begin{cases} \kappa\tau = 0 \\ g_{ss} = 2g\kappa^2 \\ \kappa g_s = -(g\kappa)_s \end{cases}. \tag{17.2}$$

The first equation shows us that in the case of a constant force the trajectory is always plane (and in particular can be a straight line). From the last two equations we obtain $\kappa g^{3/2} = \text{const}$. Since the trajectory is a plane curve, we can choose locally a *flat* coordinate system where $\boldsymbol{r} = (x, y(x), 0)$. In these coordinates we have $g = \sqrt{1 + y'^2}$, $\kappa = y''/(1 + y'^2)^{3/2}$, and it results $y'' = 0$ so the trajectory in the case of a constant force field is always a parabola.

In the case of a constant (but not uniform) magnetic field $\boldsymbol{B}$ we have from (17.1)

$$\left(\frac{v^2}{2}\right)_s \boldsymbol{t} + v^2 \kappa \boldsymbol{n} = \frac{q}{m}(v\boldsymbol{t} \times \boldsymbol{B}). \tag{17.3}$$

Since $\boldsymbol{t}$ is perpendicular on the RHS of (17.3), we have $v = v_0 = \text{const.}$ and $g = g_0 = v_0^2 = \text{const.}$, which agrees with the well-known fact that magnetic field does not change the kinetic energy of charged particles. In terms of geometric quantities (17.3) reads

$$g\kappa \boldsymbol{n} = \frac{qv}{m}(\boldsymbol{t} \times \boldsymbol{B}). \tag{17.4}$$

Also, since the metrics along the trajectory is constant, we can write

$$\boldsymbol{t}_s = \frac{q}{mv_0}(\boldsymbol{t} \times \boldsymbol{B}) = \frac{qB}{mv_0}\left(\boldsymbol{t} \times \frac{\boldsymbol{B}}{B}\right),$$

and we denote by $C(s) = qB(\boldsymbol{r}(s))/mv_0$ and $\boldsymbol{T}(s) = \boldsymbol{B}(\boldsymbol{r}(s))/B(\boldsymbol{r}(s))$ the unit tangent of the magnetic field line that intersects the path of the particle at every point $\boldsymbol{r}(s)$. With these notations (17.4) reads

$$\kappa \boldsymbol{n} = \boldsymbol{t}_s = C(\boldsymbol{t} \times \boldsymbol{T}). \tag{17.5}$$

From here we have $\kappa_{helix} = C \sin\theta$, where θ is the angle between $\boldsymbol{t}$ and $\boldsymbol{T}$. We can express the components of the magnetic field in terms of the local Serret–Frenet frame of the particle path, $\boldsymbol{B} = B_t\boldsymbol{t} + B_n\boldsymbol{n} + B_b\boldsymbol{b}$, and since $\boldsymbol{t} \times \boldsymbol{B} = B_n\boldsymbol{b} - B_b\boldsymbol{n}$ we have $B_n = 0$, or $\boldsymbol{T} \cdot \boldsymbol{n} = 0$. Equivalently, $\boldsymbol{T} = T_t\boldsymbol{t} + T_b\boldsymbol{b}$, $T_t^2 + T_b^2 = 1$. It means that the particle moves such that the unit normal to its trajectory is always in the normal plane of the field lines. It also means that the tangent to the field line is always in the rectifying plane of the trajectory. According to the definition of a generalized helix (Definition 41), the motion of the particles is always a local helix with its axis perpendicular on the normal plane to the magnetic field lines, curvature $\kappa_{helix} = C \sin\theta = -CT_b$. In other words, the particle trajectories wind locally around the magnetic field lines. If the magnetic field is uniform, or if we study the motion in a small region where the field is almost uniform, the trajectory is a cylindrical helix. We know from Definition 41 that a helix has a constant ratio between its curvature and torsion. It results that locally, if the intensity of the field increases (hence curvature increases) the torsion increases, too. It means that when a particle enters a region with increasing magnetic field its "local helix" becomes flatter and narrower, and this is the *magnetic mirror* effect. Eventually, for a critical value of the field, the torsion cancels, the trajectory becomes flat, and the particles turns around.

If we differentiate (17.5) with respect to s we obtain

$$\kappa_s \boldsymbol{n} - \kappa^2 \tau - \kappa\tau \boldsymbol{b} = \frac{C_s}{C}\kappa \boldsymbol{n} + C^2 \boldsymbol{T}(\boldsymbol{t}\cdot\boldsymbol{T}) - C^2 \boldsymbol{t} + C\kappa(\boldsymbol{t}\times\boldsymbol{T}_s).$$

If we identify in the equation above the coefficients of $\boldsymbol{b}$ on the LHS with those on the RHS we obtain a relation defining the torsion of the trajectory

$$\tan\theta = \frac{\kappa}{\tau} - \frac{K_B \cos\Psi}{\tau\cos\theta}, \tag{17.6}$$

where K_B is the curvature of the field line, and Ψ is the angle between $\boldsymbol{n}$ and the unit normal of the field line, $\boldsymbol{N}$. Equation (17.6) can also be written in the form

$$\tau^2 + \kappa^2 = C^2 + \mathcal{O}(K_B).$$

From $\boldsymbol{b} = \boldsymbol{t}\times\boldsymbol{n}$ we obtain

$$\boldsymbol{b} = -\frac{C}{\kappa}\boldsymbol{T}_b, \text{ where } \boldsymbol{T}_b = (\boldsymbol{T}\cdot\boldsymbol{b})\boldsymbol{b} = \sin\theta\boldsymbol{b}.$$

It is easy to note that if the field lines are almost rectilinear, we can neglect the term containing K_B, and then the ratio between curvature and torsion becomes a constant, i.e., the trajectory is a helix surrounding the field line. Another simple situation occurs if the magnitude of the magnetic field is constant along the particle trajectories. In this case we have

$$\kappa_s = C\cos\theta\theta_s$$
$$\boldsymbol{b}_s = C\left(\frac{\kappa_s}{\kappa}\boldsymbol{T}_b - \frac{1}{\kappa}\boldsymbol{T}_{b,s}\right) \text{ and}$$
$$\boldsymbol{T}_b\cdot\boldsymbol{T}_{b,s} = 0.$$

Consequently, we can express the angle θ as function of the curvature of the isomagnetic lines, K_B

$$(\sin\theta)_s = -K_B(\boldsymbol{N}\cdot\boldsymbol{b}).$$

This last expression allows us to obtain a simple equation for the torsion of the particle trajectory, in the isomagnetic field case

$$\tau = C\left(1 + \frac{K_B}{\kappa}\cos\psi\right),$$

where $\cos\psi = \boldsymbol{N}\cdot\boldsymbol{n}$.

An interesting question is to find the structure of the magnetic field lines to have the particles trapped inside a certain bounded region of the space. This problem is an interesting exercise for the theory of compact surfaces and closed curves. In the following we assume that the magnetic field is constant in time, and the speeds of the particles are also constant. This is of course an approximation of the real situation inside a plasma region where the magnetic field is actually perturbed by the field generated by particle motion itself. Also the field is not stationary, because there is a combination of electric and magnetic fields. Moreover, the particles collide and their speeds spread into a thermal equilibrium configuration, and relativistic dynamics may occur, too. For a general yet comprehensive treatment of the theoretical problem of hydromagnetic stability we would recommend the reader the book of Chandrasekhar [50].

We assume that each magnetic field line is a regular parametrized curve $\boldsymbol{B}(\boldsymbol{r}(s))$ of curvature $K_B(s)$. For any point $(\boldsymbol{r}_0)$, and any initial direction of the motion of a charged particle $(\boldsymbol{t}_0)$, we can predict the trajectory of the particle, $\boldsymbol{r} = \alpha(t)$ by integrating the equation of motion (17.3). The question is whether all possible particle trajectories launched in magnetic field can be organized in regular surfaces parametrized by time or arc-length, and some other parameter describing the initial conditions. For example, if the magnetic field is uniform, all trajectories are helices. All particles having initial position at points placed on a tube of constant radius along on one magnetic field line, and initial velocities (initial tangents) making the same angle with this magnetic field, move only on the surface of this tube. The space can be filled with such disjoint, coaxial tubes of different radii.

Finding the equation of such particle motion surfaces in the general case of an arbitrary magnetic field is not a typical Frobenius problem (see Theorem 5), because we do not have two given vector fields in involution to be integrated. One vector field can be the magnetic field, but the other field is not uniquely defined, since the initial velocities of particles in different points are arbitrary. We actually have a Cauchy problem defined by (17.3), and by Cauchy conditions of type 16. Specifically, we choose the Cauchy initial conditions for one particle in the form $\boldsymbol{r}_0(v)$ and $\boldsymbol{t}_0(v)$, where v is one real parameter, which labels different initial conditions. To find the particle motion integral surfaces we use Theorems 3 and 23 in Sect. 9.6, and the above initial condition to solve (17.5). This equation is an ODE for the unit tangent vector $\boldsymbol{t}(s)$ and has the general solution in the form

$$\begin{aligned} \boldsymbol{t}(s) &= \exp\left(C \int_0^s \hat{T}(s')ds' \right)\boldsymbol{t}_0 \\ \boldsymbol{t}(s)_i &= \exp\left(\mathcal{E}_{ijk} C \int_0^s \boldsymbol{T}_k(\boldsymbol{r}(s'))ds' \right)\boldsymbol{t}_{0j}, \end{aligned} \tag{17.7}$$

where the summation indices $i, j, k = 1, 2, 3$ label the Cartesian components, and $\mathcal{E}_{ijk}$ is the Levi–Civita signature tensor. The coefficient C is written in front of the integral operator because it is a constant. $\boldsymbol{T}(\boldsymbol{r}(s))$ is the unit tangent to the magnetic field along the particle path, $\boldsymbol{T} \circ \boldsymbol{r}$. This solution of the unit tangent field is actually the flow, or the exponential map, of the tensorial field $\hat{T}$, where hat means the dual of the vector $\boldsymbol{T}$. This dual is a 3×3 antisymmetric matrix associated to $\boldsymbol{T}$. The formal exponential of a 3×3 matrix A is the 3×3 matrix obtained from the series

$$\exp(A) = \sum_{i \geq 0} \frac{A^n}{n!}.$$

For example, if we have a uniform field in the direction of Oz-axis, $\boldsymbol{B} = (0, 0, B_0)$ and $\boldsymbol{T} = (0, 0, 1)$, the dual antisymmetric matrix has the form

$$(\hat{T})_{ij} = (\mathcal{E}_{ijk} \boldsymbol{T}_k) = \begin{pmatrix} 0 & 1 & 0 \\ -1 & 0 & 0 \\ 0 & 0 & 0 \end{pmatrix}.$$

The exponential has the form

$$\exp(C\hat{T}s) = \exp\begin{pmatrix} 0 & Cs & 0 \\ -Cs & 0 & 0 \\ 0 & 0 & 0 \end{pmatrix} = \begin{pmatrix} \cos(Cs) & \sin(Cs) & 0 \\ -\sin(Cs) & \cos(Cs) & 0 \\ 0 & 0 & 1 \end{pmatrix},$$

and the solution for the tangent is the well-known helix along the Oz axis

$$\boldsymbol{t}(s) = (t_{01}\cos(Cs) + t_{02}\sin(Cs), -t_{01}\sin(Cs) + t_{02}\cos(Cs), t_{03}).$$

Consequently, the general solution of the (17.3) is obtained by one more integration

$$\boldsymbol{r}(s) = \int_0^s \exp\left(C \int_0^{s'} \hat{T}(\boldsymbol{r}(s''))ds'' \right) ds' \cdot \boldsymbol{t}_0 + \boldsymbol{r}_0. \tag{17.8}$$

Equation (17.8) is actually an implicit equation for the trajectory of the particle, because the unknown function $\boldsymbol{r}(s)$ appears also in the exponent in the RHS. This inconvenience makes the problem more difficult to solve. Yet, one can check the validity of (17.8) by trying simple examples of field configurations, like the helical motion presented earlier.

A possible approach toward the closing or boundness of trajectories is to use the Bonnet Theorem 20, which provides a sufficient condition for a (complete) surface to be compact. The hypothesis is to assume that the particles describe helical trajectories around the magnetic field lines, and remain confined within tubular surfaces centered on the magnetic lines. Let us consider a set of identical particles, launched with the same initial speed (they have the same $C(s)$ function), and at the same distance from a given the magnetic field line, denoted Γ. The particles differ by only one parameter denoted v, which describes the relative position around Γ of the initial launching points. Consequently, the particle trajectories lie on a smooth surface S of equation

$$\boldsymbol{r}(s, v) = \int_0^s \exp\left(\int_0^{s'} C\hat{T}(\boldsymbol{r}(s''))ds'' \right) ds' \cdot \boldsymbol{t}_0(v) + \boldsymbol{r}_0(v). \tag{17.9}$$

The coordinate curves along this surface are

$$\begin{aligned} \boldsymbol{r}_s &= \exp\left(C \int_0^{s'} \hat{T}(\boldsymbol{r}(s''))ds'' \right) \cdot \boldsymbol{t}_0 + \boldsymbol{r}_0 \\ \boldsymbol{r}_v &= \int_0^{s} \exp\left(C \int_0^{s'} \hat{T}(\boldsymbol{r}(s''))ds'' \right) ds' \cdot \boldsymbol{t}_{0v} + \boldsymbol{r}_{0v}. \end{aligned} \tag{17.10}$$

For example, in the case of uniform parallel magnetic field we choose all particles to start their motion at same distance r from a magnetic field line, i.e., from the surface of a tube around the magnetic field line. All particles will have the same initial speed, and their initial velocities ($\boldsymbol{t}_0$) make the same angle with the magnetic field (same θ). The surface is a cylinder of radius r with the magnetic field as axis. This cylindrical surface will follow and surround the magnetic field lines, even in the case of curved field lines, if this field lines are not too much bent, i.e., if $k \gg K_B$. Let us assume such a situation when the curvature of the magnetic field is much smaller than the curvature of the particle trajectories. Let us also assume that this surface is complete and regular. This implies that we study the system for long enough time such that all trajectories can be considered a dense set in this abstract surface, and that the system does not contain any "free force" or uniform field regions. Moreover, even if the completeness condition is not fulfilled, still the surface having its Gaussian curvature bounded from below by a positive number is bounded. An example is provided by an ergodic surface winding inside asymptotically. If the field curvature K_B is smaller than that of the particles, we can approximate this surface $\boldsymbol{r}(s, v)$ with a *tube* of radius r around the curve $\Gamma(s)$ (s is the arc-length of the field line). If the field lines are closed, such a tube is homeomorphic with a torus surface. The surface equation is

$$\boldsymbol{r}(s, v) = \Gamma(s) + r(\boldsymbol{n} \cos v + \boldsymbol{b} \sin v), \tag{17.11}$$

and its first fundamental form is

$$|vec\, \boldsymbol{r}_s \times \boldsymbol{r}_v| = EG - F^2 = r^2(1 - rK_B \cos v)^2.$$

We assume that $rK_B \ll 1$ and we have the normal to this surface defined by

$$\boldsymbol{N} = -(\boldsymbol{n} \cos v + \boldsymbol{b} \sin v), \quad \boldsymbol{r}_s \times \boldsymbol{r}_v = r(1 - rK_B \cos v)\boldsymbol{N}.$$

The Gaussian curvature of the tube surface is

$$K = -\frac{K_B \cos v}{r(1 - rK_B \cos v)}. \tag{17.12}$$

and we are ready to apply Bonnet Theorem 20. If the Gaussian curvature in (17.12) is always strictly larger than a positive number δ, the tube is a compact surface. Unfortunately, in our case the Gaussian curvature has always a change of sign. This happens because, even if the magnetic curve is closed, the tube surface is homeomorphic to a torus and has also negative Gaussian curvature in some regions. We cannot apply the Bonnet theorem in this form. However, we can relax the local condition and substitute it with a global one. Indeed, we have

$$\iint_S K dA = \iint_S K\sqrt{EG - F^2} ds dv = \int_0^l \int_0^{2\pi} K_B \cos v ds dv = 2\int_0^l K_B(s) ds,$$

and we can apply the Gauss–Bonnet Theorem 20 for a certain tube radius such that the LHS of the equation above is 4π. This result conducts us to use another approach, more related to the intrinsic curve geometry. In addition, it is worth to mention that we do not need actually to prove that plasma is confined in some region, but rather to obtain the conditions under which the particles do not move too far away from the magnetic field lines.

An alternate general approach to find conditions for plasma confinement is to use the curve equivalent of the Bonnet theorem, namely the Fenchel and Fary–Milnor Theorems 14 and 15. In that, we can take profit of (17.8) and analyze its geometrical properties. If the trajectory of a charged particle is a closed and simple curve, the Fenchel Theorem 14 provides us with a necessary criterion for closeness. The Fenchel criterion for having the charged particles move along closed paths is

$$\int_0^l |k| ds > 2\pi. \tag{17.13}$$

However, it is hard to have the particle trajectories represented by simple curves, since usually the particles wind many times around the magnetic field lines. The situation can be slightly improved by taking into consideration more general curves, like knotted curves. In this case, we have Fary–Milnor Theorem 15 which increases the minimum allowed value of total curvature from 2π to 4π. Both Fenchel and Fary–Milnor theorems are valid even if the trajectories are not simple curves, see [46, Sects. 5–7]. We can require the trajectory to have no just one self-intersection, and that is the point where this trajectory will close. In this case the RHS in (17.13) has to be substituted with $2N\pi$, where N is the rotation index of the trajectory.

From (17.5) we have $|k| = C|\sin\theta|$, where $\theta(s)$ is the current angle between the tangent to the trajectory and the local direction of the magnetic field, $\cos\theta = \boldsymbol{t}(s) \cdot \boldsymbol{T}(s)$. To fulfill the closing condition we need to design the magnetic field, and to send the particle within the following constraint. We need to find a number $0 < \delta < 1$ such that $|\sin\theta(s)| < \delta$ for all the points of arc-length s along the trajectory, i.e., to fulfill the Fary–Milnor criterion. Consequently, to have closed trajectories we need to adjust the two parameters: particle velocity and maximum magnitude of the field, accordingly. The closeness condition reduces to a restriction upon θ, namely there should be a minimum angle such that $\forall s \in [0, l]$, $\theta(s) > \theta_{min}$. For example, launching a particle as parallel as possible to the field lines, or keeping the field lines straight and open is not a good idea. To find out how this criterion acts on the field configuration, we choose an arbitrary magnetic field described by $\boldsymbol{B}(\boldsymbol{r}) = B(\boldsymbol{r})\boldsymbol{T}(s)$, with $|\boldsymbol{T}| = 1$. The solution of (17.7), written in components, reads

$$t_i = \left(\exp \frac{q}{mv_0} \mathcal{E}_{ijk} \int_0^s B_k(s')ds' \right) t_{0j}. \tag{17.14}$$

The dual antisymmetric tensor associated to the unit tangent $\boldsymbol{T}(s)$ is

$$\hat{T} = \begin{pmatrix} 0 & CT_1 & CT_2 \\ -CT_1 & 0 & CT_3 \\ -CT_2 & CT_3 & 0 \end{pmatrix}. \tag{17.15}$$

We introduce the notations

$$\rho_i(s) = \int_0^s C(s')T_i(s')ds', \quad \rho_0(s) = \int_0^s C(s')ds'.$$

Since $\boldsymbol{T}$ is a unitary vector, we have

$$\rho_0(s) = \int_0^s C ds' = \frac{q}{mv_0} \int_0^s B ds' \le \frac{qB_{max} l(s)}{mv_0} < \frac{l_{max}}{R_{min}},$$

where B_{max} is the maximum value of the magnitude of magnetic field along the path of the particle (in principle can be taken the maximum value of the magnitude of magnetic field in all plasma region). Also, v_0 is the constant speed of the particle, $l(s)$ is the length of the particle trajectory at s, and R_{min} is the minimum possible radius of rotation of the particle, if it would be launched in a region with maximum magnetic field, perpendicular on the magnetic field. With these notations the matrix exponential of $\hat{T}$ from (17.15) becomes

$$\begin{pmatrix} \rho_3^2 + (\rho_1^2 + \rho_2^2) \cos \rho_0 & \rho_2 \rho_3 (1 - \cos \rho_0) + \rho_1 \sin \rho_0 & \rho_1 \rho_3 (1 - \cos \rho_0) - \rho_2 \sin \rho_0 \\ \rho_2 \rho_3 (1 - \cos \rho_0) - \rho_1 \sin \rho_0 & \rho_2^2 + (\rho_1^2 + \rho_3^2) \cos \rho_0 & \rho_1 \rho_2 (1 - \cos \rho_0) + \rho_3 \sin \rho_0 \\ \rho_1 \rho_3 (1 - \cos \rho_0) + \rho_2 \sin \rho_0 & \rho_1 \rho_2 (1 - \cos \rho_0) - \rho_3 \sin \rho_0 & \rho_1^2 + (\rho_2^2 + \rho_3^2) \cos \rho_0 \end{pmatrix}. \tag{17.16}$$

This matrix exponential has determinant 1, and hence is similar to a three-dimensional proper rotation. So, the exponential in (17.7) and (17.14) act like a rotation operator upon the initial direction of the particle.

The closeness criterion can be written

$$\left| T_i \left(\exp\left(\mathcal{E}_{ijk} \int_0^s \hat{T}(s')ds' \right) \right)_{ij} t_{0j} \right| < \delta < 1, \tag{17.17}$$

or in more condensed matrix notation

$$|\boldsymbol{T} \widehat{\exp} \boldsymbol{t}_0| < \delta < 1, \tag{17.18}$$

where $\hat{\exp}$ represents the exponential matrix in (17.17). There is no point in using the Stokes equation

$$\oint BT_k(s')ds' = \mathcal{E}_{ijk} \iint_S \left(\frac{\partial}{\partial x^i} - \frac{\partial}{\partial x^j}\right) BdA,$$

because the exponential of each of the two terms in the above transformation do not commute, so we cannot separate the exponential of the difference in a product of exponentials. However, such a transformation is useful to prover that for an axial $\boldsymbol{B} = (0, 0, B_0) =$ const. uniform field, of a pure poloidal or toroidal field, the exponent is a diagonal matrix, so the exponential matrix is also diagonal. Equation (17.18) has a maximum value of 1 if the vector $\boldsymbol{t}_0$ is an eigenvector for the matrix $\hat{\exp}$. This matrix has one real eigenvalue 1, and two complex conjugated eigenvalues. For the real eigenvalue the eigenvector is $(\rho_3/\rho_1, \rho_2/\rho_1, 1)$. So, the necessary condition for closeness of the particle trajectories is to choose the initial direction such that $|\boldsymbol{t}_0 - (\rho_3/\rho_1, \rho_2/\rho_1, 1)| > \delta > 0$.

Let us check this criterion on a toroidal geometry, for example, where we try to confine the plasma inside a torus surface. The surface of a torus of larger radius R, and smaller radius r, parametrized by the polar (v), and azimuthal (or toroidal u) angles has the form

$$\boldsymbol{r}(u, v) = ((R + r \cos u) \cos v, (R + \cos v) \sin v, r \sin u). \tag{17.19}$$

In the case of a poloidal magnetic field

$$\boldsymbol{T}_{pol} = (-r \cos v \sin u, -r \sin v \sin u, s \cos u), \tag{17.20}$$

the matrix $\hat{\exp}$ is diagonal for all s, so the trajectories will not close. The same thing happens for a toroidal field

$$\boldsymbol{T}_{tor} = (-(R + r \cos u) \sin v, (R + \cos v) cos\ v, 0). \tag{17.21}$$

Only a linear combination of toroidal and poloidal field could fulfill the criterion in (17.17).

Usually, the particles travel distances longer than their Larmor radius ($1/\kappa$), so the exponential matrix cannot be approximated with its Taylor polynomial. The smallness parameter for such an expansion would be $max_{s \in [0,l]} \rho_0 = l/R_{min}$. A Taylor expansion in this smallness order works rather in escape areas, or for weak fields, than along regular field lines. For the sake of completeness we present here such an expansion, in the case of constant magnitude of magnetic field along the path ($C = C_0 =$ const.), and valid only if the length of the trajectory is smaller than the Larmor radius ($s \ll mv_0/qB_0$). We can write

$$\cos\theta(s) = \boldsymbol{T}(s)\cdot\boldsymbol{t}_0 + C_0\int_0^s \hat{T}(s')ds' + \frac{C_0^2}{2}\int_0^s \hat{T}ds'\int_0^s \hat{T}ds'' + \ldots. \quad (17.22)$$

The general term in this expansion has the form of toroidal multipoles

$$\frac{(-1)^n C_0^n}{n!}\int_0^s\int_0^{s_1}\ldots\int_0^{s_n} \boldsymbol{T}(s)\cdot\boldsymbol{T}(s_1)\times(\boldsymbol{T}(s_2)\times\cdots\times(\boldsymbol{T}(s_n)\times\boldsymbol{t}_0))\ldots)ds_1 ds_2\ldots ds_n.$$

In the first-order approximation we have

$$\cos\theta(s) \simeq \boldsymbol{t}_0\cdot\Bigg[\boldsymbol{T}(s) - C_)\int_0^s \boldsymbol{T}(s)\times\boldsymbol{T}(s')ds'$$
$$+\tfrac{C_0^2}{2}\int_0^s\int_0^{s'} \boldsymbol{T}(s')(\boldsymbol{T}(s)\cdot\boldsymbol{T}(s'')) - \boldsymbol{T}(s)(\boldsymbol{T}(s')\cdot\boldsymbol{T}(s''))\Bigg]ds'ds'',$$

and the closeness criterion becomes

$$\int_0^l \boldsymbol{T}(s)\cdot\boldsymbol{t}_0 ds - C_0\int_0^l\int_0^s \boldsymbol{T}(s)\cdot(\boldsymbol{T}(s')\times\boldsymbol{t}_0)dsds' < \delta < 1. \quad (17.23)$$

In conclusion, (17.17) and (17.18) provide the criterion needed by the magnetic field configuration, and by the initial conditions of the particle velocity to have the trajectory confined closer to the field lines. The smaller δ in these equations, the more confinement we realize. It is interesting how theorems from differential geometry of curves and surfaces help to solve this problem. Apparently Bonnet theorem is more powerful. First, it provides a sufficient condition for confinement: if the Gaussian curvature is larger than a given positive limit, the surface carrying the particle trajectories is bounded. Second, it provides a local, differential criterion, which is more helpful than a global one. Third, it provides a quantitative criterion. If one finds a lower positive bound for the Gaussian curvature, this limit provides a measurement of the diameter of the surface (see Theorem 20). Although the equivalent theorems for curves, namely the Fenchel and Fary–Milnor ones, are only necessary conditions, they are only global (integral) conditions, and they do not provide but a qualitative result. It is also true that the result these theorems provide for curves is more restrictive than the result provided by the Bonnet theorem for surface. This is because closeness is a more specific restriction than compactness.

17.1.3 Magnetic Surfaces in Static Equilibrium

If a vessel containing plasma is placed in an uniform magnetic field $\boldsymbol{B}_0$, the plasma particles cannot reach the side walls, but they will strike the ends of the vessel. To prevent the particles from coming into contact with the material walls in this way, special types of magnetic fields configurations are introduced. One can either increase the magnetic field intensity at the ends of the container so that the particles are reflected by tandem magnetic mirror, or one can curve the magnetic filed lines to form loops, in such a way that the particles are trapped inside a *magnetic surface*. The mirror configurations (also called the linear configuration) is not quite the best because the particle collision effects render the system liable to high particles losses at the mirror points. Such systems are not being considered as potential controlled thermonuclear fusion reactors. More interesting from the geometrical point of view, there are three main types of closed magnetic surfaces configuration: *Tokamak*, *Stellarator*, and *Reversed field pinch* systems. The confinement solution consists in closing the magnetic field lines $\boldsymbol{B}(\boldsymbol{r})$ on themselves to trap the particles. In such an ideal configuration, the magnetic field lines would lie on closed surfaces, named magnetic surfaces. The magnetic field is tangent to this surface at any point and interacts with the charged particle velocity field, i.e., the plasma current [28, 60]. It can be described by the velocity field $\boldsymbol{v}(\boldsymbol{r})$, or by the density of electric current $\boldsymbol{j}(\boldsymbol{r}) = \text{curl}\boldsymbol{B}/\mu$, where μ is the magnetic permeability of plasma. In the following we provide analytical criteria for the magnetic field to create confining surfaces.

Let us have a constant magnetic field $\boldsymbol{B}(\boldsymbol{r})$ fulfilling

$$div\, \boldsymbol{B} = 0, \tag{17.24}$$

and let us assume the existence of a regular parametrized surface S of equation $\boldsymbol{r}(u, v)$, such that the magnetic field is tangent to S at any of its points, $\boldsymbol{B}(\boldsymbol{r}) \in T_{\boldsymbol{r}} S$. We can choose the parametrization of S such that it fulfills the condition

$$\boldsymbol{r}_u = \boldsymbol{B}(\boldsymbol{r}(u, v)). \tag{17.25}$$

The magnetic field lines provide natural coordinate curves on S. In addition we request that the other coordinate curves on S fulfill the differential equation

$$\boldsymbol{r}_v = \nabla \times \boldsymbol{B}(\boldsymbol{r}(u, v)). \tag{17.26}$$

In this situation, the Lorentz force acting on plasma currents

$$\boldsymbol{F}_L = \boldsymbol{j} \times \boldsymbol{B} = \frac{1}{\mu}(\nabla \times \boldsymbol{B}) \times \boldsymbol{B},$$

fulfills the equation

$$\boldsymbol{F}_L = \frac{1}{\mu}(\boldsymbol{r}_u \times \boldsymbol{r}_v) = \frac{1}{\mu}|\boldsymbol{r}_u \times \boldsymbol{r}_v| \boldsymbol{N}. \tag{17.27}$$

This configuration provides a Lorentz force parallel to the normal of the magnetic surface S. If S is oriented and closed we realized a confinement system configuration. This is because the Lorentz force acts always perpendicular on the magnetic surface, toward its inside, and hence the particles are supposed to be trapped. Even if

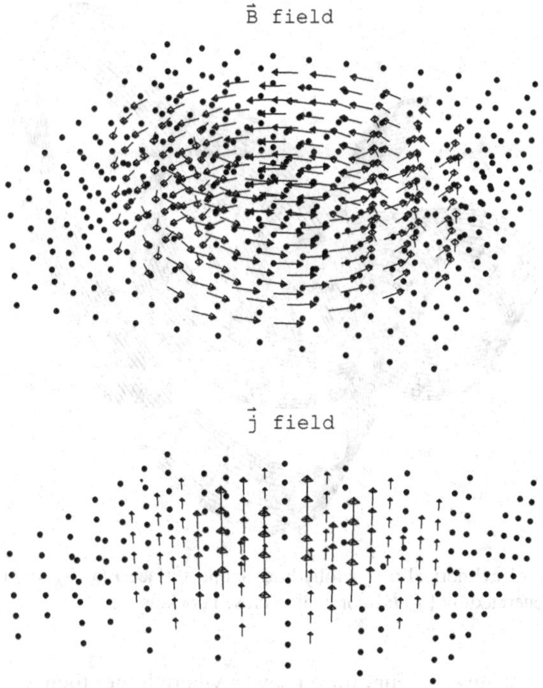

Fig. 17.1 Pure toroidal magnetic field, and corresponding axial current $\boldsymbol{j}$

(17.24)–(17.26) describe the magnetic surface, we need a criterion for its existence. The condition for the existence of an integral magnetic surface is provided by the Frobenius criterion of involution (Theorem 5) between the two vector fields

$$[\boldsymbol{B} \cdot \nabla, (\nabla \times \boldsymbol{B}) \cdot \nabla] = 0. \tag{17.28}$$

Equation (17.28) can be also written in the form

$$D_{\boldsymbol{B}}(\nabla \times \boldsymbol{B}) = D_{\nabla \times \boldsymbol{B}} \boldsymbol{B}, \tag{17.29}$$

or $D_{\boldsymbol{B}} \boldsymbol{j} = D_{\boldsymbol{j}} \boldsymbol{B}$, i.e., the directional derivatives of the magnetic field and the current with respect to one other should commute. A simple example of such a surface is provided by an axisymmetric configuration of magnetic field (Fig. 17.1), where the field is toroidal and the resulting electric current is axial. A more complicated example of open configuration is presented in Fig. 17.2. Such exact polynomial solutions are useful in providing estimates of the displacement of the magnetic boundaries with plasma flow [60].

However, such open surfaces cannot confine the particles because it is open. To have an equilibrium confinement situation we need two more criteria: one for compactness and one for closeness of the magnetic surface. The magnetic surface

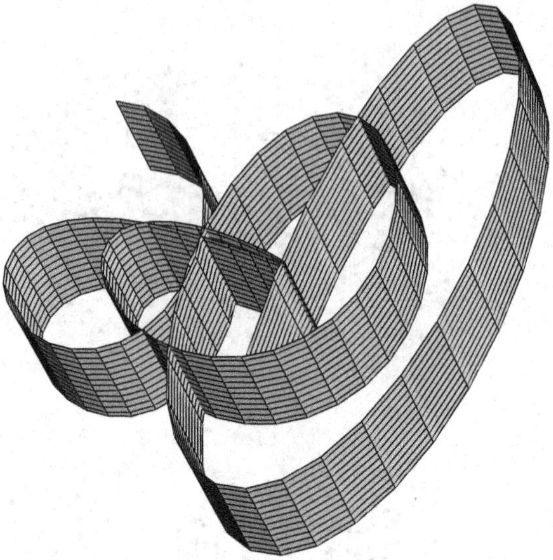

Fig. 17.2 Example of cylindrical magnetohydrodynamic surface $\boldsymbol{r}(x,y) = (10x + 30\cos y + y, 10y\sin y, 2x)$ generated by $\{\boldsymbol{j}, \boldsymbol{B}\}$ containing closed pockets

S is compact if its Gaussian curvature is everywhere larger then a positive constant $\delta > 0$ (Theorem 20). The unit normal to the magnetic surface is

$$N = \frac{\boldsymbol{B} \times curl\, \boldsymbol{B}}{|\boldsymbol{B} \times curl\, \boldsymbol{B}|},$$

and the Gaussian curvature results in a complicated expression

$$K = \left(\frac{(\boldsymbol{B} \times curl\, \boldsymbol{B}) \cdot [(\boldsymbol{B} \cdot \nabla)\boldsymbol{B}]}{B|\boldsymbol{B} \times curl\, \boldsymbol{B}|} \frac{(\boldsymbol{B} \times curl\, \boldsymbol{B}) \cdot [(curl\, \boldsymbol{B} \cdot \nabla) curl\, \boldsymbol{B}]}{|curl\, \boldsymbol{B}||\boldsymbol{B} \times curl\, \boldsymbol{B}|} - \left(\frac{(\boldsymbol{B} \times curl\, \boldsymbol{B}) \cdot (\boldsymbol{B} \cdot \nabla) curl\, \boldsymbol{B}}{B|\boldsymbol{B} \times curl\, \boldsymbol{B}|} \right)^2 \right) \cdot [B^2 curl\, \boldsymbol{B}^2 - (\boldsymbol{B} \cdot curl\, \boldsymbol{B})^2]^{-1}. \tag{17.30}$$

We can write the Bonnet condition (17.30) in a simpler form by using the notation $j\mu = |\nabla \times \boldsymbol{B}|$

$$\frac{B(N \cdot D_{\boldsymbol{B}} \boldsymbol{B})(N \cdot D_{\boldsymbol{j}} \boldsymbol{j})}{\mu j (N \cdot D_{\boldsymbol{B}} \boldsymbol{j})(N \cdot D_{\boldsymbol{j}} \boldsymbol{B})} > 1 + \delta > 1, \tag{17.31}$$

where $D_X Y$ represents the directional derivative of field Y in the direction of the field X. Equation (17.31), coupled with (17.24), represents a sufficient condition for the magnetic field to create a compact magnetic surface. It requests that a combination of directional derivatives of the magnetic field and the current projected along the unit normal fulfill a certain inequality. The second criterion (for closeness) is derived from the Gauss–Bonnet Theorem 20. If the integral of the Gaussian curvature all over the surface (the total curvature) is equal to $4\pi, 0, -4\pi, \ldots,$ then the surface is closed. This even multiplier of 2π in the RHS of the total curvature is the *Euler characteristics* $\chi(S)$ of the surface S. For a sphere $\chi = 2$ and for a torus $\chi = 0$. In conclusion, the conditions fulfilled by a magnetic field to create a stationary confinement system is to have $\delta > 0$ and $g = 0, 2, \ldots$ such that

$$K(u,v) > \delta, \text{ and } \iint_S K dA = 2\pi\chi(S) = 4\pi(1-g),$$

where $K(u,v)$ is the Gaussian curvature in the (u,v) parametrization.

Let us choose, for example, a poloidal magnetic field (Fig. 17.3) described by

$$\boldsymbol{B}(\boldsymbol{r}) = (axz, byz, c(z^2 + d - x^2 - y^2)),$$

together with its curl

$$\nabla \times \boldsymbol{B} = (-(b+2c)y, (a+2c)x, 0),$$

which is a toroidal field (Fig. 17.4). It is easy to check that the Frobenius integrability condition $[\boldsymbol{B}, \nabla \times \boldsymbol{B}] = 0$ is fulfilled for the two fields (Theorem 5). Consequently, (17.25) and (17.26) describe the coordinate curves of an integral surface

$$\boldsymbol{r}_u = \boldsymbol{B}, \text{ and } \boldsymbol{r}_v = \nabla \times \boldsymbol{B}.$$

A particular solution can be chosen with $z = z(u)$, and it results $x = \xi(u)\cos v$, $y = \xi(u)\sin v$, and $a = b$. From (17.25) we have $\xi' = a\xi z(u)$ and $z' = (\xi''\xi - \xi'^2)/(a\xi^2)$. Since $z_u = z' = c(z^2 + d - x^2 - y^2)$ we have $a(\xi''\xi - \xi'^2) = c\xi'^2 + a^2cd\xi^2 - a^2c\xi^4$. A solution is

$$\xi = \frac{\sqrt{B^2-1}}{2c(B+\cos u)}, \quad d = \frac{1}{uc^2}, \quad a = 2c,$$

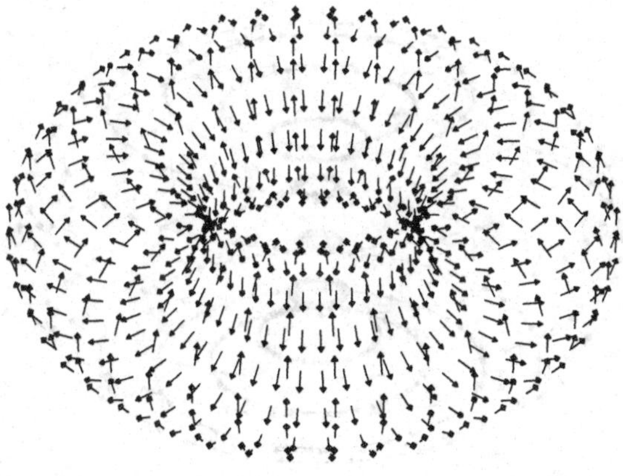

Fig. 17.3 Poloidal field

where B is an arbitrary integration constant. If we choose $c = 1/2$ we can write the integral surface, i.e., the magnetic surface, equation in the form

$$\boldsymbol{r} = \left(\frac{\sqrt{B^2 - 1}\cos v}{B + \cos u}, \frac{\sqrt{B^2 - 1}\sin v}{B + \cos u}, \frac{\sin u}{B + \cos u} \right),$$

which is actually a T_1 torus. Indeed, by denoting $B = cosh\, s$ and from $\sqrt{B^2 - 1} = \sinh s$ we can rewrite the surface equation in the toroidal coordinates (s, u, v)

$$\boldsymbol{r}(s, u, v) = \left(\frac{\sinh s\ \cos v}{\cosh s + \cos u}, \frac{\sinh s\ \sin v}{\cosh s + \cos u}, \frac{\sin u}{\cosh s + \cos u} \right).$$

Toroidal coordinates form an orthogonal three-dimensional curvilinear coordinate system (among other 11 orthogonal curvilinear coordinates in $\mathbb{R}_3$, like cartesian, cylindrical, spherical, parabolic, elliptic, hyperbolic, etc.), and are defined in the theory of separation of variables for Laplace's equation (cf. [242] and references herein in Sect. 1.3). The orthogonal coordinate surfaces are represented by concentric coaxial tori s = const. of small radius inversely proportional to s, meridian planes v = const. localized at different azimuthal angles $v \in [0, 2\pi)$, and concentric spheres u = const., of radius proportional to u (see Fig. 17.5). Expressed in toroidal coordinates, the magnetic vector field has a "flat" appearance $\boldsymbol{B} = -2\frac{\partial}{\partial u}$. The resulting Lorentz force is oriented toward the inside of the integral torus (Fig. 17.6). It is easy to verify that the total curvature of this configuration is zero.

There are several other approaches on the problem of plasma stability and confinement inside magnetic surfaces, both analytical and numerical. For example in [60] the authors use a special type of curvilinear coordinates (Boozer's flux coordinates) consisting in a normal coordinate ρ and two angular coordinates θ_B, ξ_B. The magnetic surface is parametrized by isomagnetic lines defined by B = const.,

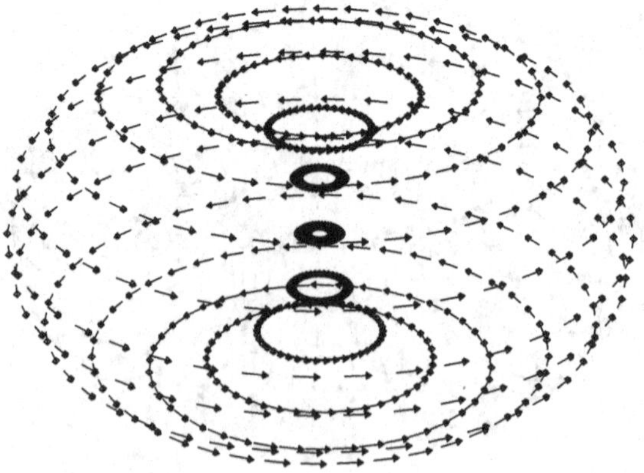

Fig. 17.4 Toroidal field

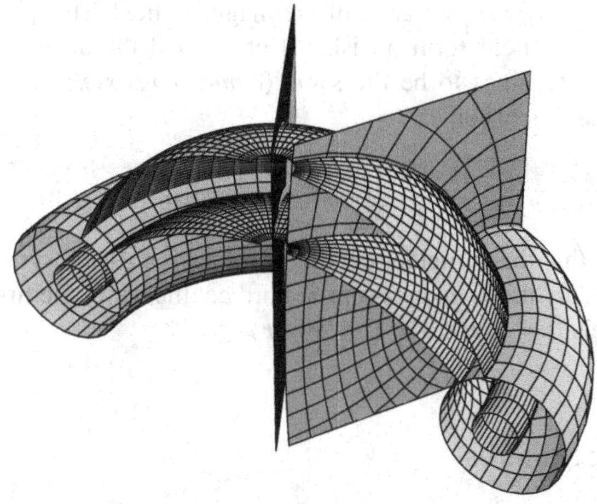

Fig. 17.5 Toroidal coordinates: $s =$ const., concentric tori, $v =$ const., meridian planes, and $u =$ const., concentric hemispheres

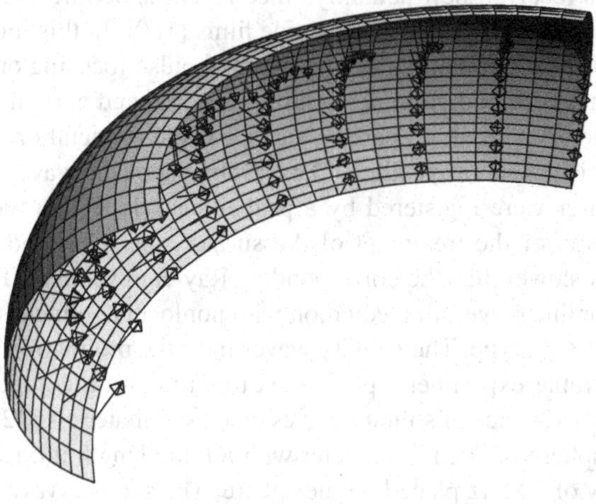

Fig. 17.6 The magnetohydrodynamic pressure $\boldsymbol{j} \times \boldsymbol{B}$ directed toward inside the closed surface, along $\boldsymbol{N}$

and another surface solenoidal isomagnetic vector, $\boldsymbol{i}_B = \nabla(\boldsymbol{B} \cdot \nabla\rho)$, such that the Frobenius criterion of integrability is fulfilled $[d/d\lambda, d/dB] = 0$. The isomagnetic lines are parametrized by a parameter λ, and their equation is

$$\frac{dB}{d\lambda} = \frac{(\nabla(\boldsymbol{B} \cdot \nabla\rho) \cdot \nabla)\boldsymbol{B}}{(\boldsymbol{B} \cdot \nabla)\boldsymbol{B}} = 0,$$

where we used the normalized isomagnetic vector $\boldsymbol{i}_B/|\boldsymbol{i}_B|$. In this approach, the condition for the regularity of the surface is related to the property of *pseudosymmetry* (or *quasisymmetry*) of the magnetic field. This property requests that the isomagnetic field form no islands on S, and the distance between two adjacent isomagnetic lines to be the same (*omnigenous systems* [28]). In vector notation this condition becomes

$$\frac{\boldsymbol{B}\cdot\nabla(\boldsymbol{B}\cdot\nabla\rho)}{\boldsymbol{B}\cdot\nabla\rho}=\text{bounded}.$$

It is interesting that the sufficient condition for such a pseudosymmetry configuration is provided by the boundness of the third coefficient of the first fundamental form of S, namely $\delta=\text{const.}>0$, $F=\boldsymbol{r}_\lambda\cdot\boldsymbol{r}_B\leq\delta$.

17.2 Elastic Spheres

A natural question inspired by the existence of solitons on the surface of shallow water is whether solitons may also propagate along the surface of a solid medium. The problem has received new actuality since recent experiments of formation of solitary elastic surface pulses on metal-oxide films [180]. In this interesting experiment, the soliton was initiated by laser-generated pulse focusing on a flat surface. To have a medium with both nonlinear elastic response and normal and anomalous dispersion, Lomonosov et al. prepared a surface made of metal or titanium nitride film coated with isotropic fused silica. The traveling acoustic waves pulse triggered by the pulsed laser were registered by a probe-beam deflection technique at two locations. Function of the treatment of the surface, the measured solitary waves traveled faster or slower than the corresponding Rayleigh velocity. The dynamics is modeled by a nonlinear evolution equation with nonlocal nonlinearity and nonlocal dispersion of the KdV type. The solitary waves have the profile of a "Mexican hat."

Another favorable experiment, performed this time on a compact surface, put into evidence the existence of solitary waves on elastic materials [322]. The authors excited a glass sphere of 80 mm diameter with a finite length ultrasonic transducer with a frequency of 1 MHz placed on the sphere. The surface waves were detected with similar PZT transducers at different points on the surface of the sphere. The surface acoustic waves were propagated along the equator of the sphere in a direction perpendicular to the line source without beam spreading. The traveling wave was both very localized (about 30° width) and propagated around for at least four round trips with a velocity very close to the corresponding Rayleigh surface wave speed in glass (3,334 m s^{-1}). Moreover, in another experiment, a signal produced on the surface at a certain point generated two twin surface wave pulses that traveled in opposite direction along the equator, intersect, interact, and return back without damping. This phenomenon is very much in favor of existence of solitary acoustic waves, even solitons, on the surface of the glass sphere.

A theoretical analysis of the existence of surface acoustic solitons was performed in [83]. The Rayleigh waves propagating along the surface (x–y plane) of an elastic medium give rise to a dynamical corrugation of the surface. The strain field produced by this corrugation decays into the bulk medium after a distance when compared with the wavelength. Consequently, for planar homogeneous media, the Rayleigh waves are nondispersive, and the balance between nonlinearity and dispersion can be obtained only by modifying the surface, by coating, grating, or just damaging the surface. In a second-order nonlinearity approach, the dynamics is governed by the equation

$$T_{\alpha\beta} = C_{\alpha\beta\mu\nu} u_{\mu\nu} + \frac{1}{2} S_{\alpha\beta\mu\nu\zeta\xi} u_{\mu\nu} u_{\zeta\xi},$$

where $\hat{T}$ is the Euler–Piola stress tensor, $u_{\alpha\beta} = \partial u_\alpha / \partial x_\beta$ are the displacement gradients, and the coefficients are the elastic moduli of the substrate, of second and second–third order, respectively. The equation of motion is

$$\rho A_\alpha = T_{\alpha\beta,\beta},$$

where A is the surface acceleration and ρ is the mass density of the substrate. The displacement field is expanded in an asymptotic series in terms of a smallness coefficient ϵ of the same order of magnitude as the depth of the layer. This asymptotic series is plugged into the boundary condition at the surface

$$T_{\alpha 3}|_{z=0} = d(D_{\alpha\beta} u_{\beta,11} - \rho_F A_\alpha)_{z=0},$$

where d is the thickness of the layer, D is another material coefficient, and ρ_F is the film material density. We expand the displacement field in plane waves

$$u = \sum_k e^{ik(x-Vt)} \frac{w(z,k)}{k} B_k,$$

where $w(z,k)$ is the depth profile of the linear Rayleigh wave, k is the wave number, and B_k are strain amplitudes. By introducing this series in the dynamic equation, and in the boundary conditions, one obtains a dynamical nonlinear recursion relations for the strain amplitudes equivalent to the Bejamin–Ono (BO) equation [83]. Numerical simulations show the existence of traveling waves very similar to the cnoidal waves of the KdV equation, or the solitons of the BO equation. Numerical tests show that these solitons are linearly stable. Moreover, same numerical procedure was used to simulate collision between two such solitons. The two models show different behavior. The BO solitons repel each other at a certain minimum distance and bounce off with unchanged shapes. On the contrary, the KdV pulses strongly contract while accelerating and radiation is shed after the collision. Consequently, the authors conclude that solitary nonlinear waves can propagate on the surface of a nonlinear homogeneously coated elastic solid. These solitary waves are stable with respect to perturbations, but they do not survive collisions with each other. A possibility to enhance the soliton character of such nonlinear waves is to use curved surfaces, and take profit both from the diffraction-free propagation along curved surfaces, and from the geometrical nonlinearities that occur in this case.

17.3 Curvature Dependent Nonlinear Diffusion on Closed Surfaces

Particles motion along a closed surface with thickness, with dynamics controlled by nonlinear diffusion equations, represents a topic of interest in protein diffusion within lipid bi-layers, cell membrane processes, patterns on animal skins, [239], etc. In these models the thickness of the physical surface (membrane, skin) is taken into account as a perturbation, but stronger effects are expected when the curvature radius becomes similar in size with the thickness.

The mathematical model for this type of thick surface diffusion consists in using normal surface coordinates and use the normal variation approach, see Sect. 10.4.1. In [239] the authors consider a physical medium described by a smooth surface Σ surface with a constant thickness ϵ. The interesting problem is to describe the regular three-dimensional diffusion of certain particles (proteins, colorants) inside this medium as an effective two-dimensional diffusion, imbedded in and along the surface Σ, whose Laplace operator and diffusion constant are curvature dependent. The effective two-dimensional diffusion field is described by the scalar field $\Phi(q1, q2, t) : \Sigma \times [0, \infty) \to \mathbf{R}$ which fulfills the two-dimensional diffusion equation

$$\frac{\partial \Phi}{\partial t} = D \Delta^{eff} \Phi, \tag{17.32}$$

where D is the diffusion coefficient, and Δ^{eff} is a *effective* two-dimensional surface Laplace operator, similar to the one defined in section 6.5.3. We need to stress that this operator is not the Beltrami operator because it contains the thickness dependence, too. With the usual notations for the surface geometry, q_j for surface coordinates, g^{ij} for the surface metric, $g = \det(g^{ij})$, Π^{ij} the second fundamental tensor of Σ, $H = g^{ij}\Pi_{ij}$ the mean curvature of Σ, and $R = 2\det(\Pi^i_j)$ the Ricci scalar curvature, with $i, j, k, \dots = 1, 2$ we can write the effective two-dimensional nonlinear diffusion equation in the form

$$-\frac{\partial \Phi}{\partial t} = \nabla_i (J^i_N + J^i_A) = g^{-1/2} \frac{\partial}{\partial q^j} g^{1/2} (J^i_N + J^i_A), \tag{17.33}$$

where

$$J^i_N = -D g^{ij} \frac{\partial \Phi}{\partial q^j}, \tag{17.34}$$

is the normal diffusion flow,

$$J_A^i = -\tilde{D}\left[(3\Pi^{im}\Pi_m^j - 2H\Pi^{ij})\frac{\partial \Phi}{\partial q_j} - \frac{1}{2}g^{ij}\frac{\partial R}{\partial q^j}\Phi\right], \tag{17.35}$$

is the anomalous flow, and $\tilde{D} = \epsilon^2 D/12$. The diffusion into the normal direction to the surface arrives to an equilibrium in a time scale $\delta t = \epsilon^2/D$. Consequently, for larger time scales $t >> \delta t$ equilibrium is assumed in the normal direction at all times. Therefore, the nonlinear diffusion equation (17.33) will be considered up to order ϵ^2.

The consequence of this type of anomalous diffusion is that the flow goes from smaller Ricci scalar points to larger Ricci scalar points, i.e. from hyperbolic or flat points to convex or concave points with positive large Ricci scalar curvature. An example of such a system is provided by the spot pattern on the skin of the Char fish, [239], which has white spots on the side parts, but labyrinth type of pattern on the dorsal parts.

17.4 Nonlinear Evolution of Oscillation Modes in Neutron Stars

The recent discovery of a millisecond pulsar binary system [48], with an orbital period of 2.2 h, brings the question of the importance of different interaction mechanisms between the stars in such close binaries. In particular, the tidal interactions have an important role in producing gravitational waves. In fact, even if the majority of the gravitational radiation in the binary systems comes from the orbital mass distribution quadrupole, the asymmetry created in the neutron star by the tidal bulge can produce certain amount of gravitational waves, and the effect is even more enhanced if the bulge can rotate fast. The neutron star systems are very layered so the surface waves induced by the binary interaction are very dispersive. On the other hand, the neutron star's oscillations, especially the so-called r-modes [175], can be highly nonlinear, being driven toward instability by gravitational radiation. All in all, it looks like such systems are appropriate for the occurrence of solitary waves on their surface, especially since long duration movement of tides have been detected. A factor that can suppress the occurrence of solitons is the existence of strong dissipative mechanisms, many of them are still completely unknown.

The nonlinear evolution of a neutron star can be modeled using Newtonian equations of motion, like the equation of continuity and the Euler equation in a compact domain D

$$\rho_t + \nabla \cdot (\rho V) = 0,$$

$$\rho(V_t + (V \cdot \nabla)V) = -\nabla P - \rho \nabla \Phi + \rho \boldsymbol{F}_{GR},$$

where V, ρ, and P are the velocity, density, and pressure of the neutron fluid, respectively; Φ is the Newtonian gravitational potential fulfilling the Poisson equation

$$\Delta \Phi = 4\pi G \rho,$$

and $\boldsymbol{F}_{GR}$ is gravitational radiation reaction force. This last term is due to the time-varying current quadrupole and can be written [175]

$$F_{GR}^{x} - iF_{GR}^{y} = -\kappa i(x+iy)[3V^{z}J_{22}^{(5)} + zJ_{22}^{(6)}],$$

$$F_{GR}^{z} = -\kappa \, \mathrm{Im}\left[(x+iy)^{2}\left(3\frac{V^{x}+iV^{y}}{x+iy}J_{22}^{(5)} + J_{22}^{(6)}\right)\right],$$

where $J_{22}^{(n)}$ represents the nth time derivative of the quadrupole moment

$$J_{22} = \int_{D} \rho r^{2} \boldsymbol{V} \cdot \boldsymbol{Y}_{22}^{B*} d^{3}x,$$

where $\boldsymbol{Y}_{22}^{B} = \boldsymbol{r} \times r\nabla Y_{22}/\sqrt{6}$ is the magnetic type vector spherical harmonics. The parameter describing the strength of the gravitational radiation force is

$$\kappa = \frac{32\sqrt{\pi}G}{45\sqrt{5}c^{7}},$$

from general relativity theory. The authors mentioned in [175] solve this complicated nonlinear evolutionary system numerically, and investigated the evolution of the so-called r-modes. These are modes specific for rotational stars, whose restoring force is the Coriolis force, and can balance the dissipative effects even for slow rotations. In time, the r-mode grows to a relative large amplitude on behalf of the gravitational radiation reaction force. However, shock waves begin to form at the leading edges of the surface of the neutron star at this point, which have as result suppressing the r-modes. The shock waves occur most likely because of the nonlinear coupling between various oscillatory modes within the star, or from *elliptic flow* instability similar to the one identified in fluid that are forced to flow along elliptical stream lines [247].

Chapter 18
Mathematical Annex

This chapter represents a mathematical annex. We briefly remember the properties of the Riccati equation and of some elliptic functions used in soliton theory. We also describe the one-soliton solutions of the KdV and MKdV equations. In the end we present a simple procedure, the so called nonlinear dispersion relation approach, through which one can find information about the relations between amplitude, half-width and speed of a soliton solution of any nonlinear equation (scalar, vector, or system, no matter of the nature of the nonlinearity) without actually solve the equation, providing such an equation admits soliton solutions. Several examples on well known cases are also given in order to illustrate how this procedure works.

18.1 Differentiable Manifolds

Definition 63. We define a d-dimensional C_p differentiable manifold $\mathcal{M} = (X, p \geq 1, d \geq 1, \{U_i, \phi_i\}_{i\in I})$ to be the set of a Hausdorff topological space X, and a family (atlas) of pairs of open sets U_i and bijective applications $\phi_i : U_i \to \phi_i(U_i) \subset \mathbb{R}^d$ fulfilling the properties:

- $\{U_i\}_{i\in I}$ is an open covering of X.
- $\forall i, j \in I, \phi_i(U_i \cap U_j) \subset \mathbb{R}^d$ is open.
- $\forall i, j \in I, \phi_j \circ \phi_i^{-1} : \phi_i(U_i \cap U_j) \to \phi_j(U_i \cap U_j)$ is a C_p diffeomorphism (i.e., bijective function of class C_p together with its inverse).

Every such set (U_i, ϕ_i) is called a chart, and $\forall x \in X$ such that $x \in U_i$, $\phi_i(x)$ are called the local coordinates of x. All the mappings $\phi : U \to \phi(U)$ are homeomorphisms. Two different atlases are *compatible* if their reunion is also an atlas. An equivalence class modulo this compatibility relation is called a differentiable structure on $\mathcal{M}$. No matter of the original topology of X, there is always a *canonical* topology induced by $\mathbb{R}^d$, where the open sets are reunion of chart domains. In that,

A. Ludu, *Nonlinear Waves and Solitons on Contours and Closed Surfaces*,
Springer Series in Synergetics, DOI 10.1007/978-3-642-22895-7_18,

the differential manifold is inheriting locally the topological properties of $\mathbb{R}^d$. Differentiable manifolds are locally compact and locally connected topological spaces. Moreover, they are connected if and only if they are path connected.

18.2 Riccati Equation

The Riccati differential equation for $f(x) : \mathbb{R} \to \mathbb{R}$ has the form

$$f' + Af + Bf^2 = C, \tag{18.1}$$

where $A, B,$ and C are differentiable functions of x. Equation (18.1) can be linearized if we perform the substitution $f = \theta'(B\theta)^{-1}$

$$\theta'' + \left(A - \frac{B'}{B}\right)\theta' - BC\theta = 0. \tag{18.2}$$

Conversely, the reduction of order from (18.2) to (18.1) is a consequence of the invariance of (18.2) under the scale transformation $(x, \theta(x)) \to (x, \lambda\theta(x))$. If $\theta_{1,2}$ are two independent particular solutions of (18.2), then the general solution of the Riccati equation depends only on one free parameter c and has the form

$$f_{gen}(x) = \frac{c\theta_1' + \theta_2'}{cB\theta_1 + B\theta_2}. \tag{18.3}$$

Another representation of the general solution of (18.1) in terms of two independent particular solutions $f_{1,2}$ of the same equation can be given in the form

$$f_{gen}(x) = \frac{f_2 - f_1 C e^{\int^x B(x')(f_1(x') - f_2(x'))dx'}}{1 - C e^{\int^x B(x')(f_1(x') - f_2(x'))dx'}}. \tag{18.4}$$

If we know just one particular solution f_p of (18.1), we still can build the general solution in the form

$$f_{gen}(x) = \frac{1}{F_{gen}} + f_1(x), \tag{18.5}$$

where $F_{gen}(x)$ is the general solution of the adjunct equation

$$F' - (A + 2Bf_1)F - B = 0, \tag{18.6}$$

for which there are quadrature formulas.

18.3 Special Functions

The solutions of all nonlinear PDEs are very much related to the Jacobi elliptic functions and Jacobi elliptic integrals. The incomplete elliptic integral of the first kind is defined as

$$F(\varphi|k) = \int_0^{\varphi} \frac{1}{\sqrt{1-k^2\sin^2\theta}} d\theta$$

and the complete elliptic integral of the first kind is $K(k) = F(\pi/2|k)$. Similarly, we define the incomplete elliptic integral of the second kind in the form

$$E(\varphi|k) = \int_0^{\varphi} \sqrt{1-k^2\sin^2\theta} d\theta,$$

and its complete elliptic integral of the second kind is $E(k) = E(\pi/2|k)$. The inverse of the elliptic integral of the first kind, i.e., if $u = F(\varphi|k)$ the $\varphi = \mathrm{am}(u|k)$, is called the amplitude for the Jacobi elliptic functions. The amplitude can generate the 12 *cnoidal functions*, among which the most used in soliton theory is the cnoidal sine function $\mathrm{sn}(u|k) = \sin(\varphi)$, cnoidal cosine function $\mathrm{cn}(u|k) = \cos(\varphi)$, and $\mathrm{dn}(u|k) = \sqrt{1-k^2\mathrm{sn}^2(u)}$. The sn and cn functions have the remarkable property of making smooth transition between periodic functions and aperiodic functions, so basically they connect the compact and noncompact structures. We have $\mathrm{sn}(u|0) = \sin(u), \mathrm{sn}(u|1) = \tanh(u), \mathrm{cn}(u|0) = \cos(u), \mathrm{cn}(u|1) = \mathrm{sech}(u)$. The cnoidal sine and cosine are double periodic functions. The real period is $T = 4K(k)$, and the imaginary one is $4iK(k)$. The cnoidal sine is the solution of the nonlinear ODE

$$(f_x)^2 = (1-f)(1-k^2 f), \tag{18.7}$$

i.e., $f(x) = \mathrm{sn}(x|k)$.

The spherical harmonics $Y_{lm}(\theta,\varphi)$, with $l = 0, 1, \ldots$ and $\mathbf{Z} \ni m \in (-l, l)$, form an orthonormal complete basis of harmonic ($\Delta_{S_2} Y_{lm} = 0$) polynomial functions defined on the unit sphere $S_2 \subset \mathbb{R}^3$. The general expression is

$$Y_{lm} = (-1)^m \sqrt{\frac{(2l+1)(l-m)!}{4\pi(l+m)!}} P_l^m(\cos\theta) e^{im\varphi},$$

where $P_l^m(x) : [-1, 1] \to \mathbb{R}$ are the associate Legendre functions defined as

$$P_l^m(x) = \frac{(1-x^2)^{\frac{m}{2}}}{2^l l!} \frac{d^{l+m}}{dx^{l+m}} (x^2-1)^l.$$

The restriction $P_l^0 = P_l$ is called Legendre polynomial. The orthonormality and closure relations are

$$\int_{S_2} Y^*_{lm} Y_{l'm'} \sin\theta d\theta d\varphi = \delta_{mm'}\delta_{ll'},$$

$$\sum_{l=0}^{\infty}\sum_{m=-l}^{l} Y^*_{lm}(\theta,\varphi)Y_{lm}(\theta',\varphi') = \frac{1}{\sin\theta}\delta(\theta-\theta')\delta(\varphi-\varphi').$$

From their definition, the spherical harmonics are the natural solutions of the Laplace equation in spherical coordinates, so any harmonic functions defined on the unit sphere can be expanded in series of spherical harmonics. The same role is played by the Legendre polynomial on the unit circle S_1. In a physical problem the angular part of the solution is usually handled by spherical harmonics, and the radial dependence is usually manipulated with the help of the spherical Bessel functions $j_l(r), n_l(r) : [0,\infty) \to \mathbb{R}$. The ODE for the spherical Bessel functions is

$$\left(\frac{1}{r}\frac{d^2}{dr^2}r + 1 - \frac{l(l+1)}{r^2}\right)q_l = 0,$$

where $q_l(r)$ is either $j_l(r)$ or $n_l(r)$. The j_l solution is regular in the origin, and the n_L one (Neumann function) is irregular in the origin. With these solutions we can also construct the Hankel functions as $h_l^{1,2}(x) = j_l \pm i n_l$. More details and proofs, integral or series representations and summations formulae, recursion formulas, and asymptotic relations about these special functions can be found in several books, among which we mention [5, 129, 238, 284].

18.4 One-Soliton Solutions for the KdV, MKdV, and Their Combination

The Korteweg–de Vries equation (KdV or $K(2,1)$) in $\eta(x,t)$

$$A\eta_t + D\eta_x + B\eta\eta_x + C\eta_{xxx} = 0 \tag{18.8}$$

with traveling solutions $\eta(x,t) = f(\xi)$, $\xi = x - Vt$ where V is a free parameter, becomes

$$-VAf' + Bff' + Cf''' + Df' = 0, \tag{18.9}$$

where $f' = \frac{df(\xi)}{d\xi}$ and A is the original coefficient of the time derivative evolutionary equation $Ad\eta/dt \to Adf/d\xi$. The

$$\eta(x,t) = \eta_0 \,\mathrm{sn}^2\left(\frac{x-Vt}{L}\middle|k^2\right) + \eta_1 \tag{18.10}$$

is the "cnoidal" sine Jacobi elliptic solution to the KdV equation, and k is the modulus of the cnoidal sine (Sect. 18.3). The solutions depend on the free parameter

$a = \eta_0$, i.e., its *amplitude*. The *half-width* is L, and the velocity V is given by

$$L = \sqrt{-\frac{12kC}{\eta_0 B}},$$

$$V = \frac{D}{A} + \frac{B}{3A}\left[\eta_0\left(1 + \frac{1}{k}\right) + \eta_1\right]. \tag{18.11}$$

In the limit $k \to 1$, the cnoidal solution approaches the one-soliton solution and the parameters become

$$\eta(x,t) = \eta_0 sech^2 \frac{x - Vt}{L} + \eta_1, \tag{18.12}$$

$$L = \sqrt{-\frac{12C}{aB}},$$

$$V = \frac{D}{A} + \frac{B}{3A}\left(\eta_0 + 3\eta_1\right). \tag{18.13}$$

We note that the amplitude η_0 is proportional to the velocity V (higher solitons run faster), and the width L is inversely proportional to the amplitude a (higher solitons are narrower).

Another typical equation is the modified KdV (MKdV or $K(3,1)$)

$$A\eta_t + D\eta_x + B\eta^2\eta_x + C\eta_{xxx} = 0, \tag{18.14}$$

which reduces for traveling solutions to

$$-VAf' + Bf^2 f' + Cf''' + Df' = 0. \tag{18.15}$$

A one-soliton solution family is

$$\eta(x,t) = a sech \frac{x - Vt}{L} \tag{18.16}$$

depending on the free a parameter, the *amplitude*. The *half-width* L and the velocity V of the shape are given by

$$L = \sqrt{\frac{6C}{a^2 B}},$$

$$V = \frac{a^2 B}{6A} + \frac{D}{A}. \tag{18.17}$$

We note that the square of the amplitude a is proportional to the velocity V (higher solitons run faster), and the width L is inversely proportional to the amplitude a (higher solitons are narrower).

Another solution of the MKdV equation is given by the topological soliton, i.e.,

$$\eta(x,t) = a Tanh \frac{x - Vt}{L}, \tag{18.18}$$

with the following relations among the parameters:

$$L = \sqrt{-\frac{6C}{Ba^2}}$$
$$V = \frac{D}{A} + \frac{Ba^2}{3A}. \tag{18.19}$$

A mixed nonlinear equation which contains both the KdV and the MKdV specific terms is always equivalent to a MKdV equation. Suppose we have

$$\eta_t + d\eta_x + a\eta\eta_x + b\eta^2\eta_x + c\eta_{xxx} = 0, \tag{18.20}$$

then this equation is equivalent with

$$f_t + \left(d + \frac{a^2}{4b}\right) f_x + bf^2 f_x + cf_{xxx} = 0,$$
$$f(x,t) = \eta(x,t) + \frac{a}{2b}. \tag{18.21}$$

18.5 Scaling and Nonlinear Dispersion Relations[1]

In this book we focus our investigations on solutions of nonlinear PDE defined on compact contours or surfaces. Before solving such a system, it is natural to look for a simple and qualitative criterion to find out if solutions could exist on such compact spaces. The idea is to extract information on simple properties of possible soliton solutions, like half-width, amplitude, and velocity, without actually solving the equation, i.e., to analyze the nonlinear dispersion relation (NLDR) associated to the system [194, 308]. We present in the table below some NLDR results for several nonlinear PDEs. In order, the equations analyzed in the first column of this table are KdV, mKdV, K(n,n), K(n,m), Burgers, a nonlinear dispersion equation, sine–Gordon, Φ^4-equation, Schrödinger cubic, Schrödinger higher nonlinearity, generalized nonlinear Schrödinger equation, vector nonlinear Schrödinger equation, and the two-dimensional KP equation.

The NLDR does provide a type of dimensional analysis of the solutions of nonlinear PDEs. The procedure is the following. For a PDE of the form

$$G(u, u_t, u_{tt}, \ldots, u_x, u_{xx}, \ldots) = 0, \tag{18.22}$$

[1] I am indebted to Dr. Panayotis Kevrekidis for the existence of this section.

where $x \in \mathbb{R}$, subscripts denote partial derivatives, and $u(x,t)$ is a real, complex, or vector-valued function, we substitute in the PDE, according to

PDE	Analytic solution and parameters ($\xi = x - Vt$)	NLDR ($\sim$ means proportional)
$u_t + 6uu_x + u_{xxx} = 0$	$u = A\ \text{sech}^2(\xi/L)$,	$L = 1/\sqrt{3A - V}$,
	$V = 2A,\ L = \sqrt{2/A}$	$V \sim A \to L \sim 1/\sqrt{A}$
$u_t + 6u^2u_x + u_{xxx} = 0$	$u = A\ \text{sech}(\xi/L)$,	$L = 1/\sqrt{2A^2 - V}$,
	$V = A^2, L = 1/A$	$V = A^2 \to L = 1/A$
$u_t + (u^n)_x + (u^n)_{xxx} = 0$	$u = [A\cos^2(\xi/L)]^{\frac{1}{n-1}}$	$L = \sqrt{\frac{1}{1-\alpha}} = \text{const.}$
	if $\lvert\xi\rvert \leq \frac{2n\pi}{n-1}$, 0 else;	
	$L = \frac{4n}{(n-1)}, V = \frac{(n+1)A^{n-1}}{2n}$	if $V = \alpha A^{n-1}$
$u_t + (u^n)_x + (u^m)_{xxx} = 0$	Unknown	$L = \sqrt{\frac{A^{n-1}}{A^{m-1} - V}}$,
	if $n \neq m$	$V \sim A^{m-1} \to L \sim A^{(n-m)/2}$
$u_t + uu_x - u_{xx} = 0$	$u = 2A\tan(A\xi + C_1) + V$,	$L = 1/(A - V)$,
	$A = \sqrt{C_2 - V^2}$	$V \sim A \to L \sim 1/A$
$u_t + a(u^m)_x - \mu(u^k)_{xx}$	Only particular cases	$cA^{\gamma-1}L^2 + (amA^{m-1} - V)L$
$+cu^{\gamma} = 0$	known	$+\mu k(2-k)A^{k-1} = 0$
$u_{tt} - u_{xx} + \sin u = 0$	$u = 4\tan^{-1} e^{\frac{\xi}{L}}$,	$L^2 = -(1 - V^2)/\cos AL$,
	$L^2 = V^2 - 1$	$AL = \text{cst.} \to L^2 = V^2 - 1$
$u_{tt} - u_{xx} - m^2u + u^3 = 0$	$u = \pm m\tanh(\xi/L)$,	$L^2 = \frac{1-V^2}{m^2 - 3A^2L^2}$
	$L^2 = 2(1 - V^2)/m^2$	if $AL \simeq \text{const.}$
		$\to L^2 \sim 1 - V^2$
$i\Psi_t + \Psi_{xx} + \lvert\Psi\rvert^2\Psi = 0$	$Ae^{i(Vx/2 + A^2t/2 - V^2t/4)}\text{sech}\left(\frac{\xi}{L}\right)$;	$L^1 = \frac{1}{A^2 + (V^2/4 - \omega)}$,
	$L = \sqrt{2}/A$	$\omega - V^2/4 \sim A^2 \to L \sim 1/A$
$i\Psi_t + \Psi_{xx} + \lvert\Psi\rvert^{\sigma}\Psi = 0$	Unknown in general	$L = \frac{1}{A}\frac{1}{\sqrt{A^{\sigma-1} + \frac{V^2}{4A^2}}}$
		if $V \sim A,\ L \sim A^{-\frac{\sigma+1}{2}}$
$i\frac{\partial u}{\partial t} + u_{xx} - \frac{\lvert u\rvert^k u}{1+\gamma\lvert u\rvert^k}$	$e^{i[\frac{V}{2}x - (\frac{V^2}{4} + \omega - \frac{4}{kL^2})t]}$.	$L = \frac{-V^2 \pm \sqrt{V^4 + 4(\frac{A^k}{1+\gamma A^k} - \omega)}}{2(\frac{A^k}{1+\gamma A^k} - \omega)}$,
$= \omega u$	$\cdot A\ \text{sech}^{\frac{2}{k}}\frac{\xi}{L}$	$V^2 \sim \frac{A^{k/2}}{(1+\gamma A^k)^{1/2}} \sim \omega^{1/2}$,
	$L = \left(\frac{2(2+k)}{k^2}\right)^{\frac{1}{2}}\left(\frac{A^k}{1+\gamma A^k}\right)^{-\frac{1}{2}}$	$L \sim \left(\frac{A^k}{(1+\gamma A^k)}\right)^{-1/2}$
$iq_t^{(1)} + q_{xx}^{(1)}$	$2\boldsymbol{P}\eta e^{-2i\zeta x + 4i(\zeta^2 - \eta^2)t - i\frac{\pi}{2}}$.	$L^{(j)} = \frac{1}{A}$,
$= -2(\lvert q^{(1)}\rvert^2 + \lvert q^{(2)}\rvert^2)q^{(1)}$,	$\cdot\text{sech}(2\eta x - 8\zeta\eta t - 2\delta_0)$	$A = \sqrt{(A^{(1)})^2 + (A^{(2)})^2}$
$iq_t^{(2)} + q_{xx}^{(2)}$	$L = \frac{1}{2\eta}$	
$= -2(\lvert q^{(1)}\rvert^2 + \lvert q^{(2)}\rvert^2)q^{(2)}$		
$(-4u_t + 6uu_x + u_{xxx})_x$	$\frac{(k_1 - k_2)^2}{2}\ \text{sech}^2\left(x\frac{k_2 - k1}{2}\right.$	$L^2 = \frac{1}{A + \alpha^2 - \frac{4}{3}V}$,
$+3u_{yy} = 0$	$\left. + y\frac{k_1^2 - k_2^2}{2} + t\frac{k_2^3 - k_1^3}{2}\right)$	$\alpha^2 - 4V/3 \sim A \to L^2 \sim 1/A$
		for $k_i << k_j$

$$u_t \to -V u_x \tag{18.23}$$

$$u \to A, \; u^{(k)}_{xx\ldots x} \to (-1)^{k-1}\frac{A}{L^k}, \; k > 0, \tag{18.24}$$

$$\int u dx \to AL, \tag{18.25}$$

where the superscript denotes the number of derivatives with respect to x. The result of the substitution is to obtain the NLDR connecting the length scale of the solution L, its speed V and amplitude A in the form

$$G\left(A, -V\frac{A}{L}, -V^2\frac{A}{L^2}, \ldots, \frac{A}{L}, -\frac{A}{L^2}, \ldots\right) = 0. \tag{18.26}$$

References

1. Y. Abe, Y. Kondo, T. Matsuse, Suppl. Prog. Theor. Phys. **68**, 68 (1980)
2. M.J. Ablowitz, P.A. Clarkson, *Solitons, Nonlinear Evolution Equations and Inverse Scattering* (Cambridge University Press, Cambridge, 1992)
3. M.J. Ablowitz, J. Hammack, D. Henderson, C.M. Schober, Phys. Rev. Lett. **84**, 887 (2000); Physica D **152**, 46 (2001)
4. R. Abraham, J.E. Marsden, *Foundations of Mechanics* (W. A. Benjamin, New York, 1967)
5. M. Abramowitz, I.A. Stegun, *Handbook of Mathematical Functions* (National Bureau of Standards Applied Mathematics Series 55, Washington, DC, 1964)
6. S. Alexander, T. Biswas, G. Caleagni, Phys. Rev. D **81**, 043511 (2010)
7. J. Angulo et al., Nonlinearity **15**, 759 (2002)
8. P. Annamalai, E. Trinh, T.G. Wang, J. Fluid Mech. **158**, 317 (1985)
9. R.E. Apfel et al., Phys. Rev. Lett. **78**, 1912 (1997)
10. R. Aris, *Vectors, Tensors, and the Basic Equations of Fluid Mechanics* (Dover, New York, 1989)
11. V.I. Arnold, B.A. Khesin, *Topological Methods in Hydrodynamics* (Springer, Berlin, 1998)
12. M. Barros, Proc. Am. Math. Soc. **126**(5), 1503 (1997)
13. O.A. Basaran, J. Fluid Mech. **241**, 169 (1992)
14. G.K. Batchelor, *An Introduction to Fluid Dynamics* (Cambridge University Press, New York, 1967)
15. D. Baye, P. Descourvement, Nucl. Phys. A **419**, 397 (1984)
16. E. Becker, W.J. Hiller, T.A. Kowalewski, J. Fluid Mech. **258**, 191 (1994)
17. T.B. Benjamin, Q. Appl. Math. **231** (July 1982)
18. T.B. Benjamin, J.L. Bona, J.J. Mahony, Philos. Trans. R. Soc. Lond. A **272**, 47 (1972)
19. M. Berger, B. Gostiaux, *Differential Geometry: Manifold, Curves, and Surfaces* (Springer, Berlin, 1988)
20. M.V. Berry, J.O. Indekeu, M. Tabor, N.L. Balazs, Physica D (Amsterdam) **11**, 1 (1984)
21. M.V. Berry, Nonlinearity **21**, T19 (2008)
22. R.B. Bird, R.C. Armstrong, O. Hassager, *Dynamics of Polymeric Liquids* (Wiley, New York, 1977)
23. M.C. Birse, Prog. Part. Nucl. Phys. **25**, 1 (1990)
24. A.I. Bobenko, U. Eitner, *Painlevé Equations in the Differential Geometry of Surfaces* Lect. Notes. Math. **1753** (Springer, Heidelberg 2000)
25. A. Bohr, Fys. Medd. K. Dan. Vidensk. Selsk. **26**, 14 (1952)
26. J.L. Bona, Phys. Fluids **23**, 438 (1983)
27. J. Bona, H. Chen, Physica D **116**, 191 (1998)
28. A. Boozer, Phys. Fluids **26**, 496 (1983)

A. Ludu, *Nonlinear Waves and Solitons on Contours and Closed Surfaces*, Springer Series in Synergetics, DOI 10.1007/978-3-642-22895-7,

29. L. Bourdieu, T. Duke, M.B. Elowitz, D.A. Winkelmann, S. Leibler, A. Libehaber, *Phys. Rev. Lett.* **75**, 176 (1995)
30. C.J. Brokaw, Cell Motil. Cytoskel. **42**, 134 (1999)
31. U. Brosa, Z. Naturforsch. A **43**, 1141 (1986)
32. R.C. Brower, D.A. Kessler, J. Koplik, H. Levine, Phys. Rev. A **29**, 1335 (1984)
33. B.M. Budak, S.V. Fomin, *Multiple Integrals, Field Theory and Series* (MIR, Moscow, 1973) (English Translation)
34. E.N. Bukina, V.M. Dubovik, Turk. J. Phys. **23**, 927 (1999)
35. S.S. Bupara, *Spontaneous Movements of Small Round Bodies in Viscous Fluids*, Ph.D. Thesis, Department of Chemical Engineering, University of Minnesota, 1964
36. F.H. Busse, J. Fluid. Mech. **142**, 1 (1984)
37. J.W. Cahn, C.M. Elliott, A. Novick-Cohen, Eur. J. Appl. Math. **7**, 287 (1996)
38. W. Cai, T.C. Lubensky, Phys. Rev. Lett. **73**, 1186 (1994)
39. A.M. Calini, T. Ivey, G. Marí-Beffa: Physica D **238**, 8, 788 (2009)
40. A.M. Calini, T.A. Ivey, Math. Comp. Sim. **55**, 341 (2001)
41. A.M. Calini, T.A. Ivey, J. Nonlin. Sci. **15**, 321 (2005); arXiv:nlin/0612065; arXiv:nlin/0411065
42. R. Camassa, Discrete Contin. Dyn. Syst., Ser. B **3**, 115 (2003)
43. E.J. Campbell, Proc. Lond. Math. Soc. **29**, 14 (1898); **34**, 347 (1902)
44. E.J. Campbell, Proc. Lond. Math. Soc. **35**, 333 (1903); **2**, 293 (1904); **3**, 24 (1904)
45. R. Capovilla, C. Chryssomalakos, J. Guven: arXiv:nlin.SI/0204049 v2 (13 June 2002)
46. M.P. do Carmo: *Differential Geometry of Curves and Surfaces* (Prentice-Hall, Englewood Cliffs, 1976)
47. F.J.C. De Carvalho, Am. Math. Mon. **92**, 202 (1985)
48. D. Chakrabarty, E.H. Morgan, Nature **394**, 364 (1998)
49. S. Chandrasekhar, *Ellipsoidal Figures of Equilibrium* (Yale University Press, New Haven, 1969)
50. S. Chandrasekhar, *Hydrodynamic and Hydromagnetic Stability* (Dover, New York, 1981)
51. S.J. Chapman, G. Richardson, SIAM J. Appl. Math. **58**, 2, (1998) 587
52. E.W. Cheney, *Applications of Fixed-Point Theorems to Approximation Theory* (Academic, New York, 1976)
53. A. Chevalley, *The Theory of Lie Groups* (Princeton University Press, Princeton, NJ, 1946)
54. A.J. Chorin, J.E. Marsden, *A Mathematical Introduction to Fluid Mechanics* (Springer, Berlin, 1992)
55. K.-S. Chou, X.-P. Zhu, *The Curve Shortening Problem* (Chapman & Hall/CRC, Boca Raton, FL, 2001)
56. N. Cindro, W. Greiner, J. Phys. G: Nucl. Part. Phys. **9**, L175 (1983)
57. S. Cohen, F. Plasil, W.J. Swiatecki, Ann. Phys. **82**, 557 (1974)
58. S. Coleman, Phys. Rev. **D15**, 2929 (1977)
59. S. Coleman, The uses of instantons, in *The Whys of Subnuclear Physics*, ed. by A. Zichichi (Plenum, New York, 1979)
60. C. Copenhaver, Phys. Fluids **26**, 2635 (1983)
61. L. Cortelezzi, A. Prosperetti, Q. Appl. Math. **38**, 375 (1981)
62. S.I.R. Costa, Proc. Am. Math. Soc. **109**, 205 (1990)
63. R. Courant, *Partial Differential Equations* (Interscience, New York, 1962)
64. R. Courant, D. Hilbert, *Methods of Mathematical Physics* (Interscience, New York, 1966)
65. G.S. Deem, N.J. Zabusky, Phys. Rev. Lett. **40**, 859 (1978)
66. J.M. Deutsch, Science **240**, 922 (1988)
67. L.A. Dickey, *Soliton Equations and Hamiltonian Systems* (World Scientific, New Jersey, 2003)
68. J. Dieudonné, *Fondements de l'analyse moderne*, vol 1 (Gauthier-Villars, Paris, 1965)
69. J. Dieudonné, *Eléments d'analyse*, vol 4, chapter XX, section 1 (Gauthier-Villars, Paris, 1971)
70. J. Dieudonné, *Eléments d'analyse*, vol 3, chapter XVII (Gauthier-Villars, Paris, 1970)

71. R.K. Dodd, J.C. Eilbeck, J.D. Gibbon, H.C. Morris, *Solitons and Nonlinear Wave Equations* (Academic, London, 1982)
72. L. Dolan, Phys. Rev. D **13**, 528 (1976)
73. A.T. Dorsey, R.E. Goldstein, Phys. Rev. B **57**, 3059 (1998)
74. B. Doubrovine, S. Novikov, A. Fomenko, *Geometrie Contemporaine* (MIR, Moscow, 1985)
75. B Dubrovin, J. Phys. A: Math. Thoeret. **43** (2010) 434002
76. J.P. Draayer, A. Ludu, G. Stoiev, Rev. Mex. Fisica **45**, 80 (1999)
77. S. Dramanyan et al., Phys. Rev. E **55**, 7662 (1997)
78. P.G. Drazin, *Nonlinear Systems* (Cambridge University Press, Cambridge, 1997)
79. P.G. Drazin, R.S. Johnson, *Solitons: An Introduction* (Cambridge University Press, Cambridge, 1996)
80. H.R. Dullin, G.A. Gottwald, D.D. Holm, Phys. Rev. Lett. **87** (2001)
81. E.A.H. Dupont, *Quotient Manifolds by Group Action* (MS Thesis, University of Copenhagen, 2001).
82. D. Ebin, J. Marsden, Ann. Math. **92**, 102 (1970)
83. C. Eckl, A.P. Mayer, A.S. Kovalev, Phys. Rev. Lett. **81**, 983 (1998)
84. J. Eggers: Rev. Mod. Phys. **69**, 865 (1997)
85. S.-I. Ei, E. Yanagida, J. Dyn. Diff. Eqs. **7**, 423 (2005)
86. U. Eichmann, A. Ludu, J.P. Draayer, J. Phys. A: Math. Gen. **35**, 6075 (2002)
87. J.M. Eisenberg, W. Greiner, *Nuclear Theory*, vol 1 (North-Hollan, Amsterdam, 1976)
88. L.P. Eisenhart, *A Treatise on The Differential Geometry of Curves and Surfaces* (Dover, New York, 1960)
89. K.A. Erb, D.A. Bromley, Phys. Rev. C **23**, 2781 (1981)
90. E. Falcon, C. Laroche, S. Fauve, *Phys. Rev. Lett.* **89**, 204501 (2002)
91. A.J. Faller, Phys. Today **10** (October 2006) and references herein
92. Y.-F. Fang, M.G. Grillakis, Commun. Part. Diff. Eqs. **21**, 1253 (1996)
93. M.J. Feigenbaum, J. Stat. Phys. **19**, 25 (1978)
94. M.J. Feigenbaum, J. Stat. Phys. **21**, 669 (1979)
95. M.D. Feit, J.A. Fleck, Jr., J. Comput. Phys. **47**, 412 (1982)
96. Z.C. Feng, Y.H. Su, Phys. Fluids **9**, 519 (1997)
97. H. Feshbach, *Theoretical Nuclear Physics: Nuclear Reactions* (Wiley, New York, 1972)
98. R.P. Feynman, Phys. Rev. **84**, 108 (1951)
99. H.J. Fink, W. Scheid, W. Greiner, J. Phys. G: Nucl. Part. Phys. **1**, L85 (1975)
100. L. Friedland, A.G. Shagalov, Phys. Rev. Lett. **85**, 2941 (2000)
101. S. Ganguli, *Fibre Bundles and Gauge Theories in Classical Physics*, Course Notes (University of California, Berkeley, 2005)
102. Y. Giga, K. Ito, *Proceedings of Nonlinear Differential Equations Applications*, vol 35 (Birkhäuser, Basel, 1999)
103. S. Giovanazzi, L. Pitaevskii, S. Stringari, Phys. Rev. Lett. **72**, 3230 (1994)
104. R. Glaser, *Biophysics* (Springer, Berlin, 1996)
105. R.A. Gherghescu, A. Ludu, J.P. Draayer, J. Phys. G: Nucl. Part. Phys. **27**, 63 (2001)
106. M. Ghomi, B. Solomon, Comment. Math. Helv. **77**, 767 (2002)
107. H. Gluck, Enseignement Math. **17**, 295 (1971)
108. R.E. Goldstein, D.M. Petrich, Phys. Rev. Lett. **67**, 3203 (1991)
109. J. Langer, R. Perline, Phys. Lett. A **239** (1998) 36; R.E. Goldstein, S.A. Langer, Phys. Rev. Lett. **75**, 1094 (1995)
110. V.P. Goncharov, V.I. Pavlov, JETP **119**, 685 (2001)
111. R.A. Granger, *Fluid Mechanics* (Dover, New York, 1995)
112. M.J. Greenberg, J.R. Harper, *Algebraic Topology* (Westwiew, Boulder, CO, 1981)
113. W. Greiner, J. Eisenberg, *Nuclear Models* (North-Holland, Amsterdam, 1987)
114. W. Greiner, J.Y. Park, W. Scheid, *Nuclear Molecules* (World Scientific, Singapore, 1995)
115. K.A. Gridnev, Z. Phys. A: Hadrons Nucl. **349**, 269 (1994)
116. M. Grigorescu, A. Sandulescu, Phys. Rev. C **48**, 940 (1993)
117. P.G. Grinevich, M.U. Schmidt, arXiv:dg-ga/9703020 v1 (26 March 1997)

118. W.T. Growers, *Further Analysis*, Course Notes, University of Cambridge, Cambridge (Lent term, 1997)
119. H.W. Guggenheimer, *Differential Geometry* (Dover, New York, 1977)
120. A. Halanay, *Ecuatii diferentiale* (Editura Didactica si pedagogica, Bucharest, 1972)
121. J.L. Hammack, H. Segur, J. Fluid Mech. **65**, 289 (1974)
122. A. Hanke, R. Metzler, J. Phys. A: Math. Gen. **36**, L473 (2003)
123. H. Hasimoto, J. Fluid Mech. **51**, 477 (1972)
124. F. Hausdorff, Ber. Verhandl. Sachs. Akad. Wiss. Leipzig, Math. Naturw. Kl. **58**, 19 (1906)
125. H.V. Helmholtz, J. für die reine und angewandte Mathematik (Crelle's Journal) **55** (1858) (see Wiss. Abh. i. 101)
126. M. Hines, J.J. Blum, Biophys. J. **25**, 421 (1979)
127. R. Hirota, Phys. Rev. Lett. 1192 (1971)
128. R. Hirota, J. Math. Phys. **14**, 805 (1973)
129. H. Hochstadt, *The Functions of Mathematical Physics* (Wiley-Interscience, New York, 1971)
130. R.G. Holt, E.H. Trinh, Phys. Rev. Lett. **77**, 1274 (1996)
131. P. Holzer, U. Mosel, W. Greiner, Nucl. Phys. A **138**, 241 (1969)
132. H. Horiuchi, K. Ikeda, *Cluster Models and Other Topics* (World Scientific, Singapore, 1986)
133. W. Hurewicz, *Lectures on Ordinary Differential Equations* (MIT, Cambridge, 1958)
134. F. Iachello, Phys. Rev. C **23**, 2778 (1981)
135. E. Infeld, G. Rowlands, *Nonlinear Waves, Solitons and Chaos* (Cambridge University Press, Cambridge, 2000)
136. V.I. Istratescu, *Introduction to Linear Operator Theory* (Dekker, New York, 1981)
137. I.B. Ivanov, *Thin Liquid Films* (Dekker, New York, 1988)
138. I.B. Ivanov, P.A. Kralchevsky, Mechanics and thermodynamics of curved thin films, in *Thin Liquid Films*, ed. by I.B. Ivanov (Dekker, New York, 1988), pp. 49–130
139. T.A. Ivey, Contemp. Math **285**, 71 (2001)
140. T.A. Ivey, *Geometry and Topology of Finite-Gap Vortex Filaments*, in *Geometry, Integrability and Quantization*, ed. by I.M. Mladenov, M. de Léon (Softex, Sofia, 2006) pp. 187
141. R. Jackiw, V.P. Nair, S.-Y. Pi, A.P. Polychronakos: arXiv:hep-ph/0407101 v1 (8 July 2004) (electronic preprint)
142. T.R.N. Jansson, M.P. Haspang, K. Jensen, P. Hersen, T. Bohr, Phys. Rev. Lett. **96**, 174502 (2006)
143. J.P. Jaric, J. Elasticity **51**, 73 (1998)
144. J. Javanainen, J. Ruostekoski, arXiv:cond-matt/0411154 (2004)
145. D. Jou, J. Casas-Vàzquez, G. Lebon, *Extended Irreversible Thermodynamics* (Springer, Berlin, 1996)
146. P.E.T. Jorgensen, R.T. Moore, *Operator Commutation Relations* (D. Reidel, Dordrecht, 1984)
147. D.D. Joseph, Y.Y. Renardy, *Fundamentals of Two-Fluid Dynamics* (Springer, Berlin, 1992)
148. V.G. Kartavenko, Sov. J. Nucl. Phys. **40**, 240 (1984)
149. V.G. Kartavenko, K.A. Gridnev, W. Greiner, Int. J. Mod. Phys. **3**, 1219 (1994)
150. V.G. Kartavenko, A. Ludu, A. Sandulescu, W. Greiner, Int. J. Mod. Phys. E **5**, 329 (1996)
151. K. Kato, Y. Abe, Prog. Theor. Phys. **80**, 119 (1988)
152. S. Kawamoto, J. Phys. Soc. Jpn. **54**, 2055 (1985)
153. Lord Kelvin (Sir William Thomson), *On Vortex Atoms*, Proc. R. Soc. (Edinburgh) **6**, 94 (1867); Edin. Trans. **25** (1869)
154. Y.G. Kevrekidis, A. Ustinov, *The 17th Annual CNLS Conference* (Los Alamos, 12–16 May 1997)
155. P.G. Kevrekidis, A. Khare, A. Saxena, Phys. Rev. E **68**, 047701 (2003)
156. D.M. Kishmatullin, A. Nadim, Phys. Rev. E **63**, 061508 (2001)
157. D.M. Kishmatullin, A. Nadim Phys. Fluids **14**, 3534 (2002)
158. S. Kobayashi, K. Nomizu, *Foundations of Differential Geometry* (Interscience, London, 1963)
159. Y. Kodama, J. Phys. A: Math. Gen. **37**, 11169 (2004)
160. A. Kolmogorov, S. Fomine, *Elèments de la thèorie des fonctions et de l'analyse fonctionelle* (Editions MIR, Moscow, 1974)

161. K. Kreutz, Phys. Rev. D **12**, 3126 (1975)
162. E. Kreyszig, *Differential Geometry* (University of Torronto Press, Toronto, 1959)
163. M.D. Kruskal, Lect. Notes Phys. **278**, 310 (1975)
164. H. Kumaon-go, *Pseudo-Differential Operators* (MIT, Cambridge, 1974)
165. C.N. Kumar, P.K. Panigrahi, solv-int/9904020
166. M. Lakshmanan, R. Myrzakulov, S. Vijayalakshmi, A.K. Danlybaeva, J. Math. Phys. **39**, 3765 (1998)
167. Sir H. Lamb, *Hydrodynamics* (Dover, New York, 1932)
168. G.L. Lamb, J. Math. Phys. **18**, 1654 (1977)
169. G.L. Lamb, Jr., *Elements of Soliton Theory* (Wiley, New York, 1980)
170. L. Landau, F. Lifchitz, *Theorie de L'elasticitè* (MIR, Moscow, 1967); A.E.H. Love, *A Treatise on Mathematical Theory of Elasticity* (Dover, New York, 1944)
171. L. Landau, E. Lifchitz, *Fluid Mechanics* (MIR, Moscow, 1971)
172. J. Langer, R. Perline, J. Nonlinear Sci. **1**, 71 (1991)
173. D. Lewis, J. Marsden, R. Montgomery, T. Ratiu, Physica D **18**, 391 (1986)
174. J. Lighthill, *Waves in Fluids* (Cambridge University Press, Cambridge, 1979)
175. L. Lindblom, J. Tohline, M. Vallisneri, Phys. Rev. Lett. **86**, 1152 (2001)
176. C.B. Lindemann, Cell Motil. Cytoskel. **29**, 141 (1994)
177. M. Lakshmanan, J. Math. Phys. **20**, 1667 (1979)
178. R. de la Llave, P. Panayotaros, J. Nonlinear Sci. **6**, 147 (1996)
179. R. Loll, arXiv: gr-qc/9701007
180. A.M. Lomonosov, P. Hess, A.P. Mayer, Phys. Rev. Lett. **88**, 076104 (2002)
181. D. Lovelock, H. Rund, *Tensors, Differential Forms, and Variational Principles* (Dover, New York, 1989)
182. H.-L. Lu, R.E. Apfel, J. Fluid Mech. **222**, 351 (1991)
183. A. Ludu, A. Sandulescu, W. Greiner, Int. J. Mod. Phys. E **1**, 169 (1992)
184. A. Ludu, A. Sandulescu, W. Greiner, Int. J. Mod. Phys. E **2**, 855 (1993)
185. A. Ludu, A. Sandulescu, W. Greiner, K. M. Källmann, M. Brenner, T. Lönnroth, P. Manngärd, J. Phys. G: Nucl. Part. Phys. **21**, L41 (1995)
186. A. Ludu, A. Sandulescu, W. Greiner, J. Phys. G: Nucl. Part. Phys. **21**, 1715 (1995)
187. A. Ludu, R.A. Ionescu, W. Greiner, Found. Phys. **26**, 665 (1996)
188. A. Ludu, A. Sandulescu, W. Greiner, J. Phys. G: Nucl. Part. Phys. **23**, 343 (1997)
189. A. Ludu, J.P. Draayer, Phys. Rev. Lett. **80**, 2125 (1998)
190. A. Ludu, J.P. Draayer, Physica D **123**, 82 (1998)
191. A. Ludu, G. Stoitcheva, J.P. Draayer, Math. Comput. Simulat. **55**, 621 (2001)
192. A. Ludu, Math. Comput. Simulat. **69**, 389 (2005)
193. A. Ludu, N. Hutchings, *Proceedings of Conference ISIS, Natchitoches, LA, 2004*, AIP Conf. Proc. **755** (AIP, Melville, 2005), pp. 91, 137, 253
194. A. Ludu, P. Kevrekidis, Math. Comput. Simulat. **74**, 229 (2007)
195. A. Ludu, N. Hutchings, Math. Comput. Simulat. **74**, 179 (2007)
196. A. Ludu, J. Nonlin. Math. Phys. **15** (2008) 157
197. N.G. Berloff, Phys. Rev. Lett. **94** (2005) 010403; A. Ludu, M. Milosevic, F.M. Peeters, Math. Comp. Sim. **51** 082903 (2010)
198. C. Cibert, A. Ludu, Math. Comp. Sim. **80** (2009) 223; J. Theoret. Biol. **265**, 2 (2010) 95–103
199. A. Ludu, M. Milosevic, A. Cuyt, J. Van Deunn, F.M. Peeters, J. Math. Phys. **51**, 1 (2010) 082903: 1–29
200. E.V. Vargas, R. Hustert, P. Gumrich, A. Ludu, A.D. Jackson, T. Heimburg, Biophysical Chem. **153** (2011) 159
201. A. Ludu, J. Geom. Symmetry Phys. **20** (2010)
202. J.C. Luke, J. Fluid Mech. **27**, 395 (1967)
203. T.S. Lundgren, N.N. Mansour, J. Fluid Mech. **194**, 479 (1988)
204. W. Magnus, Commun. Pure Appl. Math. **7**, 649 (1954)
205. V.G. Makhankov, Phys. Rep. **35**, 1 (1978)

206. H.J. Mang, Phys. Rev. **119**, 1069 (1960); R. Blendowske, A. Walliser, Phys. Rev. Lett. **61**, 1930 (1988)
207. Y. Manin, *Mathematics and Physics* (Birkhäuser, Boston, MA, 1981)
208. N. Manton, Nonlinearity **21**, T221 (2008)
209. N. Manton, P. Sutcliffe, *Topological Solitons* (Cambridge University Press, Cambridge, 2004)
210. J.E. Marsden, T.J.R. Hughes, *Mathematical Foundations of Elasticity* (Dover, New York, 1994)
211. N.S. Manton, N.A. Rink, J. Phys. A: Math. Theoret. **43** (2010) 434024
212. J.E. Marsden, S. Shkoller, *Math. Proc. Cambridge Phil. Soc.* **125**, 3 (1999) 553–575
213. J.E. Marsden, T.S. Ratiu, S. Shkoller, arXiv:math.AP/9908103 v1 (19 August 1999) (electronic preprint)
214. J.E. Marsden, S. Shkoller, arXiv:math.AP/0005033 v2 (8 May 2000) (electronic preprint)
215. J.E. Marsden, S. Pekarsky, S. Shkoller, M. West, J. Geom. Phys. **38**, 253 (2001)
216. P.L. Marston, S.G. Goosby, Phys. Fluids **28**, 1233 (1985)
217. R.I. McLachlan, H. Segur, A note on the motion of surfaces, preprint arXiv:sol-int/9306003 (1993)
218. E. Medina, M.J. Marín, Inverse Probl. **17**, 985 (2001)
219. A. Menchca-Rocha et al., Phy. Rev. E **63**, 046309 (2001)
220. R.E. Meyer, *Introduction to Mathematical Fluid Dynamics* (Dover, New York, 1971)
221. M.D. Meyerson, Am. Math. Mon. **61**, 181 (1976)
222. A.S. Mikhailov, V. Calenbuhr, *From Cells to Societies* (Springer, Berlin, 2002)
223. C.A. Miller, L.E. Scriven, J. Fluid Mech. **32**, 417 (1968)
224. L.M. Milne-Thomson, *Theoretical Hydrodynamics* (Dover, New York, 1996)
225. J.W. Milnor, Ann. Math. (2nd series) **52**, 248 (1950)
226. M. Hadzhilazova, J.-F. Ganghoffer, I. Mladenov, CRAS (Sofia) **63**, 1155 (2010)
227. V.M. Vassilev, P.A. Djondjorov, I.M. Mladenov, J. Phys. A: Math. Theor. **41**, 435201 (2008)
228. P. Djondjorov, M. Hadzhilazova, I. Mladenov, V. Vassilev, *Proceedings of the Ninth International Conference on Geometry, Integrability and Quantization* (SOFTEX, Sofia 2008), pp. 175–186
229. H.L. Morrison, A.D. Speliotopoulos, J. Nonlin. Math. Phys. **9**, 9 (2002) 464
230. C. Müller, *Spherical Harmonics* (Springer, Berlin, 1966)
231. M. Murase, *The Dynamics of Cell Motility* (Wiley, New York, 1992); C.B. Lindemann, K.S. Kanous, Cell Motil. Cytoskel. **31**, 1 (1995)
232. S. Murugesh, R. Balakrishnan, Eur. Phys. J. B **29**, 193 (2002)
233. K. Nakayama, H. Segur, M. Wadati, Phys. Rev. Lett. **69**, 2603 (1992)
234. K. Nakayama, M. Wadati, J. Phys. Soc. Japan, **62** 6, 1895 (1993)
235. C. Nash, S. Sen, *Topology and Geometry for Physicists* (Academic, London, 1988)
236. R. Natarajan, R.A. Brown, Phys. Fluids **29**, 2788 (1986)
237. R. Natarajan, R.A. Brown, J. Fluid Mech. **183**, 95 (1987)
238. A. Nikiforov, V. Ouvarov, *Elementes de la theorie des functions speciales* (MIR, Moscow, 1974)
239. N. Ogawa: Phys. Rev. E **81**, 061113 (2010)
240. K. Ohtsuka, R. Takaki, S. Watanabe, Phys. Fluids **15**, 1065 (2003)
241. J.G. Oldroyd, Proc. R. Soc. Lond. A **200**, 523 (1950)
242. P.J. Olver, *Applications of Lie Groups to Differential Equations*, 1st edn. (Springer, Berlin, 1986)
243. P.J. Olver, P. Rosenau, Phys. Rev. E **53**, 1900 (1996)
244. T.M. O'Neil, Phys. Today **52**, 29 (1999)
245. A.R. Osborne, M. Onorato, M. Serio, L. Bergamasco, Phys. Rev. Let. **81**, 3559 (1998)
246. R. Osserman, *A Survey of Minimal Surfaces* (Dover, New York, 2002)
247. S. Ou, J.E. Tohline, arXiv e-prep, astro-ph/0406037
248. E.A. Overman, N.J. Zabusky, Phys. Rev. Lett. **45**, 1693 (1980)
249. E.A. Overman, N. Zabusky, J. Fluid Mech. **125**, 7 (1982)
250. V. Yu. Ovsienko, B.A. Khesin, Funktsional'nyi Analiz i Ego Prilozheniya **21**(4), 81 (1987)

251. V.P. Palamodov: *Linear Differential Operators with Constant Coefficients* (Springer, Berlin, 1970)
252. P. Panayotaros, Near-monochromatic water waves on the sphere, in *IMACS Conference on Nonlinear Waves*, ed. by T. Taha (University of Georgia, Athens, April 1999)
253. T.W. Patzek, R.E. Brenner, O.A. Basaran, J. Comput. Phys. **97**, 489 (1991)
254. V. Penna, M. Spera, J. Geom. Phys. **27** (1998) 99
255. D.H. Peregrine, J. Fluid Mech. **27**, 815 (1967)
256. K.A. Pericak-Spector, J. Sivaloganathan, S.J. Spector, Private Communication (2000)
257. R. Perline, J. Phys. A: Math. Gen. **27**(15), 5335 (1994)
258. L.M. Pismen, J. Rubinstein, Physica D **47**(1991) 353
259. Y. Pomeau, E. Villermaux, Phys. Today **39** (March 2006)
260. L. Porcar et al., Phys. Rev. Lett. **93**, 198301 (2004)
261. T. Powers, G. Huber, R.E. Goldstein, Phys. Rev. E **65**, 041901 (2002)
262. P.B. Price, Annu. Rev. Nucl. Part. Sci. **39**, 19 (1989)
263. A. Prosperetti, Q. Appl. Math. **35**, 339 (1977)
264. A. Prosperetti, J. de Mecanique **19**, 149 (1980)
265. A. Prosperetti, J. Fluid Mech. **100**, 333 (1980)
266. E. Prugovecki, *Quantum Mechanics in Hilbert Spaces* (Academic, New York, 1971)
267. R. Rajaraman, *Solitons and Instantons* (North-Holland, Amsterdam, 1987)
268. Lord Rayleigh, Mess. Math. **9**, 177 (1880)
269. W.H. Reid, Q. Appl. Math. **18**, 86 (1960)
270. M. Rein (ed.), *Drop–Surface Interaction* (Springer, Berlin, 2002)
271. M. Remoissenet, *Waves Called Solitons* (Springer, Berlin, 1999)
272. G. Richardson, J.R. King, J. Phys. A: Math. Gen. **35**, 9857 (2002)
273. S. Richardson, J. Fluid Mech. **56**, 609 (1972)
274. F. Riesz, B. Sz.-Nagy, *Functional Analysis*, 2nd edn. (Dover, New York, 1990)
275. A. Ronveaux, *Heun's Differential Equations* (Oxford University Press, Oxford, 1995)
276. P. Rosenau, J.M. Hyman, Phys. Rev. Let. **70**, 564 (1993)
277. G. Rosensteel, J. Troupe, J. Phys. G: Nucl. Part. Phys. **25**, 549 (1999)
278. G. Rosensteel, J. Troupe, *Gauge Theory of Riemannian Ellipsoids*, arXiv:math-ph/9909031
279. G. Rosensteel, J. Phys. A: Math. Gen. **37**, 10967 (2004)
280. R.E. Rosensweig, *Ferrohydrodynamics* (Cambridge University Press, Cambridge, 1985)
281. K. Rottmann, *Mathematische Formelsammlung* (Bibliographisches Institut, Mannheim, 1960)
282. G. Royer, J. Phys. G: Nucl. Part. Phys. **26**, 1149 (2000)
283. B.M. Rush, A. Nadim, Eng. Anal. Bound. Elem. **24**, 43 (2000)
284. I.M. Ryshik, I.S. Gradstein, *Tables of Series, Products, and Integrals* (Veb Deutscher Verlag der Wissenschaften, Berlin, 1963)
285. A. Sakovich, S. Sakovich, arXiv:nlin.SI/0601019 v1 (9 January 2006)
286. R.L. Sani: *Convective Instability*, Ph.D. Thesis, Department of Chemical Engineering, University of Minnesota, 1963
287. L. Satpathy, L.P. Sarangi, A. Faessler, J. Phys. G: Nucl. Part. Phys. **6**, 201 (1986)
288. W. Scheid, W. Greiner, Z. Phys. A **226**, (1979) 364
289. W.K. Schief, C. Rogers, Proc. R. Soc. Lond. A **455**, 3163 (1999)
290. T. Schlick, W.K. Olson, J. Mol. Biol. **223**, 1089 (1992)
291. L. Schwartz, *Analyse*, 2nd edn. (Hermann, Paris, 1970) Fluid Mech. **19**, 321 (1963)
292. L.E. Scriven, Chem. Eng. Sci. **12**, 98 (1960)
293. L.E. Scriven, C.V. Sternling, J. Fluid Mech. **19**, 321 (1964)
294. J. Serrin, *Handbuch der Physik*, VIII/1, sections 58–65 (Springer, Berlin, 1966)
295. J.A. Setiah, J. Diff. Geom. **31**, 131 (1991)
296. J.A. Setiah, *Level Set Methods and Fast Marching Methods* (Cambridge University Press, Cambridge, 1999)
297. A. Shapere, F. Wilczek, Phys. Rev. Lett. **58**, 2051 (1987)
298. T. Shi, R.E. Apfel, Phys. Fluids **7**, 1545 (1995)

299. T. Shifrin: *Differential Geometry: A First Course in Curves and Surfaces* (University of Georgia, Fall, 2005)
300. A. Cannas da Silva, A. Weinstein, *Geometric Models for Noncommutative Algebras* (University of California at Berkeley, Berkeley 1998)
301. W.F. Simmons, Proc. R. Soc. Lond. A **309**, 551 (1969)
302. T.H.R. Skyrme, *Nucl. Phys.* **9**, 615 (1959)
303. V.I. Smirnov, N.A. Lebedev, *Functions of a Complex Variable* (MIT, Cambridge, 1968); S.G. Krantz, *Function Theory of Several Complex Variables* (American Mathematical Society, Providence, 2001)
304. G.F. Smith, Tensor, N.S. **19**, 79 (1968)
305. H. Sohr, *The Navier–Stokes Equations. An Elementary Functional Analytic Approach* (Birkhäuser, Basel, 2001)
306. N. Steenrod, *The Topology of Fibre Bundles* (Princeton University Press, Princeton, NJ, 1951)
307. H.A. Stone, Phys. Rev. Lett. **77**, 4102 (1996)
308. W.A. Strauss, *Partial Differential Equations* (Wiley, New York, 1992)
309. V.M. Strutinsky, Nucl. Phys. A **95**, 420 (1967)
310. D.J. Struik, *Lectures on Differential Geometry*, 2nd edn. (Addison-Wesley, Reading, MA, 1961)
311. C. Sulem, P.L. Sulem, *The Nonlinear Schrödinger Equation* (Springer, Berlin, 1999)
312. A.G. Sveshnikov, A.N. Tikhonov, *The Theory of Functions of a Complex Variable*, 2nd edn. (MIR, Moscow, 1973)
313. T. Taha (ed.), *IMACS Conference on Nonlinear Waves* (University of Georgia, Athens, April 2005)
314. J.L. Thiffeault, J. Phys. A: Math. Gen. **34**, 1 (2001); arXiv:nlin.CD/0105026; arXiv:nlin.CD/0105010
315. J.J. Thomson, *A Treatise on the Motion of Vortex Rings* (Macmillan, London, 1983) (also known as the *Adam Prize Essay*, 1884)
316. Y. Tian, R.G. Holt, R.E. Apfel, Phys. Fluids **7**, 2938 (1995)
317. A.N. Tikhonov, A.A. Samarskii, *Equations of Mathematical Physics* (Dover, New York, 1990)
318. E. Trinh, T.G. Wang, J. Fluid Mech. **122**, 315 (1982)
319. E.H. Trinh, R.G. Holt, D.B. Thiessen, Phys. Fluids **8**, 43 (1995)
320. J. Troupe, G. Rosensteel, Ann. Phys. **270**, 126 (1998)
321. J. A. Tsamopoulos, R.A. Brown, J. Fluid Mech. **127**, 519 (1983)
322. Y. Tsukahara, N. Yamanaka, Appl. Phys. Lett. **77**, 2926 (2000)
323. A.N. Varchenko, P.I. Etingof, *Why the Boundary of a Round Drop Becomes a Curve of Order Four* (American Mathematical Society, Providence, RI, 1991)
324. W. Volkmuth, R.H. Austin, Nature **358**, 600 (1992)
325. W. Volkmuth, R.H. Austin, Nature **358**, 600 (1992); T.T. Perkis, S.R. Quake, D.E. Smith, S. Chu, Science **264**, 822 (1994)
326. K. Yasui, Phys. Rev. E **58**, 471 (1998)
327. J.A. Yorke, *Selected Topics in Differential Delay Equations*, Lect. Notes Math. **243** (Springer, Berlin, 1971)
328. F.R. Young, *Sonoluminescence* (CRC, Boca Raton, FL, 2004)
329. N.J. Zabusky, M.H. Hughes, K.V. Roberts, J. Comput. Phys. **30**, 96 (1979)
330. V.E. Zakharov, A.B. Shabat, Sov. Phys. JETP **34**, 62 (1972)
331. V.E. Zakharov, E.A. Kuznetsov, Physics-Uspekhi **40**(11), 1087 (1997)
332. Z. Zapryanov, S. Tabakova, *Dynamics of Bubbles, Drops and Rigid Particles* (Kluwer, Dordrecht, 1999)
333. N.B. Zhitenev, R.J. Haug, K.V. Klitzing, K. Eberl, Phys. Rev. B **52**, 11277 (1995)
334. M. Wadati, H. Tsuru, Physica D (Amsterdam) **21**, 213 (1986)
335. H.D. Wahlquist, F.B. Estabrook, J. Math. Phys. **16**, 1 (1975)
336. D.F. Walnut, *An Introduction to Wavelet Analysis* (Birkhäuser, Boston, 2004)
337. C.E. Weatherburn, Q. J. Math. **50**, 230 (1925)

338. C.E. Weatherburn, *Differential Geometry of Three Dimensions* (Cambridge University Press, Cambridge, 1927)
339. J. Weiner, Proc. Am. Math. Soc. **67**, 306 (1977)
340. C. Wexler, A.T. Dorsey, Phys. Rev. Lett. **82**, 620 (1999)
341. C. Wexler, A.T. Dorsey, Phys. Rev. B **60**, 10971 (1999)
342. G.B. Whitham, Proc. R. Soc. Lond. A **283**, 238 (1965)
343. R.M. Wilcox, J. Math. Phys. **8**, 962 (1967)
344. S. Willard, *General Topology* (Addison-Wesley, Reading, MA, 1970)
345. D.R.M. Williams, M. Warner, J. Phys. (Paris) **51**, 317 (1990)
346. A.J. Willson, Proc. Camb. Philos. Soc. **61**, 595 (1965)

Index

A. Ludu, *Nonlinear Waves and Solitons on Contours and Closed Surfaces*,
Springer Series in Synergetics, DOI 10.1007/978-3-642-22895-7,